Chemicals in the Atmosphere

solubility, sources and reactivity

Wiley Series in Solution Chemistry

Editor-in-Chief

P. G. T. Fogg, *University of North London, UK* (retired)

Editorial Board

W. E. Acree, *University of North Texas, USA*
A. Bylicki, *Polish Academy of Sciences, Warsaw, Poland*
A. F. Danil de Namor, *University of Surrey, UK*
H. J. M. Grünbauer, *Dow Benelux NV, Ternauzen, The Netherlands*
S. Krause, *Rennselaer Polytechnic Institute, Troy, USA*
A. E. Mather, *University of Alberta, Canada*
H. Ohtaki, *Ritsumeikan University, Kusatsu, Japan*
A. D. Pelton, *Ecole Polytechnique de Montreal, Canada*
M. Salomon, *MaxPower Inc., Harleyville, PA, USA*
A. Skrzecz, *Polish Academy of Sciences, Warsaw, Poland*
R. P. T. Tomkins, *New Jersey Institute of Technology, USA*
W. E. Waghorne, *University College Dublin, Ireland*
B. A. Wolf, *Johannes-Gutenburg-Universität, Mainz, Germany*
C. L. Young, *University of Melbourne, Australia*

Volume 1
pH and Buffer Theory – A New Approach
H Rilbe
Chalmers University of Technology, Gothenburg, Sweden

Volume 2
Octanol–Water Partition Coefficients: Fundamentals and Physical Chemistry
J Sangster
Sangster Research Laboratories, Montreal, Canada

Volume 3
Crystallization Processes
Edited by
H Ohtaki
Ritsumeikan University, Kusatsu, Japan

Volume 4
The Properties of Solvents
Y Marcus
The Hebrew University of Jerusalem, Israel

Volume 5
Physical Chemistry of Non-aqueous Solutions of Cellulose and Its Derivatives
V V Myasoedova
Institute of Chemical Physics, Russian Academy of Sciences, Moscow, Russia

Volume 6
The Experimental Determination of Solubilities
Edited by
G T Hefter
Murdoch University, Murdoch, Australia
and
R P T Tomkins
New Jersey Institute of Technology, USA

Volume 7
Chemicals in the Atmosphere – Solubility, Sources and Reactivity
P G T Fogg
University of North London, UK (retired)
and
J Sangster
Sangster Research Laboratories, Montreal, Canada

Chemicals in the Atmosphere

solubility, sources and reactivity

Peter Fogg and James Sangster

With contributions from
Yin-Nan Lee
Stephen Schwartz
Peter Warneck

WILEY

Email (for orders and customer service enquiries): cs-books@wiley.co.uk
Visit our Home Page on www.wileyeurope.com or www.wiley.com

This publication is designed to provide accurate and authoritative information in regard to the subject matter covered. It is sold on the understanding that the Publisher is not engaged in rendering professional services. If professional advice or other expert assistance is required, the services of a competent professional should be sought.

Other Wiley Editorial Offices

John Wiley & Sons Inc., 111 River Street, Hoboken, NJ 07030, USA

Jossey-Bass, 989 Market Street, San Francisco, CA 94103-1741, USA

Wiley-VCH Verlag GmbH, Boschstr. 12, D-69469 Weinheim, Germany

John Wiley & Sons Australia Ltd, 33 Park Road, Milton, Queensland 4064, Australia

John Wiley & Sons (Asia) Pte Ltd, 2 Clementi Loop #02-01, Jin Xing Distripark, Singapore 129809

John Wiley & Sons Canada Ltd, 22 Worcester Road, Etobicoke, Ontario, Canada M9W 1L1

Wiley also publishes its books in a variety of electronic formats. Some content that appears in print may not be available in electronic books.

British Library Cataloguing in Publication Data

A catalogue record for this book is available from the British Library

ISBN 0-471-98651-8

Typeset in 10/12pt Times by Laserwords Private Limited, Chennai, India
Printed and bound in Great Britain by TJ International, Padstow, UK
This book is printed on acid-free paper responsibly manufactured from sustainable forestry in which at least two trees are planted for each one used for paper production.

Contents

Contents

List of Contributors

Peter Fogg
University of North London (retired)
Home address: Silver Birches, High Molewood, Hertford, UK

Yin-Nan Lee
Atmospheric Sciences Division, Brookhaven National Laboratory, Upton, NY, USA

James Sangster
Sangster Research Laboratories, Montreal, Canada

Stephen Schwartz
Atmospheric Sciences Division, Brookhaven National Laboratory, Upton, NY, USA

Peter Warneck
Max-Planck-Institut für Chemie, Abteilung Biogeochemie, Mainz, Germany

Preface

This book is published as Project 2001-037-1-500 of the IUPAC Commission on Solubility Data. This resulted from discussions between members of the IUPAC Commission on Solubility Data and the IUPAC Commission on Atmospheric Chemistry extending over several years. The book provides a broad survey and theoretical basis of many aspects of the behaviour of stable and unstable chemicals in the atmosphere. There is special emphasis on heterogeneous processes at all levels of the atmosphere. Modern techniques of investigation and of modelling such processes are discussed. Many experimental data relating to the interaction of gases with cloud droplets and with surfaces of solid particles are included.

The book includes an extensive compilation of evaluated Henry's law constants for dissolution in water and in sea water of substances which normally occur in the atmosphere or which may escape into the atmosphere.

The editors are indebted to Dr Simon Clegg for advice during the early stages of the project and to Dr Stephen Schwartz for advice during the project. They also acknowledge help from Dr Werner Hauthal, Dr Douglas Worsnop, Dr Rolf Sander, Dr Edward Edney and Dr Kenneth S. Carslaw. Contributions to the text from Dr Yin-Nan Lee, Dr Stephen Schwartz and Dr Peter Warneck are especially appreciated.

Peter Fogg
London, UK

James Sangster
Montreal, Canada
October 2002

Introduction

Peter Fogg

University of North London, UK (retired)

The behaviour and hence the possible impact of trace gases and solids in the atmosphere is now of major concern. Sources and concentrations of the principal trace gases are considered in Chapter 1 with an indication of their environmental significance.

There are many aspects of the chemistry of the atmosphere. The rest of the volume is chiefly concerned with heterogeneous processes in all layers of the atmosphere.

Of particular importance are the factors which determine the rate and extent of dissolution of gases in aqueous media. For many systems the solubility of a volatile solute in a liquid tends to become proportional to the partial pressure of the solute the lower the partial pressure of the solute. This is true whatever the units of solubility and of pressure. In practice, there is a pressure range over which the proportionality can be assumed to hold within experimental error. The pressure range over which linearity can be assumed varies from system to system. However, trace gases in the atmosphere have very low partial pressures. With the exception of gases such as ammonia which are reversibly hydrolysed in water, Henry's law as a simple proportionality between partial pressure and concentration is appropriate. Serious error can arise if solubility measurements at low pressures are extrapolated to a pressure higher than the range in which linearity can be assumed. Conversely, a solubility/pressure measurement at a pressure outside the linear region will not correspond to the ratio at low pressures.

The units for measuring Henry's law constants are discussed in detail in Chapter 2. This is followed by an explanation of important aspects of Henry's law constants in the understanding of mass transport of material in the atmosphere and other key atmospheric processes.

The measurement, prediction and thermodynamic aspects of the solubility of species which may be found in the atmosphere are considered in detail in Chapters 2–4. Prediction of values of Henry's law constants are discussed in

Chemicals in the Atmosphere – Solubility, Sources and Reactivity. Edited by P. G. T. Fogg and J. Sangster
© 2003 IUPAC   ISBN: 0-471-98651-8

Chapters 5 and 6. Evaluated values of experimental measurements of Henry's law constants for dissolution in pure water and in seawater are tabulated in Chapters 11 and 12.

The effects of electrolytes on the solubility of gases in water are explained in Chapter 7. Sea spray can cause salts from the sea to be taken up into clouds. Depending on the temperature, stratospheric clouds of liquid droplets may consist of a concentrated sulfuric–nitric acid aerosol. The physical properties of these solvents are discussed in relation to conditions in the stratosphere.

The Henry's law constant is essentially an equilibrium property of a gas. The rate of dissolution is also of great significance because gases in the atmosphere may not be in equilibrium with cloud droplets. Many gases react irreversibly with water after they dissolve. The kinetics of solution processes and any subsequent reaction are discussed in Chapter 8. Details of measurements of accommodation coefficients and of uptake coefficients are described and their significance is discussed.

The behaviour of individual gases which are unstable in the presence of liquid water is discussed in Chapter 9. Experimental measurements of accommodation coefficients and uptake coefficients are tabulated. Values of Henry's law constants are tabulated in cases where such a constant is meaningful and has been measured.

Many solids are taken into the troposphere as small particles, often termed particulates. These may arise from terrestrial surface dust. Volcanoes emit dust into the atmosphere. Soot particles are formed during combustion processes. Ice particles are formed in both the stratosphere and the troposphere. Important reactions of environmental significance can occur on the surface of these solids. Sources of these particles and heterogeneous reactions involving solid surfaces are described and discussed in Chapter 10.

Full details of readily accessible sources of additional data of importance to atmospheric chemistry are given in Appendix I. These include sources of data on homogeneous gas-phase reactions in addition to heterogeneous reactions.

Appendix II contains a brief account of the effect of climate change on oxygen isotope ratios.

CHAPTER 1

General Behaviour and Origins of Greenhouse and Significant Trace Gases

Peter Fogg
University of North London, UK (retired)

1 UNITS FOR QUANTITIES OF TRACE GASES

It is often appropriate to express global quantities of trace gases in terms of the weight of carbon, nitrogen or sulfur equivalent to the trace gas in question. One teragram, Tg, is 10^{12} grams. One petagram, Pg, is 10^{15} grams. Hence $10\,Tg\,C\,yr^{-1}$ is 10^{13} g of carbon or its equivalent per year. The abundance of a trace gas is often expressed as the ratio of the number of molecules of the trace gas to the total number of molecules present in the gas phase in parts per million (ppm) or parts per trillion (ppt). This mixing ratio may vary by several percent with the water vapour content of the air. It is therefore an advantage to express the ratio by reference to dry air. Sometimes the abundance of a trace gas is expressed as the number of molecules per unit volume. This has the disadvantage of being pressure and temperature dependent.

2 ATMOSPHERIC LIFETIMES OF GASES

The global atmospheric lifetime of a gas is a measure of the time that a particular molecule of gas is likely to stay in the atmosphere before it decomposes or is washed out by rain. It characterizes the time that it takes to turn over the global atmospheric burden of a particular gas. It is defined as the total quantity of the particular gas in the atmosphere divided by the rate at which the gas enters the atmosphere under conditions in which the total quantity in the atmosphere remains constant, i.e. when the rate of gain is equal to the rate of loss. It is usually called the lifetime of the gas but sometimes it is called the turnover time. The lifetime of carbon monoxide is $0.08–0.25\,yr$. That of tetrafluoromethane

Chemicals in the Atmosphere – Solubility, Sources and Reactivity. Edited by P. G. T. Fogg and J. Sangster
© 2003 IUPAC   ISBN: 0-471-98651-8

is $>50\,000\,\text{yr}$ and that of methane is 8.4 or 12 yr depending on the method of estimation [1].

3 GLOBAL WARMING POTENTIALS

The surface temperature of the earth depends on the balance reached between the incoming radiation from the sun and the radiation from the earth's surface back into space. Much of the radiation reaching earth from the sun is in the visible and ultraviolet region of the spectrum (short wavelength) and that radiated back into space in the infrared region (long wavelength). If the incoming radiative energy over a period of time is balanced by the outgoing radiative energy, the temperature of the earth's surface remains constant. Any factor which alters the balance between the energy from the sun and the energy radiated from the earth causes what is known as radiative forcing. An increase in dust in the upper atmosphere due to volcanic activity causes an increase in the proportion of the short wavelength radiation which is reflected back into space without reaching the surface of the earth. This is negative radiative forcing which tends to decrease the temperature of the surface of the earth. The lower the surface temperature the less radiation from the earth's surface so a new balance tends to be reached. Clouds also reflect radiation from the sun back into space. An increase in cloud cover causes negative radiative forcing.

Greenhouse gases are gases in the atmosphere which absorb the infrared radiation from the surface of the earth and act as a blanket to reduce loss of energy by the earth. An increase in the proportion of greenhouse gases therefore causes positive radiative forcing with an increase in the surface temperature of the earth.

Gases in the atmosphere vary in the extent to which they cause radiative forcing. This depends on the extent to which they absorb infrared radiation. Over a period of time the overall effect on the climate of a certain quantity of gas depends also on the lifetime of the gas in the atmosphere. The overall effect of a gas on the temperature of the earth is measured by the global warming potential (GWP). This is the radiative forcing due to an instantaneous release into the atmosphere of one kilogram of gas integrated over a period of time divided by the corresponding effect of a standard gas. By convention the standard gas is carbon dioxide. The GWP of a gas x relative to a reference gas r (i.e. carbon dioxide) is defined [2,3] as

$$\text{GWP}(x) = \frac{\displaystyle\int_0^{TH} a_x[x(t)]\,\mathrm{d}t}{\displaystyle\int_0^{TH} a_r[r(t)]\,\mathrm{d}t}$$

where a_x, a_r are the radiative efficiencies due to the increase in abundance of the gas in question (these have the units $\text{W}\,\text{m}^{-2}\,\text{kg}^{-1}$), TH is the period of time under consideration (time horizon) and $x(t)$, $r(t)$ are expressions for the decay of the gases as functions of time, t.

Values of GWP depend on the time period under consideration. In the case of gases with a shorter lifetime than carbon dioxide the GWP decreases with increase in the time period but in the case of gases with a longer lifetime the GWP increases with increases in the time period. In the case of methane, which has a shorter lifetime than carbon dioxide, the GWP is 62, 23, 7 for time periods of 20, 100 and 500 years, respectively. In the case of sulfur hexafluoride, which has a very long lifetime estimated to be 3200 years, the GWP is 15 100, 22 200 and 32 400, respectively.

Any time period could be chosen for comparing the behaviour of gases. However, a 100-year period has been recommended for comparative purposes [4]. This is the period currently (2002) chosen by the United States for policy making [5].

4   CARBON DIOXIDE; CO_2; [CAS 124-38-9]

The concentration of carbon dioxide is estimated to have been about 280 ± 10 ppm for several thousand years until about 1750 when industrialization began. There has been a continuous increase since that time. In 1999 the concentration was 367 ppm [6]. About three quarters of the increase is due to combustion of fossil fuel. Change of land use has been responsible for the rest of the increase [6]. It has been estimated that fossil fuel burning together with a minor contribution from cement manufacture was responsible for an emission of an average of $5.4 \pm 0.3 \, Pg \, C \, yr^{-1}$ in the period 1980–1989 and 6.3 ± 0.1 in 1990–1999 [6].

The carbon dioxide in the air is currently estimated to be about 730 Pg C [6]. It increased by an average of $3.3 \pm 0.1 \, Pg \, C \, yr^{-1}$ from 1980 to 1989 and by 3.2 ± 0.1 from 1990 to 1999. The increase is less than the emission because of removal of carbon dioxide by the oceans and by participation in photosynthesis.

Carbon dioxide in the sea is in equilibrium with carbonate ion and bicarbonate ions:

$$CO_2 + H_2O \rightleftharpoons H^+ + HCO_3^- \rightleftharpoons 2H^+ + CO_3^{2-}$$
$$\sim 1\% \qquad\qquad \sim 91\% \qquad\qquad \sim 8\%$$

The exact proportions depend on pH, temperature and the abundance of carbon dioxide in the air. The bulk of the carbon is in the form of bicarbonate ions. Some of the carbon is removed from the liquid phase by the formation by marine organisms of shells consisting mostly of calcium carbonate, which eventually accumulate as a component of sand or other sediment or as a coral reef. It has been estimated that the sea holds about 50 times the carbon in the atmosphere [6]. The solubility of carbon dioxide in seawater decreases with increase in temperature and is also influenced by the salinity and pH of the water.

Carbon dioxide is absorbed through the stomata of higher plants. The total quantity taken up in this way is estimated to be about $270 \, Pg \, C \, yr^{-1}$ [7]. This is more than one-third of all the carbon dioxide in the atmosphere. About $120 \, Pg \, C \, yr^{-1}$ reacts to form carbohydrates [8] and the rest diffuses out without reaction. About half of these carbohydrates undergo reactions to form plant tissue. The remainder decomposes back to carbon dioxide during respiratory processes [9,10].

Change of land use has had an important effect on the concentration of carbon dioxide. Development of agriculture in areas covered with natural vegetation usually results in a net increase in carbon dioxide in the atmosphere. This is partly due to an overall loss of plant biomass but also to increased decomposition of organic matter in the soil. Deforestation is especially significant. It has been estimated that the removal of 20 % of the world's forest area since 1850 has contributed about 90 % of the emissions due to land use change [6,11].

5   CARBON MONOXIDE; CO; [CAS 630-08-0]

Carbon monoxide has a relatively short lifetime as it is oxidized in the troposphere to carbon dioxide by oxygen in the presence of light or by direct reaction with OH. In the period 1996–1998 the global average concentration was about 80 ppb [1] with greater concentrations in winter than in summer. In high northern latitudes concentrations vary from about 60 ppb in the summer to about 200 ppb in the winter. Concentrations are lower in the southern hemisphere where there is less industry than in the northern hemisphere [1].

6   VOLATILE ORGANIC COMPOUNDS (VOC)

A wide range of organic compounds are emitted as a result of fossil fuel and biomass burning. However, the greatest portion of the emission is from vegetation. This consists mostly of isoprene and terpenes with more isoprene than terpenes. Smaller proportions of other compounds such as acetone are also emitted. The recommended estimates of quantities emitted, given in the Third Assessment Report of IPCC [1], total $571\,\mathrm{Tg\,C\,yr^{-1}}$. Of these 571, a total of 161 are from the burning of fossil fuel, 33 from biomass burning and 377 from vegetation.

Most of the VOC are readily oxidized by HO. However, photochemical reaction leading to the formation of ozone as a byproduct together with products containing carbon occurs in the presence of NO_x. Isoprenes, on oxidation, give about a 5 % yield of solid aerosols [1].

7   NITRIC OXIDE; NO; [CAS 10102-43-9] AND NITROGEN DIOXIDE; NO_2; [CAS 10102-44-0]

The bulk of these gases is a product of fossil fuel combustion.

These two oxides of nitrogen readily interconvert in the atmosphere and are often given the general formula NO_x. They are thought [1] to catalyse the formation of O_3 and OH and removal of CO through a reaction sequence such as

$$OH + CO + O_2 \longrightarrow CO_2 + HO_2$$

$$HO_2 + NO \longrightarrow NO_2 + HO$$

$$NO_2 + h\nu \longrightarrow NO + O(^3P)$$

$$O(^3P) + O_2 + M \longrightarrow O_3 + M$$

Table 1.1 Estimated quantities of NO_x from various sources reported in *climate change 2001* [1]

Source	Estimated global emission/Tg N yr^{-1}
Fossil fuel	33.0
Aircraft	0.7
Biomass burning	7.1
Soils	5.6
Lightning	5.0
Stratosphere	<0.5
Total	51.9

The overall reaction is

$$CO + 2O_2 + hv \longrightarrow CO_2 + O_3$$

Much of the NO_x is eventually oxidized by OH to give nitric acid. Some reacts with acetyl peroxide to from peroxyacetyl nitrate (PAN). This is a relatively stable compound which can be transported long distances without decomposition [1].

Current sources of NO_x are shown in Table 1.1

8 NITROUS OXIDE; N₂O; [CAS 10024-97-2]

Nitrous oxide is a powerful greenhouse gas with a global warming potential 296 times that of carbon dioxide [1]. It has a long lifetime of 120 years and reaches the stratosphere. About 90 % is eventually decomposed by photodissociation and the rest by reaction with electronically excited oxygen atoms [O(^{1}D)]. Sources of nitrous oxide are shown in Table 1.2.

Table 1.2 Estimates of the emission of nitrous oxide from various sources in 1994 reported in *climate change 2001* [1] and based of measurements given in Refs [12] and [13]

Source	Emission/Tg N yr^{-1}	Range
Ocean	3.0	1–5
Atmosphere (NH_3 oxidation)	0.6	0.3–1.2
Tropical soils:		
Wet forest	3.0	2.2–3.7
Dry savannas	1.0	0.5–2.0
Temperate soils:		
Forests	1.0	0.1–2.0
Grasslands	1.0	0.5–2.0
Total from natural sources	*9.6*	*4.6–15.9*
Agricultural soils	4.2	0.6–14.8
Biomass burning	0.5	0.2–1.0
Industrial sources	1.3	0.7–1.8
Cattle farming	2.1	0.6–3.1
Total from anthropogenic sources	*8.1*	*2.1–20.7*
Total from all sources	17.7	6.7–36.6

Table 1.3 Estimated emission of methane in 1992 by Lelieveld *et al.* [14] reported in climate change 2002 [1][a]

Source	Estimated emission/Tg CH_4 yr^{-1}
Wetlands including rice fields	225
Termites	20
Oceans	15
Decay of methane hydrates in sediments	10
Energy generation (power stations, etc.)	110
Landfills	40
Ruminants	115
Waste treatment	25
Biomass burning	40
Total	600

[a]Details of estimates by other authors are also given in [1].

9 METHANE; CH_4; [CAS 74-82-8]

Methane is a greenhouse gas with a global warming potential 23 times that of carbon dioxide over a 100-year period [2]. Currently the concentration is about 1745 ppb. This corresponds to a total burden of about 4850 Tg CH_4 [1]. About 60 % of the emission of methane is from agriculture, fossil fuel burning and waste disposal and other anthropogenic sources. The main agricultural source is the growing of rice. The main natural sources of methane is decaying vegetation in wetlands (see Table 1.3). There is little decrease in concentration with altitude in the troposphere. At higher altitudes decomposition occurs mainly as a result of oxidation by OH radicals [1]. Figure 1.2 shows the predicted variation in the concentration with altitude at high altitudes for specified conditions and time of year.

10 HYDROFLUOROCARBONS

The most abundant hydrofluorocarbons are trifluoromethane (HFC-23), fluoroethane (HFC-134a) and 1,1-difluoroethane (Table 1.4). All are products of industrialization and concentrations have increased significantly during recent years. The longest lived is fluoroethane with a lifetime of 269 years and a global warming potential 1200 times that of carbon dioxide. Like other hydrofluorocarbons and hydrochlorofluorocarbons, they are removed chiefly by reaction with OH radicals.

11 PERFLUOROCARBONS, SULFUR HEXAFLUORIDE AND TRIFLUOROMETHYL SULFUR HEPTAFLUORIDE

Tetrafluoromethane and hexafluoroethane have global warming potentials of 5700 and 11 900 relative to carbon dioxide, respectively [1]. They have lifetimes of 50 000 and 10 000 years, respectively [2]. They are very stable. Sulfur hexafluoride also has a high global warming potential (22 200) and a long lifetime of

Table 1.4 Concentrations and lifetimes of important greenhouse gases in the troposphere reported in *climate change 2001* [1]

Gas	Concentration/ ppt	Lifetime/ yr	100-yr GWP
Carbon dioxide; CO_2; [CAS 124-38-9]	367 (ppm)	– [a]	1
Methane; CH_4; [CAS 74-82-8]	1745 (ppb)	8.4–12	23
Nitrous oxide; N_2O; [CAS 10024-97-2]	314 (ppb)	120–114	296
Perfluoromethane, tetrafluoromethane; CF_4; [CAS 75-73-0]	80	>50000	5700
Perfluoroethane, hexafluoroethane; C_2F_6; [CAS 76-16-4]	3.0	10000	11900
Sulfur hexafluoride; SF_6; [CAS 2551-62-4]	4.2	3200	22200
Trifluoromethane (HFC-23); CHF_3; [CAS 75-46-7]	14	260	12000
1,1,1,2-Tetrafluoroethane (HFC-134a); $C_2H_2F_4$; [CAS 811-97-2]	7.5	13.8	1300
1,1-Difluoroethane (HFC-152a); $C_2H_4F_2$; [CAS 75-37-6]	0.5	1.40	120
Trichlorofluoromethane (CFC-11); CCl_3F; [CAS 75-69-4]	268	45	4600
Dichlorodifluoromethane (CFC-12); CCl_2F_2; [CAS 75-71-8]	533	100	10600
Chlorotrifluoromethane (CFC-13); $CClF_3$; [CAS 75-72-9]	4	640	14000
1,1,2-Trichlorotrifluoroethane (CFC-113); $C_2Cl_3F_3$; [CAS 76-13-1]	84	85	6000
1,2-Dichlorotetrafluoroethane (CFC-114); $C_2Cl_2F_4$; [CAS 374-07-2]	15	300	9800
Chloropentafluoroethane (CFC-115); C_2ClF_5 [CAS 76-15-3]	7	1700	7200
Carbon tetrachloride; CCl_4; [CAS 56-23-5]	102	35	1800
1,1,1-Trichloroethane; $C_2H_3Cl_3$; [CAS 71-55-6]	69	4.8	140
Chlorodifluoromethane (HCFC-22); $CHClF_2$; [CAS 75-45-6]	132	11.9	1700
1,1-Dichloro-1-fluoroethane (HCFC-141b); $C_2H_3Cl_2F$; [1717-00-6]	10	9.3	700
1-Chloro-1,1-difluoroethane (HCFC-142b); $C_2H_3ClF_2$; [75-68-3]	11	19	2400
Bromochlorodifluoromethane (Halon-1211); $CBrClF_2$; [353-59-3]	3.8	11	1300
Bromotrifluoromethane (Halon-1301); $CBrF_3$; [75-63-8]	2.5	65	6900
1,2-Dibromotetrafluoroethane (Halon-2402); $C_2Br_2F_4$; [124-73-2]	0.45	<20	

[a] There is currently no consensus on a value of the lifetime of carbon dioxide in the atmosphere because of the complexity of the carbon cycle.

3200 years [1,2]. Tetrafluoromethane and sulfur hexafluoride occur naturally in fluorites. Tetrafluoromethane therefore has a natural abundance of 40 ppt and sulfur hexafluoride one of 0.01 ppt. The concentrations of all these gases have increased very markedly since the 1970s [15–17]. The present concentration of tetrafluoromethane is now about 80 ppt and that of sulfur hexafluoride about 4.2 ppt. The only significant removal from the atmosphere of any of these compounds is by photolysis or reaction with ions.

Recently, there has been concern over the finding of trifluoromethyl sulfur pentafluoride in the atmosphere [18]. This has the largest radiative forcing per molecule of any of the greenhouse gases because it absorbs strongly in the infrared region. The concentration has grown from almost zero in the late 1960s to about 0.12 ppt in 1999. If the lifetime is similar to that of sulfur hexafluoride, the GWP over 100 yr would be about 18 000. However, a recent investigation of the properties of the compound by dissociative electron attachment spectrometry showed dissociation to give $SF_5^- + CF_3$ and the formation of ions CF_3^- and F^- [19]. The authors stated that the work indicated the possibility that photoinduced electron dissociation may occur when SF_5CF_3 is adsorbed on polar stratospheric ice particles. If this is indeed the case the atmospheric lifetime would be smaller than that of SF_6 and the GWP less than 18 000.

12   HYDROXYL RADICALS; HO; [CAS 3352-57-6]

Hydroxyl radicals react with greenhouse gases which contain one or more hydrogen atoms. They also oxidize carbon dioxide. The primary source of OH in the troposphere is the photodissociation of O_3:

$$O_3 + h\upsilon \longrightarrow O(^1D) + O_2$$

followed by

$$O(^1D) + H_2O \longrightarrow 2OH$$

13   OZONE; O₃; [CAS 10028-15-6]

Tropospheric ozone is a greenhouse gas and has a major effect on the climate. It is also toxic to animals and plants. The global total is estimated to be about 370 Tg O_3 or about 50 ppb. This is thought to be an increase of at least 30 ppb since the beginning of the 20th century [1]. Diffusion of ozone from the stratosphere is one source of tropospheric ozone. However, much of the ozone is produced by photochemical reactions involving NO_x and other pollutants. The principal destruction pathways involve reaction with HO_2 and reaction with $O(^1D)$, itself a product of photodissociation of O_3. Ozone is also lost to surface vegetation. The lifetime of tropospheric ozone is only a few weeks. As a consequence, it is not uniformly distributed and concentrations in different areas vary between about 10 and 100 ppb [1].

The concentration is much greater in the stratosphere where it is formed from oxygen by photochemical reaction. It absorbs harmful ultraviolet radiation. This has a small cooling effect on the earth's surface. Ozone is destroyed in the stratosphere by photolysis following absorption of UV radiation. It is also destroyed by the direct or indirect effect of breakdown products formed by photochemical reactions of CFCs and the corresponding bromine compounds [20]. These breakdown products ultimately yield chlorine or bromine, which is photodissociated to atomic chlorine or bromine. Ozone is destroyed by chain reactions involving atomic chlorine and, to a lesser extent, atomic bromine. Breakdown products from CFCs also form hydrogen chloride and chlorine nitrate. Both of these compounds are stable under cold stratospheric conditions and act as reservoirs for chlorine. They undergo a surface reaction on the surface of ice particles in polar stratospheric clouds which form over polar regions in the winter. This leads to formation of molecular chlorine, which is photochemically decomposed to chlorine atoms in springtime. Chlorine nitrate is re-formed as shown below:

$$HCl + ClONO_2 \xrightarrow{\text{ice}} HNO_3(s) + Cl_2$$

$$Cl_2 + h\upsilon \longrightarrow 2Cl$$

$$2Cl + 2O_3 \longrightarrow 2ClO + 2O_2$$

$$ClO + NO_2 \longrightarrow ClONO_2$$

Overall reaction:

$$HCl + 2O_3 + NO_3 \longrightarrow HNO_3(s) + 2O_2 + ClO$$

ClO can decompose to yield additional chlorine atoms.

Breakdown products from bromine compounds in the atmosphere lead to the formation of hydrogen bromide and bromine nitrate. These bromine reservoirs are more readily decomposed by photolysis than the corresponding chlorine compounds and yield bromine atoms. These are much more effective at removing ozone than are chlorine atoms [21] (see Section 10.3).

14 VARIATION OF THE NATURE OF THE ATMOSPHERE WITH ALTITUDE

The density of the atmosphere decreases with increase in altitude. The temperature decreases over the first 15–20 km but increases with altitude at higher to about 50 km. Concentrations of various reactive species change with altitude. Many species undergo photolysis in the upper atmosphere. The rate constant for photolysis is the first-order rate constant for the reaction

$$X \xrightarrow{h\upsilon} \text{dissociation products}$$

It is usually given the symbol J and referred as the 'J' value:

$$-d[X]/dt = J[X]$$

The constant J depends on the actinic flux $F(\lambda)$, the absorption cross-section of the molecule $\sigma(\lambda)$ and the quantum yield $\phi(\lambda)$. For a particular photolytic reaction (i) subject to radiation of wavelength λ_1 to λ_2 the value of J_i is given by

$$J_i = \int_{\lambda_1}^{\lambda_2} F(\lambda)\sigma_i(\lambda)\phi_i(\lambda)\, \mathrm{d}\lambda$$

Absorption cross-sections and quantum yields depend on wavelength and may change with change of temperature and pressure. They are specific for a particular species and reaction. The actinic flux changes with time of day and of year, longitude, latitude and altitude. It is greatly affected by the earth's surface albedo. It is also affected by species in the atmosphere which scatter or absorb radiation. Accurate prediction of photolysis rates depends on complex models [22].

Typical species distribution curves and variations of J values with altitude for local noon on March 15 at a latitude of 40° N are shown in Figures 1.1–1.7. The corresponding variations in temperature, pressure and density are shown in Figure 1.8. The following is a statement taken from JPL Publication 97-4 [23] of the assumptions that were made and procedures adopted when the original graphs were prepared:

They were generated by the LLNL 2-D model of the troposphere and stratosphere. The temperature profile is an interpolation to climatological values. Surface source gas boundary conditions are those for the year 1990 as reported in chapter 6 of the WMO/UNEP report [24]. The equatorial tropopause source gas mixing ratios

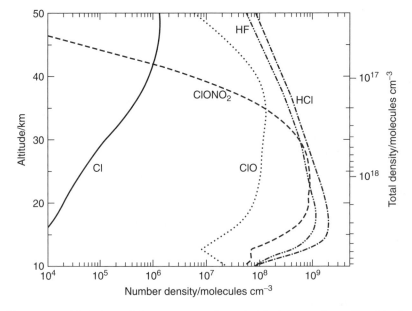

Figure 1.1 Variation of the concentration of chlorine species with altitude (March 15, local noon, 40° N)

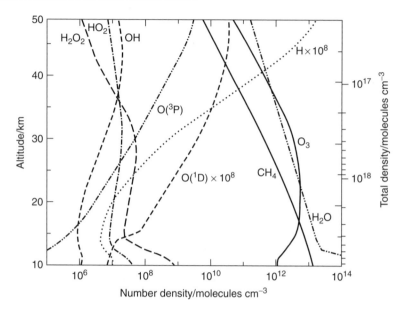

Figure 1.2 Variation of the concentration of O and H species and of CH_4 with altitude (March 15, local noon, 40° N)

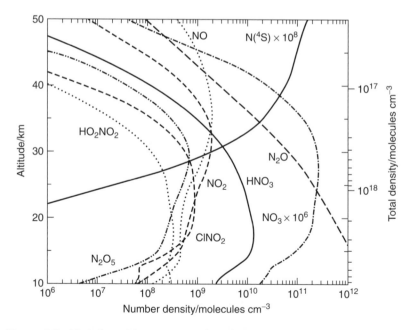

Figure 1.3 Variation of the concentration of nitrogen compounds with altitude (March 15, local noon, 40° N)

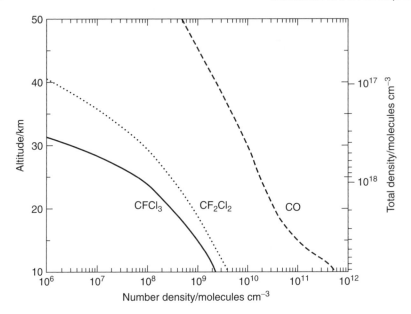

Figure 1.4 Variation of the concentrations of CFCl$_3$, CF$_2$Cl$_2$ and CO with altitude (March 15, local noon, 40° N)

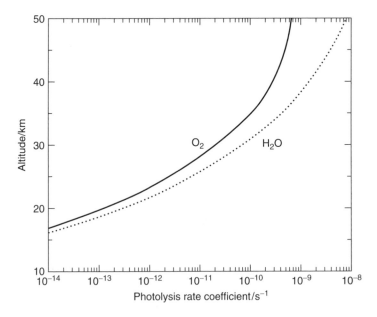

Figure 1.5 Variation with altitude of first-order photolysis rate constants (*J* values) for O$_2$ and H$_2$O (March 15, local noon, 40° N)

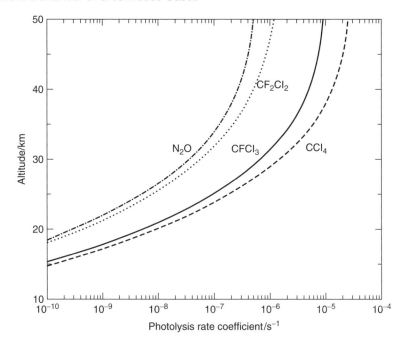

Figure 1.6 Variation of first-order photolysis rate constants (J values) for N_2O, $CFCl_3$ and CF_2Cl_2 with altitude (March 15, local noon, 40° N)

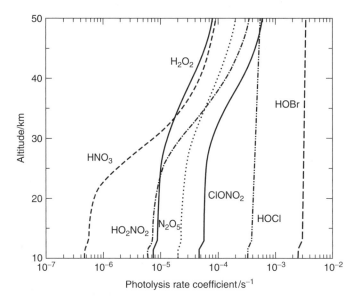

Figure 1.7 Variation of first-order photolysis rate constants (J values) for H_2O_2, HOBr, HOCl, HNO_3, HO_2NO_2, N_2O_5 and $ClONO_2$ with altitude (March 15, local noon, 40° N)

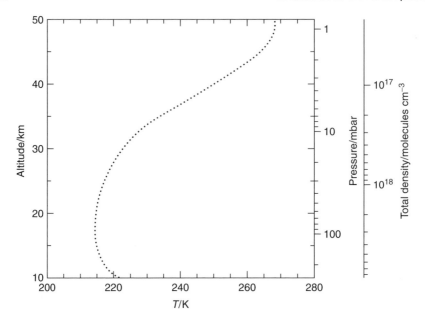

Figure 1.8 Variations in pressure, temperature and density with altitude (March 15, local noon, 40° N)

are: total chlorine 3.4 ppb, total fluorine 1.6 ppb, total bromine 18 ppt, methane 1.67 ppm, and nitrous oxide 309 ppb. The kinetic parameters used were consistent, to the extent possible, with the current recommendations. Representations of sulfate aerosol and polar stratospheric heterogeneous processes which were included are hydrolysis of nitrogen pentoxide and chlorine and bromine nitrate and reaction of hydrogen chloride with chlorine nitrate and hypochlorous acid. The model run represents a periodic steady-state atmosphere with 1990 surface abundances of source gases.

The 'J' values were calculated with a clear sky, two-stream radiative transfer model with wavelength binning of 5 nm above 310 nm and 500 cm^{-1} below. Surface reflectance includes the effect average cloudiness on the albedo. Oxygen cross-sections in the Schumann–Runge region were calculated by the method of Allen and Frederick [25], corrected for the Herzberg continuum values of Yoshino et al. [26].

(LLNL 2-D refers to a two-dimensional model developed at the Lawrence Livermore National Laboratory).

The graphs are included in the current work to give an indication of the relative concentrations at different altitudes and rates of photolysis of a selection of species in the atmosphere. It is stated in JPL Publication 97-4 that the fluxes and profiles are given to provide 'order of magnitude' values of important photochemical parameters. They are not intended to be standards or recommended values.

Graphs of estimated data for solar irradiances and fluxes are also included in JPL Publication 97-4.

REFERENCES

1. Prather, M.; Ehhalt, D.; Dentener, F.; Derwent, R.; Dlugokencky, E.; Holland, E.; Isaksen, I.; Katima, J.; Kirchhoff, V.; Matson, P.; Midgley, P.; Wang, M., Atmospheric chemistry and greenhouse gases, in *Climate Change 2001: the Scientific Basis. Contribution of Working Group I to the Third Assessment Report of the Intergovernmental Panel on Climate Change*, Houghton, J. T.; Ding, Y.; Griggs, D. J.; Noguer, M.; van der Linden, P. J.; Dai, X.; Maskell, K.; Johnson, C. A. (eds), Cambridge University Press, Cambridge, 2001, 881pp.
2. Ramaswamy, V.; Boucher, O.; Haigh, J.; Hauglustaine, D.; Haywood, J.; Myhre, G.; Nakajima, T.; Shi, G. Y.; Solomon, S., Radiative forcing of climate change, in *Climate Change 2001: the Scientific Basis. Contribution of Working Group I to the Third Assessment Report of the Intergovernmental Panel on Climate Change*, Houghton, J. T.; Ding, Y.; Griggs, D. J.; Noguer, M.; van der Linden, P. J.; Dai, X.; Maskell, K.; Johnson, C. A. (eds), Cambridge University Press, Cambridge, 2001, 881pp.
3. Houghton, J. T.; Callander, B. A.; Varney, S. K. (eds), *Climate Change 1990. The Intergovernmental Panel on Climate Change*, Cambridge University Press, Cambridge, 1990.
4. Houghton, J. T.; Meira Filho, L. G.; Callander, B. A.; Harris, N.; Kattenberg, A.; Maskell, K. (eds), *Climate Change 1995. The Science of Climate Change. Contribution of Working Group I to the Second Assessment Report of the Intergovernmental Panel on Climate Change*, Cambridge University Press, Cambridge, 1996.
5. US Environmental Protection Agency, *Global Warming Potentials*, EPA, Washington, DC, 2002; http://yosemite.epa.gov/oar/globalwarming/EmissionsNationalGlobalWarmingPotentials.htm.
6. Prentice, I. C.; Farquhar, G. D.; Fasham, M. J. R.; Goulden, M. L.; Heimann, M.; Jaramillo, V. J.; Kheshgi, H. S.; Le Quére, C.; Scholes, R. J.; Wallace, D. W. R., The carbon cycle and atmospheric carbon dioxide, in *Climate Change 2001: the Scientific Basis. Contribution of Working Group I to the Third Assessment Report of the Intergovernmental Panel on Climate Change*, Houghton, J. T.; Ding, Y.; Griggs, D. J.; Noguer, M.; van der Linden, P. J.; Dai, X.; Maskell, K.; Johnson, C. A. (eds), Cambridge University Press, Cambridge, 2001, 881pp.
7. Farquhar, G. D.; Lloyd, J.; Taylor, J. A.; Flanagan, L. B.; Syvertsen, J. P.; Hubick, K. T.; Wong, S. C.; Ehleringer, J. R., *Nature*, 1993, **363**, 439–43.
8. Cias, P.; Denning, P. A. S.; Tans, P. P.; Berry, J. A.; Randall, D. A.; Collatz, G. J.; Sellers, P. J.; White, J. W. C.; Trolier, M.; Meijer, H. A. J.; Francey, R. J.; Monfray, P.; Heimann, M., *J. Geophys. Res. Atmos.*, 1997, **102**, 5857–72.
9. Lloyd J.; Farquhar, G. D., *Funct. Ecol.*, 1996, **10**, 4–32.
10. Waring, R. H.; Landsberg, J. J.; Williams, M., *Tree Physiol.*, 1998, **18**, 129–34.
11. Houghton, R. H., *Tellus*, 1999, **51B**, 298–350.
12. Mosier, A.; Kroeze, C.; Nevison, C.; Oenema, O.; Seitzinger, S.; van Cleemput, O., *Nutrient Cycling Agroecosyst.*, 1998, **52**, 225–48.
13. Kroeze, C.; Mozier, A.; Bouwman, I., *Global Biogeochem. Cycles*, 1999, **13**, 1–8.
14. Lelieveld, J.; Crutzen, P.; Demtener, F. J., *Tellus*, 1998, **50B**, 128–50.
15. Harnisch, J.; Borchers, R.; Fabian, P.; Maiss, M., *Geophys. Res. Lett.*, 1996, **23**, 1099–102.
16. Maiss, M.; Steele, L. P.; Francey, R. J.; Fraser, P. J.; Langenfelds, R. L.; Trivett, N. B. A.; Levin, I., *Atmos. Environ.*, 1996, **30**, 1621–9.
17. Maiss, M.; Brenninkmeijer, C. A. M., *Environ. Sci. Technol.*, 1998, **32**, 3077–86.
18. Sturges, W. T.; Wallington, T. J.; Hurley, M. D.; Shine, K. P.; Sihra, K.; Engel, A.; Oram, D. E.; Penkett, S. A.; Mulvaney, R.; Brenninkmeijer, C. A. M., *Science*, 2000, **289**, 611–3.
19. Sailer, W.; Drexel, H.; Plec, A.; Grill, V.; Mason, N. J.; Illenberger, E.; Skalny, J. D.; Mikoviny, T.; Scheier, P.; Mark, T. D., *Chem. Phys. Lett.*, 2002, **351**, 71–8.

20. Molina, M. J.;   Tso, T.-L.;   Molina, L. T.;   Wang, F. C.-Y.,   *Science*,   1988,   **238**, 1253–17.
21. www.nas.nasa.gov/Services/Education/Resources/TeacherWork/Ozone/Ozone_chem. html.
22. Roselle, S. J.; Schere, K. L.; Pleim, J. E., Photolysis rates for CMAQ, in *Science Algorithms of the EPA Models*, Environmental Protection Agency Report EPA/600/R-99/030, National Exposure Research Laboratory, Research Triangle Park, NC, 1999, Chapt. 14.
23. Connell, P., *Chemical Kinetics and Photochemical Data for Use in Stratospheric Modeling*, JPL Publication 97-4, National Aeronautics and Space Administration, Jet Propulsion Laboratory, California Institute of Technology, Pasadena, CA, 1997, pp. 58–267.
24. WMO, *Scientific Assessment of Ozone Depletion: 1994*, World Meteorological Organization Global Ozone Research and Monitoring Project, Report No. 37, National Aeronautics and Space Administration, Geneva, 1994.
25. Allen, M.; Frederick, J. E., *J. Atmos. Sci.*, 1982, **39**, 2066–75.
26. Yoshino, K.; Cheung, A. S. C.; Esmond, J. R.; Parkinson, W. H.; Freeman, D. E.; Guberman, S. L.; Jenouvrier, A.; Coquart, B.; Merienne, M. F., *Planet. Space Sci.*, 1988, **36**, 1469–75.

CHAPTER 2

Presentation of Solubility Data: Units and Applications

Stephen Schwartz
Atmospheric Sciences Division, Brookhaven National Laboratory, Upton, NY, USA

The solubility of gases in water and other aqueous media such as seawater and more concentrated solutions is central to the description of the uptake and reactions of these gases in aerosols, precipitation, surface water and other aqueous media such as the intracellular fluids of plants and animals. It is also pertinent to sampling of soluble atmospheric gases in aqueous medium for analytical purposes. This book presents evaluated summaries of data pertinent to the solubility of gases in aqueous media. This chapter introduces the terminology by which this solubility is described and the pertinent units and presents examples of applications pertinent to atmospheric chemistry. As is seen below, a variety of units have been and continue to be employed for gas solubility data, so some attention must be given to this subject. As this is an IUPAC publication, every effort is made to employ units that are consistent with the International System of Units (Système International, SI) [1]. However, in IUPAC publications of solubility data it is usual to publish data in the original units in addition to SI units. The consistency of SI makes this system of units convenient for application in atmospheric chemistry and related disciplines [2]. However, as elaborated below, there are some departures from strict SI that persist in chemical thermodynamics that require special consideration.

The solubility of weakly soluble gases in liquids is one of the oldest branches of physical chemistry. The origins of this study go back to William Henry (1775–1836), whose presentation before the Royal Society and subsequent publication [3] gave rise to what has come to be known as Henry's law. That paper describes a series of experiments with several gases which remain of concern to present-day atmospheric scientists, CO_2, H_2S, N_2O, O_2, CH_4, N_2, H_2, CO and PH_3. The gas and liquid were contacted and allowed to reach solubility equilibrium at various values of applied pressure, with the amount of absorbed

Chemicals in the Atmosphere – Solubility, Sources and Reactivity.   Edited by P. G. T. Fogg and J. Sangster
© 2003 IUPAC   ISBN: 0-471-98651-8

gas determined from the decrease in gaseous volume after equilibration with the water. The generalization reached by Henry is stated in the original paper as follows:

> The results of a series of at least fifty experiments on [the several gases] establish the following general law: *that under equal circumstances of temperature, water takes up, in all cases, the same volume of condensed gas* [at higher applied pressure] *as of gas under ordinary pressure.* But as the spaces [volumes] occupied by every gas are inversely as the compressing force [pressure], it follows, *that water takes up, of gas condensed by one, two, or more additional atmospheres, a quantity which, ordinarily compressed, would be equal to twice, thrice &c. the volume absorbed under the common pressure of the atmosphere.* [italics in original]

In other words, the amount of gas absorbed, as measured by volume at ordinary pressure, is proportional to the applied pressure.

It is also clear in the presentation that Henry realized that the amount of dissolved material (extensive property) scaled linearly with the amount of water; in fact, Henry normalized his results to a standard amount of water, so that it is clear that he recognized that it is the concentration (intensive property) that is proportional to the pressure of the gas.

In a subsequent appendix, Henry [4] noted that the residuum of undissolved gas was not always entirely pure and exhibited an increasing proportion of less soluble contaminant, necessitating correction of his originally presented data. He states in this connection:

> *For, the theory which Mr. DALTON has suggested to me on this subject, and which appears to be confirmed by my experiments, is, that the absorption of gases by water is a purely a mechanical effect, and that its amount is exactly proportional to the density of the gas, considered abstractedly from any other gas with which it may accidentally be mixed. Conformably to this theory, if the residuary gas contain 1/2, 1/10, or any other proportion, of foreign gas, the quantity absorbed by water will be 1/2, 1/10, &c. short of the maximum.*

In this elaboration, Henry makes use of Dalton's newly enunciated law of partial pressures to correct his previously reported data.

This brief summary of these experiments permits us to restate Henry's law with complete fidelity to his original formulation, changing only the terminology, as follows: the equilibrium concentration of a dissolved gaseous species is proportional to the partial pressure of the gas, the proportionality constant being a property of the gas and a function of temperature, or

$$[X(aq)] = k_{H_X}(T) p_X \qquad (2.1)$$

where $[X(aq)]$ is the aqueous concentration, p_X is the gas-phase partial pressure and $k_{H_X}(T)$ is the Henry's law coefficient of the gas X in water at temperature T.

The law is readily generalized to other solvents, the Henry's law coefficient depending on both the solvent and the solute, to mixed solvents, and to solvents containing nonvolatile solutes. In the terminology of modern physical chemistry, Henry's law is a limiting law:

$$k_{H_X} = \lim_{p_X \to 0} \frac{[X(aq)]}{p_X}$$

the reason being that the properties of the solvent change with increasing concentrations of solute. This situation can be accounted for by expressing Henry's law in terms of the activities in the pertinent phases:

$$a_{X(aq)} = k_{H_X}(T) a_{X(g)}$$

We thus observe that the Henry's law coefficient is an equilibrium constant for the reaction

$$X(g) = X(aq)$$

Recognition of this leads to an intrinsic connection to chemical thermodynamics, and in particular to the fact that the Henry's law coefficient is related in the customary way to the free energy change of reaction as

$$k_{H_X} = \exp(-\Delta G^\circ / RT)$$

where in this case ΔG° is the free energy of dissolution:

$$\Delta G^\circ = \Delta_f G^\circ(X_{aq}) - \Delta_f G^\circ(X_g)$$

the $\Delta_f G^\circ$s being the standard free energies of formation of the aqueous and gaseous species. In practice, it is the usual situation that knowledge of the Henry's law coefficient allows one to determine the unknown standard free energy of formation of one of the two species from the known value for the other, rather than the Henry's law coefficient being calculated from two known standard free energies of formation. However, there are important practical exceptions in instances of highly reactive gases whose reactivity precludes establishment of the solubility equilibrium at measurable concentration and partial pressure.

The reason for focusing here on Henry's law and for quoting from Henry's original paper is in part to provide justification for the form of Henry's law employed in the present work, and advocated for universal use, namely that it is the aqueous-phase concentration (or, alternatively, molality) that is viewed as proportional to the gas-phase partial pressure, rather than the other way around, as has been a customary expression of Henry's law by many subsequent investigators in the nearly 200 years since Henry's initial report. From the perspective of the atmospheric scientist, this approach has a certain further justification as well: it is usually the gas-phase partial pressure that can be considered the independent or controlling variable, with the aqueous-phase concentration dependent on the gas-phase partial pressure. Henry's law is the basis for any consideration of exchange of material between the gas phase and natural waters, and the Henry's law coefficient is central to expressions that describe the extent and rate of such exchange, even for species whose equilibrium solubility is not adequately described by Henry's law.

It must be stressed that expressing Henry's law according to Equation (2.1) is by no means universal either in the physical chemistry literature or in the environmental sciences. Frequently it is customary to express Henry's law the other way around (gas-phase partial pressure proportional to aqueous-phase concentration); this can even be sometimes more convenient, as in considerations of the evasion of dissolved gases from solution, as from a lake to the atmosphere. In any event, it is inevitable to encounter Henry's law expressed in this reverse sense. Also, a variety of modes of expression have been employed for describing the concentration of dissolved gas, such as the volume that it would occupy at a standard pressure, or mole fraction relative to solvent, that have led to yet further variants of Henry's law and to tabulations of Henry's law coefficients in a multiplicity of units. It is thus imperative, at the present time and for the foreseeable future, to pay attention to the sense and units of Henry's law as employed by different investigators and to convert Henry's law coefficients from one set of units (and sense of the law) to another. Consequently, we address these issues in this chapter and present algorithms for and examples of such conversions. As noted above, there is a further desire to employ SI units, and the consequent necessity for making the use of these units more familiar and for dealing with special problems associated with the fact that the units conventions of chemical thermodynamics are not entirely consistent with SI.

For a detailed review of modeling interactions between gaseous species and liquid water in the atmosphere, see Sander [5].

1 UNITS

As a measure of aqueous-phase abundance we employ molality, the mixing ratio of the solute in the solution, having unit mole of solute per kg of solvent, $mol\,kg^{-1}$. Use of this measure and unit is consistent with current practice in chemical thermodynamics and is preferred to concentration, which, because of density change associated with temperature change, is not constant with temperature for a solution of a given composition. Nonetheless, concentration is commonly employed, with the most common unit being molar (moles per litre of solution, $mol\,l^{-1}$ or M). This unit suffers from the further disadvantage that the litre is not an SI unit, the SI unit of volume being the cubic meter. In practice, because the mass of $1\,l$ of water is approximately equal to $1\,kg$, the molality of a dilute aqueous solution is approximately equal numerically to the molar concentration. More precisely, for a single component solution, the concentration of a solute X, C_X, is related to the molality of the solute m_X as

$$C_X \left(\frac{mol_{solute}}{l_{solution}} \right) = m_X \left(\frac{mol_{solute}}{kg_{solvent}} \right) \times \rho_{solvent} \left(\frac{kg_{solvent}}{l_{solution}} \right) \qquad (2.2)$$

where $\rho_{solvent}$ denotes the density of the solvent in the solution:

$$\rho_{solvent} = \frac{\rho_{solution}}{1 + MW_X m_X}$$

where $\rho_{solution}$ denotes the solution density and MW_X denotes the molar mass (molecular weight) of species X, in units of $kg\,mol^{-1}$. More generally, for a multicomponent solution,

$$\rho_{solvent} = \frac{\rho_{solution}}{1 + \sum_i MW_{X_i} m_{X_i}} \tag{2.3}$$

where the summation is taken over all solutes. For dilute aqueous solutions the solvent density is nearly equal to $1\,kg\,mol^{-1}$, so the concentration and molality are numerically essentially equal, i.e.

$$C_X \left(\frac{mol_{solute}}{l_{solution}} \right) \approx m_X \left(\frac{mol}{kg_{solvent}} \right) \left\{ \times 1 \left(\frac{kg_{solvent}}{l_{solution}} \right) \right\} \tag{2.4}$$

where the quantity in braces can be omitted for numerical work but must be retained for units calculus. For more concentrated solutions of importance in atmospheric chemistry, such as aqueous clear-air aerosol, including sea salt, or for stratospheric sulfuric acid aerosol, the volume fraction of solute and the resulting departure from unit solvent density can be appreciable.

Where sufficient pressure dependence data are available to report the Henry's law coefficient as a limiting law, then that value is reported, i.e.

$$k_{H_X}(mol\,kg^{-1}\,bar^{-1}) \equiv \lim_{p_X \to 0} \frac{m_X(mol\,kg_{solvent}^{-1})}{p_X(bar)}$$

When such data are not available (which is generally the situation), the Henry's law coefficient is reported as the quotient of molality by partial pressure, with the partial pressure (range) of the measurements specified:

$$k_{H_X}(mol\,kg^{-1}\,bar^{-1}) = \frac{m_X(mol\,kg_{solvent}^{-1})}{p_X(bar)}$$

For application in atmospheric chemistry, the lack of such pressure-dependent data should make little practical difference provided that there are good reasons to assume that the measurements were made within a pressure range in which Henry's law is applicable. This means that linear extrapolation to zero pressure would correspond to zero concentration.

Solubility data are reported at the temperature(s) of measurement. Where sufficient temperature-dependence data are available, the enthalpy change of solution is reported as

$$\frac{\Delta H^\circ}{R_g} = -\frac{d\ln(k_{H_X}/1\,mol\,kg^{-1}\,bar^{-1})}{d(1/T)}$$

In principle, this quantity can also be obtained from the limiting-law solubility data, but again in practice such data are not generally available, again a situation of little practical consequence.

The unit $mol\,kg^{-1}\,bar^{-1}$ for the Henry's law coefficient is practical for many applications, generally, and specifically in atmospheric chemistry, with the obvious proviso that the partial pressure of the gas be expressed in units of bar.

We adopt the symbol k_H for this practical quantity, and likewise employ the symbol p_X to denote the partial pressure in bar of the species X. However, as these units are not products of powers of base SI units, their use leads to complications, especially in expressions in which they are combined with SI quantities. It is necessary to employ a Henry's law coefficient in SI units in such expressions, which we distinguish by a caret, $\hat{k}_H$, where $\hat{k}_H = 10^{-5}k_H$. Likewise to avoid ambiguity, when we employ partial pressure in pascals, we use the symbol $\hat{p}_X$.

Increasingly the mixing ratio x_X is gaining favour in atmospheric chemistry as the quantity for expressing local abundance; units are mol/mol (air), e.g. nmol/mol. Since Henry's law solubility depends on the partial pressure of the gas, rather than mixing ratio, it is necessary to 'convert' from mixing ratio to partial pressure. This is readily achieved as

$$p_X(\text{bar}) = p(\text{bar})x_X(\text{mol/mol})$$

so that

$$m_X(\text{mol kg}_{\text{solvent}}{}^{-1}) = k_{H_X}(\text{mol kg}^{-1}\text{bar}^{-1})p(\text{bar})x_X(\text{mol/mol})$$

In many applications, such as the evaluation of chemical kinetics and mass transport rates, the concentration is required, rather than the molality, and it is therefore often convenient to employ a Henry's law coefficient that yields the concentration directly. If we denote this concentration Henry's law coefficient H_X (units M bar^{-1}), then within the approximation that the solvent density is equal to 1 kg mol^{-1},

$$H_X\left(\frac{\text{M}}{\text{bar}}\right) \approx k_{H_X}\left(\frac{\text{mol kg}^{-1}}{\text{bar}}\right)\left\{\times 1\left(\frac{\text{kg}_{\text{solvent}}}{\text{l}_{\text{solution}}}\right)\right\} \qquad (2.5)$$

In many considerations of mass transport it is useful to employ a dimensionless Henry's law coefficient, the ratio of concentration in solution (aqueous) phase to that in the gas phase (or vice versa; extreme caution is suggested in using dimensionless Henry's law coefficients because the sense of the equilibrium cannot be inferred from the units of the quantity). We take this dimensionless Henry's law coefficient also in the sense aqueous/gaseous:

$$\tilde{H}_X \equiv \frac{C_{X(\text{aq})}(\text{mol m}^{-3})}{C_{X(\text{g})}(\text{mol m}^{-3})}$$

where the dimensionless property is denoted by the tilde over the symbol. Within the accuracy of the ideal gas law (adequate for all practical purposes in atmospheric chemistry, at least on earth), the concentration (mol m^{-3}) of a gaseous species X having partial pressure $\hat{p}_X$ (Pa) is given as

$$C_{X(\text{g})} = \hat{p}_X/R_g T$$

where R_g is the universal gas constant and T is the absolute temperature. From Equation (2.2) the aqueous-phase concentration of the dissolved gas $(\text{mol}\,\text{m}^{-3})$ is

$$C_{X(aq)}\left(\frac{\text{mol}}{\text{m}^3_{\text{solution}}}\right) = \hat{k}_{Hx}\left(\frac{\text{mol}\,\text{kg}_{\text{solvent}}^{-1}}{\text{Pa}}\right)\hat{p}_X(\text{Pa})\rho_{\text{solvent}}\left(\frac{\text{kg}_{\text{solvent}}}{\text{m}^3_{\text{solution}}}\right)$$

Both the aqueous and gaseous concentrations scale with partial pressure; the ratio of these quantities, the dimensionless Henry's law coefficient, is of course independent of partial pressure:

$$\tilde{H}_X = \hat{k}_{Hx}\left(\frac{\text{mol}\,\text{kg}_{\text{solvent}}^{-1}}{\text{Pa}}\right)\rho_{\text{solvent}}\left(\frac{\text{kg}_{\text{solvent}}}{\text{m}_{\text{solution}}^3}\right)R_g\left(\frac{\text{Pa}\,\text{m}^3}{\text{mol}\,\text{K}}\right)T(\text{K}) \quad (2.6)$$

For all quantities on the right-hand side of Equation (2.6) in strict SI units the quantity $\hat{k}_{Hx}\rho_{\text{solvent}}R_gT(\text{K})$ is a dimensionless group. This expression exhibits its simple form as a direct consequence of employing consistent units, specifically the Henry's law coefficient in units of $\text{mol}\,\text{kg}_{\text{solvent}}^{-1}\,\text{Pa}^{-1}$. Note that for dilute aqueous solution the solvent density $(\text{kg}_{\text{solvent}}\,\text{m}_{\text{solution}}^{-3})$ has a value of approximately $10^3\,\text{kg}_{\text{solvent}}\text{m}_{\text{solution}}^{-3}$.

For the Henry's law coefficient given in units of $\text{mol}\,\text{kg}_{\text{solvent}}^{-1}\text{bar}^{-1}$, a conversion factor must be applied:

$$\tilde{H}_X = 10^{-5}\left(\frac{\text{bar}}{\text{Pa}}\right)k_{Hx}\left(\frac{\text{mol}\,\text{kg}_{\text{solvent}}^{-1}}{\text{bar}}\right)\rho_{\text{solvent}}\left(\frac{\text{kg}_{\text{solvent}}}{\text{m}_{\text{solution}}^3}\right)R_g\left(\frac{\text{Pa}\,\text{m}^3}{\text{mol}\,\text{K}}\right)T(\text{K})$$

For solvent density expressed in units of $\text{kg}_{\text{solvent}}\,\text{l}_{\text{solution}}^{-1}$, this becomes

$$\tilde{H}_X = 10^{-2}k_{Hx}\left(\frac{\text{mol}\,\text{kg}_{\text{solvent}}^{-1}}{\text{bar}}\right)R_g\left(\frac{\text{Pa}\,\text{m}^3}{\text{mol}\,\text{K}}\right)T(\text{K})\rho_{\text{solvent}}\left(\frac{\text{kg}_{\text{solvent}}}{\text{l}_{\text{solution}}}\right)$$

$$\approx 10^{-2}k_{Hx}R_gT$$

where the final approximate equality holds within the approximation of the solvent density equal to $1\,\text{kg}\,\text{l}^{-1}$. The conversion factors are a direct consequence of use of inconsistent units. For this reason, use of strict SI units for the Henry's law coefficient $(\text{mol}\,\text{kg}_{\text{solvent}}^{-1}\,\text{Pa}^{-1})$ is to be preferred where this quantity is combined with others in algebraic expressions.

2   AQUEOUS-PHASE SOLUBILITY EQUILIBRIA

Before proceeding, we should note an important qualification to Henry's law, namely that it applies to sparingly soluble gases or, more precisely, to gases which are physically dissolved but not undergoing chemical reaction in solution to form a chemically different species. Here physical absorption includes solvation but does not include reaction to form a distinct chemical species, such as hydration.

For example, the dissolution of acetaldehyde can be represented by the following sequence of reactions:

$$CH_3CHO(g) \longrightarrow CH_3CHO(aq)$$

$$CH_3CHO(aq) + H_2O \longrightarrow CH_3CH(OH)_2(aq)$$

The first of these is physical dissolution, i.e. the Henry's law equilibrium; the second is a chemical reaction. The pertinent equilibrium expressions are:

$$k_{H_{CH_3CHO}} = \frac{[CH_3CHO(aq)]}{p_{CH_3CHO}}$$

and

$$k_{hyd_{CH_3CHO}} = \frac{[CH_3CH(OH)_2(aq)]}{[CH_3CHO(aq)]}$$

The total dissolved acetaldehyde concentration is evaluated as the sum of the two dissolved species:

$$[CH_3CHO(aq, tot)] = [CH_3CHO(aq)] + [CH_3CH(OH)_2(aq)]$$

$$= k_{H_{CH_3CHO}} p_{CH_3CHO}(1 + k_{hyd_{CH_3CHO}})$$

In practice, whether or not to include the hydrated species as a part of the dissolved component depends on the experiment or application and whether the dissolution process is rapid or slow compared with the hydration reaction or other possible reaction, such as reaction of H_2CO with dissolved S(IV). What is important to note for the present purpose is that the equilibrium concentration of the hydrated species is itself proportional to that of the unhydrated species, so that the total is proportional to the concentration of the unhydrated species and the overall linearity between gas-phase partial pressure and aqueous concentration is retained. One can take advantage of this linearity to define an effective Henry's law coefficient, in this instance

$$k_{H_{CH_3CHO}}{}^* \equiv k_{H_{CH_3CHO}}(1 + k_{hyd_{CH_3CHO}})$$

such that one has for the total dissolved acetaldehyde concentration an expression that is formally identical with Henry's law, viz.

$$[CH_3CHO(aq, tot)] = k_{H_{CH_3CHO}}{}^* p_{CH_3CHO}$$

It is of course crucial in any potentially ambiguous situation to specify whether the Henry's law coefficient refers to the total or only to the unhydrated species. This is important also in correlations of Henry's law coefficients with chemical structure. Physical solubility depends on properties such as polarizability, which depend on molecular structure and number of electrons; correlations based on such properties can obviously be disrupted if there is a structural change associated with the hydration.

Much important in this context is the exclusion from the rubric of Henry's law any chemical reaction that results in a solubility that is not linear in the partial pressure of the gaseous species. Consider the solubility equilibrium for the dissolution of the strong acid HCl:

$$HCl(g) \longrightarrow H^+(aq) + Cl^-(aq)$$

for which the equilibrium relation is

$$k_{eq} = \frac{[H^+][Cl^-]}{p_{HCl}}$$

For both H^+ and Cl^- deriving entirely from the dissolved hydrochloric acid, which may be considered entirely dissociated, then

$$p_{HCl} = k_{eq}^{-1}[Acid]^2$$

where [Acid] denotes the aqueous concentration of the dissolved acid, a nonlinear relation, and *ipso facto*, not Henry's law (here we have considered the aqueous acid concentration to be the independent variable, consistent with the fact that virtually all the material would be present in solution, rendering this concentration the more readily measured variable). For a variety of reasons such as description of mass transport rates it may be useful to consider the actual Henry's law constant of HCl. Formally this is related to the overall equilibrium constant k_{eq} through the sequence of reactions

$$HCl(g) \longrightarrow HCl(aq)$$

$$HCl(aq) \longrightarrow H^+(aq) + Cl^-(aq)$$

which sum to give the overall reaction

$$HCl(g) = H^+(aq) + Cl^-(aq)$$

Correspondingly, the equilibrium expressions are

$$k_{H_{HCl}} = \frac{[HCl(aq)]}{p_{HCl}}$$

$$k_a = \frac{[H^+][Cl^-]}{[HCl(aq)]}$$

and

$$k_{eq} = \frac{[H^+][Cl^-]}{p_{HCl}}$$

where $k_{H_{HCl}}$ is the Henry's law coefficient for HCl, k_a is the acid dissociation constant of aqueous HCl and k_{eq} is the equilibrium constant for the overall reaction. If the overall equilibrium constant k_{eq} can be determined, by measurement of partial pressure and aqueous activities of the two ionic species, and if, generally in a separate experiment, the acid dissociation constant K_a can be determined, then the Henry's law coefficient (equilibrium constant of physical solubility) of HCl can be evaluated as

$$k_{H_{HCl}} = k_{eq}/k_a$$

Where data are available to permit the determination of both the Henry's law coefficient and the equilibrium constant(s) of the pertinent aqueous-phase reaction(s), then all these data are presented. Frequently only the equilibrium constant for the overall reaction is known or, because of uncertainties in determination of the individual component equilibrium constants, it is known with greater accuracy. In that case the overall solubility equilibrium constant is presented, together with the enthalpy change associated with the overall reaction.

3   EFFECTIVE HENRY'S LAW COEFFICIENTS

In the case of a weakly acidic gas dissolving in an aqueous solution that may be considered well buffered relative to the incremental acidity resulting from dissolution of the gas, it may be possible to assume that the acid concentration is constant, in which case the total amount of dissolved gas is linear in the gas-phase partial pressure. Consider the dissolution of SO_2:

$$SO_2(g) \longrightarrow SO_2(aq)$$

$$k_{H_{SO_2}} = \frac{[SO_2(aq)]}{p_{SO_2}}$$

$$SO_2 + H_2O \longrightarrow H^+ + HSO_3^-$$

$$k_{a1} = \frac{[H^+][HSO_3^-]}{[SO_2(aq)]}$$

$$HSO_3^- \longrightarrow H^+ + SO_3^{2-}$$

$$k_{a2} = \frac{[H^+][SO_3^{2-}]}{[HSO_3^-]}$$

For this system it is possible and convenient to define a pH-dependent effective Henry's law coefficient for total dissolved sulfur(IV):

$$k_{H_{S(IV)}}{}^* \equiv k_{H_{SO_2}} \left(1 + \frac{k_{a1}}{[H^+]} + \frac{k_{a1}k_{a2}}{[H^+]^2} \right)$$

such that under the assumption that the aqueous solution is well buffered one obtains a Henry's law-like expression for total dissolved sulfur(IV):

$$[S(IV)(aq)] = k_{H_{S(IV)}}{}^* p_{SO_2}$$

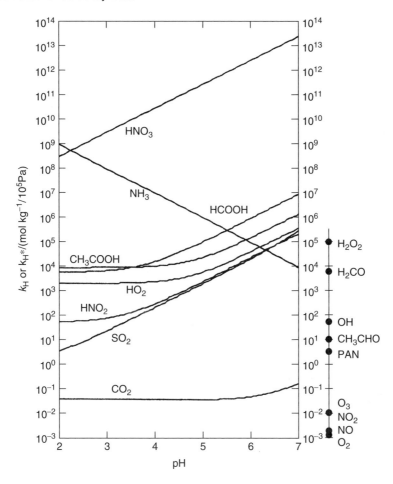

Figure 2.1 pH dependence of the effective Henry's law constant for gases which undergo rapid acid–base dissociation reactions in dilute aqueous solution, as a function of solution pH. The buffer capacity of the solution is assumed to exceed greatly the incremental concentration from the uptake of the indicated gas. Also indicated at the right of the figure are Henry's law constants for nondissociative gases. $T \approx 300\,K$ (modified from Schwartz [6]; for references, see Schwartz [7])

Expressions such as this have found much application in consideration of scavenging of gases by cloudwater and in considerations also of the coupled mass transport and chemical reaction in cloudwater. Figure 2.1 shows the pH dependence of the effective Henry's law coefficient for gases which undergo rapid acid–base dissociation reactions in dilute aqueous solution as a function of solution pH. Also shown for reference are the Henry's law coefficients of several nondissociative gases of atmospheric interest. Note the great increase in effective Henry's law coefficient resulting from the greater solubility of the ionic species than the parent neutral species.

4 PARTITION EQUILIBRIA IN LIQUID-WATER CLOUDS

The key initial characterization of the distribution of a volatile substance between gas and liquid phases in a cloud is the equilibrium distribution. This distribution depends in the first instance on the liquid water content of the cloud. Values of cloud liquid water content L are typically of the order of $1\,g\,m^{-3}$, i.e. $1 \times 10^{-3}\,kg\,m^{-3}$. (The commonly used measure of liquid water content, the liquid water volume fraction, is related to L as L/ρ_{cw}, where ρ_{cw} the density of cloudwater is ca $1\,kg\,m^{-3}$; the liquid water volume fraction is thus of the order of 1×10^{-6}.) For a volatile substance dissolving according to Henry's law, the amount of material per cubic meter in the aqueous phase is

$$n_{X(aq)}\left(\frac{mol}{m^3}\right) = L\left(\frac{kg}{m^3}\right)\hat{k}_{Hx}\left(\frac{mol\,kg^{-1}}{Pa}\right)\hat{p}_X(Pa)$$

The amount of material per cubic meter in the gas phase is

$$n_{X(g)} = \hat{p}_X(1 - L/\rho_{cw})/R_gT \approx \hat{p}_X/R_gT \text{ for } L/\rho_{cw} \ll 1$$

The partition ratio, the ratio of material in the liquid to gas phase, is thus

$$\frac{n_{X(aq)}}{n_{X(g)}} = \frac{L\hat{k}_{Hx}\hat{p}_X}{\hat{p}_X/R_gT} = L\hat{k}_{Hx}R_gT$$

This result is readily extended to more complicated solution equilibria, especially through the use of the effective Henry's law coefficient.

Similar considerations pertain to the evaluation of the equilibrium distribution between air and an aqueous scrubbing solution employed in extracting a soluble gas from the atmosphere; here the liquid water volume fraction would be replaced by the ratio of flow-rates of solution to air.

Note that we have employed the strict SI version of the Henry's law coefficient having units, $mol\,kg^{-1}\,Pa^{-1}$. Unfortunately, if we use the Henry's law coefficient in practical units, $mol\,kg^{-1}\,bar^{-1}$, we must use a units conversion factor of $1 \times 10^{-5}\,bar\,Pa^{-1}$:

$$\frac{n_{X(aq)}}{n_{X(g)}} = 10^{-5}\left(\frac{bar}{Pa}\right)L\left(\frac{kg}{m^3}\right)k_{Hx}\left(\frac{mol\,kg^{-1}}{bar}\right)R_g\left(\frac{Pa\,m^3}{mol\,K}\right)T(K)$$

The partition ratio can be readily assessed from the value of the quantity $L\hat{k}_{Hx}R_gT$. For a given situation of liquid water content and temperature it is useful to consider the value of the Henry's law coefficient k_{H50} such that 50 % of the substance is in each phase:

$$\hat{k}_{H50} = (LR_gT)^{-1} \text{ or } k_{H50} = 10^5(LR_gT)^{-1}$$

For scoping purposes, we note that for $T \approx 300\,K$, $R_gT \approx 2.5 \times 10^3\,m^3\,Pa/(mol\,kg^{-1})$, so that for $L \approx 1 \times 10^{-3}\,kg\,m^{-3}$, $\hat{k}_{H50} \approx 0.4\,mol\,kg^{-1}\,Pa^{-1}$ and $k_{H50} \approx 4 \times 10^4\,mol\,kg\,bar^{-1}$. Note that for most gases of atmospheric interest $k_H \ll k_{H50}$

so that the equilibrium distribution consists of virtually all of the substance present in the gas phase; a key exception is H_2O_2. However, the effective Henry's law coefficient approaches or in some instances substantially exceeds $k_{H_{50}}$, for example HNO_3, which at equilibrium is present virtually entirely in solution phase.

The fraction of the total volatile material that is present in the aqueous phase is

$$F_{X(aq)} = \frac{n_{X(aq)}}{n_{X(g)} + n_{X(aq)}} = \frac{10^{-5} L k_{H_X} R_g T}{1 + 10^{-5} L k_{H_X} R_g T}$$

For a weakly soluble gas such that $\hat{k}_{H_X} R_g T \ll 1$ (i.e. $10^{-5} L k_{H_X} R_g T \ll 1$),

$$F_{X(aq)} \longrightarrow \hat{k}_{H_X} R_g T = 10^{-5} L k_{H_X} R_g T$$

As noted above, the recurrence of the factor 10^5 is due to the standard state of gaseous substances being taken as 1 bar (rather than 1 Pa), which leads to the Henry's law coefficient k_{H_X} having the units $mol\,kg^{-1}\,bar^{-1}$, rather than $mol\,kg^{-1}\,Pa^{-1}$, which would be consistent with SI units.

5 EXPRESSIONS FOR ABUNDANCE OF MATERIALS IN MULTI-PHASE SYSTEMS

Frequently it is useful to compare on an equivalent basis the amounts of trace species present in a volume of air in different phases (gas, aqueous, particulate). A variety of quantities and units have been employed to express such abundances, as summarized by Schwartz and Warneck [2]. Use of mole based units is advocated whenever possible (known composition and molecular weight) because these units display chemically meaningful relationships. Because mixing ratio by volume is appropriate only for gas-phase species, this quantity is not suitable. Suitable quantities and units are concentration ($mol\,m^{-3}$), mixing ratio (mole per mole air) and equivalent partial pressure, the partial pressure that the species would exhibit if an ideal gas, $\hat{p}_X^*$ (Pa) or p_X^* (bar). Of the several quantities, mixing ratio seems most general and suitable.

Consider a species X present in cloudwater (liquid water content $L\,kg\,m^{-3}$) with molality $m_X\,mol\,kg^{-1}$. The several equivalent measures of abundance are presented in Table 2.1.

Table 2.1 Expressions for molar abundance of cloudwater dissolved species. Cloud liquid water volume fraction, L; molality of dissolved species, $m_X\,mol\,kg^{-1}$

Quantity	Unit	Symbol	Expression
Concentration	$mol\,m^{-3}$	c_X	$L m_X$
Partial pressure	Pa	$\hat{p}_X^*$	$L R_g T m_X$
Partial pressure	bar	p_X^*	$10^{-5} L R_g T m_X$
Mixing ratio	mol/mol (air)	x_X	$10^{-5} L R_g T m_X / \hat{p}_{air}(bar)$

6   MASS ACCOMMODATION AND INTERFACIAL MASS-TRANSPORT RATES

Much attention has been paid in recent years to the measurement of mass accommodation coefficients of gaseous substances on the air–water interface and to the examination of the implication of such mass accommodation coefficients on the rate of uptake and reaction of gases in cloudwater. The mass accommodation coefficient is the fraction of collisions from the gas phase on to the solution interface that result in transfer of molecules from the gas phase to the liquid phase. Here we review the pertinent derivations and implications. The Henry's law coefficient is central to both.

Consider the flux of a species X from the gas phase into a solution. The collision rate ($\mathrm{mol\,m^{-2}\,s^{-1}}$) is evaluated from the kinetic theory of gases as

$$\sigma_X{}^+ = \tfrac{1}{4}\alpha_X \bar{v}_X C_{X(g)}(0+) \qquad (2.7)$$

where $\bar{v}_X = (8R_g T/\pi M W_X)^{\frac{1}{2}}$ is the mean molecular speed of the species X, α_X is the mass accommodation coefficient and $C_{X(g)}(0+)$ is the concentration of the species at the interface, on the gas side, which may differ appreciably from the bulk concentration because of a diffusive gradient in the vicinity of the interface. This flux can be expressed equivalently in terms of the hypothetical aqueous-phase concentration $C_{X(aq)}{}^*(0+)$ that would exist in equilibrium with the gas-phase concentration $C_{X(g)}(0+)$, as

$$\sigma_X{}^+ = \tfrac{1}{4}\alpha_X \bar{v}_X C_{X(aq)}{}^*(0+)/\tilde{H}_X$$

Here it is most convenient to use the dimensionless (concentration/concentration) Henry's law coefficient. The flux given by Equation (2.7) is a gross flux, which may be offset by a return flux from the solution phase into the gas phase. This return flux is

$$\sigma_X{}^- = \tfrac{1}{4}\bar{v}_X C_{X(aq)}(0-)/\tilde{H}_X$$

where $C_{X(aq)}(0-)$ denotes the concentration of the species at the interface, on the water side. This is so because, by microscopic reversibility, $\sigma_X{}^-$ must be equal to $\sigma_X{}^+$ at phase equilibrium [i.e. when $C_{X(aq)}(0-) = \tilde{H}_X C_{X(g)}(0+)$] and because the flux scales linearly with the concentration. The net flux into solution is

$$\sigma_X \equiv \sigma_X{}^+ - \sigma_X{}^- = \tfrac{1}{4}\alpha_X \bar{v}_X [C_{X(g)}(0+) - C_{X(aq)}(0-)/\tilde{H}_X] \qquad (2.8)$$

Equation (2.8) sets an upper limit to the mass transfer rate that must be considered in any treatment of interphase mass transfer; depending on the value of α_X, this may or may not present an appreciable limitation to the overall rate.

7   COUPLED MASS TRANSFER AND AQUEOUS-PHASE REACTION

This discussion (based on Schwartz [8]) develops expressions describing the rate of gas and aqueous-phase mass transport and aqueous-phase reaction. The exchange flux of a gaseous species between the gas and aqueous phases may be described phenomenologically (e.g. Danckwerts [9]) as the product of an overall mass-transfer coefficient K_g times the difference between the gas-phase concentration $C_{X(g)}$ and the aqueous-phase concentration $C_{X(aq)}$ divided by the dimensionless Henry's law coefficient $\tilde{H}_X$ describing the equilibrium solubility of the gas in the aqueous medium:

$$\sigma_X = K_g[C_{X(g)} - C_{X(aq)}/\tilde{H}_X] \qquad (2.9)$$

where $C_{X(g)}$ and $C_{X(aq)}$ refer to concentrations in the 'bulk' of the phase, that is, at distances sufficiently far from the interface to be outside the region of strong gradient in the vicinity of the interface. For concentrations in units of mol m^{-3}, the flux has units mol m^{-2} s^{-1}. Note that here we use the dimensionless Henry's law coefficient $\tilde{H}_X$. This expression is fairly general, being applicable to a variety of situations of interest in atmospheric chemistry, for example to the uptake of a gas into cloudwater by uptake followed by aqueous-phase reaction or to dry deposition of a gas to surface water. All of the physics (and chemistry) of the mass-transfer process is embodied in the mass-transfer coefficient K_g; the subscript g denotes that this is the gas-side mass transfer coefficient. This quantity has dimension of length/time, or velocity.

It is evident from Equation (2.9) that in order for there to be a net flux, the aqueous-phase concentration $C_{X(aq)}$ must not be in equilibrium with the gas-phase concentration $C_{X(g)}$, i.e. that $C_{X(aq)} \neq \tilde{H}_X C_{X(g)}$; more specifically, for there to be a net uptake in the aqueous medium $C_{X(aq)}$ must be less than $\tilde{H}_X C_{X(g)}$. The lack of equilibrium may be due to aqueous-phase chemical reaction depleting the concentration the dissolved gas or, alternatively, in the case of dry deposition to surface water, simply to the fact that the water is undersaturated relative to the atmospheric concentration of the gas, for example because of transfer from the mixed layer of the ocean to the deep ocean.

If the aqueous-phase concentration of the depositing substance $C_{X(aq)}$ is 0 [more precisely, if $C_{X(aq)} \ll \tilde{H}_X C_{X(g)}$], then $\sigma_X = K_g C_{X(g)}$. However, the condition $C_{X(aq)} \ll \tilde{H}_X C_{X(g)}$ does not necessarily imply that the flux is equal to its maximum possible value, as governed by atmospheric mass transport only, since there may be a return flux (aqueous phase to gas phase) resulting from a near-interface aqueous-phase concentration that is substantial compared to $\tilde{H}_X C_{X(g)}$ even when $C_{X(aq)} \ll \tilde{H}_X C_{X(g)}$.

The choice of the gas-phase concentration, rather than the aqueous-phase concentration, as that to which to refer the flux is arbitrary, reflecting the gas-phase orientation of atmospheric chemists. The flux might entirely equivalently be referred to the aqueous-phase concentration, viz.

$$\sigma_X = K_l[\tilde{H}_X C_{X(g)} - C_{X(aq)}]$$

where K_l is referred to as the liquid-side mass transfer coefficient; it is seen that $K_l = K_g/\tilde{H}_X$.

Considerable progress is made in understanding and describing the overall mass-transfer process by considering it to be a sequence of processes, from the bulk gas phase to the interface, across the interface, and from the interface to the bulk aqueous phase. These fluxes are described, respectively, as

$$\sigma_X = k_g[C_{X(g)} - C_{X(g)}(0+)] \tag{2.10}$$

$$\sigma_X = \tfrac{1}{4}\alpha_X \bar{v}_X[C_{X(g)}(0+) - C_{X(aq)}(0-)/\tilde{H}_X] \tag{2.8}$$

$$\sigma_X = \beta k_L(C_{X(aq)}(0-) - C_{X(aq)}) \tag{2.11}$$

where $C_{X(g)}(0+)$ denotes the concentration of the species at the interface, on the gas side, which may differ appreciably from the bulk gas-phase concentration because of a diffusive gradient in the vicinity of the interface; likewise, $C_{X(aq)}(0-)$ denotes the concentration of the species at the interface, on the water side. The gas-phase and liquid-phase mass-transfer coefficients k_g and k_l describe *physical* mass transfer (turbulent diffusion plus molecular diffusion through the laminar boundary layer immediately adjacent to the interface); the magnitudes of these coefficients depend on the degree of physical agitation characterizing the system. The coefficient β, to be discussed later, represents enhancement of the aqueous-phase mass-transfer flux due to removal of the material by chemical reaction; when there is no enhancement, $\beta = 1$.

Under steady-state conditions the flux is constant and equal in both media and across the interface. By equating the several expressions for the flux [Equations (2.8)–(2.11)], one obtains the overall mass-transfer coefficient as the inverse sum of the mass-transfer coefficients in the two media and at the interface:

$$\frac{1}{K_G} = \frac{1}{k_G} + \frac{1}{\left(\frac{1}{4}\right)\bar{v}_X\alpha_X} + \frac{1}{\tilde{H}_X k_L \beta} \tag{2.12}$$

In the absence of any aqueous-phase chemical reaction of the dissolved gas that would enhance the rate of uptake, the enhancement coefficient β is equal to unity. This corresponds to the two-film expression commonly employed in evaluating air–sea fluxes of nonreactive gases (e.g. [10]). Under the assumption that the interfacial resistance is negligible, then there is a critical solubility $\tilde{H}_{crit} \equiv k_g/k_l$ for which the gas- and liquid-phase resistances are equal. For $\tilde{H} \ll \tilde{H}_{crit}$, liquid-phase mass transport is controlling and the uptake rate is linearly dependent on $\tilde{H}$. For $\tilde{H} \gg \tilde{H}_{crit}$, gas-phase mass transport is controlling and the uptake rate is independent of $\tilde{H}$. For reactive gases the effect of the chemical removal process is exhibited in the enhancement coefficient β, and it is $\beta\tilde{H}$ that is to be compared with $\tilde{H}_{crit}$.

To examine reactive enhancement of uptake, it is necessary to model the concurrent mass-transfer and reactive processes, and a number of models have been introduced over the years, largely in the chemical engineering literature. The most familiar such model is the so-called diffusive-film model, which posits an unstirred laminar film at the interface having thickness $D_X{}^{aq}/k_l$, where $D_X{}^{aq}$ is the molecular diffusion coefficient of the dissolved gas in aqueous solution.

The reactive gas must diffuse through this film to the bulk convectively mixed liquid. Reaction occurs in the film and/or in the bulk liquid depending on the relative rates of reaction and mass transport. Although criticized as unrealistic, this model has received widespread application in the geochemical literature. Alternative models posit systematic or stochastic transport of stagnant, near-surface parcels of liquid into the bulk, again with reaction occurring to greater or lesser extent in these unstirred parcels depending on the relative rates of reaction and mass transport. In all cases the liquid-phase mass transport is characterized by the single parameter k_l as well as by $D_X{}^{aq}$.

Rates of reactive uptake for the three models were examined and compared by Danckwerts and Kennedy [11]. For the diffusive film model the expression for β for a reversible first-order reaction of the dissolved gas can be written as

$$\beta = \frac{\eta}{1 + (\eta - 1)\left(\dfrac{\tanh \kappa^{\frac{1}{2}}}{\kappa^{\frac{1}{2}}}\right)} \tag{2.13}$$

where η is the ratio at equilibrium of the total concentration of dissolved material to the concentration given by Henry's law dissolution alone, $\eta = \tilde{H}_X{}^* / \tilde{H}_X$, and κ is a dimensionless rate coefficient for reaction,

$$\kappa = \frac{k_{aq}{}^{(1)}}{k_l^2 / D_X{}^{aq}} \cdot \frac{\eta}{\eta - 1} \tag{2.14}$$

where $k_{aq}{}^{(1)}$ is the effective first-order rate coefficient for aqueous-phase reaction of the dissolved gas.

A few words should be said about the factors $\eta/(\eta - 1)$ and $k_l^2/D_X{}^{aq}$ employed in the definition of κ. At $\eta \gg 1$, which corresponds to large equilibrium enhancement of the solubility and which is often the situation of interest for reactive uptake, $\eta/(\eta - 1) \approx 1$ and this factor can thus be neglected in evaluation of κ. At low values of η (recall that $\eta \geq 1$) the factor $\eta/(\eta - 1)$ gives rise to substantial enhancement of κ essentially because the reaction does not need to proceed very far to reach equilibrium. The quantity $k_{crit} \equiv k_l^2/D_X{}^{aq}$ has dimension of inverse time; comparison of $k_{aq}{}^{(1)}$ with k_{crit} permits the importance of reactive enhancement to the rate of uptake to be immediately assessed.

Despite somewhat different and more realistic assumptions of the surface renewal models, these models were found to yield expressions for β that differ from Equation (2.13) by no more than a few percent [11]. These models have gained substantial support in laboratory studies (e.g. [9]). Expressions for the kinetic enhancement equivalent to Equation (2.13) were later obtained in the context of atmosphere–surface water exchange by Bolin [12] and by Hoover and Berkshire [13] and are sometimes is associated with those investigators.

Although all that remains to evaluate β (and ultimately K_g) is knowledge of the pertinent parameters, it is worthwhile to examine the dependence of the enhancement of uptake on the equilibrium constant and reaction rate constant.

At low values of the argument (≤ 0.5), i.e. for low reaction rate coefficients, the hyperbolic tangent function tanh closely approximates the argument itself, so that $\tanh \kappa^{\frac{1}{2}} / \kappa^{\frac{1}{2}}$ approaches unity and in turn β approaches unity. Thus approximately (within 10%)

$$\beta \approx 1 \qquad \text{for } \kappa \leq 0.3 \qquad (12.13a)$$

In this limiting situation there is essentially no enhancement of the rate of uptake over that given by physical dissolution alone. For somewhat greater values of κ, where the enhancement becomes appreciable, β is approximated by series expansion of the tanh function (again within 10%) as

$$\beta \approx 1 + \frac{1}{3}\frac{\eta - 1}{\eta}\kappa = 1 + \frac{1}{3}\frac{k_{aq}^{(1)}}{k_{l}^{2}/D_{X}^{aq}} \qquad \text{for } \kappa \leq 2.1 \qquad (2.13b)$$

At the other extreme, for values of the argument of tanh greater than about 1.5 (i.e. at high reaction rates), tanh approaches unity, and to good approximation

$$\beta \approx \frac{\eta}{1 + \dfrac{\eta - 1}{\kappa^{\frac{1}{2}}}} \equiv \frac{\kappa^{\frac{1}{2}}}{1 + \dfrac{\kappa^{\frac{1}{2}} - 1}{\eta}} \qquad \text{for } \kappa \geq 2.4 \qquad (2.13c)$$

For values of κ such that $(\kappa^{\frac{1}{2}} - 1) \ll \eta$, this expression simplifies to

$$\beta \approx \kappa^{\frac{1}{2}} \qquad \text{for } 2.4 \leq \kappa \leq (1 + 0.1\eta)^{2} \qquad (2.13d)$$

Finally, if $\kappa^{\frac{1}{2}}$ is sufficiently great, the first term in the denominator of Equation (2.13) predominates and

$$\beta \approx \eta \qquad \text{for } \kappa \geq 2.4 \text{ and } \kappa \geq 100(\eta - 1)^{2} \qquad (2.13e)$$

In this limit the rate of uptake is equal to that for a nonreactive gas whose solubility is enhanced, owing to instantaneous chemical equilibrium, by the factor η. This limit justifies the use of an effective Henry's law coefficient $k_{H_X}^* \equiv \eta k_{H_X}$ for rapidly established equilibria and specifies the range of applicability of this treatment.

The dependence of β on the effective first-order rate coefficient for reaction $k_{aq}^{(1)}$ is illustrated in Figure 2.2 for a range of values of η. The plateauing of β for large values of $k_{aq}^{(1)}$ and intermediate values of η corresponds to the limit in Equation (2.13e). The linear region of the graphs (slope = 0.5 on the logarithmic plot) corresponds to the limit in Equation (2.13d). This is equivalent to the well known [9] expression for diffusion controlled reaction for $\eta \gg 1$:

$$\beta k_l \approx [k_{aq}^{(1)} D_X^{aq}]^{\frac{1}{2}} \qquad \text{for } 2.4 \leq \kappa \leq (1 + 0.1\eta)^{2} \qquad (2.15)$$

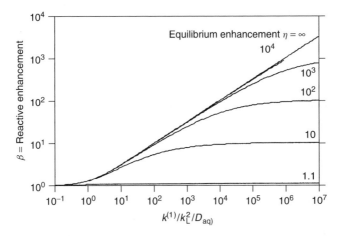

Figure 2.2 Dependence of the kinetic enhancement factor β of the aqueous-phase mass-transfer coefficient k_L as a function of the effective first-order rate constant for reaction, $k^{(1)}$, normalized to $k_L{}^2/D_{aq}$, where D_{aq} is the aqueous-phase molecular diffusion coefficient, for indicated values of the equilibrium solubility enhancement factor η (adapted from Schwartz [8])

8 RATES OF MULTI-PHASE REACTIONS IN LIQUID WATER CLOUDS

It is desired to evaluate the rate of aqueous-phase reactions in cloudwater under the assumption of solution equilibrium of the dissolved reacting gas and to express this reaction rate in units such that it can be compared with rates of gas-phase reactions. We take the cloud liquid water content L such that the liquid water volume fraction $L/\rho_{cw} \ll 1$) and assume Henry's law equilibrium; this assumption can be examined as outlined below by consideration of mass-transfer rates relative to the rate of reaction.

For a first-order aqueous-phase reaction having rate coefficient $k_{aq}{}^{(1)}$ (s^{-1}), the rate of aqueous-phase reaction (mol kg^{-1} s^{-1}) of a species X having molality $m_{X(aq)}$ (mol kg^{-1}) is

$$R_{X(aq)} = k_{aq}{}^{(1)} m_{X(aq)} = k_{H_X} k_{aq}{}^{(1)} p_X \tag{2.16}$$

Equivalent expressions for this rate are readily obtained from Table 2.1. In particular, the rate of reaction expressed in terms of partial pressure of the reagent gas p_X (Pa) is

$$\left(-\frac{d\hat{p}_X}{dt}\right)_{aq\ rxn} = L R_g T R_{X(aq)} = L R_g T k_{aq}{}^{(1)} \hat{k}_{H_X} \hat{p}_X$$

from which the effective first-order loss rate coefficient for this process referred to the total amount of material X in the volume $\gamma_{X,aq}$ (unit: s^{-1}) is

$$\gamma_{X,aq} \equiv \frac{1}{(\hat{p}_X + \hat{p}_X{}^*)} \left(-\frac{d\hat{p}_X}{dt}\right)_{aq\ rxn} = L R_g T \hat{k}_{H_X} k_{aq}{}^{(1)} \frac{\hat{p}_X}{\hat{p}_X + \hat{p}_X{}^*}$$

For a gaseous reagent species predominantly present in the gas phase, $\hat{p}_X \gg \hat{p}_X{}^*$ and we obtain the result

$$\gamma_{X,\text{aq}} \equiv \frac{1}{(p_X + p_X{}^*)}\left(-\frac{dp_X}{dt}\right)_{\text{aq rxn}} = L R_g T \hat{k}_{H_X} k_{\text{aq}}{}^{(1)} = 10^{-5} L R_g T k_{H_X} k_{\text{aq}}{}^{(1)}$$

This important quantity may be readily evaluated for known aqueous-phase first-order reaction rate coefficient $k_{\text{aq}}{}^{(1)}$. $\gamma_{X,\text{aq}}$ scales linearly with liquid water content, Henry's law coefficient, and aqueous-phase rate constant. It may be readily and immediately compared with the rate coefficient for gas-phase reaction of species X and thus provides the means of comparing rates of gas- and aqueous-phase reactions of a given species.

For other than first-order reaction, $k_{\text{aq}}{}^{(1)}$ is replaced by the effective first-order reaction rate coefficient, evaluated, for example, as the product of second-order reaction rate coefficient and concentration of other reacting species.

9 DEPARTURE FROM PHASE EQUILIBRIUM AT AIR–WATER INTERFACE

For a given flux into the aqueous phase, corresponding to an assumed reaction rate, the fractional departure from equilibrium at the interface can be evaluated to examine for mass-transport limitation to the rate of reaction as a function of measured or specified mass accommodation coefficient and reaction rate [13]. This fractional departure is

$$\frac{\tilde{H}_X C_{X(g)}(0+) - C_{X(\text{aq})}(0-)}{\tilde{H}_X C_{X(g)}(0+)} = \frac{\sigma_X}{\alpha_X \bar{v}_X C_{X(g)}(0+)/4} = \frac{\sigma_X}{\alpha_X \bar{v}_X \hat{p}_{X(g)}(0+)/4 R_g T} \tag{2.17}$$

Note that here the partial pressure of the gas is in units of Pa. Particular attention must be paid here to consistency of units or else errors can result (e.g. Ref. 9 p. 69); a major strength of SI is that such consistency is intrinsic.

Following Schwartz [14], we may use Equation (2.17) to test whether the rate of reaction in a spherical drop is appreciably limited by the rate of interfacial mass transport. For departure from phase equilibrium not to exceed a criterion, arbitrarily taken as 10 %,

$$\sigma_X \le 0.1 \frac{\alpha_X \bar{v}_X \hat{p}_{X(g)}(0+)}{4 R_g T}$$

For a spherical drop of radius a, the flux σ_X corresponds to an uptake rate $4\pi a^2 \sigma_X$ or a mean volumetric reaction rate $\overline{R}_{X(\text{aq})} = (4\pi a^2 \sigma_X)/(4\pi a^3/3) = 3\sigma_X/a$, whence the criterion that must be satisfied for mass transport limitation not to exceed 10 % is

$$\overline{R}_{X(\text{aq})} \le 0.1 \frac{\alpha_X \bar{v}_X \hat{p}_{X(g)}(0+)}{(4/3)a R_g T}$$

For the aqueous-phase reaction rate given by Equation (2.16) this criterion becomes

$$\hat{k}_{H_X} k_{aq}^{(1)} \leq 0.1 \frac{\alpha_X \bar{v}_X}{(4/3) a R_g T} \quad \text{or} \quad k_{H_X} k_{aq}^{(1)} \leq 10^4 \frac{\alpha_X \bar{v}_X}{(4/3) a R_g T}$$

Note that here, as in general where the Henry's law coefficient appears in expressions, we must use the Henry's law coefficient in strict SI units, i.e. $mol\, kg^{-1}\, Pa^{-1}$. In the second form of the criterion this Henry's law coefficient has been replaced by that in $mol\, kg^{-1}\, bar^{-1}$ units.

A similar expression has been given by Schwartz [14] as a criterion for absence of limitation to the rate of multi-phase reaction due to the finite rate of gas-phase diffusion, viz.

$$\hat{k}_{H_X} k_{aq}^{(1)} \leq 0.1 \frac{3 D_g}{a^2 R_g T} \quad \text{or} \quad k_{H_X} k_{aq}^{(1)} \leq 10^4 \frac{3 D_g}{a^2 R_g T}$$

10   DRY DEPOSITION OF GASES TO SURFACE WATER

As presented above, the flux of dry deposition of atmospheric gases to surface water, or more generally the exchange flux between the atmosphere and surface water, may be evaluated for known values of the pertinent mass-transfer coefficients in the two phases and the Henry's law coefficient, taking into account any reactive enhancement. Dry deposition is generally expressed as the product of the atmospheric concentration times a deposition velocity v_d:

$$\sigma_X = v_d C_{X(g)}$$

whence $v_d = K_g$ provided that $C_{X(aq)} \ll \tilde{H}_X C_{X(g)}$.

Here we examine the dependence of deposition velocity on the several parameters. An approximate value for k_g is 0.13 % of the wind speed, for a reference height of 10 m; the exact proportionality coefficient depends somewhat on the reference height and on the atmospheric stability [15]. Thus, for typical wind speeds of $3-15\, m\, s^{-1}$, values of k_g range from 4 to $20\, mm\, s^{-1}$. An expression for k_l given by Liss and Merlivat [16], which exhibits three linear regions of dependence on wind speed (increasing slope with increasing wind speed), has gained substantial support (e.g. [17]). That expression gives $k_l = 1.4 \times 10^{-6}$, 1.3×10^{-5} and $1.1 \times 10^{-4}\, m\, s^{-1}$ for wind speeds of 3, 5 and $10\, m\, s^{-1}$, respectively; the actual values depend somewhat on the identity of the transported gas and on temperature. We take these values as representative of the range that must be considered for the present purpose of identifying factors governing dry deposition of gases to surface waters. However, it must be stressed that although these values are representative, the problem of specifying the precise values pertinent to a given environmental situation is by no means solved.

Figure 2.3 shows the overall mass-transfer coefficient K_g in the absence of reactive enhancement as a function of Henry's law coefficient for representative values of k_l and k_g. Also shown are values of Henry's law and effective Henry's law coefficients for several gases of atmospheric interest. The points marked by ● represent the equilibrium solubility after the reaction (acid−base dissociation,

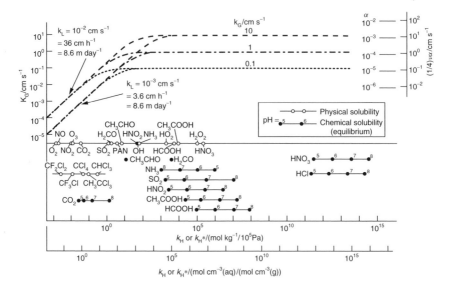

Figure 2.3 Overall mass-transfer coefficient K_G (in the absence of interfacial mass-transport limitation) for nonreactive gases ($\beta = 1$) as a function of solubility for values of k_G and k_L indicated on the asymptotes of the curves. The scale at the right permits comparison of interfacial and overall conductance for indicated values of the mass-accommodation coefficient α. Also shown are Henry's law coefficients k_H (open circles) and effective Henry's law coefficients k_H^* (filled circles) of a number of atmospheric gases; the intermediate abscissa scale gives k_H and k_H^* in units of $\text{mol kg}^{-1}/(10^5 \text{ Pa})$ (adapted from Schwartz [8])

pH dependent, or aldehyde hydrolysis) has gone to completion; the ratio to the physical solubility (indicated by ○) gives the equilibrium enhancement factor η. The thrust of the figure is that without reactive enhancement there are few gases that are sufficiently soluble in water that for representative values of k_g and k_l their deposition velocity would be controlled by atmospheric mass transport. This figure demonstrates the important influence of reactive enhancement on k_l.

A further feature of the figure is the device at the upper right, which displays the interfacial conductance, evaluated as $(1/4)\bar{v}\alpha$, for indicated values of the mass-accommodation coefficient α. Here the mean molecular speed, which depends on molecular weight, has been taken as $4 \times 10^2 \text{ m s}^{-1}$. Interfacial resistance to mass transport should be taken into consideration if interfacial conductance is comparable to K_g evaluated with only the gas- and liquid-phase resistances. Interfacial resistance is not limiting to the rate of dry deposition for values of $\alpha \gtrsim 10^{-4}$ but, depending on the value of K_g evaluated with only the gas- and liquid-phase resistances, might become limiting for lower values of α.

11 CONCLUSIONS

Exchange of volatile species between the gas phase of the atmosphere and liquid water is a key process in the evolution of materials in the earth's environment system. The driving force for this exchange is the Henry's law solubility equilibrium together with rapid equilibria in aqueous solution. The Henry's law constant, or the effective Henry's law constant that takes these rapid equilibria into account, is thus a fundamental property of volatile atmospheric species in the earth's environment that must be well known in order to describe the extent and rate of these phenomena under circumstances of interest. The Henry's law constant (or effective Henry's law constant) ranges fairly widely for substances of interest to the earth's atmospheric environment, by some 18 orders of magnitude. Because of this wide range of solubilities, the rates and extents of various processes, such as the distribution between gas and liquid phase in clouds, likewise vary substantially. In many situations it is sufficient to identify limiting cases applicable to sparingly soluble or highly soluble gases. For example, the zeroth-order question is where the bulk of the material resides, and often knowledge of this is sufficient for the task at hand, for which an order of magnitude estimate of the solubility is often sufficient. Clearly insoluble gases such as nitrogen and oxygen are present essentially entirely in the gas phase. Very soluble material, such as nitric acid, would be expected to be present essentially entirely in the liquid phase, at least for clouds of sufficient liquid water content. In some instances it is necessary to know the amount of material present in the lesser compartment, for example in the calculation of the rate of aqueous phase reactions in cloudwater. These calculations require rather precise knowledge of the Henry's law solubility.

This chapter has presented formalism to describe several applications of Henry's law solubility to atmospheric chemistry. It is hoped that it provides a sense of the pertinence and use of solubility data in atmospheric chemistry and will stimulate interest in the evaluated Henry's law data presented in this volume.

REFERENCES

1. Mills, I.; Cvitaš, T.; Homann, K.; Kallay, N.; Kuchitsu, K., *Quantities, Units and Symbols in Physical Chemistry* ('The Green Book'), Blackwell Scientific Publications, Oxford, for International Union of Pure and Applied Chemistry, 1993.
2. Schwartz, S. E.; Warneck, P., Units for use in atmospheric chemistry, *Pure Appl. Chem.*, 1995, **67**, 1377–1406.
3. Henry, W., Experiments on the quantity of gases absorbed by water, at different temperatures, and under different pressures, *Philos. Trans. R. Soc. London*, 1803, **93**, 29–43.
4. Henry, W., Appendix to Mr. William Henry's paper, on the quantity of gases absorbed by water, at different temperatures, and under different pressures, *Philos. Trans. R. Soc. London*, 1803, **93**, 274–6.
5. Sander, R., Modeling atmospheric chemistry: interactions between gas-phase species and liquid cloud/aerosol particles, *Surv. Geophys.*, 1999, **20**, 1–31.
6. Schwartz, S. E., in *The Chemistry of Acid Rain: Sources and Atmospheric Processes*, Johnson, R. W.; Gordon, G. E. (eds), American Chemical Society, Washington, DC, 1987, pp. 93–108.
7. Schwartz, S. E., Mass-transport considerations pertinent to aqueous-phase reactions of gases in liquid-water clouds, in *Chemistry of Multiphase Atmospheric Systems*, Jaeschke, W. (ed.), Springer, Heidelberg, 1986, pp. 415–71.

8. Schwartz, S. E., Factors governing dry deposition of gases to surface water, in *Precipitation Scavenging and Atmosphere–Surface Exchange*, Schwartz, S. E.; Slinn, W. G. N. (coordinators) Hemisphere, Washington, DC, 1992.

9. Danckwerts, P. V., *Gas–Liquid Reactions*, McGraw-Hill, New York, 1970.

10. Liss, P. S.; Slater, P. G., Flux of gases across the air–sea interface, *Nature*, 1974, **247**, 181–4.

11. Danckwerts, P. V.; Kennedy, B. E., Kinetics of liquid-film process in gas absorption. Part I: models of the absorption process, *Trans. Inst. Chem. Eng.*, 1954, **32**, S49–S52.

12. Bolin, B., On the exchange of carbon dioxide between the atmosphere and the sea, *Tellus*, 1960, **12**, 274–81.

13. Hoover, T. E.; Berkshire, D. C., Effects of hydration on carbon dioxide exchange across an air–water interface, *J. Geophys. Res.* 1969, **74**, 456–64.

14. Schwartz, S. E., Chemical conversions in clouds, in *Aerosols: Research, Risk Assessment and Control Strategies*, Lee, S. D.; Schneider, T.; Grant, L. D.; Verkerk, P. J. (eds), Lewis, Chelsea, MI, 1986, pp. 349–75.

15. Liss, P. S.; Merlivat, L., Air–sea gas exchange rates: introduction and synthesis, in *The Role of Air–Sea Exchange in Geochemical Cycling*, Buat-Ménard, P. (ed), Reidel, Dordrecht, 1986, pp. 113–27.

16. Hicks, B. B.; Liss, P. S., Transfer of SO_2 and other reactive gases across the air–sea interface, *Tellus*, 1976, **28**, 348–54.

17. Watson, A. J.; Upstill-Goddard, R. C.; Liss, P. S., Air–sea gas exchange in rough and stormy seas measured by a dual-tracer technique, *Nature*, 1991, **349**, 145–47.

CHAPTER 3

Thermodynamic Aspects of Henry's Law

James Sangster
Sangster Research Laboratories, Montreal, Canada

1 HENRY'S LAW AND RAOULT'S LAW

Henry's law and Raoult's law are both commonly used expressions in equilibrium thermodynamics. They have not always been well understood or used properly. Carroll [1–3] and Näfe [4] present a succinct explanation of their relationship. A sketch of the fundamental principles is given here.

It has been found convenient, in chemical thermodynamics, to describe the properties of a solution in terms of deviation from an agreed-upon model or reference solution. This is the so-called 'ideal' solution, i.e. one for which both the volume change and heat of mixing are zero. Although no real mixing process produces an ideal solution, the concept has proven useful because deviations are generally small, that is, the ideal solution is a good zeroth approximation.

The relation between Henry's and Raoult's laws may be presented in slightly different, though equivalent, ways. We use here the approach of Carroll [3].

In defining an ideal solution, for the purposes of deducing excess or mixing properties, we need to specify a reference state for each component comprising the mixture. The reference state for Raoult's law is the pure component, and the reference fugacity is that of the pure substance at the temperature of interest (as before, we assume fugacity and pressure to be equivalent). The fugacity of a component in a real solution is f, given by

$$f = \gamma x f^{\circ} \tag{3.1}$$

where x is its mole fraction, f° is the reference fugacity (vapour pressure) and γ is an activity coefficient, which takes into account all deviations from ideality.

If we choose the Henry's law convention for a component, its fugacity will be

$$f = \gamma^* x f^* \tag{3.2}$$

Chemicals in the Atmosphere – Solubility, Sources and Reactivity. Edited by P. G. T. Fogg and J. Sangster
© 2003 IUPAC   ISBN: 0-471-98651-8

where f^* is a proportionality constant, a hypothetical 'vapour pressure' and γ^* is the activity coefficient in the Henry's law convention. Note that f° is independent of the other component, but that f^* does depend on the second component, which must be specified.

The quantities γ^* and γ are distinct and should not be confused. They have the properties

$$\lim_{x \to 1} \gamma = 1 \qquad (3.3)$$

and

$$\lim_{x \to 0} \gamma^* = 1 \qquad (3.4)$$

It is seen that, in the Henry's law convention, the reference state for the component is the infinitely dilute solution.

Since the fugacity of a component is a physically measurable quantity and is independent of the conventions chosen, we have, from Equations (3.1) and (3.2)

$$\gamma^* f^* = \gamma f^\circ \qquad (3.5)$$

In the case of infinitely dilute solution, $\gamma^* = 1$ and $\gamma = \gamma^\infty$, i.e. the infinite dilution activity coefficient in the Raoult's law convention. These relationships are illustrated in Figure 3.1, where, as an example, $\gamma^\infty = 1.7$, $f^\circ = 100\,\text{kPa}$ and $f^* = 170\,\text{kPa}$.

The form of Henry's law constant adopted in this book is

$$k_{\text{H}} = \lim_{x \to 0} (x/f) \qquad (3.6)$$

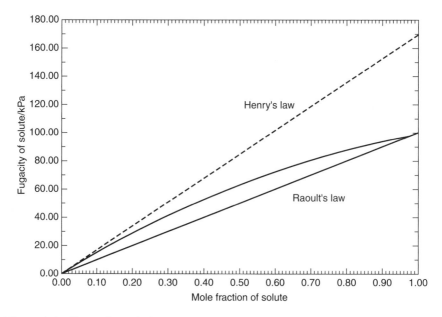

Figure 3.1   Illustration of Henry's and Raoult's laws for a solute whose vapour pressure is 100 kPa and for which $\gamma^\infty = 1.7$

Table 3.1 Forms of Henry's law constant currently in use

	'Solubility' form atmospheric chemists, environmental chemists	'Volatility' form environmental chemists, chemical engineers, physical chemists
(A) Using concentration and pressure:		
mole fraction	x/p	p/x
molality	m/p	p/m
molarity	M/p	p/M
(B) Using concentrations only (partition coefficient):		
	$[X]_{aq}/[X]_{gas}$	$[X]_{gas}/[X]_{aq}$

or, equivalently,

$$k_H = \lim_{x \to 0}(\partial x/\partial f) \tag{3.7}$$

From Equations (3.2) and (3.6),

$$k_H = \lim_{x \to 0}(x/f) = 1/f^* \tag{3.8}$$

which relates Henry's law constant to the hypothetical quantity f^*. Also,

$$\ln k_H = \lim_{x \to 0}(RT \ln x - RT \ln f) \tag{3.9}$$

$$= -\Delta_{sol}\overline{G} \tag{3.10}$$

where $\Delta_{sol}\overline{G}$ is the partial molar Gibbs energy change for the formation of a solution for which solubility is proportional to pressure within the limits of measurement. Equation (3.6) is the form of Henry's Law adopted in this book. The reader should be aware, however, that other forms are currently in use. This is shown in Table 3.1.

2 THERMODYNAMIC QUANTITIES FROM HENRY'S LAW CONSTANTS

Equation (3.10) shows that $\ln k_H$ represents the Gibbs energy change for the transfer of the solute from one phase to the other. From usual thermodynamic relationships we have [5]

$$[\partial(\Delta_{sol}\overline{G})/\partial T]_P = -\Delta_{sol}\overline{H}/RT^2 \tag{3.11}$$

where $\Delta_{sol}H$ is the corresponding enthalpy change. This is commonly called the 'enthalpy of solution'. The enthalpy and Gibbs energy are related by

$$\Delta_{sol}\overline{G} = \Delta_{sol}\overline{H} - T\Delta_{sol}\overline{S} \tag{3.12}$$

where $\Delta_{sol}\overline{S}$ is the entropy change.

Although Equation (3.12) is thermodynamically valid, the quantities $\Delta_{sol}\overline{H}$ and $\Delta_{sol}\overline{S}$ cannot usually be found from Henry's law constant data in practice.

This is because most measurements of k_H are not sufficiently precise to allow meaningful results upon differentiation of Equation (3.11). State-of-the-art calorimetric techniques can be used to measure directly the heat of solution of gases in liquids. For liquids in liquids, current calorimetric techniques routinely permit very precise measurement of the heat of solution, ΔH(liquid $\rightarrow$ aq). The gas $\rightarrow$ aq enthalpy can then be found from the heat of vaporization [6]:

$$\Delta H(\text{gas} \rightarrow \text{aq}) = \Delta H(\text{liquid} \rightarrow \text{aq}) - \Delta H(\text{liquid} \rightarrow \text{gas}) \qquad (3.13)$$

3 HOW DILUTE IS 'DILUTE'?

We have seen that Henry's law is defined to be operative when the solute concentration approaches zero, Equation (3.4). However, Equation (3.4) represents a limit and says nothing about what 'infinitely dilute' means in practice.

In strict theoretical terms, of course, there is no cut-off concentration below which Henry's law holds, and above which it does not, just as there is no solvent concentration which defines the limits of Raoult's law. Instead, one can speak of concentration limits *for practical purposes*, that is, concentration limits defined by the current precision of measurement. An informal benchmark may be taken, for example the solubility of atmospheric gases at low pressure (one atmosphere) at ordinary temperatures in water [5]. The concentrations in these cases are approximately 10^{-3}–10^{-4} mole fraction.

In phenomenological terms, one can imagine that the solute has entered the Henry's law region when solute molecules are sufficiently far apart that they no longer 'see' each other, i.e. there are no solute–solute interactions. This clearly depends on the nature of both solute and solvent.

Attempts have been made to find limits of applicability of Henry's law. A particularly focused example is the work of Koga *et al.* [7–11]. The total pressure of *tert*-butanol aqueous solutions was measured very precisely in dilute solutions between 20 and 30 °C. The lowest concentration of solutions was 7×10^{-5} mole fraction. The heat of mixing was also measured at 20–60 °C. The lowest concentration of solute in these experiments was 4×10^{-4} mole fraction. The data are plotted in Figures 3.2 and 3.3.

Westh *et al.* [12] found similar results for a number of aqueous and nonaqueous systems: hexane–cyclohexane, glycerol–water, urea–water and 2-butoxyethanol–water.

Tucker and Christian [13] measured Henry's law for benzene in water at 35 °C, and found that k_H/mol kg^{-1} Pa^{-1} increased from 6.8×10^{-6} at 3×10^{-4} mole fraction to 7.1×10^{-6} at infinite dilution. They ascribed this increase to benzene dimerization. This finding is consistent with model calculations using the osmotic (McMillan–Mayer) second virial coefficient [14].

Sanemasa *et al.* [15,16], in a series of elegant experiments, 'titrated' a given quantity of water with various amounts of solute vapour from a separate reservoir. After equilibration, both liquid and vapour were analysed for solute. The final solute concentration was varied from unsaturation to saturation at various temperatures (5–45 °C). The solutes were benzene, alkylbenzenes and condensed aromatics. In all cases, Henry's law was obeyed within a few percent.

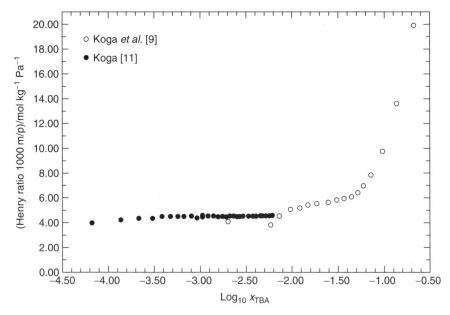

Figure 3.2 Illustrating the approach to Henry's law constant in dilute solutions of *tert*-butyl alcohol (TBA) in water at 25 °C

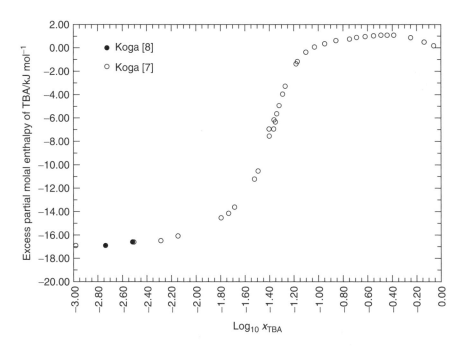

Figure 3.3 Illustrating the approach to Henry's law from the variation of excess partial molar enthalpy of *tert*-butyl alcohol (TBA) in dilute solutions in water at 27 °C

One can conclude that, for liquids and solids in water, Henry's law is obeyed up to saturation according to current methods of measurement. These methods are described in Chapter 4; they are not of the highest precision. In particular, the calculation method known as 'vapour pressure/aqueous solubility' is valid for all but the most accurate data needs.

4 HENRY'S LAW CONSTANT AND OTHER MEASURES OF SOLUBILITY

Solubility has been reported in the scientific literature in many ways, as reviewed by Battino *et al.* [17]. In the present instance, a number of these (encountered in Chapter 11) are defined here and related to Henry's law constant as adopted in this book. Equivalences used, both here and in Chapter 11, are given in Table 3.2. It is assumed that the solute concentration is low enough for the solubility to be taken to be proportional to pressure and that gases may be treated as ideal.

(a) Mole fraction, x
For dilute aqueous solutions, the solute mole fraction is effectively $m/55.51$. If the partial pressure of solute above the solution is p, then $k_H = 55.51x/p$.

(b) Concentration (molarity), M
For dilute aqueous solutions, m and M are numerically equal and thus $k_H = M/p$.

(c) Ostwald coefficient, L
L is defined as the ratio of the volume of gas absorbed (V_g) to the volume of water (V_w), all measured at the same temperature:

$$L = V_g/V_w \qquad (3.14)$$

L is therefore unitless. Under the assumptions of dilute solution and gas ideality, L is independent of the gas partial pressure. The number of moles of gas

Table 3.2 Definitions and equivalences useful in Henry's law constant conversions[a]

Conversion factors:
$1\,atm = 760\,mmHg = 101.325\,kPa$
$1\,bar = 100\,kPa = 10^5\,Pa$
$1\,litre = 1000\,cm^3 = 1\,dm^3$
$1\,mol\,m^{-3} = 10^{-3}\,M$
Other:
Density of water $= 1\,g\,cm^{-3}$
Molecular weight of water $= 18.015\,g\,mol^{-1}$
1 litre of water contains 1 kg
molality (m) = moles of solute per kg of water
molarity (M) = moles of solute per litre of solution

[a]These equivalences are used in Chapter 11 and are valid for practical calculations. They may not be valid for the most precise calculations.

represented by V_g is pV_g/RT, and the mass of water represented by V_w is ρV_w, where ρ is the density. Henry's law constant is then given by m/p, or

$$k_H = V_g/\rho RT V_w = L/\rho RT \qquad (3.15)$$

(d) Bunsen coefficient, α
This is the volume of gas, reduced to STP, which is absorbed by unit volume of water at the temperature of measurement under a partial pressure of 1 atm. Using the same volume symbols as for the Ostwald coefficient, we have

$$\alpha = (V_g/V_w)(273.15/T) \qquad (3.16)$$

The Ostwald and Bunsen coefficients are thus related. Following the same argument as in (c),

$$k_H = \alpha/(273.15\rho R) \qquad (3.17)$$

(e) Water–air partition coefficient, k_{WA}
The water–air partition coefficient is the ratio of the concentration of the solute (X) in the aqueous phase to that in the gas phase:

$$k_{WA} = [X]_{aq}/[X]_{gas} \qquad (3.18)$$

Using the usual approximations, we have $[X]_{gas} = p/RT$ and $[X]_{aq} = m$. Thus

$$k_H = m/p = [X]_{aq}/[X]_{gas} RT = k_{WA}/RT \qquad (3.19)$$

5 HENRY'S LAW CONSTANT AND THE INFINITE DILUTION ACTIVITY COEFFICIENT, γ^∞

The infinite dilution activity coefficient is a quantity familiar to physical chemists and chemical engineers. It is, in fact, closely related to Henry's law constant and is often used for the same purposes.

The relation between these two quantities may be derived from Equations (3.1) and (3.6):

$$f = \gamma x f^\circ \qquad (3.1)$$

$$k_H = \lim_{x \to 0} (x/f) \qquad (3.6)$$

From Equation (3.1), the ratio x/f is identically $1/\gamma f^\circ$. In the limit, as in Equation (3.6),

$$k_H = \lim_{x \to 0} (x/f) = 1/(\gamma^\infty f^\circ) \qquad (3.20)$$

6 USES OF HENRY'S LAW CONSTANT AND INFINITE DILUTION ACTIVITY COEFFICIENT

The water solubility and vapour pressure are two fundamental properties of pure substances and are used extensively in many areas of chemistry and engineering.

As usually understood, water solubility is expressed as a concentration at a given temperature. The solubility is therefore an upper limit of how much solute the solvent can contain. The vapour pressure of a liquid or solid is a measure of how much of the substance can be found in the gas phase at a given temperature.

For some applications, knowledge of water solubility and vapour pressure, *by themselves*, is not sufficient. Henry's law constant, combining as it does both pressure and concentration, is a type of 'limiting' solubility. It can also be thought of as the vapour pressure of a 'pure' component in the state of infinite dilution in a solvent. Two solutes can have very different properties, but their Henry's law constants may be comparable. This is shown for acetone and 2,2′-PCB (Table 3.3). Acetone is completely miscible and has a relatively high vapour pressure; the pesticide is, by comparison, an involatile and insoluble solid. Yet their Henry's law constants are not widely different.

Examples of the practical use of Henry's law constants are described here.

(a) Comparative olfactometry
The odours, fragrances and aromas of foods, wines, perfumes, etc., arise from the presence of volatile components. The prediction and correlation of olfactometric properties need data on the pure compounds, of which Henry's law constant is a principal example [18,19].

(b) Migration of pollutants in the environment
It was a recent news item that organochlorines have been found in Arctic marine biota [20]. Although the Canadian Arctic is geographically distant from pollutant sources (southern Canada and the USA), this migration has been well documented [21–23]. The pollutants had migrated through the atmosphere, i.e. the gas phase. This may be initially surprising, since these substances are commonly considered to be involatile and insoluble.

In fact, the transport of contaminants has been modelled adequately [24,25]. The model contains as parameters such quantities as temperature, chemical degradability, distance, transport and physicochemical properties. The migration of pollutants from warm to cold waters is principally explained by the temperature dependence of Henry's law constant. Mackay and Wania [24] illustrated this fact with the example of hexachlorobenzene. The Henry's law constant of hexachlorobenzene is given by

$$\log(k_H/\text{mol kg}^{-1}\,\text{Pa}^{-1}) = 2560/(T/\text{K}) - 12.98 \qquad (3.21)$$

The equilibrium model postulates that the fugacity of the pollutant is the same in warm and cold environments. By definition, $k_H = m/f$. Thus in the warm

Table 3.3   Properties of acetone and 2,2′-PCB at 25 °C

Compound	Water solubility	Vapour pressure/Pa	$k_H/\text{mol kg}^{-1}\text{Pa}^{-1}$
Acetone	∞	30 710	2.5×10^{-4}
2,2′-PCB	$x = 8.1 \times 10^{-8}$	0.26	2.6×10^{-5}

environment (25 °C)

$$f(25\,^\circ C) = m(25\,^\circ C)/k_H(25\,^\circ C) \qquad (3.22)$$

and in the cold environment (0 °C)

$$f(0\,^\circ C) = m(0\,^\circ C)/k_H(0\,^\circ C) \qquad (3.23)$$

Since by hypothesis $f(25\,^\circ C) = f(0\,^\circ C)$,

$$m(0\,^\circ C)/m(25\,^\circ C) = k_H(0\,^\circ C)/k_H(25\,^\circ C) = 6.3 \qquad (3.24)$$

Without claiming that this simple calculation is a complete account of the matter, we can understand that there is a thermodynamic tendency for condensed phase pollutants to 'distill' from warmer to colder environments.

(c) Infinite dilution activity coefficients in chemical engineering
Sandler [26] summarized some applications of Henry's law constants (in the form of γ^∞) in chemical engineering. In certain cases, a knowledge of γ^∞ for both components of a binary mixture allows one to obtain values of parameters in a two-parameter activity coefficient model that can be used to make phase equilibrium predictions over the whole concentration range.

Infinite dilution activity coefficients are very commonly used in the design of air-stripping towers or scrubbers, for example in the treatment of wastewater [27]. The principle of a scrubber is illustrated by Figure 4.1 in the next Chapter 4, i.e. it is a bubble tower for 'purging' a solution. In such a stripping design, air is bubbled through the solution and the pollutant passes from the solution to the gas phase and is removed. The equilibrium concentration of pollutant in the gas phase is governed by γ^∞, and the effectiveness of the operation depends also upon flow rate, interphase transport, etc.

In liquid–liquid extraction, a desired product is prepared in one solvent and extracted with another. The separation factor β [28] can be approximated by

$$\beta_{12} = \gamma^\infty(1)/\gamma^\infty(2) \qquad (3.25)$$

for the same solute in immiscible solvents 1 and 2.

REFERENCES

1. Carroll, J. J., *Chem. Eng. Prog.*, 1991, **87**(9), 48–51.
2. Carroll, J. J., *Chem. Eng. Prog.*, 1992, **88**(8), 53–8.
3. Carroll, J. J., *Chem. Eng. Prog.*, 1999, **95**(1), 49–56.
4. Näfe, H., *Ber Bunsenges. Phys. Chem.*, 1994, **98**, 1281–6.
5. Wilhelm, E.; Battino, R.; Wilcock, R. J., *Chem. Rev.*, 1977, **77**, 219–62.
6. Abraham, M. H., *J. Chem. Soc., Faraday Trans. 1*, 1984, **80**, 153–81.
7. Koga, Y., *Can. J. Chem.*, 1986, **64**, 206–7.
8. Koga, Y., *Can. J. Chem.*, 1988, **66**, 1187–93.
9. Koga, Y.; Siu, W. W. Y.; Wong, T. Y. H., *J. Phys. Chem.*, 1990, **94**, 7700–6.
10. Koga, Y.; Wong, T. Y. H.; Siu, W. W. Y., *Thermochim. Acta*, 1990, **169**, 27–38.
11. Koga, Y., *J. Phys. Chem.*, 1995, **99**, 6231–3; 12370.

12. Westh, P.; Haynes, C. A.; Koga, Y., *J. Phys. Chem.*, 1998, **102**, 4982–7.
13. Tucker, E. E.; Christian, S. D., *J. Phys. Chem.*, 1979, **83**, 426–7.
14. Rossky, P. J.; Friedman, H. L., *J. Phys. Chem.*, 1980, **84**, 587–9.
15. Sanemasa, I.; Araki, M.; Deguchi, T.; Nagai, H., *Bull. Chem. Soc. Jpn.*, 1982, **55**, 1054–62.
16. Sanemasa, I.; Ishibashi, K.; Kumamaru, M.; Deguchi, T., *Bull. Chem. Soc. Jpn.*, 1989, **62**, 2908–12.
17. Battino, R.; Clever, H. L.; Fogg, P. G. T.; Young, C. L., in *Carbon Dioxide in Water and Aqueous Electrolytes*, Solubility Data Series, Vol. 62, Scharlin, P. (ed.), Oxford University Press, Oxford, 1996, pp. vi–xiv.
18. Buttery, R. G.; Ling, L. C.; Guadagni, D. G., *J. Agric. Food Chem.*, 1969, **17**, 385–9.
19. Amoore, J. E.; Buttery, R. G., *Chem. Senses Flavour*, 1978, **3**, 57–71.
20. Macdonald, R.; McLaughlin, F.; Adamson, L., *Can. Chem. News*, 1997, **49**(8), 28–9.
21. Hung, H.; Halsall, C. J.; Blanchard, P.; Li, H. H.; Fellin, P.; Stern, G.; Rosenberg, B., *Environ. Sci. Technol.*, 2001, **35**, 1303–11.
22. Patton, G. W.; Hinckley, D. A.; Walla, M. D.; Bidleman, T. F., *Tellus*, 1989, **41B**, 243–55.
23. Hinckley, D. A.; Bidleman, F.; Rice, C. P., *J. Geophys. Res. C4*, 1991, **96**, 7201–13.
24. Mackay, D.; Wania, F., *Sci. Total Environ.*, 1995, **160–161**, 25–38.
25. Wania, F.; Mackay, D., *Sci. Total Environ.*, 1995, **160–161**, 211–32.
26. Sandler, S. I., *Fluid Phase Equilib.*, 1996, **116**, 343–53.
27. Berglund, R. L.; Whipple, G. M., *Chem. Eng. Prog.*, 1987, **83**(11), 46–54.
28. Wisniak, J.; Segura, H.; Reich, R., *Phys. Chem. Liquids*, 1996, **32**, 1–24.

CHAPTER 4

The Experimental Measurement of Henry's Law Constant

James Sangster

Sangster Research Laboratories, Montreal, Canada

As explained in Chapter 3, Henry's law constant, and also the infinite dilution activity coefficient, are derived from measured quantities related to the solute at equilibrium between liquid and vapour (gaseous) phases. All the techniques for determining these two quantities to be described here are therefore necessarily based on the same principle, viz. the establishment of solute equilibrium between two phases. In other words, the same *method* is used, but the particular apparatus and procedure (the *technique*) vary from case to case.

The purpose of this chapter is not to be absolutely exhaustive in coverage and detail, but to give the reader an overview of the experimental measurement techniques mentioned in Chapter 11, which is a compilation/evaluation of reported Henry's law constant data. The abbreviations used in Chapter 11 are mentioned in the title of each technique (e.g., Direct Measurement = 'DIRECT').

Two recommended reviews of experimental measurement of Henry's law constants are those by Staudinger and Roberts [1] and Bamford *et al.* [2].

Techniques usually associated with Henry's law constants are discussed first. Those associated with limiting activity coefficients follow second.

1   HENRY'S LAW CONSTANT

Vapour Pressure–Aqueous Solubility ('VP/AS')

The mole fraction definition of Henry's law constant, from Chapter 1, is

$$k_H = x/p \tag{4.1}$$

where x is the mole fraction of solute in the aqueous solution and p is the partial pressure of solute above the solution (ideal gases are assumed). It is taken for

Chemicals in the Atmosphere – Solubility, Sources and Reactivity.   Edited by P. G. T. Fogg and J. Sangster
© 2003 IUPAC   ISBN: 0-471-98651-8

granted also that x is sufficiently small that the ratio x/p is at its limiting value in an infinitely dilute solution. This definition is applicable to gaseous, liquid and solid solutes.

For gaseous solutes, the measurement of solubility is directly a measurement of Henry's law constant. Thus if, for example, the solubility of nitrogen is measured at a partial pressure of 1 bar, the data for x and p may be inserted directly into Equation (4.1).

For liquid and solute solutes, the calculation of Henry's law constant from aqueous solubility and vapour pressure is strictly not an experimental determination in most cases, although the quantities used are experimental.

Consider a system of water and an 'insoluble' liquid organic solute at constant temperature. There will be three phases at equilibrium, viz. a vapour phase (vap), an aqueous phase (aq) and an organic phase (org). In the following analysis, it is assumed that Henry's law applies to solute concentrations up to and including saturation. Since vapour and aqueous phases are in equilibrium, the partial pressure p^{vap} in Equation (4.1) will be given by

$$p^{\text{vap}} = x^{\text{aq}} \gamma^{\text{aq}} p^{\circ,1} \tag{4.2}$$

where $p^{\circ,1}$ is the vapour pressure of pure liquid solute at the temperature of measurement and x^{aq} and γ^{aq} are, respectively, the mole fraction solubility and activity coefficient of solute at saturation. Substituting Equation (4.2) in Equation (4.1), we have

$$k_{\text{H}} = 1/(\gamma^{\text{aq}} p^{\circ,1}) \tag{4.3}$$

In this equation, γ^{aq} is usually not known. In this case, however, it can be easily found from known quantities. Since pure liquid solute is in equilibrium with the aqueous solution,

$$p^{\text{vap}} = x^{\text{aq}} \gamma^{\text{aq}} p^{\circ,1} = x^{\text{org}} \gamma^{\text{org}} p^{\circ,1} \tag{4.4}$$

Assuming that water is negligibly soluble in solute, we have $x^{\text{org}} = \gamma^{\text{org}} = 1$, following the usual Raoult's law convention. From Equation (4.4), then,

$$\gamma^{\text{aq}} = 1/x^{\text{aq}} \tag{4.5}$$

Substituting Equation (4.5) in Equation (4.3),

$$k_{\text{H}} = x^{\text{aq}}/p^{\circ,1} \tag{4.6}$$

Equation (4.6) is the basis for calculating k_{H} from vapour pressure and aqueous solubility for a liquid solute.

For example, the solubility of CCl_4 in water at 25 °C is $x^{\text{aq}} = 9.529 \times 10^{-5}$ [3]. Its vapour pressure at this temperature is $p^{\circ,1} = 1.525 \times 10^4$ Pa [4]. Thus,

$$k_{\text{H}} = 9.529 \times 10^{-5}/1.525 \times 10^4 = 6.25 \times 10^{-9}\,\text{Pa}^{-1} \tag{4.7}$$

or

$$k_{\text{H}} = 3.47 \times 10^{-7}\,\text{mol}\,\text{kg}^{-1}\,\text{Pa}^{-1} \tag{4.8}$$

using the conversion factor from Chapter 3.

For a solute which is solid at the measurement temperature T, the analysis is similar. Equation (4.1) is again the starting point. If one knows both the aqueous solubility of a solid solute and the vapour pressure of the solid at the measurement temperature, one can insert these data directly into Equation (4.1) to obtain k_H. In the more common eventuality, the solid solubility may be known, but its vapour pressure may not.

For the case of a solid solute, we have three phases in equilibrium: a vapour phase (vap), an aqueous phase (aq) and a solid organic phase (cr). Since all three phases are in equilibrium,

$$p^{\text{vap}} = x^{\text{aq}}\gamma^{\text{aq}}p^{\circ,\text{l}} = x^{\text{cr}}\gamma^{\text{cr}}p^{\circ,\text{cr}} \tag{4.9}$$

where $p^{\circ,\text{l}}$ is the vapour pressure of pure supercooled liquid solute and $p^{\circ,\text{cr}}$ is the vapour pressure of pure solid solute at the temperature of measurement. If, as before, water is negligibly soluble in solid solute,

$$x^{\text{cr}} = \gamma^{\text{cr}} = 1 \tag{4.10}$$

Substituting Equation (4.10) into Equation (4.9), we have

$$x^{\text{aq}}\gamma^{\text{aq}}p^{\circ,\text{l}} = p^{\circ,\text{cr}} \tag{4.11}$$

or

$$\gamma^{\text{aq}} = p^{\circ,\text{cr}}/x^{\text{aq}}p^{\circ,\text{l}} \tag{4.12}$$

If this expression for γ^{aq} is substituted into Equation (4.3), we obtain

$$k_H = (x^{\text{aq}}/p^{\circ,\text{l}})(p^{\circ,\text{l}}/p^{\circ,\text{cr}}) \tag{4.13}$$

(thus k_H is equivalently $x^{\text{aq}}/p^{\circ,\text{cr}}$). It can be seen that Equation (4.13) is similar to Equation (4.6), with the addition of a 'correction factor' $(p^{\circ,\text{l}}/p^{\circ,\text{cr}})$.

Now the ratio $(p^{\circ,\text{l}}/p^{\circ,\text{cr}})$ is not a directly accessible experimental quantity. It can, however, be calculated by a thermodynamic argument from the thermal properties of the solute.

The Gibbs energy of fusion at T $(\Delta_{\text{fus}}G_T{}^{\circ})$ is related to the fugacities (vapour pressures) of pure supercooled liquid and pure solid:

$$\Delta_{\text{fus}}G_T{}^{\circ} = RT \ln(p^{\circ,\text{l}}/p^{\circ,\text{cr}}) \tag{4.14}$$

We let the fusion properties be $\Delta_{\text{fus}}H^{\circ}$, $\Delta_{\text{fus}}S^{\circ}$ and T_{fus}. By definition, $\Delta_{\text{fus}}H^{\circ} = T_{\text{fus}}\Delta_{\text{fus}}S^{\circ}$. The heat capacities are $C_p{}^{\circ}(\text{l})$ and $C_p{}^{\circ}(\text{cr})$ and $\Delta C_p{}^{\circ} = C_p{}^{\circ}(\text{l}) - C_p{}^{\circ}(\text{cr})$. From thermodynamics,

$$\Delta_{\text{fus}}G_T{}^{\circ} = \Delta_{\text{fus}}H^{\circ}(1 - T/T_{\text{fus}}) + \Delta C_p{}^{\circ}[T - T_{\text{fus}} + T \ln(T_{\text{fus}}/T)] \tag{4.15}$$

or

$$\Delta_{\text{fus}}G_T{}^{\circ} = (T_{\text{fus}} - T)(\Delta_{\text{fus}}S^{\circ} - \Delta C_p{}^{\circ}) + \Delta C_p{}^{\circ}T \ln(T_{\text{fus}}/T) \tag{4.16}$$

If one knows all the quantities on the right-hand side of Equation (4.15) or (4.16), one can calculate $\Delta_{fus}G_T{}^\circ$ and hence $(p^{\circ,l}/p^{\circ,cr})$.

Usually, not all these thermochemical quantities are known for the solute. Equations (4.15) and (4.16) can be simplified in a number of ways; a specific example will illustrate the effects of various approximations. Good thermodynamic data are available [5] for ethyl 4-aminobenzoate (benzocaine):

$$T_{fus} = 87.9\,^\circ C (362.9\,K) \tag{4.17}$$

$$\Delta_{fus}H^\circ = 22\,300\,J\,mol^{-1} \tag{4.18}$$

$$\Delta_{fus}S^\circ = 61.45\,J\,mol^{-1}\,K^{-1} \tag{4.19}$$

$$C_p{}^\circ(cr) = 18.95 + 0.724(T/K)\,J\,mol^{-1}\,K^{-1} \tag{4.20}$$

$$C_p{}^\circ(l) = 183.8 + 0.906(T/K)\,J\,mol^{-1}\,K^{-1} \tag{4.21}$$

At T_{fus}, $C_p{}^\circ(cr) = 281.7\,J\,mol^{-1}\,K^{-1}$ and $C_p{}^\circ(l) = 453.9\,J\,mol^{-1}\,K^{-1}$, so that $\Delta_{fus}C_p{}^\circ = 172.2\,J\,mol^{-1}\,K^{-1}$.

In a first approximation (let us call it case 'a'), we can set $\Delta C_p{}^\circ = 0$ in Equation (4.15) or (4.16), assuming a knowledge of $\Delta_{fus}H^\circ$ (or of $\Delta_{fus}S^\circ$ and T_{fus}). Hence

$$\Delta_{fus}G_T{}^\circ = \Delta_{fus}H^\circ(1 - T/T_{fus}) \tag{4.22}$$

or

$$\Delta_{fus}G_T{}^\circ = \Delta_{fus}S^\circ(T_{fus} - T) \tag{4.23}$$

In another approximation (case 'b'), neither $\Delta_{fus}H^\circ$ nor $\Delta_{fus}S^\circ$ is known. Yalkowsky [6] showed that $\Delta_{fus}S^\circ$ for many organic compounds was approximately $56\,J\,mol^{-1}\,K^{-1}$. It can more closely be approximated from structure by considering molecular rotational symmetry and molecular flexibility [7,8]. In case 'b,' $\Delta_{fus}S^\circ = 56\,J\,mol^{-1}\,K^{-1}$ is used in Equation (4.23). In a closer approximation (case 'c'), we assume we know $\Delta_{fus}H^\circ$ and $\Delta_{fus}C_p{}^\circ$ and use Equation (4.15). For the best approximation (case 'd'), we use all the data for benzocaine above, and derive

$$\Delta_{fus}G_T{}^\circ = -49\,508 + 1141.1\,T/K - 0.091(T/K)^2$$
$$- 164.85(T/K)\ln(T/K)\,J\,mol^{-1} \tag{4.24}$$

The final results for $\Delta_{fus}G_T{}^\circ$ and $(p^{\circ,l}/p^{\circ,cr})$ are given in Table 4.1 for the four approximations (in increasing order of accuracy). Assuming that the most accurate result is that of case 'd,' we see that the introduction of various approximations changes the result appreciably. Also, in this example, $\Delta T = T_{fus} - T = 65\,^\circ C$. In many environmental applications, the solid is much higher melting, i.e. $\Delta T > 65\,^\circ C$, and the approximations become less accurate still.

The calculation of k_H from vapour pressure and aqueous solubility, Equation (4.6), has been found to be accurate for nonpolar and slightly polar liquid

Table 4.1 Gibbs energy of fusion of benzocaine and its reference vapour pressure
ratio at 25 °C

Method of calculation	$\Delta_{fus}G_{298}°$ (J mol^{-1})	$p^{°,l}/p^{°,cr}$
(a) Eq. (4.22), $\Delta C_p° = 0$, $\Delta_{fus}H° = 22\,300$ J mol^{-1}	3976	4.97
(b) Eq. (4.23), $\Delta_{fus}S° = 56$ J mol^{-1} K^{-1}	3623	4.31
(c) Eq. (4.15), $\Delta C_p° = 172.2$ J mol^{-1} K^{-1}	2918	3.24
$\Delta_{fus}H° = 22\,300$ J mol^{-1}		
(d) Full expression, Equation (4.24) (see text)	2585	2.84

organic compounds, provided good data are available for both quantities. This is
demonstrated in Chapter 3 for many compounds.

Direct Measurement ('DIRECT')

This technique is called 'direct' here because it follows straightforwardly from
the definition of Henry's law constant. It explicitly requires the analysis of both
aqueous and vapour phases at equilibrium.

In the simplest arrangement, sometimes called the 'static equilibrium' approach, one adds a gas or vapour and water to a flask at constant temperature,
waits (with stirring) for equilibrium, then analyses both phases. The procedure
of Tancrède and Yanagisawa [9] is a good example of this technique, in which a
tiny amount of liquid solute (below the saturation limit) was added to the water
beforehand, and the solution allowed to equilibrate with the vapour phase.

In a modified version of this 'flask' technique, air or an inert gas is bubbled
through a dilute solution containing the solute (Figure 4.1), by means of a metal
or glass frit. This is sometimes called the 'bubble column' variant of the direct
technique. Often the gas phase is analysed by gas chromatography. The bubble
column technique was used by Leighton and Calo [10], Yin and Hassett [11],
Murphy *et al.* [12], Zhou and Mopper [13] and Kames *et al.* [14].

Note that, in this technique, the concentration of solute in the aqueous phase
is assumed to remain constant. This technique should not be confused with the
'PURGE' technique (described next).

Mention should be made here of the procedure adopted by the Sanemasa group
[15,16]. In an elegant application of the direct technique, the water was 'titrated'
by solute vapour to successively increasing pressures up to saturation. Both liquid
and solid solutes were used to supply vapour.

Purge or Gas Stripping ('PURGE')

This has also been called the 'dilution technique' [17]. The apparatus is again
that shown in Figure 4.1, with the difference that, instead of assuming that the
aqueous solute concentration is constant, the solute is deliberately purged from
the aqueous solution. The principle is described by Burnett [18], Leroi *et al.* [17]
and Mackay *et al.* [19].

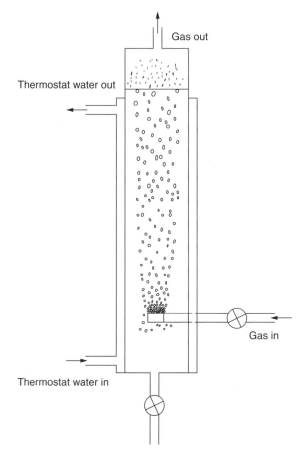

Figure 4.1 Bubble column for DIRECT and PURGE measurement techniques for Henry's law constant

Mackay *et al.* [19] discuss the conditions under which Henry's law constant may be simply related to the time rate of depletion of solute from the aqueous phase. If these conditions are satisfied, then [17,19]

$$dC/dt = -GC/(k_H VRT) \qquad (4.25)$$

where C = solute aqueous concentration (moles per unit volume)
G = flow rate of inert gas (volume per unit time)
V = volume of water in the bubble column
R = gas constant
T = absolute temperature
k_H = Henry's law constant (here, in units of concentration per unit pressure).

We can integrate this equation from $t = 0$ to $t = t$:

$$\ln(C/C_0) = -Gt/(k_H VRT) \tag{4.26}$$

so that a plot of logarithm of concentration versus time is linear with a slope of $-(G/k_H VRT)$. If solute concentration is measured by absorption spectrophotometry, then the logarithm of absorbance can be plotted against time; it is not necessary to know the initial solute concentration [19].

The purge technique has been used successfully for both liquid and solid solutes [20–24].

Equilibrium Partitioning in Closed Systems ('EPICS')

This technique is based upon a comparison of mass balances in two similar systems [25,26]. It requires no special apparatus. Two identical flasks, 1 and 2, containing different volumes of water, are spiked with solute. The flasks are capped and allowed to come to equilibrium. It can be shown [25,26] that

$$k_{AW} = [V_w(2) - rV_w(1)]/[rV_g(1) - V_g(2)] \tag{4.27}$$

where k_{AW} = dimensionless partition coefficient
V_w = liquid volume
V_g = gas volume
$r = [C_g(1)/C_g(2)]/(M_2/M_1)$
C_g = concentration of solute in the gas phase
M_2/M_1 = ratio of solute masses added to the flasks.

The best results are obtained when water volumes $V_w(1)$ and $V_w(2)$ are very different, and the value of k_{AW} sought lies in the interval $0.1 < k_{AW} < 1$ [25,27]. The EPICS technique has been used by Ashworth et al. [28], Hansen et al. [29], DeWulf et al. [27], Kondoh and Nakajima [30] and Park et al. [31].

Wetted-Wall Column ('WET WALL')

This is really a variant on the direct technique described above. In the present case (Figure 4.2), an aqueous solution flows down the inside wall of a glass tube as a thin film, where it equilibrates with a concurrent flow of inert gas [32]. The aqueous and gas phases are collected separately at the bottom of the tube and analysed for solute. The results are usually expressed as k_{AW}. The technique has been used for solutes in gas, liquid or solid form [32–34], notably for PCBs and pesticide solutes.

Fog Chamber ('FOG')

This is again a variant on the direct technique. Aqueous and gas phases are aspirated together, producing a spray (fog) in a chamber (Figure 4.3). The droplet size is about $40\,\mu m$ [35]. A small portion of the fog droplets are collected in the

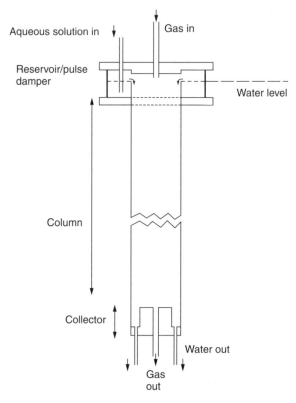

Figure 4.2 Apparatus for the wetted-wall column measurement of Henry's law constant

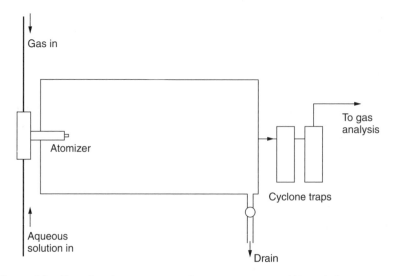

Figure 4.3 Fog chamber apparatus for measurement of Henry's law constant

cyclone traps and analysed. The gas phase is analysed downstream. The results are usually expressed as k_{AW}. The technique was used for pesticide solutes [35].

Multiple Equilibrium Headspace Chromatography ('HEADSPACE')

In this technique, only one phase is analysed for solute. The basic procedure and equations were first suggested by McAuliffe [36]. Apparatus was described by this author and Munz and Roberts [37].

Typically, the container is a gas-tight graduated syringe. Let its volume be V_{tot}. An appropriate volume V_1 of an aqueous solution of solute is introduced ($V_1 < V_{tot}$). Let the initial quantity of solute introduced be N_1 mol. The remaining volume $V_g = V_{tot} - V_1$ is filled with an inert gas, and the two phases equilibrate. The N_1 mol distribute themselves between water (L_1 mol) and gas (G_1 mol); this is the first equilibration. Hence

$$N_1 = G_1 + L_1 \tag{4.28}$$

All the gas is then expelled by manipulating the plunger and the quantity G_1 is determined by gas chromatography. An identical volume V_g of fresh gas is introduced and the phases equilibrate a second time.

From the definition of k_{AW},

$$k_{AW} = (G_1/V_g)/(L_1/V_1) \tag{4.29}$$

$$= G_1/(L_1 r) \tag{4.30}$$

where $r = V_g/V_1$, kept constant throughout the procedure. From Equations (4.28) and (4.30),

$$N_1 = G_1 + L_1 = G_1 + G_1/(rk_{AW}) \tag{4.31}$$

$$= G_1[1 + 1/(rk_{AW})] \tag{4.32}$$

From Equation (4.32),

$$G_1 = rN_1 k_{AW}/(1 + rk_{AW}) \tag{4.33}$$

In the second equilibration, $(N_1 - G_1) = N_2$ mol distribute themselves between the two phases. At the second equilibrium stage, therefore,

$$G_2 = rk_{AW}(N_1 - G_1)/(1 + rk_{AW}) \tag{4.34}$$

similar to Equation (4.27). The gas volume V_g is expelled and analysed as before. The solute is 'extracted' from the liquid in a series of steps.

Equation (4.33) is substituted into Equation (4.34). After simplification,

$$G_2 = rk_{AW}N_1/(1 + rk_{AW})^2 \tag{4.35}$$

From Equations (4.33) and (4.35),

$$k_{AW} = [(G_1/G_2) - 1]/r \tag{4.36}$$

Hence k_{AW} may be deduced from just two equilibrations. For a series of n steps,

$$G_n = r k_{AW} N_1 / (1 + r k_{AW})^n \tag{4.37}$$

$$\therefore \log G_n = an + b \tag{4.38}$$

where

$$a = -\log(1 + r k_{AW}) \tag{4.39}$$

$$b = \log(r k_{AW} N_1) \tag{4.40}$$

Thus a semilogarithmic plot of G (or chromatographic peak area) versus n is linear, and k_{AW} can be found from the slope.

Since r is kept constant, the ratio G_n / L_n is always the same, from Equation (4.30). Alternatively, the liquid may be analysed instead of the gas, and a relation similar to Equation (4.37) holds [37].

This technique was used by Kolb *et al.* [38] and more recently in a fully automated version [39].

A variant of the headspace technique was described by Robbins *et al.* [40]. Imagine a flask, into which is put a volume V_L of aqueous solution of concentration C. The number of moles of solute is $N = C V_L$, which then equilibrates between liquid volume V_L and gas volume V_G. The concentrations at equilibrium are C_L and C_G, corresponding to N_L and N_G mol of solute. The governing relations are

$$C V_L = N = N_L + N_G \tag{4.41}$$

Now

$$k_{AW} = C_G / C_L \tag{4.42}$$

by definition. Dividing Equation (4.41) by V_L,

$$N / V_L = N_L / V_L + N_G / V_L \tag{4.43}$$

or

$$C = C_L + N_G / V_L \tag{4.44}$$

$$= C_L + (N_G / V_L)(V_G / V_G) \tag{4.45}$$

$$= C_L + C_G (V_G / V_L) \tag{4.46}$$

Using Equation (4.42), we have

$$C = C_G / k_{AW} + C_G (V_G / V_L) \tag{4.47}$$

$$= C_G (1 / k_{AW} + V_G / V_L) \tag{4.48}$$

Rearranging Equation (4.48),

$$1 / C_G = 1 / C k_{AW} + (V_G / V_L) / C \tag{4.49}$$

The same equilibration is performed with a number of identical flasks, using different gas-to-liquid volume ratios, but with the same initial aqueous concentration C. By plotting $1/C_G$ versus (V_G/V_L), one obtains a straight line, and k_{AW} can be found from the ratio slope/intercept.

Critical Evaluation ('CRIT EVAL')

In a few cases in the evaluations in Chapter 11, the data source is indicated as an external critical evaluation. Occasionally such data were used to supplement other tabulated data; sometimes they were judged to be of sufficiently high quality as to merit a RECOMMENDED value themselves. The work of Mackay and Shiu [41] is an example.

Evaluation of Techniques for Determining Henry's Law Constants

There have been few attempts to compare systematically the accuracy and applicability of the various techniques mentioned above [28,35,42].

The strengths and weaknesses are sometimes readily apparent. The VP/AS technique is accurate particularly for liquids of low or zero polarity, for which good aqueous solubility and vapour pressure data are available. For high-melting solids, the technique requires properties of the supercooled liquid, which cannot be directly measured. Nevertheless, for some purposes, it may be sufficiently accurate [41].

An analysis of the factors influencing the accuracy of the EPICS technique was given by DeWulf *et al.* [27]. The PURGE technique is particularly convenient for both liquid and solid solutes. Its accuracy is primarily dependent on the degree of approach to equilibrium, i.e. the extent to which the vapour and liquid at the top of the bubble column are in equilibrium [19]. One of the limiting factors in the HEADSPACE procedure is the range over which the ratio (V_G/V_L) can be varied [40] or the number of times the solute can be extracted by gas [36] before concentrations become too small to be determined accurately.

For environmental purposes, where ultimate accuracy is not required, the various convenient and rapid techniques above produce data of comparable quality. For the highest accuracy, of course, an appropriate DIRECT technique is required (including attention to many details of equilibrium, analysis and manipulation). Properly used, the VP/AS calculation is also accurate, as mentioned above.

2   ACTIVITY COEFFICIENT AT INFINITE DILUTION (γ^∞)

Since Henry's law constant and γ^∞ are directly related through the solute vapour pressure, a technique for measuring the former is automatically one for determining the latter.

Techniques more specifically used for measuring γ^∞ have been reviewed [43–47].

Gas Chromatography ('GC')

The theory of the gas–liquid chromatographic technique for measuring γ^∞ has been summarized by Eckert *et al.* [43], Tiegs *et al.* [44] and Khalfaoui and Newsham [48]. Corrections must be applied for non-ideality of the gas mixture. Greatly simplified, the relation becomes [43,44]

$$\gamma^\infty = RT\phi_2 Z/(V_g{}^\circ M_1 \phi_2{}^\circ p_2{}^\circ) \tag{4.50}$$

where T = the absolute temperature
        ϕ = fugacity coefficient of solute at column temperature
        Z = compressibility factor of the gas mixture
        $V_g{}^\circ$ = solute specific retention volume
        M_1 = molecular weight of the solvent
        $\phi_2{}^\circ$ = fugacity coefficient of pure solute at its vapour pressure at column temperature
        $p_2{}^\circ$ = vapour pressure of solute at column temperature.

Conventional Vapour–Liquid Equilibrium

A vast quantity of data has been obtained by isothermal and isobaric methods of vapour–liquid equilibrium of binary systems. Results for aqueous binary systems have been compiled in the DECHEMA series [49–51]. The raw data are reduced to parameters for activity coefficients according to various models (Wilson, NRTL, Van Laar, Margules, UNIQUAC equations). The γ^∞ data are obtained by setting the mole fraction of the solute to zero. This is a method in which data obtained at intermediate compositions are extrapolated to zero solute concentration. The accuracy of the final results naturally depends upon the fitting equation used and its ability to represent faithfully the variation of activity coefficient with concentration.

Direct Vapour–Liquid Equilibrium Measurement of Dilute Solutions

If one measures the total pressure P of a dilute aqueous solution as a function of solute concentration $(x \to 0)$ at constant temperature, the value of γ^∞ may be found from the equation (highly simplified)

$$\gamma^\infty = P/P^\circ + (\partial P/\partial x)_{T,x\to0}/P^\circ \tag{4.51}$$

where P° is the vapour pressure of pure solute at T [44,46]. This method was used by Fischer and Gmehling [52] by measuring the absolute total pressure of aqueous butanol solutions $(x \geq 0.99)$ at $50\,^\circ\text{C}$. It can be appreciated that special attention must be paid to purity of solute, precision of pressure measurement and temperature control.

    The total pressure can be measured relative to pure solvent, in which case the method is called the 'differential static method.'

The Ebulliometric Method ('EBULLIOMETRY')

Alternatively, one can measure the boiling point T of a dilute aqueous solution ($x \rightarrow 0$) at a given pressure P. The required quantity is then $(\partial T/\partial x)_{P,x\rightarrow0}$ [53]. The governing equation analogous to Equation (4.51) is

$$\gamma^{\infty} = P_{w}^{\circ}/P^{\circ} - (\mathrm{d}P_{w}^{\circ}/\mathrm{d}T)[\partial T/\partial x]_{P,x\rightarrow0} \qquad (4.52)$$

where P_{w}° and P° are the vapour pressures of solvent and solute, respectively [44,54]. As in the static case, it is more convenient to measure the temperature relative to the boiling point of pure solvent ('differential ebulliometry').

The Dew Point Method

This method was designed for systems in which the solute is less volatile than the solvent [55,56]. What is measured is the change in dew temperature of the solvent as a function of small additions of solute. The simplified relation is

$$\gamma^{\infty} = (P_{w}^{\circ})^{2}/P^{\circ}[P_{w}^{\circ} + (\partial T/\partial y)(\partial P_{w}^{\circ}/\partial T)] \qquad (4.53)$$

where y is the solute mole fraction in the vapour.

The Rayleigh Distillation Method

The apparatus needed for this method is simpler than all those described above. A dilute solution, of known total mass and solute concentration, is subjected to a one-stage flask-to-flask distillation (in a modified version by Dohnal and Horakova [57], the distillation is effected by passing inert gas through, collecting the transferred solution quantitatively and determining the total mass and composition). As the solutions are dilute, the mass balance equations are simplified, and the governing equation is (ideal gases assumed)

$$\gamma^{\infty} = P_{w}^{\circ}[1 + \ln(x_{f}/x_{i})/\ln(m_{f}/m_{i})]/P^{\circ} \qquad (4.54)$$

where x_{i} and x_{f} are solute mole fractions in the initial and transferred solutions, respectively, and m_{i} and m_{f} are the corresponding total masses. Although the method has not been reported used on aqueous solutions, it is surprisingly accurate, considering its experimental simplicity [57].

Liquid–Liquid Chromatography ('INDIRECT')

In this technique, the solute partitions between two immiscible liquid phases (mobile and stationary) in a column. The retention time is related to the ratio of the γ^{∞}s of the solute in the two phases. In order to determine γ^{∞} in one phase, the datum in the other phase must be known. If it has not been otherwise previously determined, it can be found by, e.g., gas–liquid chromatography using the same stationary phase at the same temperature.

Evaluation of Techniques for Measuring γ^{∞}

This topic has been well investigated and can be technically detailed. For present purposes, it is sufficient to point out that Kojima *et al.* [47] presented a thorough summary, with a good bibliography.

REFERENCES

1. Staudinger, J.; Roberts, P. V., *Crit. Rev. Environ. Sci. Technol.*, 1996, **26**, 205–97.
2. Bamford, H. A.; Baker, J. E.; Poster, D. L., *Review of Methods and Measurements of Selected Hydrophobic Organic Contaminant Aqueous Solubilities, Vapor Pressures and Air–Water Partition Coefficients*, NIST Special Publication No. 928, US Department of Commerce, Washington, DC, 1998.
3. Horvath, A. L.; Getzen, F. W., *Solubility Data Series, Vol. 60, Halogenated Methanes with Water*, Oxford University Press, Oxford, 1995.
4. Gmehling, J.; Onken, U.; Arlt, W., *Chemistry Data Series, Vol. I/8, Vapor–Liquid Equilibrium Data Collection*, Deutsche Gesellschaft für Chemisches Apparatewesen, Frankfurt/Main, 1984.
5. Neau, S. H.; Flynn, G. L., *Pharm. Res.*, 1990, **7**, 1157–62.
6. Yalkowsky, S. H., *Ind. Eng. Chem. Fundam.*, 1979, **18**, 108–11.
7. Dannenfelser, R. M.; Yalkowsky, S. H., *Ind. Eng. Chem. Res.*, 1996, **35**, 1483–6.
8. Dannenfelser, R. M. Estimation of the entropy of melting from structure, *PhD Thesis*, University of Arizona, 1997.
9. Tancrède, M. V.; Yanagisawa, Y., *J. Air Waste Manage. Assoc.*, 1990, **40**, 1658–63.
10. Leighton, D. T.; Calo, J. M., *J. Chem. Eng. Data*, 1981, **26**, 382–5.
11. Yin, C.; Hassett, J. P., *Environ. Sci. Technol.*, 1986, **20**, 1213–17.
12. Murphy, T. J.; Mullin, M. D.; Meyer, J. A., *Environ. Sci. Technol.*, 1987, **21**, 155–62.
13. Zhou, X.; Mopper, K., *Environ. Sci. Technol.*, 1990, **24**, 1864–9.
14. Kames, J.; Schweighoefer, S.; Schurath, U., *J. Atmos. Chem.*, 1991, **12**, 169–80.
15. Sanemasa, I.; Araki, M.; Deguschi, T.; Nagai, H., *Bull. Chem. Soc. Jpn.*, 1982, **55**, 1054–62.
16. Sanemasa, I.; Ishibashi, K.; Kumamaru, M.; Deguschi, T., *Bull. Chem. Soc. Jpn.*, 1989, **62**, 2908–12.
17. Leroi, J.-C.; Masson, J.-C.; Renon, H.; Fabries, J.-F.; Sannier, H., *Ind. Eng. Chem. Process Des. Dev.*, 1977, **16**, 139–44.
18. Burnett, M. G., *Anal. Chem.*, 1963, **35**, 1567–70.
19. Mackay, D.; Shiu, W. Y.; Sutherland, R. P., *Environ. Sci. Technol.*, 1979, **13**, 333–7.
20. Kucklick, J. R.; Hinckley, D. A.; Billeman, T. F., *Mar. Chem.*, 1991, **34**, 197–209.
21. ten Hulscher, Th. E. M.; van der Velde, L. E.; Bruggeman, W. A., *Environ. Toxicol. Chem.*, 1992, **11**, 1595–603.
22. Alaee, M.; Whittal, R. M.; Strachan, W. J., *Chemosphere*, 1996, **32**, 1153–64.
23. Shiu, W.-Y.; Mackay, D., *J. Chem. Eng. Data*, 1997, **42**, 27–30.
24. Hovorka, S.; Dohnal, V., *J. Chem. Eng. Data*, 1997, **42**, 924–33.
25. Lincoff, A. H.; Gossett, J. M., The determination of Henry's constant for volatile organics by equilibrium partitioning in closed systems, in *Gas Transfer at Water Surfaces*, Brutsaert, W.; Jirka, G. H., (eds), Reidel, Dordrecht, 1984, pp. 17–25.
26. Gossett, J. M., *Environ. Sci. Technol.*, 1987, **21**, 202–8.
27. DeWulf, J.; Drijvers, D.; van Langenhove, H., *Atmos. Environ.*, 1995, **29**, 323–31.
28. Ashworth, R. A.; Howe, G. B.; Mullins, M. E.; Rogers, T. N., *J. Hazard. Mater.*, 1988, **18**, 25–36.
29. Hansen, K. C.; Zhou, Z.; Yaws, C. L.; Aminabhavi, T. M., *J. Chem. Eng. Data*, 1993, **38**, 546–50.
30. Kondoh, H.; Nakajima, T., *Kankyo Kagaku*, 1997, **7**, 81–9.
31. Park, S.-J.; Han, S.-D.; Ryu, S.-A., *Hwahak Konghak*, 1997, **35**, 915–20.

32. Fendinger, N. J.; Glotfelty, D. E., *Environ. Sci. Technol.*, 1988, **22**, 1289–93.
33. Brunner, S.; Hornung, E.; Santl, H.; Wolff, E.; Piringer, O. G.; Altschuh, J.; Brügge-mann., R., *Environ. Sci. Technol.*, 1990, **24**, 1751–4.
34. Rice, C. P.; Chernyak, S. M.; McConnell, L. L., *J. Agric. Food Chem.*, 1997, **45**, 2291–8.
35. Fendinger, N. J.; Glotfelty, D. E.; Freeman, H. P., *Environ. Sci. Technol.*, 1989, **23**, 1528–31.
36. McAuliffe, C., *Chem. Technol.*, 1971, **1**(1), 46–51.
37. Munz, C.; Roberts, P. V., *J. Am. Water Works Assoc.*, 1987, **79**(5), 62–9.
38. Kolb, B.; Welter, C.; Bichler, C., *Chromatographia*, 1992, **34**, 235–40.
39. Zhu, J.; Chai, X., *Am. Laboratory*, 1999, **31**(17), 28C–30C.
40. Robbins, G. R.; Wang, S.; Stuart, J. D., *Anal. Chem.*, 1993, **65**, 3113–8.
41. Mackay, D.; Shiu, W. Y., *J. Phys. Chem. Ref. Data*, 1981, **10**, 1175–99.
42. Sagebiel, J. C.; Seiber, J. N.; Woodrow, J. E., *Chemosphere*, 1992, **25**, 1763–8.
43. Eckert, C. A.; Newman, B. A.; Nicolaides, G. L.; Long, T. C., *AIChE J.*, 1981, **27**, 33–40.
44. Tiegs, D.; Gmehling, J.; Medina, A.; Soares, M.; Bastos, J.; Alessi, P.; Kikic, I., *Chemistry Data Series, Vol. IX/1, Activity Coefficients at Infinite Dilution*, Deutsche Gesellschaft für Chemisches Apparatewesen, Frankfurt/Main, 1986.
45. Bergmann, D. L.; Eckert, C. A., *ACS Symp. Ser.*, 1992, **509**, 218–28.
46. Sandler, S. I., *Fluid Phase Equilib.*, 1996, **116**, 343–53.
47. Kojima, K.; Zhang, S.; Hiaki, T., *J. Phase Equilib.*, 1997, **131**, 145–79.
48. Khalfaoui, B.; Newsham, D. M. T., *J. Chromatogr. A*, 1994, **673**, 85–92.
49. Gmehling, J.; Onken, U., *Chemistry Data Series, Vol. I/1 , Vapor–Liquid Equilibrium Data Collection*, Deutsche Gesellschaft für Chemisches Apparatewesen, Frankfurt/Main, 1977.
50. Gmehling, J.; Onken, U.; Arlt, W., *Chemistry Data Series, Vol. I/1a , Vapor–Liquid Equilibrium Data Collection*, Deutsche Gesellschaft für Chemisches Apparatewesen, Frankfurt/Main, 1981.
51. Gmehling, J.; Onken, U.; Rarey-Nies, J. R., *Chemistry Data Series, Vol. I/1b , Vapor–Liquid Equilibrium Data Collection*, Deutsche Gesellschaft für Chemisches Apparatewesen, Frankfurt/Main, 1988.
52. Fischer, K.; Gmehling, J., *J. Chem. Eng. Data*, 1994, **39**, 309–15.
53. Raal, J. D., *AIChE J.*, 2000, **46**, 210–20.
54. Dallinga, L.; Schiller, M.; Gmehling, J., *Chem. Eng. Data*, 1993, **38**, 147–55.
55. Trampe, D. B.; Eckert, C. A., *AIChE J.*, 1993, **39**, 1045–50.
56. Suleiman, D.; Eckert, C. A., *J. Chem. Eng. Data*, 1994, **39**, 692–6.
57. Dohnal, V.; Horakova, I., *Fluid Phase Equilib.*, 1991, **68**, 173–85.

CHAPTER 5

Calculation of Henry's Law Constants and Infinite Dilution Activity Coefficients – An Overview

James Sangster

Sangster Research Laboratories, Montreal, Canada

The terms 'calculation,' 'prediction' and 'correlation' have often been used interchangeably. In what follows, 'calculation' will be taken to mean a deduction of k_H or γ^∞ for organic solutes in water by means of quantities more or less distanced from actual vapour–liquid equilibrium measurements. The procedures of deriving k_H from vapour pressure and aqueous solubility ('VP/AS' – see Chapter 4) and γ^∞ from solubility measurements (also see Chapter 4) are here considered to be experimental and not calculation methods.

As pointed out in Chapter 4, k_H and γ^∞ are thermodynamically related quantities and hence are interconvertible. We have [1]

$$\Delta_{vap} G = -RT \ln p^\circ \tag{5.1}$$

$$\Delta_{hyd} G = -RT \ln k_H \quad (k_H = \text{pressure}^{-1}) \tag{5.2}$$

$$\Delta_{sol} G = RT \ln \gamma^\infty \tag{5.3}$$

$$\Delta_{hyd} G + \Delta_{vap} G = \Delta_{sol} G \tag{5.4}$$

where p° is the vapour pressure of pure solute at T and "vap", "hyd" and "sol" refer to processes of vaporization, hydration and solution respectively. The calculation of k_H or γ^∞ is therefore equivalent to calculating the Gibbs energy change in hydration or solution. Depending upon their starting points of convenience, calculation methods may focus primarily on Gibbs energy changes or on the limiting quantities themselves.

The well-known predictive methods for k_H and γ^∞ need databases of good quality experimental data for a large number of aqueous solutes. A list of such

Chemicals in the Atmosphere – Solubility, Sources and Reactivity. Edited by P. G. T. Fogg and J. Sangster
© 2003 IUPAC   ISBN: 0-471-98651-8

Table 5.1 Compilations of experimental Henry's law constants of aqueous solutes

Types of compounds (No. of compounds)	Ref.
General organic (292)	2
General organic (1656)	3
General organic (150)	4
Ketones (38)	5
Inorganic (17), C_1–C_4 organic (16)	6
Hydrocarbons, halocarbons (172)	7
PAHs[a] (13), chlorinated hydrocarbons (28), PCBs[a] (108), agrochemicals (25)	8
Pesticides (92)	9
General organic (408)	1,10,11
Inorganic (63), general organic (850)	12
General organic (44)	13
General organic (37)	14
PCBs (73)	15
PCBs (69)	16
General organic (62), pesticides (15)	17
PCBs (42)	18

[a]PAHs = polycyclic aromatic hydrocarbons; PCBs = polychlorinated biphenyls.

Table 5.2 Compilations of experimental infinite dilution activity coefficients of aqueous solutes

Types of compounds (No. of compounds)	Ref.
General organic (404)	19
General organic (427)	20
General organic (326)	21
General organic from VLE[a] (148)	22–24
General organic from γ^∞ measurements (107)	25
General organic (34)	26
General organic (325)	27
General organic ('large')	28
General organic ('large')	29

[a]VLE = vapour–liquid equilibrium.

compilations appears in Tables 5.1 and 5.2. Special mention should be made here of the Solubility Data Series (International Union of Pure and Applied Chemistry), a series of volumes containing critically evaluated solubility data (from 1979). Those volumes concerned with gases in water are listed in Table 5.3.

Since k_H and γ^∞ are closely related, the present survey is organized according to the fundamental approach taken by investigators, rather than to the final quantities themselves. Without being absolutely exhaustive, the following discussion is intended to be wide, rather than restricted.

Table 5.3 Volumes in the IUPAC Solubility Data Series concerning low-pressure gases in water

Volume	Gases	Ref.
1	He, Ne	30
2	Kr, Xe, Rn	31
4	Ar	32
5/6	H_2, D_2	33
7	O_2, O_3	34
8	Nitrogen oxides	35
9	Ethane	36
10	Ne	37
12	SO_2, Cl_2, F and Cl oxides	38
24	Propane, butane, isobutane	39
27/28	Methane	40
32	H_2S, D_2S, H_2Se	41
43	CO	42
57	Ethene	43
60	Halogenated methanes	44
62	CO_2	45
76	Ethyne	46

1 CALCULATION FROM FIRST PRINCIPLES

This method is often called 'Monte Carlo,' 'molecular dynamics' or 'computer simulation.' It is an attempt to calculate Gibbs energies *ab initio*. Some [47–49] adopt the so-called 'test-particle method,' in which a solute particle is inserted into the solvent. The residual chemical potential is related to the ensemble average of the system, using an appropriate expression for the potential energy. Other approaches use the Kirkwood coupling parameter method, in which the test molecule is gradually 'grown,' rather than inserted, into the system [50–52]. Hu *et al.* [53] used an equation-of-state method with chemical equilibrium constants to take account of hydrogen bonding. Chen and Chen [54,55] have developed a method using the hard convex body expansion. These computer simulation methods are mentioned here for the sake of completeness; the attempt to represent interactions between solutes of practical interest and a hydrogen-bonded solvent introduces intractable mathematical difficulties, and more or less drastic simplifications have to be made.

2 CALCULATION FROM 'WHOLE MOLECULE' PROPERTIES

There are a number of methods which use correlations of molecular parameters, either computed or experimental or structural. Perhaps the best known of these is the Linear Solvation Energy Relationships (LSER) approach [21,56,57]. In the original LSER model, the molecular variables were identified as 'solvatochromatic' parameters. In the present applications, the properties concerned are molecular volumes, polarizabilities and hydrogen-bond acidities and basicities.

Table 5.4   Summary of calculation methods based on molecular parameters

Quantity	Molecular parameters used	Types of compounds	Ref.
k_H	Connectivity indices	PCBs (W)[a]	58
k_H	Total surface area	PCBs (W)	59
k_H	Moment of inertia connectivity indices, polarizabilities	PCBs (W)	60
k_H	Infrared spectroscopy	Organic esters (W)	61
k_H	Connectivity indices, polarizability, hydrogen bonding index	General organic (W)	62
k_H	Heavy atom number, atomic charges, surface area	General organic (W)	63
k_H	Thermodynamic hydration parameters	4 hydrocarbons (W)	64
k_H	Connectivity indices, Cl atom number	PCBs (W)	16
k_H	12 physicochemical parameters	General organic (W)	65
γ^∞	Critical properties, molar volume	General organic	66
γ^∞	Carbon number, surface area, connectivity, electronic energy, acentric factor, dipole moment	Hydrocarbons (W)	67
γ^∞	Molar refraction	Aromatic hydrocarbons (W)	68
γ^∞	Connectivity, atomic charges, surface area, hydrogen bond acidity and basicity, polarizability	General organic (W)	27
γ^∞	Molar volume, dispersion and dipolar parameters, hydrogen bond acidity and basicity	General organic	69,70

[a](W) indicates water as solvent is included in the systems studied.

Other methods based on molecular parameters use a variety of molecular properties. These are summarized in Table 5.4.

3   METHODS BASED ON SCALED PARTICLE THEORY

The solubility of a gas in a liquid can be approached through the statistical mechanical theory of fluids, based upon the properties of radial distribution functions. An approximate expression can be obtained for the reversible work required to introduce a spherical particle into a fluid of spherical particles. Pierotti [71] used this reasoning to deduce the equation

$$\ln(1/k_H) = G_c RT + G_i/RT + \ln(RT) \quad (k_H \equiv \text{concentration/pressure}) \quad (5.5)$$

where G_c = partial molar Gibbs energy of creating a cavity in the solvent

$\quad\quad\ G_i$ = partial molar Gibbs energy for the interaction of 1 mol of solute molecules with the solvent.

Of the two Gibbs energies in Equation (5.5), G_c is easier to evaluate [72]. If the solvent is considered to be a fluid of hard sphere particles, G_c is a function of the solvent density and hard sphere diameters.

The quantity G_i is not so easily calculated. Now G_c corresponds to the insertion of a solute molecule into the solvent cavity without any solute–solvent interaction. The interaction forces are then 'switched on,' and the Gibbs energy of this step is G_i. In simple cases, such as the Lennard–Jones (6–12) potential, this is not particularly difficult. In other cases, one must consider dispersion, induction and dipole–dipole interactions [73].

In a somewhat more realistic approach [74], the molecules were assumed to be 'soft' rather than 'hard' spheres. This involved determining fitting parameters for types of solute–solvent pairs (nonpolar–nonpolar, nonpolar–polar, water–alcohol, etc.).

4 METHODS BASED ON SOLUBILITY PARAMETERS

The Scatchard–Hildebrand solubility parameter was developed from regular solution theory [75]. Although the regular solution is an idealized concept (a mixture in which the excess entropy and excess volume are zero), the solubility parameter itself is defined from measured quantities. Barton [76] provides a comprehensive overview of the both the theory and practice of this parameter.

The definition of the solubility parameter of a liquid is based on the quantity *molar cohesive energy* (the molar potential energy of the liquid relative to the ideal vapour at the same temperature). For a liquid well below its normal boiling point, this quantity may be written as $(\Delta_{vap}H - RT)$, where $\Delta_{vap}H$ is the enthalpy of vaporization at T. If V is the molar volume of the liquid, then the *cohesive energy density* is defined as $(\Delta_{vap}H - RT)/V$; the solubility parameter, usually written as δ, is defined as

$$\delta = [(\Delta_{vap}H - RT)/V]^{0.5} \tag{5.6}$$

The unit for δ is thus seen to be (energy per unit volume)$^{0.5}$ or (pressure)$^{0.5}$. The quantity known as the *internal pressure* π of a liquid is defined thermodynamically as $\pi = (\partial U/\partial V)_T$, where U is the internal energy. For nonpolar or weakly polar liquids it has been found that $\pi \approx \delta^2$. (This is not true for strongly polar or hydrogen-bonded liquids.) The solubility parameter has been reasonably, although not totally, successful as a means of representing the thermodynamic interactions among many organic substances, including polymers and paints, etc. [76].

The use of the solubility parameter for calculating k_H and γ^∞ has been developed only recently. According to regular solution theory, one can write the infinite dilution activity coefficient of a solute (subscript 2) in a solvent (subscript 1) as

$$\ln \gamma_2^\infty = V_2(\delta_2 - \delta_1)^2/RT \tag{5.7}$$

This expression is completely inadequate for mixtures containing polar or hydrogen-bonded molecules. This was recognized by a number of investigators [77–83], and a modified expression was proposed by Thomas and Eckert [28]:

$$\ln \gamma_2^{\infty} = V_2[(\lambda_2 - \lambda_1)^2 + (\tau_2 - \tau_1)^2 + (\sigma_1 - \sigma_2)(\tau_1 - \tau_2)$$
$$+ (\alpha_1 - \alpha_2)(\beta_1 - \beta_2)]/RT + \ln(V_2/V_1) + 1 - (V_2 - V_1) \quad (5.8)$$

where λ = molecular polarizability
τ = molecular polarity
σ = dipole induction
α = hydrogen bond acidity
β = hydrogen bond basicity.

The last three terms in Equation (5.8) take into account the fact that solute and solvent molecules may differ greatly in size. This equation is referred to as the original MOSCED (Modified Separation of Cohesive Energy Density) equation. It was further improved incrementally by Howell et al. [84] and Wu et al. [85]. In a slightly different approach, Poe et al. [86] used different forms of the MOSCED equation to create synthetic data matrices to be analysed by factor analysis. The factors from these matrices were used to target test the different model parameters. Much simplified versions of the MOSCED equation were suggested by Padovani and Suleiman [87] and Vilcu and Puricel [88].

The solubility parameter has also been used to derive an expression for the Henry's law constant [89]. This is part of a rather ambitious attempt to include not only k_H, but also the octanol–water partition coefficient, aqueous solubility and chromatographic retention, in a single scheme. The result is called SOFA (Solubility Parameters for Fate Analysis). In SOFA, k_H is equated to the quotient of water solubility and vapour pressure (this is described in Chapter 4 as the VP/AS method). The equation for k_H in SOFA may be written as

$$\ln k_H = -\ln p_2^{\circ} - V_2[\delta_2^2 + \delta_1^2 - 2c_1\delta_1\delta_2(c_2\delta_1 + \delta_2)/(\delta_1 + c_2\delta_2)]/RT$$
$$+ d_2 + \ln(V_2/V_1) - (V_2/V_1) + C \quad (k_H \cong \text{concentration/pressure}) \quad (5.9)$$

where p_2° = vapour pressure of pure solute
V_1 and V_2 = molar volumes as before
c_1, c_2 and d_2 = fitting parameters
C = a constant for a given solvent.

In this equation, d_2 is related to that part of the entropy of mixing other than the combinatorial contribution. The burden of SOFA has been concerned principally with organic pollutants as solutes, e.g. methylbenzenes [90], methylchlorobenzenes [89], chlorinated dibenzofurans [91–93] and chlorinated dibenzodioxins [91–94]. Many of these substances are solid at ordinary temperatures, and these authors have developed methods for predicting the needed quantities enthalpy of vaporization and liquid molar volume, from structure.

Jonah [95] showed that, for a few systems, the temperature dependence of the Henry's law constant could be correlated through the entropy of vaporization of the solvent.

5   UNIQUAC, UNIFAC, ASOG, NRTL AND WILSON EQUATIONS

Historically, it has been found easier to account for structural and thermodynamic properties of gases and crystalline solids, rather than liquids. In gases, molecules are far apart, motion is random and there is very little effect of intermolecular forces. In the crystalline solid, order is maximum and there is very little movement in the crystal lattice. The liquid state, however, is truly an intermediate one, having a minimum degree of order (because it is a condensed state), which simultaneously allows a certain amount of molecular 'wandering.'

With respect to organic liquid mixtures, the deviations from Raoult's law ideality have customarily been assigned to an activity coefficient (γ). Now γ is a global correction factor, representing an undoubted combination of interaction effects in the solution. At present there is no satisfactory general way of deriving activity coefficients from more fundamental properties.

At an early date, it was thought that, if there were molecular interactions in solution, one might postulate an 'association' of molecules. Essentially, the idea was that, if the true species in solution could be identified, then thermo-dynamic properties could be deduced from consideration of these true species mixing ideally. Specifically, in an A–B mixture, there might be pairs such A–A, B–B and A–B. Dolezalek was among the first to propose this formally. In the system acetone–chloroform [96] the negative deviations from ideality could be accounted for by an association between chloroform and acetone molecules. Indeed, in a low-temperature solid–liquid equilibrium study [96], the existence of a 1:1 compound was discovered. Later, the same was found for the diethyl ether–chloroform system [97].

This postulate was later called a 'quasi-chemical' theory. It is 'quasi' because no covalent bonds are broken in the formation of pairs. Perhaps the most suc-cessful application of the quasi-chemical model is the work of Pelton *et al.* [98]. In this application, the theory is the basis of a model for the excess Gibbs energy which is able to represent the excess properties of molten salt, alloy and oxide slag mixtures. In these systems there are very strong solution interactions.

In organic solutions, the interactions are weaker, but new factors appear which are absent in inorganic systems, viz. molecules differing greatly in size and hydrogen bonds. Chemical engineers have invented a quasi-chemical approach for organic molecules. It was decided to adopt the concept that, viewed micro-scopically, an organic liquid mixture is not homogeneous. The composition at one point in the liquid is not necessarily the same as that at another point. Alter-natively, in an A–B mixture, the nearest neighbours to A and B molecules are not the same. The 'local composition' was taken as the primary variable.

The models indicated by the acronyms in the title to this section derive from the local composition concept. They are described briefly by Acree [99]. Of these, the UNIQUAC, NRTL and Wilson equations (Group A) bear a family resemblance; ASOG and UNIFAC (Group B) may also profitably be considered together.

Group A may be said to have a 'molecular' orientation. For example, Wil-son [100] proposed that the probability of finding nearest neighbours of a certain type be given by an expression using Boltzmann factors. There are two fitting parameters. Wilson [101] also used an alternative equation with a third parameter for strongly polar hydrogen-bonded systems.

The Non-Random Two-Liquids (NRTL) model of Renon and Prausnitz [102] uses the two-liquid theory of Scott [103]. There are three parameters, of which one is called a 'nonrandomness' parameter generally in the range 0.2–0.5; it increases with the complexity of intermolecular interaction.

The Universal Quasi-chemical (UNIQUAC) equation introduces local area fractions as the primary concentration variable [104]. Although there are only two fitting parameters for a binary mixture, the equation assumes given volume and area parameters for each component. These are derivable from molecular dimensions. The resulting equations are rather cumbersome. It was modified later for particular reasons [105,106]. The UNIQUAC parameters were determined by molecular mechanics calculations for aqueous solutions of diols, glycerol and glucose [107].

The equations for γ^∞ of these models are summarized in Table 5.5. The relatively simple expressions of Margules and Van Laar are included for comparison. Abrams and Prausnitz [104] pointed out that, through various simplifications, the UNIQUAC equation reduces to the five others listed in Table 5.5.

Group B methods, in contrast to those of Group A, may be said to have 'group' orientation. The seminal ideas were early exemplified by Pierotti *et al.* [108,109].

Table 5.5 Activity coefficient expressions at infinite dilution in binary systems for different solution models

Equation	$\ln \gamma_1^\infty = \cdots$ $\ln \gamma_2^\infty = \cdots$	Fitting parameters
Margules	A_{12} A_{21}	A_{12}, A_{21}
Van Laar	A_{12} A_{21}	A_{12}, A_{21}
Wilson[a]	$1 - \ln \Lambda_{12} - \Lambda_{21}$ $1 - \ln \Lambda_{21} - \Lambda_{12}$	$(\lambda_{12} - \lambda_{11}), (\lambda_{21} - \lambda_{22})$
NRTL[b]	$\tau_{21} + \tau_{12}\exp(-\alpha_{12}\tau_{12})$ $\tau_{12} + \tau_{21}\exp(-\alpha_{12}\tau_{21})$	$(g_{12} - g_{22}), (g_{21} - g_{11})$ α_{12} (assigned)
UNIQUAC[c]	$\ln(r_1/r_2) + q_1[5\ln(q_1 r_2/q_2 r_1)] - \ln\tau_{21}$ $\quad + 1 - \tau_{12}] + l_1 - r_1 l_2/r_2$ $\ln(r_2/r_1) + q_2[5\ln(q_2 r_1/q_1 r_2)]$ $\quad - \ln\tau_{12} + 1 - \tau_{21}] + l_2 - r_2 l_1/r_1$	$(u_{12} - u_{22}), (u_{21} - u_{11})$

[a] $\Lambda_{12} = (V_2/V_1)\exp[-(\lambda_{12} - \lambda_{11})/RT]$
$\Lambda_{21} = (V_1/V_2)\exp[-(\lambda_{21} - \lambda_{22})/RT]$
$V_1, V_2 =$ liquid molar volumes.
[b] $\tau_{12} = (g_{12} - g_{22})/RT$
$\tau_{21} = (g_{21} - g_{11})/RT$
$\alpha_{12} =$ non-randomness parameter.
[c] $\tau_{12} = \exp[-(u_{12} - u_{22})/RT]$
$\tau_{21} = \exp[-(u_{21} - u_{11})/RT]$
$l_1 = 5(r_1 - q_1) - (r_1 - 1)$
$l_2 = 5(r_2 - q_2) - (r_2 - 1)$
$r_1, r_2 =$ relative van der Waals volumes
$q_1, q_2 =$ relative van der Waals surface areas.

Wilson and Deal [110] put forward the first practical equation based on the starting point for the activity coefficient:

$$\ln \gamma = \ln \gamma^{S} + \ln \gamma^{G} \tag{5.10}$$

where γ^{S} is the contribution associated with molecular size differences and γ^{G} the contribution due to interactions of groups. 'Groups' here refers to recognizable chemical groups of the usual kind, such as CH_2, OH, C=C, etc. The detailed equations are too long to be given here. These equations represent the Analytical Solution of Groups (ASOG) model [111,112].

Fredenslund *et al.* [113] incorporated the group-solution concept into the basic UNIQUAC description of solution nonideality. The groups possess both surface and volume parameters. Again, the equations are too long to be presented here. The particular attraction of the UNIFAC method (UNIQUAC Functional-Group Activity Coefficients) is that it can be used to predict activity coefficients for systems not yet studied experimentally. In order to be successful in this, a large database of group-interaction parameters is needed [114]. These are obtained from vapour–liquid equilibrium measurements on binary systems [115,116,117a, 118,119].

In recent years, the prediction of Henry's law constants and limiting activity coefficients has become particularly important in environmental concerns. A number of publications indicate interest in aqueous solutions specifically [114a,115a, 117,118a,120,121a,122a,123a].

6 ADDITIVE/CONSTITUTIVE BOND AND GROUP CONTRIBUTION METHODS

The unitless water–air partition coefficient k_{WA} of an organic compound was called by Hine and Mookerjee [2] its 'intrinsic hydrophilicity.' They presented a scheme for estimating k_{WA} based on the chemical groups in the molecule, namely CH_3, OH, NH_2, C(doubly bonded), etc. They also simultaneously presented a scheme based on the covalent bonds in the molecule, e.g., C–H, N–H, C(aromatic), etc. They listed 34 bond contributions and 71 group contributions; the group contribution scheme included corrections for intramolecular effect due to functional groups found on the same carbon skeleton.

This initiative was taken up later by Meylan and Howard [119a] and considerably expanded [3]. The latest version (HENRYWIN) incorporates many new k_{WA} data, including pesticides and organic pollutants not considered by Hine and Mookerjee. HENRYWIN is the only estimation method available as a dedicated computer package suitable for personal computer use.

7 OTHER METHODS OF CALCULATION

For the sake of completeness, a few methods are mentioned here which do not fit obviously into categories discussed so far. Lin and Sandler [120a] adopted a model similar to the starting point of the scaled particle theory, viz. a cavity is created in the solvent, the solute molecule is introduced into the cavity without solute–solvent interaction and the interactions are then 'switched on.' The procedure for calculating the Gibbs energy of cavity formation was based on UNIQUAC, to take into account different molecular shapes and solute–solvent interactions. The interaction Gibbs energy was calculated by the Effective Hamiltonian/Continuum Distribution method. Water was one solvent among those considered.

The UNIFAC model was modified by Antunes and Tassios [121] by inclusion of solubility parameters, a free volume term and one fitting parameter. It was applied to atmospheric gases and alkanes in water. Bader and Gasem [122] contributed a modified regular solution equation using molar volumes, connectivity and dipole moment. Included in this paper was an equation-of-state method using critical properties and molar volumes. Water was not included in the system examples. A correlation for small molecules in water proposed by Harvey [123] was also based on critical properties. A pseudo-critical method was given by Joffe [124] for nonaqueous systems.

Chow *et al.* [125] used a neural network approach for aqueous limiting activity coefficients of polychlorinated biphenyls (PCBs). A generalized method for computing Henry's law constants, along with octanol–water partition coefficients and aqueous solubility, was given by Drefahl and Reinhard [126]. The proposed procedure, called the Group Interchange Method, was essentially one of computing the required datum of the unknown molecule from the measured datum of a closely related compound.

Vapour–liquid equilibrium data at intermediate concentrations may be used to generate γ^{∞} values by means of common excess Gibbs energy fitting equations such as are used in the DECHEMA compilation [22–24]. This is an extrapolation method, and the results depend on the model chosen. An extrapolative method which uses the 'weighted volatility function' avoids this arbitrariness [127]. The weighted volatility function ω, for a two-component system, is

$$\omega = x_1 \ln(y_1/x_1) + x_2 \ln(y_2/x_2) \tag{5.11}$$

where x indicates liquid mole fraction and y indicates vapour mole fraction. In practice, all *PTxy* data for isothermal data sets are used, together with the vapour pressure of the pure components. The general utility of this method was demonstrated for 95 systems, most of them nonaqueous.

8 CRITIQUE OF ESTIMATION METHODS

The estimation methods described above are various, and differ greatly in complexity, applicability and accuracy. Although these methods have been much compared and discussed (Table 5.6), there has been no overall systematic evaluation, even of the principal candidates. Apart from customary scientific special interests, there are some factors which frustrate such a systematic evaluation. First, the equations used in a particular approach may be refined, corrected, expanded or otherwise changed over time. Second, the number of 'known' data

Table 5.6 Reviews and comparisons of various methods of calculating limiting properties of solutions

Methods discussed[a]	Ref.
ASOG, UNIFAC	114,29
	128a
	129a
	115a
MOSCED, UNIFAC	130
VLE, MOSCED, WVF	127
VP/AS, H&M, M&H	17
VP/AS, H&M, M&H, LSER, N&S	128
N&S, H&M	129
VLE, Suzuki	130a
ASOG, UNIFAC, SPACE	70
H&M, N&S, M&H, Russell, Suzuki, UNIFAC	131

[a]ASOG = Analytical Solution of Groups [111].
H&M = Hine and Mookerjee [2].
LSER = Linear Solvation Energy Relationships [21].
M&H = Meylan and Howard [119].
MOSCED = Modified Separation of Cohesive Energy Density [28].
N&S = Nirmalakhandan and Speece [62].
Russell = Russell *et al.* [63].
SPACE = Solvatochromic Parameters for Activity Coefficients [69].
Suzuki = Suzuki *et al.* [65].
UNIFAC = UNIQUAC Functional-Group Activity Coefficients [113].
VLE = vapour–liquid equilibrium.
VP/AS = vapour pressure/aqueous solubility.
WVF = Weighted Volatility Function [127].

available changes, as do the chemical systems represented. Third, the known data on which fitting parameters are ultimately based must be critically evaluated before use. This obvious consideration is easier said than done. Careful investigators realized this [1,3,28,86].

A detailed evaluation of all methods cannot be attempted here. There are, however, some general comments which may serve as guidelines. It should be stated immediately that the prediction of the limiting properties for solutes in aqueous solution poses a severe test for any method described above. This is true for several reasons:

- the components of the mixture sometimes differ greatly in size;
- the solute may be strongly polar or hydrogen-bonded;
- the aqueous system sometimes exhibits liquid–liquid immiscibility;
- the solute molecule may contain neighbouring functional groups, whose nearness changes their interaction with the solvent, considered separately.

For these reasons, the 'one size fits all' methods are not recommended for accurate work, e.g. SOFA [89,126]. Inevitably, accuracy in detail is sacrificed for generality. In fact, methods using solubility parameters need to be rather drastically modified, since regular solution theory cannot predict an activity coefficient less than unity: see Equation (5.7).

The group contribution methods (Section 5) represent a different attempt to represent the vapour–liquid equilibria of organic mixtures. ASOG and UNIFAC in particular set out to deduce a set of group interaction parameters, which will be much smaller than the total number of possible molecular components. These methods were designed to be straightforwardly applicable to multicomponent systems. In this, they have achieved a certain measure of success; as the number of different group interaction parameters grows, the usefulness and accuracy increase. The equations themselves are relatively complicated, but this is not a fundamental difficulty with computer calculation. In the calculation of limiting properties, however, these equations are extrapolating data measured at intermediate concentrations. As mentioned above, the results of extrapolation depend sensitively on the excess Gibbs energy model. The weighted volatility function [127] is an attempt to bypass this sensitivity. As Kehiaian [128a] pointed out, these methods divide the deviation from ideality into two parts – see Equation (5.10). The two parts are sometimes called the 'combinatorial' and 'interaction' contributions. The combinatorial contribution has usually been based on the Flory–Huggins or Staverman–Guggenheim approaches [132]; they are not equivalent. In the interaction term, these methods adopt different rules for using the contact area of groups, in addition to imposing a constant coordination number. When mixture components are not too polar, these differences or inconsistencies tend not to be critical. For systems of interest in aqueous solution chemistry, the difficulties have not been satisfactorily overcome, even today.

The methods based on molecular properties (Table 5.4) exemplify the truism that, if one restricts oneself to one or a few solute types, successful correlations are easier to obtain. In correlations of this kind, the choice of molecular parameters becomes a central concern. Some of these methods suggest a 'pick and mix' basis; the choice is also influenced by the availability of convenient computer packages for calculating parameters not based on directly measured properties. In some respects, the LSER formalism is the most successful of molecular property methods [11].

The additive/constitutive bond and group contribution methods [119] have been made available in a convenient computer package, HENRYWIN (1999). This is a user-friendly product. The program identifies the molecule to be considered through its SMILES notation (Simplified Molecular Input Line Entry System [133]. The user may also, however, import a number of other structures (ChemDraw, HyperChem, MDL, Tripos, etc.). Temperature dependence is estimated from experimental data. For those interested, a list of experimental data used (with bibliography) is included. An 'experimental value adjusted' option is available if the user wishes to do a calculation for a molecule using a selected Henry's law constant value for a similar compound.

REFERENCES

1. Abraham, M. H., *J. Chem. Soc. Faraday Trans. 1*, 1984, **80**, 153–81.
2. Hine, J.; Mookerjee, P. K., *J. Org. Chem.*, 1975, **40**, 292–8.

3. Meylan, W. M.; Howard, P. H., *HENRYWIN 3.05*, Syracuse Research Corporation, Syracuse, NY, 1999; http://esc.syrres.com/interkow/Henry.htm.
4. Mackay, D.; Shiu, W. Y., *J. Phys. Chem. Ref. Data*, 1981, **10**, 1175–99.
5. Yaws, C. L.; Sachin, D. S.; Han, M., *Pollut. Eng.*, 1998, **30**(2), 44–6.
6. Yaws, C. L.; Hopper, J. R.; Wang, X.; Rathinsamy, A. K.; Pike, R. W., *Chem. Eng. (N. Y.)*, 1999, **106**(6), 102–5.
7. Yaws, C. L.; Yang, H. C.; Pan, X., *Chem. Eng. (N. Y.)*, 1991, **98**(11), 179–185.
8. Bamford, H. A.; Baker, J. E.; Poster, D. L., *Review of Methods and Measurements of Selected Hydrophobic Organic Contaminant Aqueous Solubilities, Vapor Pressures and Air–Water Partition Coefficients*, NIST Special Publication No. 928, US Department of Commerce, Washington, DC, 1998.
9. Suntio, L. R.; Shiu, W. Y.; Mackay, D.; Seiber, J. N.; Glotfelty, D., *Rev. Environ. Contam. Toxicol.*, 1988, **103**, 1–59.
10. Abraham, M. H.; Whiting, G. S.; Fuchs, R.; Chambers, E. J., *J. Chem. Soc., Perkin Trans. 2*, 1990, 291–300.
11. Abraham, M. H.; Andonian-Haftvan, J.; Whiting, G. S.; Leo, A.; Taft, R. S., *J. Chem. Soc., Perkin Trans. 2*, 1994, 1777–91.
12. Sander, R., *Compilation of Henry's Law Constants for Inorganic and Organic Species of Potential Importance in Environmental Chemistry, Version 3*, http://www.mpch-mainz.mpg.de/~sander/res/henry.html, 1999.
13. Ashworth, R. A.; Howe, G. B.; Mullins, M. E.; Rogers, T. N., *J. Hazard. Mater.*, 1988, **18**, 25–36.
14. Amoore, J. E.; Buttery, R. G., *Chemical Senses Flavour.*, 1978, **3**, 57–71.
15. Murphy, T. J.; Mullin, M. D.; Meyer, J. A., *Environ. Sci. Technol.*, 1987, **21**, 155–62.
16. Brunner, S.; Hornung, E.; Santl, H.; Wolff, E.; Piuringer, O. G.; Altschuh, J.; Brüggemann, R., *Environ. Sci. Technol.*, 1990, **24**, 1751–4.
17. Altschuh, J.; Brüggemann, R.; Santl, H.; Eichinger, G.; Piringer, O. G., *Chemosphere*, 1999, **18**, 1871–7.
18. Shiu, W. Y., Mackay, D., *J. Phys. Chem. Ref. Data*, 1986, **15**, 911–29.
19. Hwang, Y.-L.; Olson, J. D.; Keller, G. E., *Ind. Eng. Chem. Res.*, 1992, **31**, 1759–68 (Supplementary Material).
20. Kojima, K.; Zhang, S.; Hiaki, T., *Fluid Phase Equilib.*, 1997, **131**, 145–79.
21. Sherman, S. R.; Trampe, D. B.; Bush, D. M.; Schiller, M.; Eckert, C. A.; Dallas, A. J.; Li, J.; Carr, P. W., *Ind. Eng. Chem. Res.*, 1996, **35**, 1044–58.
22. Gmehling, J.; Onken, U., *Chemistry Data Series, Vol. I/1, Vapor–Liquid Equilibrium Data Collection, Aqueous–Organic Systems*, Deutsche Gesellschaft für Chemisches Apparatewesen, Frankfurt/Main, 1977.
23. Gmehling, J.; Onken, U., *Chemistry Data Series, Vol. I/1a, Vapor–Liquid Equilibrium Data Collection, Aqueous–Organic Systems, Supplement 1*, Deutsche Gesellschaft für Chemisches Apparatewesen, Frankfurt/Main, 1981.
24. Gmehling, J.; Onken, U.; Rarey-Nies, J. R., *Chemistry Data Series, Vol. I/1b, Vapor–Liquid Equilibrium Data Collection, Aqueous–Organic Systems, Supplement 2*, Deutsche Gesellschaft für Chemisches Apparatewesen, Frankfurt/Main, 1988.
25. Gmehling, J.; Menke, J.; Schiller, M., *Chemistry Data Series, Vol. IX/4, Activity Coefficients at Infinite Dilution*, Deutsche Gesellschaft für Chemisches Apparatewesen, Frankfurt/Main, 1994.
26. Tewari, Y. B.; Miller, M. M.; Wasik, S. P.; Martire, D. E., *J. Chem. Eng. Data*, 1982, **27**, 451–4.
27. Mitchell, B. E.; Jurs, P. C., *J. Chem. Inf. Comput. Sci.*, 1998, **38**, 200–9.
28. Thomas, E. R.; Eckert, C. A., *Ind. Eng. Chem., Process Des. Dev.*, 1984, **23**, 194–209.
29. Gmehling, J.; Fischer, K.; Li, J.; Schiller, M., *Pure Appl. Chem.*, 1993, **65**, 919–26.

30. Clever, H. L., *Helium and Neon*, Solubility Data Series, Vol. 1, Pergamon Press, Oxford, 1979.
31. Clever, H. L., *Krypton, Xenon and Radon*, Solubility Data Series, Vol. 2, Pergamon Press, Oxford, 1979.
32. Clever, H. L., *Argon*, Solubility Data Series, Vol. 4, Pergamon Press, Oxford, 1980.
33. Young, C. L., *Hydrogen and Deuterium*, Solubility Data Series, Vol. 5/6, Pergamon Press, Oxford, 1981.
34. Battino, R., *Oxygen and Ozone*, Solubility Data Series, Vol. 7, Pergamon Press, Oxford, 1981.
35. Young, C. L., *Oxides of Nitrogen*, Solubility Data Series, Vol. 8, Pergamon Press, Oxford, 1981.
36. Hayduk, W., *Ethane*, Solubility Data Series, Vol. 9, Pergamon Press, Oxford, 1982.
37. Battino, R., *Nitrogen and Air*, Solubility Data Series, Vol. 10, Pergamon Press, Oxford, 1982.
38. Young, C. L., *Sulfur Dioxide, Chlorine, Fluorine and Chlorine Oxides*, Solubility Data Series, Vol. 12, Pergamon Press, Oxford, 1983.
39. Hayduk, W., *Propane, Butane and Isobutane*, Solubility Data Series, Vol. 24, Pergamon Press, Oxford, 1986.
40. Clever, H. L.; Young, C. L., *Methane*, Solubility Data Series, Vol. 27/28, Pergamon Press, Oxford, 1987.
41. Fogg, P. G. T.; Young, C. L., *Hydrogen Sulfide, Deuterium Sulfide and Hydrogen Selenide*, Solubility Data Series, Vol. 32, Pergamon Press, Oxford, 1988.
42. Cargill, R. W., *Carbon Monoxide*, Solubility Data Series, Vol. 43, Oxford University Press, Oxford, 1990.
43. Hayduk, W., *Ethene*, Solubility Data Series, Vol. 57, Oxford University Press, Oxford, 1994.
44. Horvath, A. L.; Getzen, F. W., *Halogenated Methanes and Water*, Solubility Data Series, Vol. 60, Oxford University Press, Oxford, 1995.
45. Scharlin, P., *Carbon Dioxide in Water and Aqueous Electrolyte Solutions*, Solubility Data Series, Vol. 62, Oxford University Press, Oxford, 1996.
46. Fogg, P. G. T., IUPAC–NIST Solubility Data Series, 76, Solubility of Ethyne in Liquids, *J. Phys. Chem. Ref. Data*, 2002, **30**, 1693–1876.
47. Shing, K. S.; Gubbins, K. E., *Mol. Phys.*, 1982, **48**, 1109–28.
48. Shing, K. S.; Gubbins, K. E., *Mol. Phys.*, 1983, **49**, 1121–38.
49. Shing, K. S., *Chem. Phys. Lett.*, 1985, **119**, 149–51.
50. Haile, J. M., *Fluid Phase Equilib.*, 1986, **26**, 103–27.
51. Lazaridis, T.; Paulitis, M. E., *AIChE J.*, 1993, **39**, 1051–60.
52. Slusher, J. T., *Fluid Phase Equilib.*, 1998, **153**, 45–61.
53. Hu, Y.; Azevedo, E.; Lüdecke, D.; Prausnitz, J., *Fluid Phase Equilib.*, 1984, **17**, 303–21.
54. Chen, Y.-C.; Chen, Y.-P., *Fluid Phase Equilib.*, 1985, **86**, 63–85.
55. Chen, Y.-C.; Chen, Y.-P., *Fluid Phase Equilib.*, 1993, **86**, 87–109.
56. He, Y.; Wang, Y.; Wu, C.; Wang, L., *Huanjing Kexue Xuebao*, 1997, **17**, 227–31.
57. Famini, G. R.; Benyamin, D.; Kim, C.; Veerawat, R.; Wilson, L. Y., *Collect. Czech. Chem. Commun.*, 1999, **64**, 1727–47.
58. Sabljic, A.; Güsten, H., *Chemosphere*, 1989, **19**, 1503–11.
59. Hawker, D. W., *Environ. Sci. Technol.*, 1989, **23**, 1250–3.
60. Dunnivant, F. M.; Elzerman, A. W.; Jurs, P. C.; Hasan, M. N., *Environ. Sci. Technol.*, 1992, **26**, 1567–73.
61. Collette, T. W., *Trends Anal. Chem.*, 1997, **16**, 24–36.
62. Nirmalakhandan, N. N.; Speece, R. E., *Environ. Sci. Technol.*, 1988, **22**, 1349–57.
63. Russell, C. J.; Dixon, S. L.; Jurs, P. C., *Anal. Chem.*, 1992, **64**, 1350–5.
64. Majer, V., *Wiss. Ber. Forschungszent. Karlsruhe*, 1999, **FZKA.6271**, 145–50.
65. Suzuki, T.; Ohtaguchi, K.; Kojima, K., *Comput. Chem.*, 1992, **16**, 41–52.

66. Lobien, G. M.; Prausnitz, J. M., *Fluid Phase Equilib.*, 1982, **8**, 149–60.
67. Medir, M.; Giralt, F., *AIChE J.*, 1982, **28**, 341–3.
68. Dutt, N. V. K.; Prasad, D. H. L., *Fluid Phase Equilib.*, 1989, **45**, 1–5.
69. Hait, M. J.; Liotta, C. L.; Eckert, C. A.; Bergmann, D. L.; Karachewski, A. M.; Dallas, A. J.; Eiken, D. I.; Li, J. J.; Carr, P. W.; Poe, R. B.; Rutan, S. C., *Ind. Eng. Chem. Res.*, 1993, **32**, 2905–14.
70. Eckert, C. A.; Sherman, S. R., *Fluid Phase Equilib.*, 1996, **116**, 333–42.
71. Pierotti, R. A., *J. Phys. Chem.*, 1963, **67**, 1840–5.
72. Pierotti, R. A., *Chem. Rev.*, 1976, **76**, 717–26.
73. Schulze, G.; Prausnitz, J. M., *Ind. Eng. Chem. Fundam.*, 1981, **20**, 175–7.
74. Brandani, S.; Brandani, V.; Di Giacomo, G., *Ind. Eng. Chem. Res.*, 1992, **31**, 420–4.
75. Hildebrand, J. H.; Prausnitz, J. M.; Scott, R. L., *Regular and Related Solutions*, Van Nostrand-Reinhold, Princeton, NJ, 1970.
76. Barton, A. F. M., *CRC Handbook of Solubility Parameters and Other Cohesion Parameters*, CRC Press, Boca Raton, FL, 1983.
77. Weimer, R. F.; Prausnitz, J. M., *Hydrocarbon Process. Petrol. Refiner*, 1965, **44**(9), 237–242.
78. Hansen, C. M., *J. Paint Technol.*, 1967, **39**, 505–510.
79. Helpinstill, J. G.; Van Winkle, M., *Ind. Eng. Chem. Process Des. Dev.*, 1968, **7**, 213–20.
80. Hsieh, C. K., *PhD Thesis*, University of Illinois, Urbana, IL, 1973.
81. Karger, B. L.; Snyder, L. R.; Eon, C., *J. Chromatogr.*, 1976, **125**, 71–88.
82. Tijssen, R.; Billet, H. A. H.; Schoenmakers, P. J., *J. Chromatogr.*, 1976, **122**, 185–203.
83. Newman, B. A., *PhD Thesis*, University of Illinois, Urbana, IL, 1977.
84. Howell, W. J.; Karachewski, A. M.; Stephenson, K. M.; Eckert, C. A.; Park, J. H.; Carr, P. W.; Rutan, S. C., *Fluid Phase Equilib.*, 1989, **52**, 151–60.
85. Wu, Y.; Wang, S.; Hwang, D.; Shi, J., *Can. J. Chem. Eng.*, 1992, **70**, 398–402.
86. Poe, R. B.; Rutan, S. C.; Hait, M. J.; Eckert, C. A.; Carr, P. W., *Anal. Chim. Acta*, 1993, **277**, 223–38.
87. Padovani, A. M.; Suleiman, D., *AIChE J.*, 1997, **43**, 3271–3.
88. Vilcu, R.; Puricel, E., *Rev. Roum. Chim.*, 1998, **43**, 1121–32.
89. Govers, H. A. J., *J. Chem. Soc., Faraday Trans.*, 1993, **89**, 3751–9.
90. Govers, H. A. J.; Evers, E. H. G., *Chemosphere*, 1992, **24**, 453–64.
91. Govers, H. A. J.; van der Wielen, F. W. M.; Olie, K., *J. Chromatogr. A*, 1995, **715**, 267–78.
92. Govers, H. A. J.; Krop, H. B., *Organohalogen Compd.*, 1996, **28**, 5–10.
93. Govers, H. A. J.; Krop, H. B., *Chemosphere*, 1998, **37**, 2139–52.
94. Govers, H. A. J.; Luijk, R.; Evers, E. H. G., *Chemosphere*, 1990, **20**, 287–94.
95. Jonah, D. A., *Adv. Chem. Ser.*, 1983, **204**, 395–422.
96. Dolezalek, F., *Z. Phys. Chem. Stöch. Verwandt.*, 1908, **64**, 727–47.
97. Dolezalek, F.; Schulze, A., *Phys. Chem. Stöch. Verwandt.*, 1913, **83**, 45–78.
98. Pelton, A. D.; Degterov, S. A.; Eriksson, G.; Robelin, C.; Dessureault, Y., *Met. Mater.Trans.B*, 2000, **31B**, 651–9.
99. Acree, W. E., *Thermodynamic Properties of Nonelectrolyte Solutions*, Academic Press, New York, 1984.
100. Wilson, G. M., *J. Am. Chem. Soc.*, 1964, **86**, 127–30.
101. Wilson, G. M., *J. Am. Chem. Soc.*, 1964, **86**, 133–7.
102. Renon, H.; Prausnitz, J. M., *AIChE J.*, 1968, **14**, 135–44.
103. Scott, R. L., *J. Chem. Phys.*, 1956, **25**, 193–205.
104. Abrams, D. S.; Prausnitz, J. M., *AIChE J.*, 1975, **21**, 116–28.
105. Anderson, T. F.; Prausnitz, J. M., *Ind. Eng. Chem., Process Des. Dev.*, 1978, **17**, 552–61.
106. Nagata, I.; Katoh, K., *Fluid Phase Equilib.*, 1980–81, **5**, 225–44.

84 Chemicals in the Atmosphere

107. Jonsdottir, S. O.; Klein, R. A., *Fluid Phase Equilib.*, 1997, **132**, 117–37.
108. Pierotti, R. A.; Deal, C. H.; Derr, E. L.; Porter, P. E., *J. Am. Chem. Soc.*, 1956, **78**, 2989–98.
109. Pierotti, G. J.; Deal, C. H.; Derr, E. L., *Ind. Eng. Chem.*, 1959, **51**, 95–102.
110. Wilson, G. M.; Deal, C. H., *Ind. Eng. Chem. Fundam.*, 1962, **1**, 20–3.
111. Derr, E. L.; Deal, C. H., *Proc. Int. Symp. Distill.*, 1969, **3**, 40–51.
112. Wilson, G. M., *AIChE Symp. Ser.*, 1974, **70**, 120–9.
113. Fredenslund, A.; Jones, R. L.; Prausnitz, J. M., *AIChE J.*, 1975, **21**, 1086–99.
114. Gmehling, J.; Tiegs, D.; Knipp, U., *Fluid Phase Equilib.*, 1990, **54**, 147–65.
114a. Zhang, S.; Hiaki, T.; Kojima, K., *Fluid Phase Equilib.*, 1998, **149**, 27–40.
115. Gmehling, J.; Rasmussen, P.; Fredenslund, A., *Ind. Eng. Chem. Res.*, 1982, **21**, 118–27.
115a. Zhang, S.; Hiaki, T.; Hongo, M.; Kojima, K., *Fluid Phase Equilib.*, 1998, **144**, 97–112.
116. Macedo, E. A.; Weidlich, U.; Gmehling, J.; Rasmussen, P., *Ind. Eng. Chem., Process Des. Dev.*, 1983, **22**, 676–8.
117. Wienke, G.; Gmehling, J., *Toxicol. Environ. Chem.*, 1998, **65**, 57–86.
117a. Tiegs, D.; Gmehling, J.; Rasmussen, P.; Fredenslund, A., *Ind. Eng. Chem. Res.*, 1987, **26**, 159–61.
118. Bastos, J. C.; Soares, M. E.; Medina, A. G., *Ind. Eng. Chem. Res.*, 1988, **27**, 1269–77.
118a. Mullins, M.; Rogers, T.; Loll, A., *Fluid Phase Equilib.*, 1998, **150–151**, 245–53.
119. Ashraf, S. M.; Ramakrishna, M.; Prasad, D. H. L.; Rao, M. B., *Chem. Eng. J.*, 1999, **72**, 31–6.
119a. Meylan, W. M.; Howard, P. H., *Environ. Sci. Technol.*, 1991, **10**, 1283–93.
120. Barr, R. S.; Newsham, D. M. T., *Fluid Phase Equilib.*, 1987, **35**, 207–15.
120a. Lin, S.-T.; Sandler, S. I., *AIChE J.*, 1999, **45**, 2606–18.
121. Antunes, C.; Tassios, D., *Ind. Eng. Chem. Process Des. Dev.*, 1983, **22**, 457–62.
121a. Li, A.; Doucette, W. J.; Andren, A. W., *Chemosphere*, 1994, **24**, 657–69.
122. Bader, M. S. H.; Gasem, K. A. M., *Can. J. Chem. Eng.*, 1998, **76**, 94–103.
122a. Örnektekin, S.; Paksoy, H. Ö.; Dermirel, Y., *Thermochimica. Acta*, 1996, **287**, 251–9.
123. Harvey, A. H., *AIChE J.*, 1996, **42**, 1491–4.
123a. Tabai, S.; Solimando, R.; Rogalski, M., *Fluid Phase Equilib.*, 1997, **139**, 37–46.
124. Joffe, J., *J. Chem. Eng. Data*, 1977, **22**, 348–50.
125. Chow, H.; Chen, H.; Ng, T.; Myrdal, P.; Yalkowsky, S. H., *J. Chem. Inf. Comput. Sci.*, 1995, **35**, 723–8.
126. Drefahl, A.; Reinhard, M., *J. Chem. Inf. Comput. Sci.*, 1993, **33**, 886–95.
127. Wisniak, J.; Segura, H.; Reich, R., *Phys. Chem. Lett.*, 1996, **32**, 1–24.
128. Brennan, R. A.; Nirmalakhandan, N.; Speece, R. E., *Water Res.*, 1998, **32**, 1901–11.
128a. Kehiaian, H. V., *Fluid Phase Equilib.*, 1983, **13**, 243–252.
129. Schüürmann, G.; Rothenbacher, C., *Fresenius Environ. Bull.*, 1992, **1**, 10–15.
129a. Vera, J. H.; Vidal, J., *Chem. Eng. Sci.*, 1984, **39**, 651–61.
130. Park, J. H.; Carr, P. W., *Anal. Chem.*, 1987, **59**, 2596–602.
130a. Saxena, P.; Hildemann, L. M., *J. Atmos. Chem.*, 1996, **24**, 57–109.
131. Staudinger, J.; Roberts, P. V., *Crit. Rev. Environ. Sci. Technol.*, 1996, **26**, 205–97.
132. Kikic, I.; Alessi, P.; Rasmussen, P.; Fredenslund, A., *Can. J. Chem. Eng.*, 1980, **58**, 253–8.
133. Weininger, S., *J. Chem. Inf. Comput. Sci.*, 1988, **28**, 31–6.

CHAPTER 6

Details of Specific Methods of Predicting the Solubility of Organic Compounds in Water

Peter Fogg

University of North London, UK (retired)

1 GENERAL PRINCIPLES

There have been numerous attempts to link Henry's law constants with properties of molecules [1–11]. This is partly due to efforts to find reliable methods of predicting values for compounds for which the constant has not been measured experimentally. It is also due to efforts to determine which physical properties are the most important in determining solubility in water. Cronin and co-workers [1] listed the following factors: volatilization, size, entropy, H-bonding, enthalpy of solvation and, intrinsic hydrophobicity. These are not entirely independent parameters.

Phase equilibria in aqueous systems can be predicted by molecular dynamics and by Monte Carlo simulation methods. Calculations of the solubility of inert gases from molecular dynamics have been described by Swope and Andersen [12] and Watanabe and Andersen [13]. Various models have been proposed for Monte Carlo simulation [14–26]. Some of these models are based on two-body potentials, others on more complex many-body potentials. The latter are more accurate in predicting properties but require more computing time. The current situation has been reviewed recently by Economou [27], who published graphs showing the relationship between experimental and predicted values of Henry's law constants over temperature ranges for methane, ethane, butane, hexane, cyclohexane and benzene [28–32]. The difficulties associated with this technique make it unlikely that this method can be used to make accurate predictions for compounds with more complex structures. Both the molecular dynamics and Monte

Chemicals in the Atmosphere – Solubility, Sources and Reactivity. Edited by P. G. T. Fogg and J. Sangster
© 2003 IUPAC ISBN: 0-471-98651-8

Carlo simulation methods are limited because of the approximate nature of predicted intermolecular force fields of complex molecules. There has therefore been much effort to derive simpler models which can be used for all compounds.

If an aqueous solution of a compound is in equilibrium with the solid or liquid compound, then the fugacity of the compound in the vapour phase over the solution can be assumed to be equal to the fugacity in the gas phase over the pure compound, provided that the solubility of water in the pure compound is negligible. A good approximation is to equate the fugacity with vapour pressure, thus providing a simple and reliable method of estimating a ratio of vapour pressure (VP) over concentration in an aqueous phase (AS). Provided that vapour pressures and concentrations are fairly low this ratio may be equated with a Henry's law constant in units of pressure/concentration in the aqueous layer:

$$k_H (\text{mol dm}^{-3}\,\text{bar}^{-1}) = \text{AS/VP}$$

This is the simplest and usually the most reliable of predicting the Henry's law constant if reliable solubility and vapour pressure data are available.

2 ESTIMATION FROM WATER–HEXADECANE PARTITION COEFFICIENTS

Abraham et al. [10] published a simple relationship for the ratio of concentrations of gas in the aqueous and gaseous phases. This is equivalent to the Ostwald coefficient in the form of the ratio of volume of liquid to volume of gas absorbed if it is assumed that absorption of gas does not cause significant change in the volume of the liquid phases and that the gas approximates to ideal behaviour:

$$c_w/c_g = (c_{16}/c_g)/P$$

where c_w = concentration of the solute in aqueous solution/concentration in the gas phase

   c_g = concentration of solute in the gas phase

   c_{16} = concentration of the solute in hexadecane solution/concentration in the gas phase

   P = water–hexadecane partition coefficient = c_w/c_{16}.

As an alternative, P can be taken to be any other water–alkane partition coefficient but this leads to somewhat poorer agreement with experimental values.

For 14 solutes, using the water–hexadecane partition coefficient the agreement between experimental and calculated values was 0.08 log units. For 45 solutes using other water–alkane partition coefficients, the agreement was 0.15 log units.

If it can be assumed that the gas approximates to ideal behaviour and the partial pressure is low, then it can be assumed that

$$c_w/c_g = RT k_H (\text{mol dm}^{-3}\,\text{bar}^{-1})$$

where k_H is Henry's law constant, T is the absolute temperature and R the gas constant in units of dm^3 bar K^{-1} mol^{-1}, i.e. $R = 8.31448 \times 10^{-2} dm^3$ bar K^{-1} mol^{-1}.

3 LINEAR SOLVATION ENERGY RELATIONSHIPS

Many of the models for correlation of physical properties of molecules depend on the development of quantitative structure–property relationships (QSPR). In the case of Henry's law constants the model is based on the assumption that the free energy change on dissolution is a linear function of values of other molecular properties (linear solvation energy relationship, LSER). There is an exponential relationship between the ratio of concentrations in the two states (solution, gas) and the free energy change when molecules are dissolved. An LSER can therefore be written in the form

$$\log(c_w/c_g) = aA + bB + cC + \cdots + xX \qquad (6.1)$$

where A is the value of property **A**, B is the value of property **B**, X the value of property **X**, etc., and a, b, c, etc., are coefficients for the properties under consideration and for an ideal model would be the same for each compound. In practice, however, A, B, C, etc., vary from compound to compound. The properties **A**, **B**, **C**, etc., are called *descriptors*.

Thus, for compounds 1, 2, 3, etc.:

$$\log(c_w/c_g)_1 = aA_1 + bB_1 + cC_1 + \cdots xX_I \ldots$$

$$\log(c_w/c_g)_2 = aA_2 + bB_2 + cC_2 + \cdots + xX_2$$

etc.

If N compounds are under consideration, then N simultaneous equations can be envisaged. This is likely to be much greater than the number of coefficients a, b, c, etc. In practice, however, the exact values of the coefficients are likely to vary from one compound to another. If the model is a good model, then the variation is small. Computer programs are available into which can be input values of c_w/c_g based on direct measurement of gas solubility or calculated from vapour pressure and solubility of a solid or liquid in water together with values of the various properties **A**, **B**, **C**, etc. These programs evaluate the best values of a, b, c, etc., which give the best fit between values of c_w/c_g from Equation (6.1) and experimental values of c_w/c_g. Equation (6.1) can then be applied to a compound for which an experimental value of c_w/c_g is not available. Provided that values of c_w/c_g correspond to low partial pressures and gases can be assumed to behave ideally, then

$$c_w/c_g = RTk_H(mol\,dm^{-3}\,bar^{-1})$$

as explained above.

Often the compounds with known values of c_w/c_g and properties **A**, **B**, **C**, etc., are divided into two unequal groups for development of a model. The best values of a, b, c, ..., x, etc., are determined using the larger group. These values are then tested against the smaller group.

The simplest properties common to a large group of compounds are the identities of each sort of atom in the molecule. It would be possible to associate each type of atom X with a value of X, and make the number of these atoms in the molecule the value of X. However, such a model gives unsatisfactory predictions with no differentiation between different substances with the same empirical formula.

4 THE METHODS DEVELOPED BY HINE AND MOOKERJEE

A slightly more complex model was developed by Hine and Mookerjee [4] in which each type of bond of a particular type was assumed to make the same contribution to the solubility. Some groups were treated as single atoms. These included the cyano group (–CN), the nitro group (–NO$_2$) and the carbonyl group (–CO). The contribution of –C–CO therefore included the contribution of the double bond and that of –C–CN the contribution of the triple bonds. Ethenic, ethynic and aromatic carbon were differentiated from other carbon atoms by the symbols C_d, C_t, and C_{ar}. The contribution of a C_d–X (e.g. C_d–H) bond was taken to include a quarter of the contribution of the –C=C– bond and that of a C_t–X bond that of half a –C≡C– bond. However the C_{ar}–H contribution did not allow for C_{ar} to C_{ar} bonds. These were denoted as –C_{ar}=C_{ar}– and separate allowance was made for them.

Solubilities were expressed as the ratio of the concentration in the aqueous phase to that in the gas phase, c_w/c_g. Bond contributions, based on 263 solubilities based on direct measurements, were expressed as contributions to $\log(c_w/c_g)$ (see Table 6.1)

Logarithmic values of the solubility values based on the coefficients given above differed from logarithmic values based directly on experimental data with a standard deviation of 0.41.

Table 6.1 Bond contributions to $\log(c_w/c_g)$ at 298.15 K used by Hine and Mookerjee [4]

Bond	Contribution	Bond	Contribution	Bond	Contribution
C–H	−0.11	C–C$_d$	0.15	C$_d$–H	−0.15
C–F	−0.50	C–C$_t$	0.64	C$_d$–Cl	0.16
C–Cl	0.30	C–C$_{ar}$	0.11	C$_d$–C$_d$	0.48
C–Br	0.87	C$_{ar}$–H	−0.21	C$_d$–CO	2.42
C–I	1.03	C$_{ar}$–Cl	−0.14	C$_t$–H	0.00
C–CN	3.28	C$_{ar}$–Br	0.21	CO–H	1.19
C–NO$_2$	3.10	C$_{ar}$–NO$_2$	1.83	CO–O	0.28
C–O	1.00	C$_{ar}$–O	−0.74	O–H	3.21
C–S	1.11	C$_{ar}$–S	0.53	S–H	0.23
C–N	1.35	C$_{ar}$–CO	1.14	N–H	1.34
C–C	0.04	C$_{ar}$=C$_{ar}$	0.33		
C–CO	1.78	C$_{ar}$=N$_{ar}$	1.64		

On the basis of this model, the solubility of ethane, C_2H_6, is given by

$$\log(c_w/c_g) = (6 \times -0.11) + 0.04 = -0.62$$

$$c_w/c_g = 0.24$$

Experimental value $= 0.046$
The prediction for 1-hexyne, $CH_3-CH_2-CH_2-CH_2-C\equiv CH$, is

$$
\begin{aligned}
\log(c_w/c_g) = 0.04 \times 3 & \qquad \text{C–C bonds} \\
- 0.11 \times 9 & \qquad \text{C–H bonds} \\
+ 0.64 & \qquad \text{C–C}_t \text{ bond} \\
+ 0.00 & \qquad \text{C}_t\text{–H bond} \\
= -0.23 &
\end{aligned}
$$

$$c_w/c_g = 0.59$$

Experimental value $= 0.62$
The prediction for acetonitrile, CH_3-CN, is

$$
\begin{aligned}
\log(c_w/c_g) = 3.28 & \qquad \text{C–CN bond} \\
- 0.11 \times 3 & \qquad \text{C–H bonds} \\
= 2.95 &
\end{aligned}
$$

$$c_w/c_g = 891$$

Experimental value $= 708$
Hine and Mookerjee [4] described a second model based on groups in an organic molecule, rather than bonds. In general a group was characterized by the nature of the atoms to which it is attached in addition to the atoms of which it is comprised. For example, the methylene group in propane, $CH_3-CH_2-CH_3$ is different from the methylene group in ethanol, CH_3-CH_2-OH. Both differ from the group in ethylphenyl ether, $C_6H_5-O-CH_2-CH_3$. The values of the contributions to log (c_w/c_g) were estimated from the available experimental data for 292 compounds as for the model based on bond contributions. Some groups were present in a very small number of compounds or just one compound and estimations of their contributions were of little reliability. The predicted solubilities of the 292 compounds containing groups for which reliable estimates of group contributions were available were correlated with the original experimental data. The standard deviation between the two sets was 0.12, a value considered to be comparable to the error in the experimental values themselves.

Significant deviations occurred with some compounds. In many cases this could be ascribed to interaction between polar bonds in separate groups. Such interactions within a group would not cause deviations as the interactions would be accounted for in the estimation of the contribution of the group. In most cases a particular type of interaction was found in only one compound, so no averaging method was possible. However, experimental data were available for a sufficient number of pyrazines for an average interaction of two nitrogen atoms to be calculated. The average interaction of two alkoxy groups attached to adjacent saturated carbon atoms could also be found. In other cases interactions which were specific to a particular compound such as that between the bromine atoms in p-dibromobenzene could be calculated but not correlated with similar interactions in other compounds.

5 DEVELOPMENTS BY MEYLAN AND HOWARD

Meylan and Howard [5] developed Hine and Mookerjee's bond contribution model [4] using, as their basis, a data set of values of a Henry's law constants at 298.15 K for 345 compounds. A total of 218 had been measured directly and a further 127 calculated from solubility/v.p. Some of the data used by Hine and Mookerjee were updated in the light of more accurate experimental measurements. Some of the groups were again treated as single atoms. The bond contribution values which were obtained are given in Table 6.2.

Predicted values for some classes of compounds were either consistently high or low. The authors derived correction factors where appropriate. These are listed in Table 6.3.

Table 6.2 Bond contributions to $\log(c_w/c_g)$ at 298.15 K used by Meylan and Howard [5]

Bond	Value	Bond	Value	Bond	Value
C–H	−0.1197	C_d–Cl	0.0426	C_{ar}–CO	1.2387
C–C	0.1163	C_d–CN	2.5514	C_{ar}–Br	0.2454
C–C_{ar}	0.1619	C_d–O	0.2051	C_{ar}–NO$_2$	2.2496
C–C_d	0.0635	C_d–F	−0.3824	CO–H	1.2102
C–C_t	0.5375	C_t–H	0.0040	CO–O	0.0714
C–CO	1.7057	$C_t\equiv C_t$ [a]	0.0000	CO–N	2.4261
C–N	1.3001	C_{ar}–H	−0.1543	CO–CO	2.4000
C–O	1.0855	C_{ar}–C_{ar} [b]	0.2638	O–H	3.2318
C–S	1.1056	C_{ar}–C_{ar} [c]	0.1490	O–P	0.3930
C–Cl	0.3335	C_{ar}–Cl	−0.0241	O–O	−0.4036
C–Br	0.8187	C_{ar}–OH	0.5967	O=P	1.6334
C–F	−0.4184	C_{ar}–O	0.3473	N–H	1.2835
C–I	1.0074	C_{ar}–N_{ar}	1.6282	N–N	1.0956
C–NO$_2$	3.1231	C_{ar}–S_{ar}	0.3739	N=O	1.0956
C–CN	3.2624	C_{ar}–O_{ar}	0.2419	N=N	0.1374
C–P	0.7786	C_{ar}–S	0.6345	S–H	0.2247
C=S	−0.0460	C_{ar}–N	0.7304	S–S	−0.1891
C_d–H	−0.1005	C_{ar}–I	0.4806	S–P	0.6334
C_d=C_d [a]	0.0000	C_{ar}–F	−0.2214	S =P	−1.0317
C_d–C_d	0.0997	C_{ar}–C_d	0.4391		
C_d–CO	1.9260	C_{ar}–CN	1.8606		

[a] Assigned arbitrary values of zero.
[b] Intra-ring aromatic carbon to aromatic carbon.
[c] External aromatic carbon to aromatic carbon as in biphenyl.

Table 6.3   Correction factors for certain classes of compounds

Linear or branched alkane with no substituents except alkyl groups	−0.75
Cyclic alkane with no substituents except alkyl groups	−0.28
Alkene with only one double bond and with no substituents except alkyl groups	−0.20
Cyclic alkene with only one double bond and with no substituents except alkyl groups	+0.25
Linear or branched aliphatic alcohol with no substituents except alkyl groups	−0.20
Adjacent aliphatic ether functions (−C−O−C−O−C−)	−0.70
Cyclic monoether	+0.90
Epoxide	+0.50
Each additional aliphatic alcohol function (−OH) above one	−3.00
Each additional aromatic nitrogen above one within a single ring	−2.50
Monofluoroalkane	+0.95
Monochloroalkane	+0.50
Totally chlorinated alkane	−1.35
Totally fluorinated alkane	−0.60
Totally fluorinated halogenated halofluoroalkane	−0.90

The model was tested using a second set containing logarithms of solubility data for 74 compounds, none of which had been used in determination of the bond contributions listed above. The data in this validation set correlated with the predicted values with a correlation coefficient (r^2) of 0.965, a mean error of 0.31 and a standard deviation of 0.457. This standard deviation in logarithmic values corresponds to a factor of 2.86 as far as values of c_w/c_g and hence of k_H are concerned. The values of Henry's law constant for compounds in the validation set and defined as c_w/c_g span a range of 10 orders of magnitude. It must be borne in mind that this, and similar models, are only useful in predicting the order of magnitude of Henry's law constant. They do not give precise values.

On the basis of this model the value of $\log(c_w/c_g)$ for methyl benzoate, $C_6H_5COOCH_3$, is predicted to be

5×-0.1543	$C_{ar}-H$ bonds
1.2387	$C_{ar}-CO$ bond
0.0714	$CO-O$ bond
1.0855	$C-O$ bond
3×-0.1197	$C-H$ bonds
6×0.2638	$C_{ar}-C_{ar}$ bonds

Total   2.8478

No correcting factors; predicted value of $\log(c_w/c_g) = 2.8478$

Value based directly on experimental data $= 2.838$

Difference $= 0.0098$

This corresponds to a 2 % difference between predicted and experimental values of c_w/c_g and of k_H.

The value for 3-hexanol, $CH_3–CH_2–CH_2–CH(OH)–CH_2–CH_3$, is predicted to be

13×-0.1197	C–H bonds
5×-0.1163	C–C bonds
1.0855	C–O bond
3.2318	O–H bond

Total   3.3427
Correcting factor -0.20
Predicted value of $\log(c_w/c_g) = 3.1427$
Value based directly on experimental data $= 2.757$
Difference $= 0.3857$
This corresponds to an error in c_w/c_g by a factor of 2.43, i.e. an error of 143 %.

Meylan and Howard recommended that a modified use of the bond contribution method for estimating Henry's law constant should be used for a compound **A** with a complex structure if the constant were known for a compound **B** with a similar structure. One should estimate what bonds would need to be added to or subtracted from **B** to make **A**. The corresponding contributions of these bonds should then be added or subtracted from the value of $\log(c_w/c_g)$ for **B**. This would give an estimated value of the value of $\log(c_w/c_g)$ for **A** which is likely to be more reliable than a value based on summation over all the bonds in **A**.

An example given by Meylan and Howard is the estimation of c_w/c_g for 2-isobutyl-3-methoxypyrazine. If the contributions for all the bonds and the correcting factor are summed, the predicted value of $\log(c_w/c_g)$ is 4.701, which compares badly with the value of 2.699 obtained by direct measurement. A better method is to derive the value from that of 2-ethyl-3-methoxypyrazine for which $\log(c_w/c_g)$ is 3.221. To convert an ethyl group to an isobutyl group requires the subtraction of a C–H bond and addition of two C–C bonds and five C–H bonds.

The estimated value of $\log(c_w/c_g)$ for 2-isobutyl-3-methoxypyrazine is therefore given by

$$\log(c_w/c_g) = 3.221 - (-0.1197) + (2 \times 0.1163) + (5 \times -0.1197)$$

$$= 2.975$$

This is much closer to the experimental value of 2.699.

Meylan and Howard extended their model and incorporated it into the commercially available program HENRYWIN published by Syracuse Research Corporation (see Appendix I). The group contribution model of Hine and Mookerjee was also incorporated into the program so that there is a choice of estimation methods. Contributions for more types of bond and a greater variety of groups have been included. The correction factor list has also been extended. The bond contribution method is more widely applicable but the group method is usually preferable when contributions for each fragment of the molecule is known. It is also possible to use the program to estimate Henry's law constant for one compound from that of another of similar structure by subtraction or addition of contributions of bonds as described above. Structures are input using the SMILES notation.

The latest version of HENRYWIN (version 3) is supplied with a database of experimental values of Henry's law constants for 1656 compounds.

The authors have provided a careful estimate of the reliability of the bond and group methods. A major source of estimation of the error of the method is due to values of k_H which are less than 1×10^{-8} atm m^3 mol^{-1}. However, any compound which has a value of k_H less than a value of about 3×10^{-7} atm m^3 mol^{-1} can be considered to be involatile from water and errors in the estimations of such compounds are of little practical significance. The authors considered that the errors in values greater than 1×10^{-8} atm m^3 mol^{-1} were of greater significance in estimating the reliability of the method than errors in lower values.

When all data points having an estimated value of k_H less than 1×10^{-8} atm m^3 mol^{-1} are removed from statistical consideration, the statistical accuracy (in terms of standard deviation and mean error) of the bond method improves significantly, as shown here:

	Entire database	Excluding values less than 1×10^{-8} atm m^3 mol^{-1}
No. of compounds	1656	1293
Correlation coefficient (r^2)	0.814	0.845
Standard deviation	1.426	0.87
Absolute mean error	0.819	0.51

Estimates by the bond method considered above correspond to 25 °C. Most of the data in the experimental database was obtained at 25 °C (1346 compounds). However, there are 251 values at 20 °C. The error in comparing these values with values predicted by the bond method at 25 °C is greater than the error given above for the whole database.

	20 °C database	Excluding values less than 1×10^{-8} atm m^3 mol^{-1}
No. of compounds	251	147
Correlation coefficient (r^2)	0.66	0.67
Standard deviation	1.832	1.294
Absolute mean error	1.222	1.022

Therefore, inclusion of these experimental values at 20 °C decreases the average accuracy of the predictions. If the experimental data at 25 °C are analysed with exclusion of other data, the average accuracy is significantly improved:

	25 °C database	Excluding values less than 1×10^{-8} atm m³ mol⁻¹
No. of compounds	1346	1096
Correlation coefficient (r^2)	0.832	0.86
Standard deviation	1.326	0.79
Absolute mean error	0.694	0.46

The group method is not appropriate for some compounds. This method could only be used for 960 of the 1656 compounds in the database. For this group of compounds the bond method gives a better prediction:

	Group method	Bond method
No. of compounds	960	960
Correlation coefficient (r^2)	0.847	0.919
Standard deviation	0.962	0.639
Absolute mean error	0.534	0.399

The bond method has the advantage when appropriate correction factors are available. If these are not available then the group method usually gives the better prediction.

The database supplied with HENRYWIN 3 also contains equations for variation of Henry's law constant with temperature based on experimental measurements over a temperature range for 297 compounds. Where possible equations for variation of c_w/c_g with temperature were taken directly from the literature. In other cases data were fitted to equations of the type

$$\ln(c_w/c_g) = A - B/T$$

where A and B are constants and I is the absolute temperature.

The equations were correlated with chemical structure. Similar classes of compounds were found to have similar values of the constant B with similar slopes

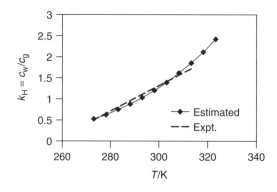

Figure 6.1 Estimated and experimental values of k_H for 2-methylpropane

when $\ln(c_w/c_g)$ was plotted against $1/T$. Average values of B were then evaluated for each class of compound – general aliphatic, general aromatic ester, nitrate, etc. The program uses these values of the slope with values of c_w/c_g at 298.15 K to calculate values of c_w/c_g and hence of k_H at 5 K intervals from 273.15 to 323.5 K (0–50 °C). When more than one type of group is contained in a molecule, the highest slope value is used by the program.

The calculated values for 2-methylpropane and values based on direct experiment are shown in Figure 6.1.

6 OTHER METHODS DEPENDENT UPON LINEAR QUANTITATIVE STRUCTURE–PROPERTY RELATIONSHIPS

Several other quantitative structure–property relationship studies have been published [2,6–11]. Some have involved correlation with physical properties in addition to structural properties. Most have involved the derivation of linear solvation energy relationships for $\log k_H$. Although some have given more reliable correlations with certain groups of compounds, none is as easy to use as the Meylan and Howard method, which remains the standard against which comparisons are made.

Abraham et al. [10] derived the following equation:

$$\log(c_w/c_g) = -0.994 + 0.557 R_2 + 2.549 \pi_2^H + 3.813 \Sigma \alpha_2^H + 4.841 \Sigma \beta_2^H$$
$$- 0.869\ V_x$$

where R_2 = excess molar refraction calculated from the experimental molar refraction

        π_2^H = dipolarity/polarizability determined experimentally

        $\Sigma \alpha_2^H$ = effective hydrogen-bond acidity

        $\Sigma \beta_2^H$ = effective hydrogen-bond basicity

        V_x = McGowan characteristic volume [33] calculated from tabulated atomic increments.

For a group of 408 gaseous compounds, the standard deviation of values of $\log(c_w/c_g)$ was 0.15.

This method suffers from the disadvantage that experimentally determined data are required.

Katritzky and Mu [9] investigated the relationship between Henry's law constant and properties related to molecular structures. They used the 408 compounds investigated by Abraham et al. with the elimination of sulfur hexafluoride and triethyl phosphate. More than 600 descriptors calculated from the structures of the molecules were investigated. In addition to the number of atoms of bonds of different types they included geometric and topological factors. A variety of quantum mechanical descriptors such as energy levels were also included.

They chose the most significant five descriptors to give the equation

$$\log(c_w/c_g) = (2.65 \pm 0.22) + (42.37 \pm 1.11)HDCA(2) + (0.65 \pm 0.02)$$
$$\times [N(O) + 2N(N)] + (-0.16 \pm 0.02)(E_{HOMO} - E_{LUMO})$$
$$+ (0.12 \pm 0.01)PCWT^E + (0.82 \pm 0.06)N_{rings}$$

where $HDCA(2)$ is a hydrogen bonding-related descriptor defined by the equation

$$HDCA(2) = \sum \frac{q_D \sqrt{S_D}}{\sqrt{S_{tot}}}$$

where q_D is the partial charge on the hydrogen bonding donor atom or atoms (H), S_D is the exposed surface of this atom or atoms and S_{tot} is the total surface area of the molecule. $N(O) + 2N(N)$ is a linear combination of the number of nitrogen atoms $N(N)$ and the number of oxygen atoms $N(O)$. N_{rings} is the number of rings. $(E_{HOMO} - E_{LUMO})$ is the energy gap associated with the dispersion energy of polar solutes in solution [34,35] $PCWT^E$ is the partial charge weighted topological electronic index discussed by Osmialowski et al. [36]. This is defined by the equation

$$PCWT^E = \frac{1}{Q_{min}} \sum_{i<j} \frac{|q_i - q_j|}{r_{ij}^2}$$

where q_i and q_j are the Zefirov partial charges of the bonded atoms [37], Q_{min} is the most negative partial charge and r_{ij} is the corresponding bond length.

The correlation coefficient (r^2) between experimental and predicted values of H was 0.9420. The standard deviation was 0.52 and the average absolute error 0.42.

Dearden et al. [7] published the following seven-parameter relationship based on a training set of 294 molecules:

$$\log(c_g/c_w) = 2.31^4 X_p^v - 1.14MR - 1.00HB_I + 0.304E_{LUMO} - 1.76\alpha$$
$$+ 0.137B_R + 1.09 \log P - 0.44$$

where 　$^4X_p^v$ = fourth-order cluster connectivity
　　　　MR = molar refractivity
　　　　HB_I = sum of the indicator variables for H-bond donation and acceptance
　E_{LUMO} = the energy of the lowest unoccupied molecular orbital
　　　　α = Kamlet's H-bonding parameter
　　　　B_R = the number of rotatable bonds in the molecule
　　HB_N = total number of hydrogen bonds that a molecule can form with water
　　　　P = octanol–water partition coefficient.

The equation was tested with 48 compounds and gave good correlations between predicted and experimental values of c_g/c_w. However, cyclohexadec-8-en-1-one gave poor agreement, probably because of neglect of the effects of flexibility.

When the results for this compound were disregarded, the predicted and experimental data for the remaining 47 compounds could be fitted to the equation

$$\log(c_g/c_w)_{\text{expt}} = 0.984 \log(c_g/c_w)_{\text{pred}} - 0.120$$

$$r^2(\text{adj.}) = 0.928 \qquad s = 0.702 \qquad F = 576.8$$

A 10-parameter LSER relationship for prediction of Henry's law coefficient has been published by Sixt [11]:

$$\log(c_g/c_w) = 0.75(\pm 0.10) - 0.18(\pm 0.01)N_{\text{ar}} - 5.4(\pm 0.3)U_q$$
$$- 3.2(\pm 0.2)\alpha_{\text{hb,max}} + 0.11(\pm 0.01)N_{\text{H,m}} + 5.0(\pm 0.4)Q_C$$
$$+ 3.7(\pm 0.4)Q_F - 2.8(\pm 0.3)\beta_{\text{hb,max}} + 2.7(\pm 0.3)Q_O$$
$$+ 2.2(\pm 0.3)Q_{\text{OxO}} - 0.17(\pm 0.03)\mu_m$$

where
N_{at} = No. of atoms in the compound
U_q = total charge interaction
$\alpha_{\text{hb,max}}$ = hydrogen bonding acidity
$N_{\text{H,m}}$ = No. of hydrophobic hydrogen atoms
Q_C = sum of the total charges on carbon atoms
Q_F = sum of the total charges on fluorine atoms
$\beta_{\text{hb,max}}$ = hydrogen bonding basicity
Q_O = sum of the total charges on oxygen atoms
Q_{OxO} = sum of charges on OXO oxygen atoms
μ_m = dipole moment.

English and Carroll [38] derived two linear regression models using descriptors to account for the influences of connectivity, charge distribution, charged surface area, hydrogen bonding characteristics and nature of groups on the Henry law coefficient for solubilities in water at 25 °C. Initially they worked with 29 descriptors. For the first model an optimum group of 10 descriptors was chosen. An additional two descriptors were used for the second model. These descriptors took into account effects due to hydrogen bonding. The models were based on data for 357 compounds. These linear models were developed using 303 of the compounds and tested with the remaining 54 compounds.

7 USE OF NEURAL NETWORKS TO PREDICT HENRY'S LAW CONSTANTS

The methods described above depend on the assumption that there are linear relationships between physical properties and solubility. The disadvantage of

Table 6.4 Independent test set performance indices for
models developed by English and Carroll [38]

Model	r^2	Standard error (log units)
Model 1, linear	0.925	0.533
Model 1, neural	0.979	0.281
Model 2, linear	0.980	0.285
Model 2, neural	0.985	0.237

this linear method of building a model is that there is no provision for nonlinear relationships between values of k_H and properties of molecules. Artificial neural networks have found wide applications in many other branches of chemistry. Types of artificial neural networks and their applications have been described by Gasteiger and Zupan [39]. It is possible to apply artificial neural networks to this problem but little work has been published. However, Sixt [11] has described the effect of inputting 15 descriptors into a 15–10–1 network using a Stuttgart Neural Network Simulator.

The training data set contained 305 compounds and it was tested with 76 compounds. The training data gave results with a correlation coefficient (r^2) of 0.987 and a root mean standard deviation of 0.24 for $\log(c_g/c_w)$. The test data gave a root mean standard deviation of 0.31. Thiobencarb, methylparathion and 2,3,6-trichloroanisole were three compounds for which there were large discrepancies between measured and predicted values of $\log(c_g/c_w)$; the differences were 1.66, 1.07 and 1.01, respectively.

English and Carroll [38] derived two models based on neural networks using the same compounds and groups of 10 or 12 descriptors. A training set of 261 and a validation set of 42 was chosen from the 303 compounds used for developing the linear models mentioned above. The neural models were tested with the 54 compounds used to test the linear models. In each case the neural versions of the models performed better than the corresponding linear versions (Table 6.4). The models which used 12 descriptors (models 2) performed significantly better than the models based on 10 descriptors (models 1).

Each of the models produced outliers with significant differences between experimental and predicted solubilities. Some compounds were outliers in the case of more than one model. In the case of some compounds a particular model gave a significantly better prediction than other models. Predictions were often poor for chloro and iodo compounds. Both of the models 2 were better than models 1 in the case of nitriles, cyclic ketones, morpholines and 1,4-dioxane but less reliable in the case of cyano compounds.

Comparisons of the predictions of the solubilities of individual compounds with experimental values are given in the supporting information available from the American Chemical Society via the Internet at http://pubs.acs.org.

REFERENCES

1. Dearden, J.; Cronin, M.; Lai, N.; Rowe, P., *QSAR Prediction of Henry's Law Constant*, www.documentarea.com/qsar/cronin.pdf.

2. Örnektekin, S.; Paksoy, H. O.; Demirel, Y., *Thermochim. Acta*, 1996, **287**, 251–9.
3. Kühne, R.; Ebert, R.-U., Kleint, F.; Schmidt, G.; Schüürmann, G., *Chemosphere*, 1995, **30**, 2061–77.
4. Hine, J.; Mookerjee, P. K., *J. Org. Chem.*, 1975, **40**, 292–8.
5. Meylan, W. M.; Howard, P. H., *Environ. Toxicol. Chem.*, 1991, **10**, 1283–93.
6. Dearden, J. C.; Cronin, M. T. D.; Sharra, J. A.; Higgins, C.; Boxall, A. B. A.; Watts, C. D., in *QSAR in Environmental Sciences – VII*, Chen, C.; Schüürmann, G. (eds), SETAC Press, Pensacola, FL 1997, pp. 135–42.
7. Dearden, J. C.; Ahmed, S. A.; Cronin, M. T. D.; Sharra, J. A., in *Molecular Modelling and Prediction of Bioactivity*, Gundertoffe, K.; Jorgensen, F. S. (eds), Plenum Press, New York, 2000, pp. 273–4.
8. Suzuki, T.; Ohtaguchi, K.; Koide, K., *Comput. Chem.*, 1992, **16**, 41–52.
9. Katritsky, A. R.; Mu, L., *J. Chem. Inf. Comput. Sci.*, 1996, **36**, 1162–8.
10. Abraham, M. H.; Andonian-Haftvan, J.; Whiting, G. S.; Leo, A.; Taft, R. S., *J. Chem. Soc., Perkin Trans. 2*, 1994, 1777–91.
11. Sixt, S. *Dissertation*, Naturwissenschaftlichen Fakultäten der Universität Erlangen-Nürnberg, Erlangen, 1998.
12. Swope, W. C.; Andersen, H. C., *J. Phys. Chem.*, 1984, **88**, 6548–56.
13. Watanabe, K.; Andersen, H. C., *J. Phys. Chem.*, 1986, **90**, 795–802.
14. Bernal, J. D.; Fowler, R. H., *J. Chem. Phys.*, 1933, **1**, 515.
15. Stillinger, F. H.; Rahman, A., *J. Chem. Phys.*, 1974, **60**, 1545.
16. Jorgensen, W. L., *J. Chem. Phys.*, 1982, **77**, 4156.
17. Jorgensen, W. L.; Chandrasekhar, J.; Madura, J. D.; Impley, R. W.; Klein, M. L., *J. Chem. Phys.*, 1983, **79**, 926.
18. Berendsen, H. J. C.; Postma, J. P. M.; van Gunsteren, W. F.; Hermans, J; in *Intermolecular Forces*, Pullman, B. (ed.), Reidel, Dordrecht, 1981.
19. Berendsen, H. J. C.; Grigera, J. R.; Straatsma, T. P., *J. Phys. Chem.*, 1987, **91**, 6269.
20. Ahlstrom, P.; Wallqvist, A.; Engström, S.; Jönsson, B., *Mol. Phys.*, 1989, **68**, 563.
21. Caldwell, J.; Dang, L. X.; Kollman, P. A., *J. Am. Chem. Soc.*, 1990, **112**, 9144.
22. Wallqcist, A.; Berne, B. J., *J. Chem. Phys.*, 1993, **97**, 13–841.
23. Chialvo, A. A.; Cummings, P. T., *J. Chem. Phys.*, 1996, **105**, 8274.
24. Boulougouris, G. C.; Economou, I. G.; Theodorou, D. N., *J. Phys. Chem. B*, 1998, **102**, 1029.
25. Errington, J. R.; Panagiotopoulos, A. Z., *J. Phys. Chem. B*, 1998, **102**, 7470.
26. Mackie, A. D.; Fernández-Cobos, J.; Vega, L. F., *J. Chem. Phys.*, 1999, **111**, 2103.
27. Economou, I. G., presented at the 14th Symposium on Thermophysical Properties, Boulder, CO, 25–30 June, 2000.
28. Boulougouris, G. C.; Errington, J. R.; Economou, I. G.; Panagiotopoulos, A. Z.; Theodorou, D. N., *J. Phys. Chem. B*, 2000, **104**, 4958.
29. Martin, M. G.; Siepmann, J. I., *J. Phys. Chem. B*, 1998, **102**, 2569.
30. Errington, J. R.; Boulougouris, G. C.; Economou, I. G.; Panagiotopoulos, A. Z.; Theodorou, D. N., *J. Phys. Chem. B*, 1998, **102**, 8865.
31. Slusher, J. T., *J. Phys. Chem. B*, 1999, **103**, 6075.
32. Tsonopoulis, C.; Wilson, G. M., *AIChE J.*, 1983, **29**, 990.
33. Abraham, M. H.; McGowan, J. C., *Chromatographia*, 1987, **23**, 243.
34. Rosch, N.; Zerner, M. C., *J. Phys. Chem.*, 1994, **98**, 5817–23.
35. Amos, A. T.; Burrows, B. L., *Quantum Chem.*, 1973, **7**, 289–313.
36. Osmialowski, K.; Halkiewicz, J.; Kaliszan, R., *J. Chromatogr.*, 1986, **361**, 63–9.
37. Zefirov, N. S.; Kirpichenok, M. A.; Izmailov, F. F.; Trofimov, M. I., *Dokl. Akad. Nauk (English Version)*, 1987, **296**, 883–7.
38. English, N. J.; Carroll, D. G., *J. Chem. Inf. Comput. Sci.*, 2001, **41**, 1150–61.
39. Gasteiger, J.; Zupan, J., *Angew. Chem. Int. Ed. Engl.*, 1993, **32**, 503–27.

CHAPTER 7

Solubility of Gases in Strong Electrolyte Solutions

Peter Fogg
University of North London, UK (retired)

1 DISTRIBUTION OF STRONG ELECTROLYTES

Cloud water contains dissolved ions. The concentrations of major ions in cloud and fog waters from different sources have been summarized by Warneck [1] as indicated in Table 7.1.

Cloud and fog droplets also contain particulate solid matter which has served as nuclei for condensation. According to Warneck [1], coarser particles above about $1\,\mu m$ in diameter consist largely of salts of alkali and alkaline earth metals. The solid material picked up over the sea is mostly NaCl and $MgSO_4$. Salts of Na, K, Ca and Mg are the main cations in the coarser particles picked up over land areas. In general, soils contain only a relatively small proportion of chloride and of sulfate ions. However, sulfuric acid is formed in the atmosphere by oxidation of sulfur dioxide and is partially neutralized by ammonia in the atmosphere. Ammonium sulfate forms a large proportion of the fine particles below $1\,\mu m$ which act as condensation nuclei.

Dissolution of condensation nuclei in cloud droplets is an important factor in determining the ionic content of cloud water.

2 SOLUBILITY OF GASES IN SALT SOLUTIONS

This solubility is often characterized by the Setchenov constant, which can be defined in different ways depending on the way in which concentration is measured [6]. One form of the relationship is

$$\log(c_0/c_g) = kc_s$$

where c_0 is the solubility of the gas in the pure solvent, c_g is the solubility of the gas in the solution of the electrolyte, c_S is the concentration of the salt and

Chemicals in the Atmosphere – Solubility, Sources and Reactivity.   Edited by P. G. T. Fogg and J. Sangster
© 2003 IUPAC   ISBN: 0-471-98651-8

Table 7.1 Concentration of the principal ions in cloud and fog waters/μmol dm^{-3} (reproduced with permission from W. Jaeschke (Ed.), *Chemistry of Multiphase Atmospheric Systems*, NATO ASI Series, Vol. 96. Copyright Springer-Verlag, 1986)

Ion	Russia (Europe) [2]		S. England [3]	Germany [4]	Pasadena, CA USA [5]
	Precipitating	Non-precipitating	Non-precipitating	Mountain fog	Urban fog
SO_4^{2-}	29	117	40	387	240–472
Cl^-	22.2	76.8	94	205	480–730
NO_3^-	3.2	16.0	18.6	450	1220–3250
HCO_3^-	11.5	16.4	–	–	–
NH_4^+	28.8	11.1	22.1	710	1290–2380
Na^+	17.0	29.8	95.2	295	320–500
K^+	5.1	20.4	12.5	85	33–53
Ca^{2+}	10.0	29.3	33.2	110	70–265
Mg^{2+}	12.3	40.0	12.3	–	45–160
H^+	5.0	94.4	0.06–40	12.6	14–1200
pH	5.3	4.0	4.4–7.2	5.1	4.85–2.92
No. of samples	125	194	23	19	4

See also P. Warneck, *Chemistry of the Natural Atmosphere*, Academic Press, 2000.

Table 7.2 Parameters for the Schumpe model for the Setchenov constant [8]

Cation	Cation parameter h_i/mol^{-1} dm^3	Anion	Anion parameter h_i/mol^{-1} dm^3	Gas	Gas parameter h_G/mol^{-1} dm^3
H^+	0.0000[a]	OH^-	0.0918	H_2	−0.0176
Li^+	0.0687	F^-	0.1058	He	−0.0368
Na^+	0.1079	Cl^-	0.0381	Ne	−0.0147
K^+	0.0929	Br^-	0.0266	Ar	−0.0026
Rb^+	0.0749	I^-	0.0233	Kr	−0.0085
Cs^+	0.0582	NO_2^-	0.0726	Xe	0.0094
NH_4^+	0.0555	NO_3^-	0.0136	Rn	0.0170
Mg^{2+}	0.1576	ClO_3^-	0.1390	CH_4	0.0028
Ca^{2+}	0.1543	ClO_4^-	0.0549	C_2H_2	−0.0174
Ba^{2+}	0.1710	BrO_3^-	0.1158	C_2H_4	0.0014
Fe^{2+}	0.1694	IO_3^-	0.0956	C_2H_6	0.0115
Co^{2+}	0.1582	IO_4^-	0.1469	C_3H_8	0.0461
Ni^{2+}	0.1556	HCO_3^-	0.1019	C_4H_{10}	0.0552
Cu^{2+}	0.1490	HSO_3^-	0.0601	CO_2	−0.0183
Mn^{2+}	0.1474	$H_2PO_4^-$	0.1009	N_2	0.0002
Zn^{2+}	0.1588	ϕCH_2COO^-	−0.0089	NH_3	−0.0506
Cd^{2+}	0.2028	CN^-	0.0722	NO	0.0060
Al^{3+}	0.2192	SCN^-	0.0670	N_2O	−0.0110
Fe^{3+}	0.0957	$HCrO_4^-$	0.0408	O_2	0.0000[a]

Table 7.2 (*continued*)

Cation	Cation parameter h_i/mol^{-1} dm^3	Anion	Anion parameter h_i/mol^{-1} dm^3	Gas	Gas parameter h_G/mol^{-1} dm^3
Cr^{3+}	0.0578	HPO$_4^{2-}$	0.1559	H$_2$S	−0.0341
		CO$_3^{2-}$	0.1558	SO$_2$	−0.0832
		S$_2$O$_3^{2-}$	0.1268	SF$_6$	0.0108
		SO$_3^{2-}$	0.1357		
		SO$_4^{2-}$	0.1164		
		PO$_4^{3-}$	0.2243		

[a]Arbitrary zero.

k is the Setchenov constant, which depends on the gas, the electrolyte and the temperature. Usually the addition of electrolyte causes a reduction in the solubility of a gas and the constant k is positive when the equation is written as above. In the literature, concentrations of electrolyte have been expressed in terms of molarity or molality. The solubility of the gas has been given as molarity, molality or mole fraction. Solubilities have also been expressed as Bunsen coefficients or Ostwald coefficients. Clever has shown that there are small but significant differences between the parameters derived from different concentration units [6,7].

There are several empirical models for estimation of the Setchenov constant [8]. Schumpe [9] published a model in which the Setchenov constant is

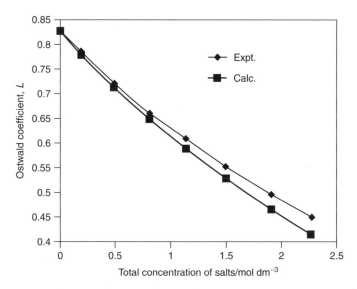

Figure 7.1 Solubility of carbon dioxide at 298.15 K in solutions containing sodium chloride (50 mol% of total electrolyte), magnesium sulfate (25 mass% of total electrolyte) and potassium nitrate (25 mass% of total electrolyte) [10]. Expt., experimental data [10]; Calc., calculated data using Schumpe coefficients

considered to be made up of ion-specific and gas specific parameters:

$$\log(c_0/c_g) = \Sigma(h_i + h_G)c_i$$

These parameters were derived from available experimental data (Table 7.2). This model has been extended by Hermann et al. [8] by using 319 sets of solubility data for various salt solutions and gases and deriving parameters for 45 ions and 22 gases at 298.15 K. These parameters can be used to estimate the corresponding Setchenov constant for a particular salt-gas system. The standard deviation of the calculated Setchenov constants from the experimental values is 0.026, with reactive gases tending to show the greatest deviation.

The values given are for 298.15 K but the authors suggest that the parameters can be used over a temperature range of about ±5 K. Figure 7.1 shows a comparison of experimental and calculated values of the Ostwald coefficient for carbon dioxide in solutions containing sodium chloride, magnesium sulfate and potassium nitrate. The Ostwald coefficient is defined as the ratio of the volume of gas absorbed to the volume of the absorbing liquid.

3   THERMODYNAMIC METHODS OF ESTIMATING THE SOLUBILITY OF GASES IN ELECTROLYTE SOLUTIONS

Estimation of the solubility of a gas in an electrolyte solution by use of Setchenov coefficients is unsatisfactory if the ionic strength is high. Prediction of the behaviour of solutions containing electrolytes requires a realistic estimate of activity coefficients of charged species. The simple Debye–Hückel expression is usually satisfactory for solutions of concentration less than about 0.1 mol dm^{-3} but is unsuitable for the treatment of concentrated electrolyte solutions. More expressions containing extra terms applicable to higher concentrations were developed by Guggenheim [11], Scatchard et al. [12] and others. These extra terms depended on the concentration of the electrolyte and are often satisfactory for estimation of the activity of electrolyte in a binary solution [13–15].

There are two general methods of estimating the properties of mixed electrolyte solutions. One method is by use of a so-called *mixing rule*. The activities of individual components in the solution are estimated by a particular method of averaging the activity coefficients of components in binary solutions, usually at the ionic strength of the multicomponent solution under investigation. A mixing rule was developed by Kusik and Meissner [16] from the Brønsted specific interaction principle and Harned's rule [17] for mixed activity coefficients at constant ionic strength. This was adapted by Tabazadeh et al. [18] to model solute behaviour in stratospheric systems [19].

A recent example of the use of the modified Huron–Vidal second-order mixing rule [20] was described by Vanderbeken et al. [21]. In this case the solubility of methane and carbon dioxide in brines at high temperature and pressure was predicted.

An alternative method of estimating activity coefficients involves detailed consideration of interactions between ions. Pitzer, in 1973, published a paper [22] describing the development of equations to extend the simple Debye–Hückel model of electrolyte systems. He recognized the ionic strength dependence of the effect of short-range forces in binary interactions. A second virial coefficient to allow for binary interactions had been included in the earlier Guggenheim–Scatchard modifications of the Debye–Hückel equation. Pitzer modified this term to allow for the effect of ionic strength. In particular, he recognized an ionic strength dependence of the effect of short-range forces in binary interactions. Previous equations contained a virial expansion and Pitzer modified the second virial coefficient introduced by other workers to take account of these binary interactions. He published equations which agreed with experimental measurement of activity coefficients in solutions of concentrations of several $mol\,dm^{-3}$ rather than about $0.1\,mol\,dm^{-3}$. The equations were subsequently developed by Pitzer and co-workers and published in a series of papers [23–31].

The Pitzer equations were originally developed for single electrolyte solutions. However, Pitzer and Simonson [32] later derived equations which could be applied to systems which contained an unlimited number of singly charged ionic species or uncharged species. They derived equations for thermodynamic quantities and showed how interaction parameters for more complex systems could be estimated from interaction parameters for binary systems.

Subsequently, Clegg *et al.* [33] showed how the equations could be generalized to include ions of arbitrary charge. These formed the basis of a thermodynamic model of phase equilibria since a component present in two phases in equilibrium must have the same chemical potential in each phase. The model can be correlated with experimental data and used to predict conditions of equilibria for situations where experimental data are unreliable or uncertain. The solubility of gases or the conditions under which a solid phase will precipitate from solution may be estimated with a reasonable degree of precision.

This model is based on an expression for the excess Gibbs free energy which contains an extended Debye–Hückel term to account for long-range effects and a Margules expansion which accounts for short-term effects. Equations for activities and activity coefficients are derived from this expression. Interaction parameters between species in solution must be based ultimately on experimental measurements. In the case of a solution of a single electrolyte three parameters are needed, B_{MX}, $W_{1,MX}$ and $U_{1,MX}$, to account for the short-range interaction between the two ions, M^+ and X^-, and also between molecules of the solvent (1) and the ions. These parameters may be found by comparing predicted values of activity coefficients with values from experimental measurements of physical properties.

In solutions containing more than two kinds of ions it is necessary to consider three-body interactions of the type anion(1)–anion(2)–cation or cation(1)–cation(2)–anion. These become more significant with increase in concentration. Additional interaction parameters are required which must be based on available experimental data.

Clegg and co-workers have shown that the model [29] is particularly useful in the prediction of the behaviour of concentrated acid solutions such as are present as stratospheric aerosols. At low partial pressures the solubility of hydrogen halides and nitric acid was considered in terms of the equilibrium

$$HY(g) \rightleftharpoons H^+(aq) + Y^-(aq)$$

on the assumption that there was an insignificant amount of undissociated acid in the liquid phase.

A constant analogous to a Henry's law constant was defined as

$$k_H = x_H f_H^* x_Y f_Y^* / P_{HY}$$

where x_i is the mole fraction of ionic species i defined as a fraction of the total number of species which includes both ions and neutral molecules and f_i^* is the activity coefficient of species i on the basis that $f_i^* \to 1$ as $x_i \to 0$.

Clegg and Brimblecombe [34] measured activity coefficients and vapour pressures of aqueous solutions of HCl at 298.15 K and used heat capacity data and the value of the enthalpy of solution of HCl to derive the following expression for the value of k_H of HCl over a temperature range:

$$\ln(k_H/\text{atm}^{-1}) = 6.4954 - 9.0027 \times 10^3[1/(T_r/K) - 1/(T/K)] - 65.346$$
$$\times [T_r/T - 1 + \ln(T/T_r)] + 0.078178[(T_r/T - 1)T_r + T - T_r]/K$$

where T_r is equal to 298.15 K

Total pressures over solutions containing 9.45 and 10.51 mol kg^{-1} HCl were measured over the temperature range 199.5–289.4 K using a capacitance manometer. Equilibrium partial pressures were also measured by mass spectrometry from 234.6 to 274.8 K. The partial pressure of HCl as a fraction of the total pressure ranged from about 0.09 to 0.38. The total pressures estimated from the equation given above plus the values of the vapour pressure of water are up to about 3 % higher than experimental values from 289.4 to 250 K. At lower temperatures the calculated values are up to 10 % higher than experimental values.

A similar equation for the solubility of HBr was derived by Carslaw et al. [35]. A revised version of the equation based on a more experimental data has been published by Massucci et al. [36]

$$\ln(k_H/\text{atm}^{-1}) = 12.5062 - 10.239 \times 10^3[1/(T_r/K) - 1/(T/K)] + 41.948$$
$$\times [T_r/T - 1 + \ln(T/T_r)] - 0.10482[(T_r/T - 1)T_r + T - T_r]/K$$

A revised equation for the solubility of nitric acid given by Carslaw et al. [35] is

$$\ln(k_H/\text{atm}^{-1}) = 385.972199 - 3020.3522/(T/K) - 71.001998 \ln(T/K)$$
$$+ 0.131442311 \ T/K - 0.420928363 \times 10^{-4}(T/K)^2$$

Experimental measurements of the partial pressure of HNO$_3$ over solutions containing 7.82, 15.73 and 35.99 mol kg^{-1} of acid were scattered but the general trend was consistent with the equation over the full temperature range of 294.6–224.6 K.

The expression for the activity coefficient of a species in a three-component system contains binary terms used in expressions for the corresponding binary systems. It also contains terms for three different particles close together. These parameters have to be found by adjusting to the available experimentally determined thermodynamic properties of the ternary system.

The equations for Henry's law coefficients discussed above are a measure of the partial vapour pressure of components in binary aqueous solutions. This, and other information on the binary systems, were used by Carslaw et al. [35] to derive thermodynamic models for the ternary system $HCl–H_2SO_4–H_2O$. When the model was fitted to experimental values of the solubility of HCl, it was found that inclusion of ternary terms did not affect the accuracy of the fit. Ternary terms were therefore not used in this case.

Carslaw et al. [35] found it convenient to express results in terms of an effective Henry's law constant k_H^* with units $mol\,dm^{-3}\,atm^{-1}$. In the case of hydrogen chloride this is the concentration of the chloride ions in solution in units of $mol\,dm^{-3}$ divided by the partial pressure of hydrogen chloride in units of atm. The variation of k_H^* with temperature in this system as predicted by the model and experimental values of k_H^* are shown in Figure 7.2.

Carslaw et al. [35] also worked on the systems $HNO_3–H_2SO_4–H_2O$, $HBr–H_2SO_4–H_2O$, $HCl–HNO_3–H_2O$ and $HBr–H_2SO_4–H_2O$.

A comparison for the values of k_H^* for hydrogen bromide in aqueous sulfuric acid is shown in Figure 7.3. This model was later revised by Massucci et al. [36].

A model of the quaternary system $HCl–HNO_3–H_2SO_4–H_2O$ on a similar basis has been published by Carslaw et al. [35].

Clegg and Brimblecombe [40] investigated the thermodynamic aspects of the solubility of ammonia in electrolyte solutions. The dissolution of ammonia

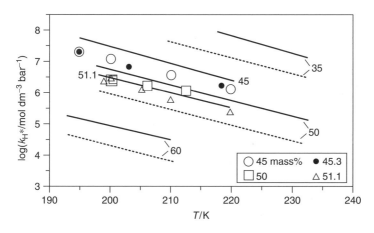

Figure 7.2 Measured and calculated effective Henry's Law constants for HCl in aqueous H_2SO_4. Data points correspond to experimental data from Hanson and Ravinshankara [38]. Numbers on the figures are mass% of H_2SO_4. Full lines represent data from model published by Carslaw et al. [35] Dotted lines from model of Zhang et al. [38a]

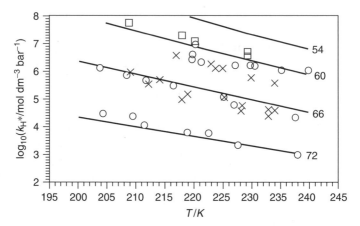

Figure 7.3 Measured and calculated effective Henry's Law constants for HBr in aqueous H_2SO_4. Data points correspond to experimental data from Williams *et al.* [39]. Numbers on the figures are mass% of H_2SO_4. Circles correspond to kinetic uptake measurements at 60, 66 and 72 mass% H_2SO_4. Crosses correspond to equilibrium vapour pressure measurements at 60 and 66 mass% H_2SO_4. Squares correspond to vapour pressure measurements at 54 mass% H_2SO_4. Full lines represent data from the model published by Carslaw *et al.* [35]

involves five related equilibria:

$$NH_3(g) \rightleftharpoons NH_3(aq) \tag{7.1}$$

$$NH_3(aq) + H_2O \rightleftharpoons NH_4^+(aq) + OH^-(aq) \tag{7.2}$$

$$H^+ + OH^- \rightleftharpoons H_2O \tag{7.3}$$

$$NH_3(g) + H^+(aq) \rightleftharpoons NH_4^+ \tag{7.4}$$

$$NH_4^+ \rightleftharpoons H^+(aq) + NH_3(aq) \tag{7.5}$$

The first equilibrium is controlled by a Henry's law equilibrium constant defined as

$$k_H = \gamma_{NH_3} m_{NH_3} / p_{NH_3}$$

and the second by a base dissociation constant

$$K_b = \gamma_{NH_4^+} m_{NH_4^+} \gamma_{OH^-} m_{OH^-} / (\gamma_{NH_3} m_{NH_3} a_w)$$

Values of activity coefficients of the dissolved components are needed to calculate the total quantity of ammonia going into solution from the partial pressure of ammonia or the partial pressure from composition of the solution. These activity coefficients depend on all the electrolytes present in the solution and may be calculated by the ion interaction model developed by Pitzer. They also depend on interaction between neutral species and on interaction of neutral species with charged species. These interactions are characterized by interaction parameters.

At concentrations of ammonia above $0.1 \, \text{mol kg}^{-1}$ the degree of ionization of ammonia can be neglected [41]. Hence in the absence of other sources of ions the activity coefficient of ammonia depends only on interaction between pairs of neutral molecules. Three-body interactions are not significant for concentrations of ammonia below $25 \, \text{mol kg}^{-1}$. Hence

$$p_{NH_3} = (m_{NH_3}/\text{mol kg}^{-1})\gamma_{NH_3}/K_H$$

$$\gamma_{NH_3} = \exp[(2m_{NH_3}/\text{mol kg}^{-1})\lambda_{N,N}]$$

where $\lambda_{N,N}$ is the NH_3-NH_3 interaction parameter. Clegg and Brimblecombe [41] gave the following expression for $\lambda_{N,N}$:

$$\lambda_{N,N} = 0.033161 - 21.12816/(T/K) + 4665.1461/(T/K)^2$$

They also derived ion–NH_3 interaction parameters from partial pressure, salt solubility and partitioning data for a wide range of ions and found that there was a simple relationship between parameters and ion charge and size. These parameters allow the estimation from Pitzer's equations of the activity of ammonia in a wide variety of multicomponent solutions containing a variety of ionic species. Depending on the nature of the system, the dissociation of ammonia to give ammonium ions may have to be taken into account.

Clegg and Brimblecombe gave equations for the two constants k_H and K_b of the form

$$\ln K_x = a + b/(T/K) + cT/K$$

These equations are valid for 0–40 °C. Values of the constant a, b and c are given below.

Dissociation constant K_b (based on data from Bates and Pinching [42]):

$K_b/\text{mol dm}^{-3}$ at 298.15 K	a	b	c
1.774×10^{-5}	16.9732	−4411.025	−0.044

Henry's law type constant, k_H (based on experimental data published by Chen et al. [43]):

$k_H/\text{mol dm}^{-3} \text{ atm}^{-1}$ at 298.15 K	a	b	c
60.72	−8.09694	3917.507	−0.003

Clegg and Brimblecombe also gave a value for k_H of $57.64 \, \text{mol dm}^{-3} \text{ atm}^{-1}$ calculated from thermodynamic data published by Wagman et al. [44]. Eigen and Schoen [45] published a value of $91.0 \, \text{mol dm}^{-3} \text{ atm}^{-1}$, which compares well with a value of 90.12 from the equation above.

The equilibrium constant for Equation (7.4) may be written as

$$K = x_{NH_4^+} f_{NH_4^+} x_{H^+} f_{H^+} p_{NH_3}$$

Clegg *et al.* [46] gave the following equation:

$$\ln(K/\text{atm}^{-1}) = 25.393 + 10373.6[1/(T_r/\text{K}) - 1/(T/\text{K})]$$
$$+ 4.131\{T_r/T - [1 + \ln(T_r/T)]\}$$

This equation is based upon the following data:

experimental value of k_H at 298.15 K = 60.72 mol kg^{-1} atm^{-1} [43];
dissociation constant, K_a, of the ammonium ion at 298.15 K according to Equation (7.5) = 5.6937 × 10^{-10} mol kg^{-1} [47];
enthalpy change $\Delta H°$ at 298.15 K in reaction (7.4) = −86.25 kJ mol^{-1} [44,48];
heat capacity change in reaction (7.4) at constant pressure, $\Delta C_p° = 34.25$ J mol^{-1} K^{-1} [48,49] (assumed to be constant).

The authors stated that the corresponding model of the behaviour of ammonia in electrolyte solutions is valid from 328 to <200 K.

Following the development of earlier models of simpler systems, Clegg *et al.* [50] developed a model for the system H$^+$–NH$_4^+$–Na$^+$–SO$_4^{2-}$–NO$_3^-$–Cl$^-$–H$_2$O at 298 K. This was based on extensive literature data for salt solubilities, electromotive forces, osmotic coefficients and vapour pressures. It is valid for the whole range of proportions of the solid phases and to 40 mol kg^{-1} for liquid acid sulfate systems.

Solubility of HOCl in Water and Aqueous H$_2$SO$_4$

Huthwelker *et al.* [51] derived a model for the dissolution of hypochlorous acid in water and sulfuric acid under atmospheric conditions.

In the atmosphere, HOCl is normally produced in the gas phase by the reaction

$$\text{HO}_2 + \text{ClO} \longrightarrow \text{HOCl} + \text{O}_2$$

It is destroyed by the photolysis or reaction with OH:

$$\text{HOCl} + h\nu \longrightarrow \text{Cl} + \text{OH}$$

$$\text{HOCl} + \text{OH} \longrightarrow \text{ClO} + \text{H}_2\text{O}$$

At low temperature it is produced by the heterogeneous reaction
aerosol

$$\text{ClONO}_2 + \text{H}_2\text{O} \xrightarrow{\text{aerosol}} \text{HOCl} + \text{HNO}_3$$

HOCl concentrations are never very high. Above about 195 K it is a few parts per trillion by volume. Below this temperature it is a few parts per billion.

Hypochlorous acid is a weak acid. The equilibrium between gas and liquid phases may be represented by

$$\text{HOCl(g)} \rightleftharpoons \text{HOCl(aq)}$$

A thermodynamic Henry's law constant may be defined as

$$k_H' = a_{HOCl}/p_{HOCl} = m_{HOCl}f_{HOCl}/p_{HOCl}$$

where a_{HOCl}, m_{HOCl} and f_{HOCl} are the activity, molality and activity coefficient in the liquid phase, respectively, and p_{HOCl} is the partial pressure of HOCl in the gas phase.

An effective Henry's law constant, k_H^*, may be defined as

$$k_H^* = m_{HOCl}/p_{HOCl}$$

The variation of k_H' with temperature can be expressed as

$$\ln[k_{H(T)}'/\text{mol kg}^{-1}\,\text{bar}^{-1}] = \ln[k_{H(T^\circ)}'/\text{mol kg bar}^{-1}] + (\Delta H^\circ/R)[1/(T^\circ/\text{K})$$
$$- 1/(T/\text{K})] + \Delta C_p^\circ/R[(T^\circ/T) - 1 - \ln(T^\circ/T)]$$

where ΔH° and ΔC_p are the enthalpy change and heat capacity change associated with the reference temperature T° and $k_{H(T^\circ)}'$ is Henry's law constant at the reference temperature. This reference temperature was taken to be 298.15 K. It was assumed that changes in ΔC_p with change in temperature could be disregarded. In practice, it was found that the all terms involving ΔC_p could be neglected.

Huthwelker et al. [51] derived a simplified equation for the variation of k_H^* with temperature and concentration of sulfuric acid taking into account the effect of the sulfuric acid on the activity coefficient of the hypochlorous acid. This may be written in the form

$$\ln[k_H^*/\text{mol kg}^{-1}\,\text{bar}^{-1}] = \ln\,[k_{H(T^\circ)}'/\text{mol kg}^{-1}\,\text{bar}^{-1}] - m_{H_2SO_4}\,k_1$$
$$+ (\Delta H^\circ/R)[1/(T^\circ/\text{K}) - 1/(T/\text{K})]$$

where $k_1 = -0.04107 + 54.56/(T/\text{K})$.

Huthwelker et al. [51] claimed that the equation can be used for predicting solubility in solutions of H_2SO_4 from 40 to 60 mass% at 200–235 K in the range of experimental data published by Hanson and Ravinshankara [38]. It also can be used outside this range 'within reasonable limits.' However, they pointed out that the reliability of the expressions depends on the accuracy of the estimation of diffusion coefficients, D, needed to calculate values of k_H^* from experimental values of $k_H^*\sqrt{D}$ measured by Hanson and Ravinshankara. These diffusion coefficients were estimated using a cubic cell model [52,53], which could lead to an uncertainty in values of k_H^* said to be within a factor of about two.

Huthwelker et al. [51] also tabulated experimental values of $\ln k_H'$ and $\ln k_H^*$ from various sources (Table 7.3). The variations of k_H^* with temperature for HOCl in pure water and in sulfuric acid of various concentrations are shown in Figure 7.4.

Table 7.3 Values of Henry's law constants of HOCl in pure water and in sulfuric acid solutions

T/K	Conc. $H_2SO_4/mol\,kg^{-1}$	Ln $(k'_H/mol\,kg^{-1}\,bar^{-1})$	Ln $(k_H*/mol\,kg^{-1}\,bar^{-1})$	Ref.
283.15	0	8.048		54
293.15	0	6.976		54
293.15	0	6.576		55
293.15	0	6.817		56
313.15	0	5.503		55
200	15.05		12.64	38
204	15.05		12.07	38
215	15.05		11.26	38
200	14.38		12.53	38
215	14.38		10.95	38
230	14.38		9.18	38
205	12.78		12.77	38
205	12.78		12.57	38
215	12.78		11.74	38
225	12.78		10.39	38
200	10.41		13.67	38
210	10.41		12.36	38
220	10.41		11.52	38
220	10.41		11.14	38
230	10.41		10.23	38
200	8.69		14.33	38
200	8.69		14.31	38
204	8.69		13.45	38
205	8.69		13.09	38
210	8.69		12.89	38
210	8.69		12.74	38
210	8.69		12.71	38
211	8.69		12.55	38
218	8.69		12.13	38
220	8.69		11.81	38
230	8.69		10.46	38

Application to Atmospheric Conditions

The application of thermodynamic models to atmospheric conditions depends on the assumption that thermodynamic equilibria are established. This condition may not always be fulfilled.

The electrolyte concentration in atmospheric aerosols depends on the location. Electrolytes in sea spray are incorporated into clouds. Sulfur compounds are absorbed over industrial areas. There is much interest in the composition of stratospheric aerosols. The current state of the modelling of the composition of these aerosols has been well reviewed by Carslaw *et al.* [57].

Stratospheric aerosols are now known from sample analysis to consist largely of concentrated sulfuric acid droplets even at temperatures below about 190 K [35]. It has been predicted that concentrations of sulfuric acid vary from about 80 mass% at 240 K in the mid-latitude stratosphere to about 40 mass%

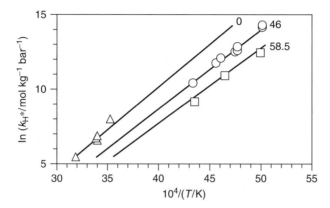

Figure 7.4 The variation of the effective Henry's Law constant k_H^* of HOCl in water and aqueous H_2SO_4 with change of temperature. The lines correspond to the model [51]. Numbers by the lines give the concentration of sulfuric acid in units of mass%. 46 mass% = 8.69 mol kg^{-1}; 58.5 mass% = 14.38 mol kg^{-1} (see Table 7.3)

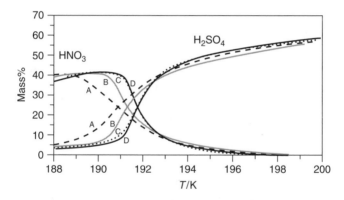

Figure 7.5 A Comparison of liquid aerosol compositions predicted by different groups. A, Beyer *et al.* [61]; B, Tabazadeh *et al.* [62]; C, Luo *et al.* [63]; D, Carslaw *et al.* [31] (based on Figure 12 in Carslaw *et al.* [19])

at 200 K in polar regions [58]. Various gases dissolve in the sulfuric acid and liquid-phase reactions may lead to formation of species such as chlorine which destroy ozone.

Pure sulfuric acid freezes at 283.5 K. The acid forms a series of hydrates which freeze at decreasing temperatures the greater the proportion of water. The highest hydrate, the octahydrate, freezes at 211 K. It follows that sulfuric acid–water mixtures should not be liquid at a temperature much below the water–octahydrate eutectic temperature even if the system were supercooled [59].

Nitric acid and gases such as ClONO$_2$, which may decompose to form nitric acid when dissolved in water–sulfuric acid mixtures, are present in the upper atmosphere. The dissolution of nitric acid lowers the freezing point of aqueous solutions of sulfuric acid. Models indicate that the liquid aerosols contain more nitric acid than sulfuric acid at temperatures below about 192 K [57].

There are various models for distribution of HBr, HCl and other species between the gas phase and liquid $H_2O-H_2SO_4$ or $H_2O-H_2SO_4-HNO_3$ in addition to that published by Carslaw *et al.* [35] discussed above. Various other programs have been reviewed by Pilinis [60]. Models from different authors may show significant differences (Figure 7.5).

Tropospheric clouds do not contain high concentrations of sulfuric acid. Much of the sulfuric acid is neutralized by ammonia to form ammonium sulfate. Sodium chloride is a significant component in clouds generated over the sea.

4 THE AEROSOL INORGANICS MODEL PROJECT

The Aerosol Inorganics Model (AIM) Project devised by Clegg, Brimblecombe and Wexler is intended to develop equilibrium models of systems containing H^+, NH_4^+, Na^+, SO_4^{2-}, NO_3^-, Cl^-, Br^- and H_2O and is available on a Web site for general use [64]. When the composition of a mixture is specified, the distribution of water and ions between liquid, solid and vapour phases at a particular temperature and relative humidity may be calculated and the result displayed. The nature of solid phases and the activity coefficients of components may also be found.

The working of the models depends upon reliable calculation of activities of water and ions present. These activities are calculated using the equations of Pitzer and co-workers [32,33]. The equilibrium state of a system is found by minimization of the Gibbs free energy. In each case the thermodynamic consistency of the result is tested.

The theoretical basis of each of these models has been published [64]. They have been developed from earlier models [31,33,35,61].

Each model may be used in several ways. In the simplest application, the user specifies the relative humidity and temperature. The user also specifies the number of moles of each ion which would be present if no chemical reaction took place to form HSO_4^-. The model then calculates the composition of the liquid, vapour and any solid phases. It is possible to specify that certain solid phases are not formed so that properties of metastable systems may be investigated.

More complex applications involve specifying total amounts of water, ions and trace gases per cubic metre of atmosphere. The model then calculates the equilibrium state of the system. It is also possible to calculate thermodynamic properties such as activities of individual species from the composition of a mixture at a specified temperature. The models may be used for batch calculations of up to 100 similar problems. There is also facility for finding the effect of varying one of the quantities specifying the system. Possible variables are temperature, reactive humidity or total water and chemical composition. Another application is the calculation of the effects of an adiabatic change in pressure over a pressure range on the state of a parcel of air containing water and ionic material. The output when one of the parameters is varied can be presented in tabular form or graphically.

Three models are available at present:

Model I H^+, SO_4^{2-}, NO_3^-, Cl^-, Br^-, H_2O
This is based on Carslaw *et al.* [35] as updated by Massucci *et al.* [36]. It is particularly applicable to stratospheric systems.

Model II H^+, NH_4^+, SO_4^{2-}, NO_3^-, H_2O
This is based on Clegg *et al.* [46] and is particularly applicable to tropospheric systems.

Model III H^+, NH_4^+, Na^+, SO_4^{2-}, NO_3^-, Cl^-, H_2O
This is based on Clegg *et al.* [50]. At present it is restricted to 298.15 K. The possible formation of up to 19 solid phases is taken into account.

The Web site includes a series of tutorials which illustrate the very diverse applications of the models. A brief outline of these tutorials is as follows:

1. **Properties of a single salt solution** – the relationship between the water content of a particle, the relative humidity and the electrolytes present.
2. **Deliquescence of a single salt** – the behaviour of ammonium nitrate.
3. **Gas–aerosol partitioning in a simple system** – the behaviour of a mixture containing water, ammonium ions and nitrate ions.
4. **Properties of solutions containing multiple salts** – the uptake of water by a mixture of salts and predictions by the Zdanovskii–Stokes–Robinson equation [65–67].
5. **Deliquescence of mixtures of salts** – mutual deliquescence points.
6. **Gas–aerosol partitioning in mixed aerosols** – the dependence on chemical composition.
7. **Variation of solution properties with temperature** – variations of water activity coefficient and water content and the effect of electrolytes.
8. **Variation of deliquescence with temperature** – comparison of the behaviour of ammonium nitrate and of ammonium sulfate.
9. **Variation of gas–aerosol partitioning with temperature** – partitioning in the hydrogen ion, ammonium ion, nitrate and water system.

Examples of Use of the AIM Models

Example 1

Aerosol Inorganics Model III

Arbitrary input data

T = 298.15 K
Pressure = 1 atm
Relative humidity = 0.7

Species	H	NH_4	Na	SO_4	NO_3	Cl
Amount/mol	10	10	10	13	2	2

Output results as presented on the screen

System pressure $= 1.00000$ atm
Volume $= 1.000$ m**3
T $= 298.15$ K
RH $= 0.70000$[pH$_2$O $= 0.21878$E-01 atm]
(The relative humidity above is the input value. The equivalent partial pressure
is also given)

****
** LIQUID PHASE **
****

Species	Moles	Grams	Molality	Mole Frac.	Act. Coeff.
H(aq)	0.15539E+01	0.1566E+01	0.8438E+00	0.1104E$-$01	0.4077E+01
NH$_4$(aq)	0.10000E+02	0.1804E+03	0.5430E+01	0.7104E$-$01	0.5956E+00
Na(aq)	0.10000E+02	0.2299E+03	0.5430E+01	0.7104E$-$01	0.1019E+01
HSO$_4$(aq)	0.84461E+01	0.8198E+03	0.4587E+01	0.6000E$-$01	0.9037E+00
SO$_4$(aq)	0.45539E+01	0.4374E+03	0.2473E+01	0.3235E$-$01	0.7044E$-$02
NO$_3$(aq)	0.20000E+01	0.1240E+03	0.1086E+01	0.1421E$-$01	0.3394E+00
Cl(aq)	0.20000E+01	0.7091E+02	0.1086E+01	0.1421E$-$01	0.1528E+01
Br(aq)	0.00000E+00	0.0000E+00	0.0000E+00	0.0000E+00	$-$
H$_2$O(l)	0.10222E+03	0.1841E+04	0.5551E+02	0.7261E+00	0.9640E+00

****
** GASES **
****

No gases were equilibrated
– Calculated partial pressures (over liquid) –

Pressure (atm)	Species	Equiv. Moles
0.2188E$-$01	H$_2$O(g)	$-$
0.2544E$-$06	HNO$_3$(g)	0.1040E$-$04
0.1476E$-$05	HCl(g)	0.6033E$-$04
0.8814E$-$11	NH$_3$(g)	0.3603E$-$09
0.5395E$-$18	H$_2$SO$_3$(g)	0.2205E$-$16

****
** SOLIDS **
****

There are no solid phases
– Calculated saturation ratios of solids (>0.01 only) –

Sat. Ratio	Species
0.6687E−01	$(NH_4)_2SO_4$
0.3609E−01	$(NH_4)_3H(SO_4)_2$
0.4000E−01	NH_4HSO_4
0.5148E−01	NH_4NO_3
0.1629E+00	NH_4Cl
0.4160E+00	Na_2SO_4
0.9662E−01	$Na_2SO_4 \cdot 10H_2O$
0.1376E+00	$Na_3H(SO_4)_2$
0.1126E+00	$NaHSO_4 \cdot H_2O$
0.3128E−01	$NaHSO_4$
0.1685E+00	$Na_2SO_4 \cdot (NH_4)_2SO_4 \cdot 4H_2O$
0.8992E−01	$NaNO_3$
0.6863E−01	$NaNO_3 \cdot Na_2SO_4 \cdot H_2O$
0.1265E+00	$NaCl$

Example 2

Aerosol Inorganics Model III

Arbitrary input data

$T = 298.15$ K
Pressure $= 1$ atm
Relative humidity $= 0.4$

Species	H	NH_4	Na	SO_4	NO_3	Cl
Amount/mol	10	10	10	13	2	2

Output results as presented on the screen

Volume $= 1.000$ m**3
$T = 298.15$ K
RH $= 0.40000$[pH2O $= 0.12502$E−01 atm]
(The relative humidity above is the input value. The equivalent partial pressure is also given)

****
** LIQUID PHASE **
****

Species	Moles	Grams	Molality	Mole Frac.	Act. Coeff.
H(aq)	0.94613E+00	0.9536E+00	0.2355E+01	0.2070E−01	0.3820E+02
NH_4(aq)	0.85102E+01	0.1535E+03	0.2118E+02	0.1862E+00	0.7225E+00
Na(aq)	0.30464E+01	0.7004E+02	0.7582E+01	0.6665E−01	0.2210E+01
HSO_4(aq)	0.67115E+01	0.6515E+03	0.1670E+02	0.1468E+00	0.1437E+01

Species	Moles	Grams	Molality	Mole Frac.	Act. Coeff.
SO_4(aq)	0.16037E+01	0.1540E+03	0.3991E+01	0.3509E−01	0.1438E−02
NO_3(aq)	0.20000E+01	0.1240E+03	0.4978E+01	0.4376E−01	0.1670E+00
Cl(aq)	0.58391E+00	0.2070E+02	0.1453E+01	0.1278E−01	0.3280E+01
Br(aq)	0.00000E+00	0.0000E+00	0.0000E+00	0.0000E+00	–
H_2O(l)	0.22303E+02	0.4018E+03	0.5551E+2	0.4880E+00	0.8197E+00

****
** GASES **
****

No gases were equilibrated
– Calculated partial pressures (over liquid) –

Pressure (atm)	Species	Equiv. Moles
0.1250E−01	H_2O(g)	–
0.6772E−05	HNO_3(g)	0.2768E−03
0.5005E−04	HCl(g)	0.2046E−02
0.1595E−11	NH_3(g)	0.6521E−10
0.3688E−16	H_2SO_4(g)	0.1507E−14

****
** SOLIDS **
****

No.	Moles	Grams	Species
1	0.2456E−01	0.6072E+01	$(NH_4)_3H(SO_4)_2$
2	0.1416E+01	0.7575E+02	NH_4Cl
3	0.2318E+01	0.6075E+03	$Na_3H(SO_4)_2$

– Calculated partial pressures (over solids) –

Pressure (atm) or Product	Reaction
pHCl × pNH$_3$ = 0.7985E−016,	$NH_4Cl = HCl(g) + NH_3(g)$
(pNH$_3$)^3 × (pH$_2$SO$_4$)$_2$	$(NH_4)_3H(SO_4)_2 = 3NH_3(g) + 2H_2SO_4(g)$
= 0.5524E−068,	

– Calculated saturation ratios of solids (>0.01 only) –

Sat. Ratio	Species
0.1498E+00	$(NH_4)_2SO_4$
0.1000E+01	$(NH_4)_3H(SO_4)_2$
0.4949E+00	NH_4HSO_4
0.2480E+00	NH_4NO_3

Sat. Ratio	Species
0.1686E−01	$2NH_4NO_3 \cdot (NH_4)_2SO_4$
0.2071E+00	$NH_4NO_3 \cdot NH_4HSO_4$
0.1000E+01	NH_4Cl
0.3817E+00	Na_2SO_4
0.1000E+01	$Na_3H(SO_4)_2$
0.5094E+00	$NaHSO_4 \cdot H_2O$
0.2478E+00	$NaHSO_4$
0.3692E−01	$Na_2SO_4 \cdot (NH_4)_2SO_4 \cdot 4H_2O$
0.2774E+00	$NaNO_3$
0.1110E+00	$NaNO_3 \cdot Na_2SO_4 \cdot H_2O$
0.4971E+00	$NaCl$

Example 3

Aerosol Inorganics Model III

Arbitrary input data

Species	H^+	NH_4^+	Na^+	SO_4^{2-}	NO_3^-	Cl^-
Conc/mol kg^{-1}	2	1	1	1	1	1

Output Results as presented on the Screen

System pressure = 1.00000 atm
Volume = 1.000 m**3
T = 298.15 K
RH = 0.89353[pH$_2$O = 0.27926E−01 atm]
(The relative humidity above is the equilibrium value, equivalent to the liquid phase water activity)

****
** LIQUID PHASE **
****

Species	Moles	Grams	Molality	Mole Frac.	Act. Coeff.
H(aq)	0.13158E+01	0.1326E+01	0.2128E−01	0.1316E+01	0.1072E+01
NH$_4$(aq)	0.10000E+01	0.1804E+02	0.1000E+01	0.1617E−01	0.4723E+00
Na(aq)	0.10000E+01	0.2299E+02	0.1000E+01	0.1617E−01	0.6967E+00
HSO$_4$(aq)	0.68417E+00	0.6641E+02	0.6842E+00	0.1107E−01	0.1080E+01
SO$_4$(aq)	0.31583E+00	0.3034E+02	0.3158E+00	0.5109E−02	0.1940E−01
NO$_3$(aq)	0.10000E+01	0.6201E+02	0.1000E+01	0.1617E−01	0.5712E+00
Cl(aq)	0.10000E+01	0.3545E+02	0.1000E+01	0.1617E−01	0.9369E+00
Br(aq)	0.00000E+00	0.0000E+00	0.0000E+00	0.0000E+00	−
H$_2$O(l)	0.55509E+02	0.1000E+04	0.5551E+02	0.8978E+00	0.9952E+00

```
*************
** GASES **
*************
```

No gases were equilibrated
– Calculated partial pressures (over liquid) –

Pressure (atm)	Species	Equiv. Moles
0.2470E−06	$HNO_3(g)$	0.1010E−04
0.5220E−06	$HCl(g)$	0.2134E−04
0.3141E−11	$NH_3(g)$	0.1284E−09
0.6027E−19	$H_2SO_4(g)$	0.2463E−17

```
*************
** SOLIDS **
*************
```

There are no solid phases
– Calculated saturation ratios of solids (>0.01 only) –

Sat. Ratio	Species
0.1781E−01	NH_4NO_3
0.2053E−01	NH_4Cl
0.1170E−01	$Na_2SO_4 \cdot 10H_2O$
0.2683E−01	$NaNO_3$
0.1375E−01	$NaCl$

Example 4

Aerosol Inorganics Model I
Figures 7.6–7.9 depict graphs as downloaded directly from the AIM program.
Figures 7.10 and 7.11 are based on eight graphs, each showing the variation with
temperature of a different single property, which were downloaded from the AIM
program. In each case the proportions of different components which were input
was arbitrary as these graphs were intended to show some of the features of the
program and not to correspond to proportions found in the atmosphere.

5   OTHER PROGRAMS AVAILABLE FOR PREDICTING EQUILIBRIA IN ELECTROLYTE SOLUTIONS

Numerous programs have been developed for predicting equilibria in solutions
containing electrolytes in systems of geological or industrial significance. These
include the ChemCad program [68], which includes a choice of either Pitzer or
MNRTL methods of calculating activities. The National Physical Laboratory has

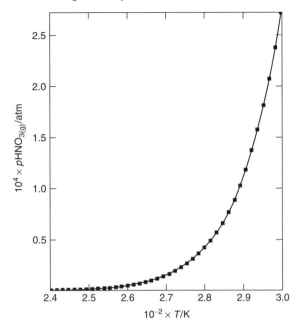

Figure 7.6 Variation of the vapour pressure of nitric acid over a solution containing 1 mol H_2SO_4, 1 mol HNO_3 and 0.5 mol HCl at a constant relative humidity of 0.4

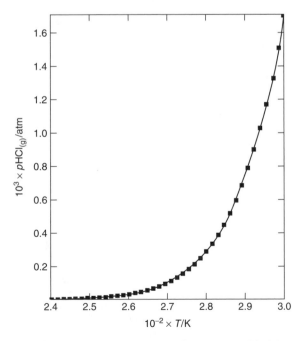

Figure 7.7 Variation of the vapour pressure of hydrogen chloride over a solution containing 1 mol H_2SO_4, 1 mol HNO_3 and 0.5 mol HCl at a constant relative humidity of 0.4

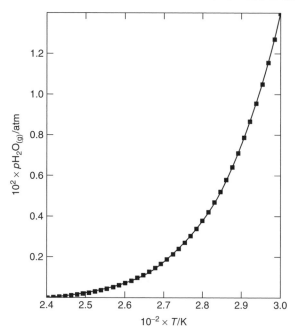

Figure 7.8 Variation of the vapour pressure of water over a solution containing 1 mol H_2SO_4, 1 mol HNO_3 and 0.5 mol HCl at a constant relative humidity of 0.4

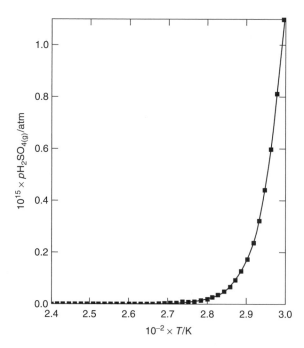

Figure 7.9 Variation of the vapour pressure of sulfuric acid over a solution containing 1 mol H_2SO_4, 1 mol HNO_3 and 0.5 mol HCl at a constant relative humidity of 0.4

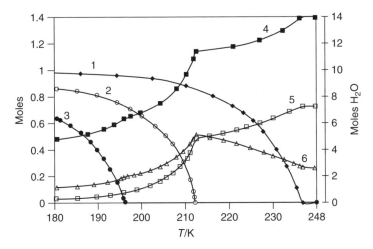

Figure 7.10 Variation with temperature of the quantities of various solid phases and of sulfate ions in the liquid phase in an arbitrary mixture of nitric acid (1 mol), sulfuric acid (1 mol), hydrogen chloride (0.5 mol) and water (14 mol) as predicted by Aerosol Inorganics Model I. 1, $HNO_3 \cdot 3H_2O$ (solid); 2, $H_2SO_4 \cdot 6 \cdot 5H_2O$ (solid); 3, H_2O (solid); 4, H_2O (liq.); 5, HSO_4^- (liq.); 6, SO_4^{2-} (liq.). The right-hand scale corresponds to the quantity of H_2O in the liquid phase and the left-hand scale to the quantities of other species

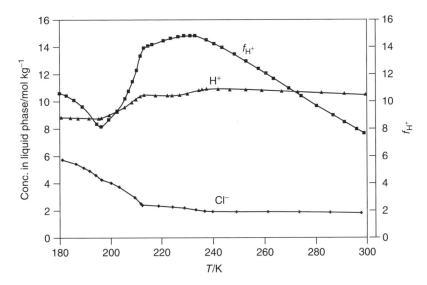

Figure 7.11 Variation with temperature of the concentrations of hydrogen ions and chloride ions and the variation of the activity coefficient of the hydrogen ions in an arbitrary mixture of nitric acid (1 mol), sulfuric acid (1 mol), hydrogen chloride (0.5 mol) and water (14 mol) as predicted by Aerosol Inorganics Model I

developed the MTDATA program [69]. A number of programs for groundwater modelling have been developed by or are available from the US Geological Survey [70]. The PHRQPITZ program [71], which can be downloaded free of charge from this source, may be used to calculate partial pressures of carbon dioxide over aqueous solutions containing a range of ions. The SOLMINEQ.88 [72] program from the same source has been used to model the solubility of carbon dioxide, methane and nitrogen in natural brines [73].

REFERENCES

1. Warneck, P., in *Chemistry of Multiphase Atmospheric Systems*, Jaeschke, W. (ed.), Springer, Berlin, 1986, pp. 473–499.
2. Petrenchuk, O. P.; Selezneva, E. S., *J. Geophys. Res.*, 1970, **75**, 3629–34.
3. Oddie, B. C. V., *Q. J. R. Meteorol. Soc.*, 1962, **88**, 535–8.
4. Mrose, H., *Tellus*, 1966, **18**, 266–70.
5. Munger, J. W.; Jacob, D. J.; Waldmann, J. M.; Hoffmann, M. R., *J. Geophys. Res.*, 1983, **88**, 5109–21.
6. Clever, H. L. Solubility Data Series, Vol. 62, *Carbon Dioxide in Water and Aqueous Electrolyte Solutions*, Scharlin, P. (ed.), Oxford University Press, Oxford, 1996, p. 84.
7. Clever, H. L., *J. Chem. Eng. Data*, 1983, **28**, 340–3.
8. Hermann, C.; Dewes, I.; Schumpe, A., *Chem. Eng. Sci.*, 1995, **50**, 1673–75.
9. Schumpe, A., *Chem. Eng. Sci.*, 1993, **48**, 153–8.
10. Yasunishi, A.; Tsuji, M.; Sada, E., *Adv. Chem. Ser.*, 1979, **177**, 189–203.
11. Guggenheim, E. A., *Philos. Mag.*, 1935, **19**, 588–643.
12. Scatchard, G.; Rush, R. M.; Johnson, J. S., *J. Phys. Chem.*, 1970, **74**, 3786–96.
13. Hamer, W. J.; Wu, Y.-C., *Phys. Chem. Ref. Data*, 1972, **1**, 1047–99.
14. Rard, J. A.; Habenschuss, A.; Spedding, F. H., *J. Chem. Eng. Data*, 1976, 374–9.
15. Goldberg, R. N., *J. Phys. Chem. Ref. Data*, 1981, **10**, 671–764.
16. Kusik, C. L.; Meissner, H. P., *AIChE J. Symp. Ser.*, 1978, **173**, 14–20.
17. Harned, H. S.; Owen, B. B., *The Physical Chemistry of Electrolytic Solutions*, Reinhold, New York, 1958.
18. Tabazadeh, A.; Turco, R. P.; Jacobson, M. Z., *J. Geophys. Res.*, 1994, **99**, 12897–914.
19. Carslaw, K. S.; Peter, T.; Clegg, S., *Rev. Geophys.*, 1997, **35**, 125–54.
20. Vidal, J., *Chem. Eng. Sci.*, 1972, **33**, 787–93.
21. Vanderbeken, I.; Ye, S.; Bouyssière, B.; Carrier, H.; Xans, P., *High Temp. High Press.*, 1999, **31**, 653–63.
22. Pitzer, K. S., *J. Phys. Chem.*, 1973, **77**, 268.
23. Pitzer, K. S.; Mayorga, G., *J. Phys. Chem.*, 1973, **77**, 2300.
24. Pitzer, K. S.; Mayorga, G., *J. Solution Chem.*, 1974, **3**, 539.
25. Pitzer, K. S.; Kim, J. J., *J. Am. Chem. Soc.*, 1974, **96**, 5701.
26. Pitzer, K. S., *J. Solution Chem.*, 1975, **4**, 249.
27. Pitzer, K. S.; Silvester, L. F., *J. Solution Chem.*, 1976, **5**, 269.
28. Pitzer, K. S.; Roy, R. N.; Silvester, L. F., *J. Am. Chem. Soc.*, 1977, **99**, 4930.
29. Silvester, L. F.; Pitzer, K. S., *J. Phys. Chem.*, 1977, **81**, 1822.
30. Pitzer, K. S.; Peterson, J. R.; Silvester, L. F., *J. Solution Chem.*, 1978, **7**, 45.
31. Pitzer, K. S., *Acc. Chem. Res.*, 1977, **10**, 371.
32. Pitzer, K. S.; Simonson, J. M., *J. Phys. Chem.*, 1986, **90**, 3005–9.
33. Clegg, S. L.; Pitzer, K. S.; Brimblecombe, P., *J. Phys. Chem.*, 1992, **96**, 9470–9; 1994, **98**, 1368.
34. Clegg, S. L.; Brimblecombe, P., *Atmos. Environ.*, 1986, **20**, 2483–5.

35. Carslaw, K. S.; Clegg, S. L.; Brimblecombe, P., *J. Phys. Chem.*, 1995, **99**, 11557–74.
36. Massucci, M.; Clegg, S. L.; Brimblecombe, P., *J. Phys. Chem. A.*, 1999, **103**, 4209–26
38. Hanson, D. R.; Ravinshankara, A. R., *J. Phys. Chem.*, 1993, **97**, 12309–19.
38a. Zhang, R. Y.; Wooldridge, P. J.; Molina, M. J., *J. Phys. Chem.*, 1993, **97**, 8541–48.
39. Williams, L. R.; Golden, D. M.; Huestis, D. L., *J. Geophys. Res.*, 1995, **100**, 7329–35.
40. Clegg, S. L.; Brimblecombe, P., *J. Phys. Chem.*, 1989, **93**, 7237–48.
41. Clegg, S. L. .; Brimblecombe, P., *J.Phys. Chem.*, 1989, **93**, 7237–48.
42. Bates, R. G.; Pinching, G. D., *J. Res. Natl. Bur. Stand.*, 1949, **42**, 419.
43. Chen, C.; Britt, H. I.; Boston, J. F.; Evans, L. B., *AIChE J.*, 1979, **24**, 820.
44. Wagman, D. D.; Evans, W. H.; Parker, V. B.; Schumm, I. H.; Bailey, S. M.; Churney, K. L.; Nuttall, R. L., *J. Phys. Chem. Ref. Data*, 1982, **11**, 1.
45. Eigen, M. ,; Schoen, J., *Z. Phys. Chem.*, 1955, **3**, 126–30.
46. Clegg, S. L.; Brimblecombe, P.; Wexler, A. S., *J. Phys. Chem. A*, 1998, **102**, 2137–54.
47. Clegg, S. L.; Whitfield, M., *Geochim. Cosmochim. Acta*, 1995, **59**, 2403–21.
48. Stull, D. R.; Prophet, H., *JANAF Thermochemical Tables*, Report NSRDS-NBS-37, US Government Printing Office, Washington, DC, 1971.
49. Roux, A.; Musbally, G. M.; Perron, G.; Desnoyers, J. E.; Singh, P. P.; Woolley, E. M.; Hepler, L. G., *Can. J. Chem.*, 1978, **56**, 24–8.
50. Clegg, S. L.; Brimblecombe, P.; Wexler, A. S., *J. Phys. Chem. A*, 1998, **102**, 2155–71.
51. Huthwelker, T.; Peter, Th.; Luo, B. P.; Clegg, S. L.; Carslaw, K. S.; Brimblecombe, P., *J. Atmos. Chem.*, 1995, **21**, 81–95.
52. Houghton, G., *J. Chem. Phys.*, 1964, **40**, 1628–31.
53. Luo, B. P.; Clegg, S. L.; Peter, Th.; Muller, R.; Crutzen, P. J., 1994, *Geophys. Res. Lett.*, 1994, **21**, 49–52.
54. Ourisson, J.; Kastner, M., *Bull. Soc. Chim. Fr., Ser. 5*, 1939, **6**, 1307–11.
55. Holzwarth, G.; Balmer, R. G.; Soni, L., *Water Res.*, 1984, **18**, 1421–27.
56. Blatchley, E. R.; Johnson, R. W.; Alleman, J. E.; McCoy, W. F., *Water Res.*, 1992, **26**, 99–106.
57. Carslaw, K. S.; Peter, T.; Clegg, S. L., *Rev. Geophys.*, 1997, **35**, 125–154; *http://grytviken.leeds.ac.uk/%7Ecarslaw/papers/review.html*.
58. Steele, H. M.; Hamill, P., *J. Aerosol Sci.*, 1981, **12**, 517–28.
59. Lide, D. R. (ed.), *Handbook of Chemistry and Physics*, CRC Press, Boca Raton, FL, 1992.
60. Pilinis, C., *Global Nest Int. J.*, 1999, **1**, 5–13.
61. Beyer, K. D.; Seago, S. W.; Chang, H. Y.; Molina, M. J., *Geophys. Res. Lett.*, 1994, **21**, 871–4.
62. Tabazadeh, A.; Turco, R. P.; Jacobson, M. Z., *J. Geophys. Res.*, 1994, **99**, 12897–914.
63. Luo, B. P.; Carslaw, K. S.; Peter, Th.; Clegg, S. L., *Geophys. Res. Lett.*, 1995, **22**, 247–50.
64. Clegg, S. L.; Brimblecombe, P.; Wexler, A. S., http://www.hpc1.uea.ac.uk/~e770/aim.html.
65. Zdanovskii, A. B., *Tru. Solyanoi Lab. Akad. Nauk SSSR*, 1936, **6**, 1.
66. Stokes, R. H.; Robinson, R. A., *J. Phys. Chem.*, 1966, **70**, 2126–30.
67. Chen, H.; Sangster, J.; Teng, T. T.; Lenzi, F., *Can. J. Chem. Eng.*, 1973, **51**, 234–41.
68. http://www.chemcad.com.
69. http://www.npl.co.uk/npl/cmmt/mtdata/mtdata.htm.
70. http://wwwbrr.cr.usgs.gov/projects/GWC_coupled/.

71. Plummer, L. N.; Parkhurst, D. L.; Fleming, G. W.; Dunkle, S. A., *US Geological Survey Water Resources Investigations Report 88-4153*, US Geological Survey, Mento Park, CA, 1988.
72. Kharaka, Y. K.; Gunter, W. D.; Aggarwal, P. K.; Perkins, E. H.; DeBraal, J. D. , *US Geological Survey Water Resources Investigation Report 88-4227*, US Geological Survey, Menlo Park, CA, 1988.
73. Seibt, A.; Naumann, D.; Hoth, P., *Geothermie Report 99-1*, STR99/04, Geo Forschungs Zentrum, Postdam, 1999.

CHAPTER 8

Details and Measurement of Factors Determining the Uptake of Gases by Water

Peter Fogg
University of North London, UK (retired)

The state of the atmosphere is determined by many complex chemical and physical processes. During recent years much work has been carried out at various centres throughout the world towards greater understanding of the chemical changes in the troposphere both in the laboratory and in the field. There have been significant developments in techniques of measurement [1].

Chemical species in the gaseous phase of the atmosphere may take part in homogeneous reactions or in heterogeneous reactions involving cloud droplets or the surface of solids. The heterogeneous processes may involve interaction with water molecules to form hydrates or hydrolysis products. Alternatively, other chemical species in solution may be involved.

A parcel of air of realistic size for study of can rarely be considered to be at equilibrium or the gas phase to be homogeneous. Various kinetic processes have to be considered in the development of useful models. Many kinetic data are now available for chemical and physical changes of hundreds of species in the atmosphere. The IUPAC Subcommittee on Gas Kinetic Data Evaluation for Atmospheric Chemistry is preparing a very extensive database of evaluated rate constants, much of which has been published in hard copy form. Summaries are available on the Web [2].

These data will play an important part in the development of better numerical models of the behaviour of the atmosphere. The scope and limitations of currently available models have been discussed by Granier *et al.* [3] and Pilinis [4].

Chemicals in the Atmosphere – Solubility, Sources and Reactivity. Edited by P. G. T. Fogg and J. Sangster
© 2003 IUPAC   ISBN: 0-471-98651-8

1 THE UPTAKE OF GAS MOLECULES BY CONDENSED PHASES

A consideration of the rate at which various processes occur is essential for the understanding of cloud systems. Mass transfer of gases occurs across interfaces between gas and condensed phases. The fraction of collisions between free gas molecules and the interface which lead to molecules crossing the interface could influence the rate at which gases are absorbed or adsorbed.

Gas molecules must first diffuse to a liquid interface and collide with the interface. They may then rebound from the interface or reversibly pass into the interface. In the case of a liquid surface they may reversibly diffuse into the bulk of the liquid and, in many cases, react in the liquid phase. In the case of a solid surface adsorption may occur, which may be followed by subsequent reaction with other adsorbed species or with the solid surface.

A liquid phase in a cloud system contains water in different proportions depending on conditions of formation of the cloud. Various dissolved species are likely to be present at low concentrations in the case of clouds in the troposphere. At high altitudes the liquid phase is mostly concentrated sulfuric acid. One or more reaction processes may occur after dissolution of a species, i.e. solvation, ionization and chemical reaction to give a new species. It is convenient to divide the process of adsorption or absorption of the gas by droplets of the condensed phase into various stages. The overall uptake of a gas from the gas phase into a liquid phase involves the processes listed below which are described in greater detail in various recent publications [5–7]:

1. diffusion of the gas to the interface;
2. accommodation of the gas at the interface;
3. dissolution;
4. diffusion into the bulk of the liquid and/or reaction; reaction can also occur in the interfacial region.

Dissolution and accommodation at the interface are both reversible processes. Gas molecules can evaporate and diffuse back into the bulk of the gas phase.

The overall process is represented in Figure 8.1.

The overall rate of uptake depends on all these processes and a rigorous mathematical representation is complex [8–10]. However, a simplified mathematical

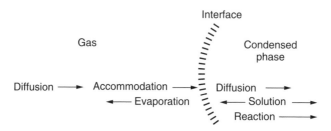

Figure 8.1 Representation of processes which occur when molecules in the gas phase dissolve and react in droplets of a liquid

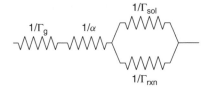

Figure 8.2   Resistances in series and parallel analogous to processes which occur when molecules in the gas phase dissolve and react in aqueous solution

model, the resistance model, useful in many situations, has been developed [8] (Figure 8.2). Various processes affecting the rate of uptake can be considered analogous to electrical resistances. In this model the differential equations describing the different processes are considered to be decoupled. It also assumes that a steady state is reached. The model is especially useful for situations in which chemical reactions occur after absorption.

In the case of a liquid condensed phase, the rate at which a gas is removed from the gas phase depends on

1. rate of diffusion to the surface dependent on the gas transport coefficient Γ_g;
2. rate of passage through the interface dependent on a coefficient α;
3. rate of diffusion into the bulk of the liquid dependent on a coefficient Γ_{sol};
4. rate of any subsequent reaction in the liquid phase dependent on a coefficient Γ_{rxn}.

Processes 3 and 4 take place simultaneously after 1 and 2, which take place sequentially. The reciprocals of Γ_g, α, Γ_{sol} and Γ_{rxn} can be thought of as resistances to the overall removal of gas from the gas phase. The overall dissolution process can be visualized as represented by four resistances connected together with two in series and two in parallel as in Figure 8.2. The overall rate of uptake of gas is dependent on a function γ_{meas} which can be measured experimentally. An equation analogous to that for a system of electrical resistances in series and parallel can be written:

$$\frac{1}{\gamma_{meas}} = \frac{1}{\Gamma_g} + \frac{1}{\alpha} + \frac{1}{\Gamma_{sol} + \Gamma_{rxn}} \tag{8.1}$$

If the gas is unreactive then the term Γ_{rxn} is omitted.

The mass accommodation coefficient, α, defines the rate at which molecules cross the interface, in one direction, from the gas to the condensed phase. It is defined as

$$\alpha = \frac{\text{number of gas molecules passing through the interface into the liquid phase}}{\text{number of gas molecules which collide with the interface with the liquid}}$$

It has also been called the *sticking coefficient*. The value of α lies between 0 and 1.

The uptake coefficient, γ_{meas}, is defined in a similar way to the mass accommodation coefficient but refers to the net rate of disappearance of gas from the gas phase. It is usual to omit the subscript and denote the quantity by γ. In the case of reactive gases it includes the loss due to chemical processes as well as physical solution with diffusion away from the interface:

$$\gamma_{meas} = \frac{\text{number of gas molecules removed by the condensed phase in unit time}}{\text{number of gas molecules striking the interface in unit time}}$$

The value of γ_{meas} also lies between 0 and 1. It is a quantity which may be measured.

The net flux of gas into the condensed phase, J, is given by

$$J = n_g \gamma_{meas} \bar{c}/4 \qquad (8.2)$$

where n_g is the gas concentration and $\bar{c}$ is the mean molecular velocity. If the units of n_g are molecules cm^{-3} and those of $\bar{c}$ are $cm\,s^{-1}$ then the units of J are molecules $cm^{-2}\,s^{-1}$. As defined, the value of γ_{meas} represents the probability that a molecule which strikes the surface will be absorbed or change by reaction. The coefficients Γ_g, Γ_{sol} and Γ_{rxn} are defined to represent rates rather than probabilities and can have values greater than unity.

The gas transport coefficient, Γ_g, is given approximately by the relationship

$$\frac{1}{\Gamma_g} = \frac{\bar{c}d}{8D_g} - \frac{1}{2} \qquad (8.3)$$

The condensed phase is assumed to be in the form of a sphere of diameter d. D_g is the gas diffusion coefficient and $\bar{c}$ the mean molecular velocity. The term $\frac{1}{2}$ is introduced to allow for the effect on the collision rate when there is net gas uptake at the surface. In many situations the diameter of the particles of condensed phase is large enough to make this second term negligible compared with the first term.

In the absence of chemical reaction to remove absorbed species, the ability of the condensed phase to dissolve a net amount of gas decreases with time until thermodynamic equilibrium is reached and the rate of desorption becomes equal to the rate of absorption. The solubility limited uptake coefficient, Γ_{sol}, varies with the time of exposure to the gas. In this model it is given by the relationship

$$\Gamma_{sol} = \frac{4k_H RT}{\pi^{\frac{1}{2}}\bar{c}} \left(\frac{D_l}{t}\right)^{\frac{1}{2}} \qquad (8.4)$$

where D_l = liquid phase diffusion coefficient of the dissolved substance
  $\bar{c}$ = mean molecular velocity in the gas phase
  k = pseudo-first-order reaction rate constant
  k_H = Henry's law constant with the units $mol\,dm^{-3}\,atm^{-1}$
  t = time during which the gas has been exposed to the liquid
  R = gas constant.

The coefficient Γ_{sol} is also a pure number but its value is not restricted to numbers less than one.

The reactive uptake coefficient, Γ_{rxn}, for a non-reversible reaction is given by the relationship

$$\Gamma_{rxn} = \frac{4k_H RT}{\bar{c}} (D_1 k_1)^{\frac{1}{2}} \tag{8.5}$$

where k_1 is the first-order rate constant of the reacting gas. In the case of a second-order reaction with another component X this term is replaced by $k_2[X]$, where k_2 is the second-order rate constant. Reaction occurs in a surface layer whose thickness is approximately $(D_1 k_1)^{\frac{1}{2}}$. The radius of the condensed phase must be greater than this if Equation (8.5) is to hold. This coefficient is also a pure number which can have any value.

In theory, the equations given above are applicable to absorption by solid condensed phases as well as liquids. However, diffusion coefficients for gases in solid phases are very much lower than for gases in liquids. In the former case it is of the order of 10^{-12} cm^2 s^{-1} and in the latter case 10^{-5} cm^2 s^{-1}. Of greater interest is the consideration of reactions in the atmosphere on the surface of solid particles such as ice, soot and dust. In these cases it is the overall rate of diffusion in the gas phase, adsorption and possible reaction on the surface which are important. These are discussed in Chapter 10.

There are various situations in which the expression for γ_{meas} may be simplified. When Γ_g and $(\Gamma_{sol} + \Gamma_{rxn})$ are all very large, the value of γ_{meas} can be taken to be equal to that of α.

If the solubility is low or the exposure time has been long so that the solution is saturated, then Γ_{sol} may be neglected and the equation for γ_{meas} becomes

$$\frac{1}{\gamma_{meas}} = \frac{1}{\Gamma_g} + \frac{1}{\alpha} + \frac{1}{\Gamma_{rxn}} \tag{8.6}$$

If Γ_g and α are known for a system, then Γ_{rxn} can be calculated from measured values of γ_{meas}. The rate constant k_1 can be calculated from Γ_{rxn} if k_H and D_1 are known. If only D_1 is known then the product $k_H k_1^{\frac{1}{2}}$ may be calculated.

If the rate of diffusion of gas to the surface of liquid is large, then $1/\Gamma_g$ may also be neglected.

If the solubility is low and there is insignificant reaction, then the term involving Γ_{rxn} can be neglected. The expression for γ_{meas} becomes

$$\frac{1}{\gamma_{meas}} = \frac{1}{\Gamma_g} + \frac{1}{\alpha} + \frac{1}{\Gamma_{sol}} \tag{8.7}$$

This is equivalent to

$$\frac{1}{\gamma_{meas}} = \frac{1}{\Gamma_g} + \frac{1}{\alpha} + \frac{\pi^{\frac{1}{2}} \bar{c}}{4k_H RT} \frac{[t]^{\frac{1}{2}}}{[D_1]} \tag{8.8}$$

A plot of $1/\gamma_{meas}$ against $t^{\frac{1}{2}}$ gives a straight line with slope proportional to $k_H \sqrt{D_1}$ and intercept of $1/\Gamma_g + 1/\alpha$.

Sometimes only one of the four stages discussed above is the significant rate-determining step and it is then possible to assume that

$$\gamma_{meas} \approx \Gamma_g \text{ or } \alpha \text{ or } \Gamma_{rxn} \text{ or } \Gamma_{sol}$$

The simple theory assuming a steady state and the corresponding equations discussed above are satisfactory for many practical purposes. More rigorous approaches have been published [5–7].

Shi *et al.* [11] have published a more comprehensive treatment of interfacial reactions which includes the formation of interfacial complexes. Taylor *et al.* [12] attempted to interpret the mass accommodation of ethanol by a water droplet. They used molecular dynamics computer simulations together with statistical mechanical perturbation theory. However, their calculated free energy barrier was not consistent with the experimentally observed mass accommodation coefficient. Aikin and Pesnell [13] have shown that account should be taken of the electronic charge on aerosol droplets. This effects the concentration of ions in the vicinity of the surface of a droplet.

Experimental values of uptake coefficients depend upon the exact conditions including the time that the surface under test has been exposed to the gas under test. Uptake coefficients have been published as part of the work of the IUPAC Subcommittee on Gas Kinetic Data Evaluation. These are available from a Web site [2]. The following symbols are used in the tables:

γ_{ss} = experimentally determined uptake coefficient obtained from measurements at a steady state when the uptake coefficient does not change with time on the time-scale of the experimental measurements.

γ_0 = experimentally determined uptake coefficient immediately after the surface is exposed to the gas under test.

This may be obtained by extrapolating back to zero time the experimental measurements over a period of time. This quantity is often taken to be close to the mass accommodation coefficient provided gas diffusion is not the rate limiting factor. Many authors, however, use this symbol to indicate values of the uptake coefficient extrapolated to zero concentration of trace gas under study.

The symbol γ without a subscript is used in the tables for the experimental values of the uptake coefficient corrected for gas-phase diffusion, liquid-phase diffusion and also the effects due to limitations of solubility expressed as a Henry's law constant.

Using the terminology used earlier in this chapter, this corrected value of an uptake coefficient is therefore given by the relationship

$$1/\gamma = 1/\alpha + 1/\Gamma_{rxn}$$

The uptake coefficient defined in this way is often called the reactive uptake coefficient and is not expected to vary with time.

If $\alpha \gg \Gamma_{rxn}$ then $\gamma \approx \Gamma_{rxn}$. Under these conditions, the measured value of γ is often referred to as the reactive uptake coefficient.

The IUPAC Committee on Gas Kinetic Data [2] also uses three other symbols for uptake coefficients:

γ_{obs} = the observed value of the uptake coefficient under defined conditions.
γ_{max} = the maximum value of the uptake coefficient under defined conditions.
γ_X = the reactive uptake coefficient of species X when more than one species is exposed to the surface under test.

Mass accommodation coefficients for interaction of a variety of gaseous species striking the surface of pure water, solutions of salts or solutions of acids have been measured by various methods. Values for gases striking solids have also been published. Available data from various sources have been published by Müller [14]. Experimental values range from 1×10^{-7} to unity. Most values are subject to appreciable uncertainty because of the difficulties in measurement.

2 EXPERIMENTAL METHODS OF MEASURING THE UPTAKE OF GASES

A wide variety of experimental techniques have been developed in recent years for investigating systems of significance. Some types of apparatus may be used for either absorption studies on liquid surfaces or adsorption studies on solid surfaces.

Some liquid surfaces become saturated with the gas under test with the uptake coefficient decreasing with time of exposure of a surface to the gas. Under these conditions, Equation (8.4) may be utilized to find a value of Henry's law constant k_H. This equation may also be applied to flat surfaces and to surfaces of droplets.

According to this equation, there will be a linear decrease in the measured uptake coefficient with a decrease in the value of $t^{-\frac{1}{2}}$ if the liquid tends to become saturated. The function t is the time of exposure of a surface of liquid to the gas under test.

The uptake of HCl on a supercooled aqueous solution containing 59.6 wt% sulfuric acid at 228 K investigated by Abbatt [15] shows this behaviour (see Figure 8.3). According to Equation (8.4), the slope of the line is equal to

$$\frac{4k_H RT D_l^{\frac{1}{2}}}{\pi^{\frac{1}{2}} \bar{c}}$$

from which $k_H D_l^{\frac{1}{2}}$ may be calculated. The diffusion coefficient in the liquid phase, D_l, can be estimated and hence values of Henry's law constant, k_H, can also be estimated. Within the limits of experimental error, linear extrapolation of the experimental points passes through the origin. This indicates the uptake tends to zero as the time increases and that reactive uptake is insignificant.

Uptake of HOBr under similar conditions depends on both the rate of solution and the rate of chemical reaction. Abbatt [15] found that the measured uptake coefficient initially varies linearly with $t^{-\frac{1}{2}}$ but an extrapolation of experimental data points does not pass through the origin (Figure 8.4). The measured uptake

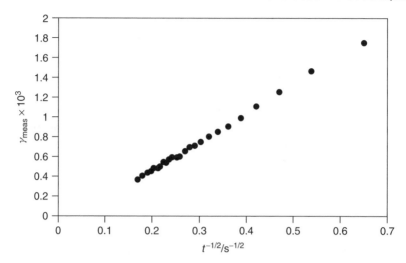

Figure 8.3 Variation with time, t, of the measured uptake coefficient of HCl on supercooled 59.6 mass% sulfuric acid film at 220 K

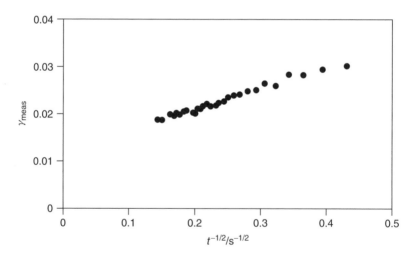

Figure 8.4 Variation with time, t, of the measured uptake coefficient of HOBr on supercooled 69.8 mass% sulfuric acid film at 228 K

tended to reach a constant finite value determined by the rate of chemical change of HOBr in the sulfuric acid solution. Abbatt suggested that this reaction was the following bimolecular reaction:

$$HOBr + HOBr \longrightarrow Br_2O + H_2O$$

In some types of apparatus a fresh layer of liquid is continuously exposed to the gas so that saturation of the surface layer is not evident.

Most of the techniques involve the following:

1. Measurement of the rate at which a trace gas under test is removed from the gas phase to the liquid phase. This usually requires analysis of the gas phase and measurement of the rate of relative mass movement of the liquid and gas phases.
2. Measurement or estimation of the surface area of the condensed phase. In the case of dissolution of gas in droplets of liquid this requires generation of droplets of uniform and measurable diameters.
3. Estimation of the rate of diffusion of gas to the liquid surface and calculation of the rate at which molecules under test reach the liquid surface.
4. Measurement of the time during which any part of the liquid phase is in contact with the gas under test.

All techniques lead to data with a wide margin of error.

2.1   Bubble Column

Shorter *et al.* have described a bubble column apparatus [16]. The way in which this apparatus was used gives some indication of the scope of other types of apparatus for measuring uptake. In this particular apparatus bubbles of carrier gas containing a known proportion of the gas under test pass through a column of liquid (Figure 8.5). The gas is analysed after it emerges from the liquid. Contact times can be varied by adjustment of the height of liquid through which the bubbles pass. The surface area of a bubble can be controlled by vibration of the orifice through which the gas enters the liquid. This technique gives Henry's law constants and is suitable for measuring low uptake coefficients between 10^{-4} and 10^{-8}.

When a substance in the gas phase does not react with water or reacts very slowly with water, the uptake is limited by the solubility of the substance absorbed. According to the model for the behaviour of the system the value of Γ_{sol} may be assumed to be given by the relationship

$$\Gamma_{sol} = \frac{8 k_H RT}{\pi^{\frac{1}{2}} \bar{c}} \left(\frac{D_l}{t} \right)^{\frac{1}{2}} \qquad (8.9)$$

The parameters are defined above after Equations (8.4) and (8.5).

In this type of apparatus, gas transport takes place from the inside of spherical bubbles to the liquid phase outside the bubble. This expression differs by a factor of two from the comparable expression for transport of gas into spherical liquid droplets.

When the substance reacts rapidly with water the model indicates that the uptake coefficient may be assumed to be given by

$$\Gamma_{rxn} = \frac{4 H RT}{\bar{c}} (D_l k_{rxn})^{\frac{1}{2}} \qquad (8.10)$$

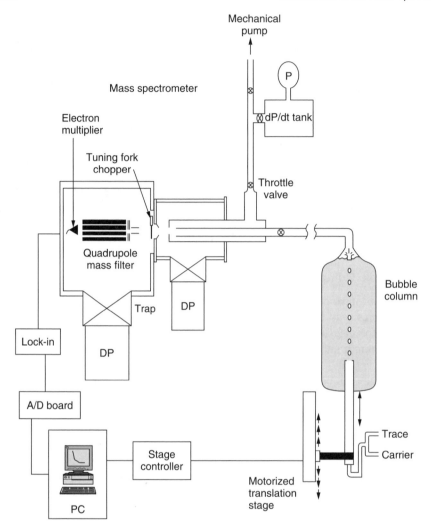

Figure 8.5 Schematic diagram of a bubble column apparatus described by shorter *et al.* [16] (reprinted with permission from J. A. Shorter *et al.*, *Environ. Sci. Technol.*, 1995, **29**, 1171–1178. Copyright 1995 American Chemical Society)

This is the same expression as that which applies for absorption of gas by droplets of water [Equation (8.4)].

The apparatus and model were tested with the following substances of known Henry's law constants which could be considered to have negligible reactivity: O_3, CH_3Br, H_2S, Cl_3Cl, CF_2ClH, COS, DMS and SO_2. The sulfur dioxide was tested at low pH to reduce hydrolysis. Values of Henry law constants using this technique agreed within about 20 % with measurements by other methods.

In the case of reactive species the bubble column gives values of $k_H k^{\frac{1}{2}}$. If a value of k_H is known from another source or can be estimated then k may be calculated. In the same way, a value of k_H may be found if a value of k is available. If neither of these options is available then it is possible to estimate the upper limit to the magnitude of k_H. The way in which this is done is to consider that both k_H and k are variable parameters rather than the true reaction rate constant and Henry's law constant for the system. It follows that

$$\log(k_H k^{\frac{1}{2}}) = \log k_H + \tfrac{1}{2} \log k$$

If the value of k is greater than about $100\,\mathrm{s}^{-1}$ the model predicts that Equation (8.2) becomes more applicable and that a plot of $\log k_H$ against $\log k$ tends to become linear. A maximum value of $k_H k^{\frac{1}{2}}$ can be estimated. At very low values of k, as k tends to zero Equation (8.1) tends to become valid. An estimate of a lower limit for values of k_H can also be made by assuming that k is about $10^4\,\mathrm{s}^{-1}$.

Shorter *et al.* [16] measured the reactive uptake of O_3 by a solution of a ferrous salt at concentration $0.06\,\mathrm{mol\,dm}^{-3}$. The plot of k_H against k based on the model and the experimental determination of the uptake coefficient is shown in Figure 8.6.

The value of k_H for very low values of k is about $0.23\,\mathrm{mol\,dm}^{-3}\,\mathrm{bar}^{-1}$. This gives a maximum value of k_H i.e. $k_H \leq 0.23\,\mathrm{mol\,dm}^{-3}\,\mathrm{bar}^{-1}$. However, experimental values for the system from other sources indicate that $k_H = 0.0112\,\mathrm{mol\,dm}^{-3}\,\mathrm{bar}^{-1}$ [17] and $k = 10\,200\,\mathrm{s}^{-1}$ [18].

Various corrections need to applied to the simple model characterized by the limiting equations above which were derived on the assumption that the system is stationary [19]. Each bubble is surrounded with a layer of liquid which moves with the bubble. Account has to be taken of the convective mixing which occurs in this layer as it moves through the bulk of the liquid. This depends on the speed

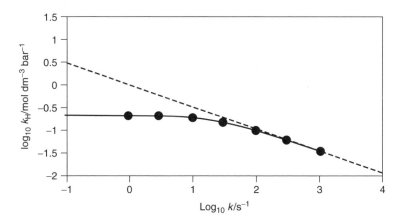

Figure 8.6   Reaction of O_3 with Fe^{2+} solution ($0.006\,\mathrm{mol\,dm}^{-3}$) investigated with a bubble train apparatus [16]

with which the bubble moves through the liquid which changes as the bubble moves towards the surface. The authors derived an empirical equation containing nine constants to make corrections due to convective mixing. Optimum values of these constant were obtained from measurements on gases for which Henry's law coefficients had been determined by other methods.

The authors state that the apparatus can measure Henry's law constants of nonreactive species in the range $0.01-2.0 \, mol \, dm^{-3} \, atm^{-1}$ to an accuracy of about 20%. For reactive species for which the Henry's law constant is known from other sources the pseudo-first-order rate constants in the range $50-10^5 \, s^{-1}$ can be measured to an accuracy of about 50%. The range of measurable $k_H k^{\frac{1}{2}}$ values is about $0.08-100 \, mol \, dm^{-3} \, atm^{-1} \, s^{-\frac{1}{2}}$.

2.2 Knudsen Cell

A Knudsen cell operates under conditions in which the mean free path of gaseous molecules is large enough to result in the majority of collisions are with the walls of the vessel or with a surface under test. The behaviour of gaseous molecules emerging through a small orifice into an evacuated vessel was studied by Knudsen in the early 1900s.

One form of the Knudsen cell reactor [9] consists essentially of two chambers separated by a valve as shown in Figure 8.7. In one of the chambers a solid or liquid surface can be prepared and cooled to low temperatures if stratospheric conditions need to be reproduced. The system operates at low pressures and gas under test can be introduced into the upper chamber. Traces of the this gas leak through a small orifice and are monitored by a mass spectrometer. When the valve between the chambers is opened the signal due to the gas under test is reduced due to absorption and perhaps reaction at the liquid or solid surface under test. Products of reaction can also be monitored by the mass spectrometer.

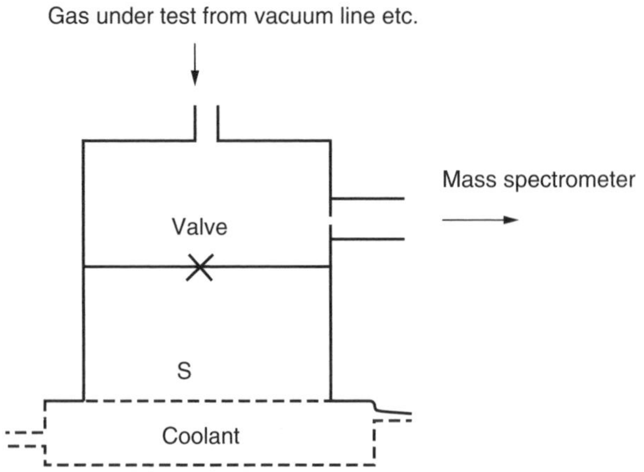

Figure 8.7 Schematic diagram of a form of Knudsen cell described by Golden *et al.* [9]

The rate at which molecules under test are lost to the surface under test is measured by the difference between the number of molecules per second reaching the mass spectrometer before the valve is opened (F_0) and the number per second after opening (F), i.e. $F_0 - F$.

Because diffusion effects due to gas–gas collisions are negligible the collision frequency with the surface under test can be calculated very accurately. The total collision frequency ω at very low pressures is given by the kinetic theory of gases as

$$\omega = (\bar{c}/4)A_s(N/V) \tag{8.11}$$

where $\bar{c}$ is the average molecular velocity, A_s the area of the surface under test and N/V the reactant gas number density, i.e. number of molecules per unit volume. In the case of a Knudsen cell the number density is related to the reactant flux F and the area of the escape aperture A_h by the equation

$$N/V = 4F/A_h\bar{c} \tag{8.12}$$

The uptake coefficient, γ_{meas}, is given by

$$\gamma_{meas} = \frac{\text{rate of loss of molecules from the gas phase to the surface}}{\text{rate of gas–surface collisions}} \tag{8.13}$$

Combining Equations (8.11)–(8.13) and remembering that the rate of loss to the surface is $F_0 - F$ gives

$$\gamma_{meas} = \frac{A_h}{A_s}\frac{F_0 - F}{F}$$

The signal for a particular molecule from the mass spectrometer is proportional to the flux from the Knudsen cell. The ratio of signals before and after opening the escape valve therefore gives F_0/F from which can be calculated $(F_0 - F)/F$ since $(F_0/F) - 1 = (F_0 - F)/F$.

The relative sizes of the escape aperture, A_h, and of the surface under test, A_s, may be varied to extend the range of uptake coefficients which may be measured. This range is between about 1×10^{-4} and 1.

The technique has certain disadvantages: it is not ideally suited to surfaces with a high vapour pressure because of the need to maintain low pressure in the apparatus, and the time-scale of several seconds may be a disadvantage.

Other forms of Knudsen cell [20] are essentially low-pressure flow reactors operating under molecular flow conditions (see Figure 8.8). There may be provision for covering the sample with a movable plunger during control experiments. There may also be provision for spectroscopic analysis of the gas port not shown in Figure 8.8. Not shown in the figure are provisions for maintaining low pressures and for heating or cooling the surface under test.

2.3 Wetted Wall Flow Tube

In this method [21,22], the inside wall of a vertical tube is completely covered with a slow moving film of the liquid under test. It is important to maintain laminar flow. Turbulent flow results in ripples which can affect the diffusion of gas

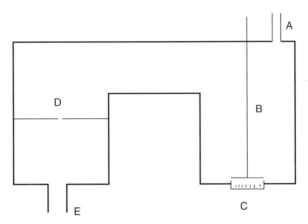

Figure 8.8 Schematic diagram of a Knudsen cell as used by Caloz *et al.* [20]. A = capillary inlet of gas under test; B = plunger to cover reaction chamber; C = reaction chamber containing liquid of solid surface under test; D = variable-size orifice; E = to mass spectrometer

to the liquid surface [5]. Carrier gas flows down the tube over the liquid surface. Samples of gas under test can be injected into the carrier gas stream. Changes in concentration of this gas at different carrier gas flow rates are monitored by optical methods or by mass spectrometry. The flow tube can be horizontal if the liquid film is stationary.

The uptake coefficient γ can be calculated for the rate of removal of the gas under test if a value of the gas-phase diffusion coefficient is known. The method is suitable for values of γ_{meas} from 10^{-6} to about 10^{-1}. An apparatus described by Rudich *et al.* [23] consisted of a vertical wetted wall flow tube, a moveable injector through which the gas under test could be introduced, a long-path optical absorption cell and a stabilized diode laser system. The reaction tube was 85 cm long with internal diameter 1.9 cm. Temperature and water vapour pressure equilibration were achieved in the upper part of the tube. Uptake coefficients were measured in the lower 25 cm of the tube where the injector tube could be moved through a distance of 18 cm. The rate of flow of the liquid through the tube could be controlled.

Similar flow tube apparatus can be used to measure the uptake of gas on films of ice or other solids.

2.4 Droplet Train Flow Tube

In this apparatus [24], a controlled train of droplets falls through a tube through which flows a stream of carrier gas containing traces of the gas under test (Figure 8.9). The total gas pressure is 6–20 Torr. Water vapour is added to the carrier gas as it enters the flow tube to prevent evaporation of the droplets as they fall and hence change in surface area. The uptake of the gas under test is monitored by optical methods or by mass spectrometry after exposure to the droplets. Droplets of uniform size and spacing are formed at a vibrating orifice.

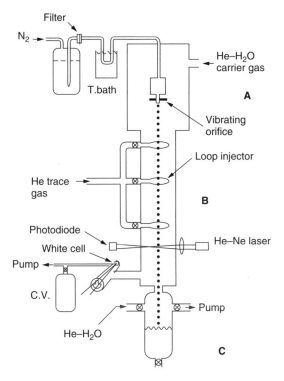

Figure 8.9 Schematic diagram of a droplet train flow tube apparatus [24]. A, Droplet generation region; B, droplet/gas interaction flow tube; C, droplet collection chamber; C. V., calibrated volume; T.bath, thermal bath (reprinted with permission from D. R. Worsnop *et al.*, *J. Phys. Chem.*, 1989, **93**, 1159–1172. Copyright 1989 American Chemical Society)

Change in the size and spacing changes the total surface area of liquid exposed to the gas. The uptake coefficient can be calculated from the change in the signal of the detector with change in surface area. The contact time between liquid surface and gas can be varied from about 2 to 20 ms, permitting measurements of change of uptake coefficients with time over a very small time-scale.

2.5 Bubble Train

The bubble train reactor [25] has been developed from the bubble column reactor. In this case bubbles pass horizontally through the liquid as shown in Figure 8.10. The advantage over bubble columns is that there is better control over the size and behaviour of the bubbles and it is applicable for a greater range of contact times.

2.6 Entrained Aerosol Flow Tube

In this technique, an aerosol of known properties is injected into a stream of carrier gas containing traces of the gas under test. Absorption is monitored

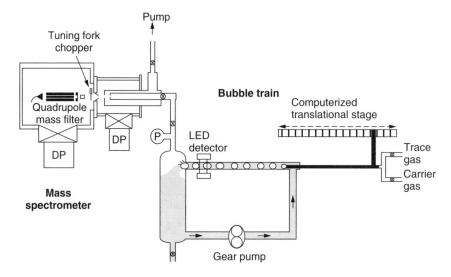

Figure 8.10 Schematic diagram of a bubble train apparatus described by Kolb *et al*. [25] (reproduced from C. E. Kolb *et al*., *Pure Appl. Chem*., 1997, **69**, 959–968, by permission of The International Union of Pure and Applied Chemistry)

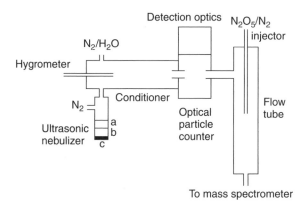

Figure 8.11 Schematic diagram of the entrained aerosol flow tube apparatus used by Hu and Abbatt [29]. a, Solution under test; b, coupling medium; c, piezoelectric vibrator

either by analysis of the aerosol or of the gas stream emerging from the reactor tube. The method has been used to measure uptake of NH_3 by sulfuric acid aerosols [26–28]. Another use has been to measure the uptake of HO_2 and N_2O_5 by nitrate, bisulfate and sulfuric acid aerosols [16,22,24]. A schematic diagram of apparatus used by Hu and Abbatt [29] for investigation of the hydrolysis of N_2O_5 on sulfuric acid and on ammonium sulfate aerosols is shown in Figure 8.11.

2.7   Aerosol Chamber Techniques

In this technique, aerosols are introduced into a chamber containing the trace gas under test. The removal of the trace gas is monitored. The uptake of N_2O_5 on NaCl aerosols and that of Cl_2CO on $NaNO_3$ aerosols have been investigated by this technique [30–32]. Behnke *et al.* used this technique to investigate reaction of N_2O_5 on NaCl solutions [33,34].

2.8   Liquid Jet

Mass accommodation and uptake coefficients can be found from the amount of gas absorbed by a liquid jet passing through a gas under test. Such a technique has been reported by Carstens and Schurath [35].

2.9   Impinging Flow

An impinging flow method has been described by Shimono and co-workers. A stream of carrier gas containing the gas under test impinged coaxially on to a stream of water. The variation of the concentration of the gas under test in the vicinity of the surface of the water was monitored by optical methods [36–38] (see also Section 9.1).

2.10   Cavity Resonances in Microparticles

Reid has reported that the chemistry of microdrops can be investigated by making use of cavity resonances [39]. The chemical composition of a droplet can be probed at different distances from its surface by using laser light of different wavelengths. This gives a method of investigating mass accommodation and chemical reaction in the interfacial region.

3   THE MASS ACCOMMODATION OF WATER VAPOUR ON PURE WATER AND AQUEOUS SOLUTIONS

If the mass accommodation coefficient α is very small then its value can significantly effect the value of γ_{meas} in Equation (8.1) and small changes in α can cause large changes in γ_{meas}. The effect is more marked the smaller are the water droplets.

There has been much interest in the mass accommodation coefficient for condensation of water vapour on water droplets because of its significance in the modelling of cloud formation. In general models are very sensitive to the exact assignment below 0.1 of a numerical value to α [40,41].

According to Li *et al.* [40], measurements from 1925 to 1985 have ranged between about 0.001 and 1. Values reported between 1987 and 2000 have ranged between 0.01 and 1. More recently, Li *et al.* determined values from the uptake of water in the gas phase which was labelled with ^{17}O. Conditions were near equilibrium. A droplet train flow reactor apparatus was used. The data are very consistent

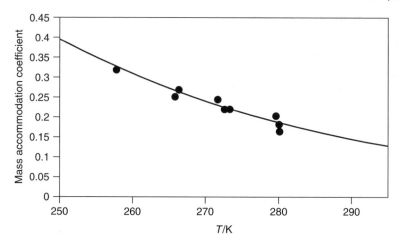

Figure 8.12 The mass accommodation coefficient of water vapor on liquid water as a function of temperature. The heavy line corresponds to the model used by the authors [40]

and show a decease of the mass accommodation coefficient with increase in temperature. The value is 0.32 ± 0.04 at 258 K and 0.17 ± 0.03 at 280 K.

The variation of the mass accommodation coefficient with temperature is shown in Figure 8.12.

As shown by Davidovits *et al.* [42], the mass accommodation coefficient can be expressed as

$$\frac{\alpha}{1 - \alpha} = \exp\left(\frac{-\Delta G_t}{RT}\right)$$

where ΔG_t is the free energy of formation of the transition state when a molecule of water is absorbed from the gas phase. The solid curve is consistent with the values of thermodynamic functions for the formation of a transition state of $\Delta H_t = -20.1 \pm 2.1 \text{ kJ mol}^{-1}$ and $\Delta S_t = 84.9 \pm 7.5 \text{ J mol}^{-1} \text{ K}^{-1}$.

Measurement of the mass accommodation coefficient of water vapour by sulfuric acid solutions by the same technique has been reported by Gershenzon *et al.* [43]. This was also carried out using isotopically labelled water vapour in a droplet flow apparatus.

Measurements were made in the temperature range 252–295 K and at acid concentrations of 50 and 70 mass% H_2SO_4. The mass accommodation coefficient was again found to decrease with increase in temperature. For 50 mass% H_2SO_4 the coefficient ranged from 0.40 ± 0.05 at 278 K to 0.55 ± 0.06 at 250 K. For 70 mass% H_2SO_4 the values ranged from 0.54 ± 0.03 at 295 K to 0.76 ± 0.05 at 252 K. The authors stated that the maximum rate of gas to liquid mass transport in the atmosphere is limited by mass accommodation if water droplets are less than about 0.5 μm in diameter.

Mass accommodation coefficients for interaction of a variety of gaseous species striking the surface of pure water, solutions of salts or solutions of acids have been measured by several different methods. Values for gases striking solids have also been published. Available data from various sources has been published by Müller [14]. Experimental values range from 1×10^{-7} to unity. Most values are subject to appreciable uncertainty because of the difficulties in measurement. Evaluated values are tabulated in JPL Publication 00-003 [44]. Selected data for interaction of gases with aqueous systems are given in Table 9.40.

REFERENCES

1. Tyndall, G. S.; Winker, D.; Anderson, T.; Eisele, F., *Advances in Measurement Methods for Atmospheric Chemistry*, Chapt. 8, http://medias.obs-mip.fr:8000/igac/html/book/chap8/chap8.html, 2000.
2. IUPAC Subcommittee for Gas Kinetic Data Evaluation, *Evaluated Kinetic Data*, http/:www.iupac-kinetic.ch.cam.ac.uk/, 2001.
3. Granier, C., *et al.*, Draft document, October 18, 2000, www-as.harvard.edu/chemistry/trop/publications/jacob1999/text.html.
4. Pilinis, C., *Global Nest Int. J.*, 1999, **1**, 5–13.
5. Worsnop, D. R.; Zahniser, M. S.; Kolb, C. E. ; Davidovits, P.; Keyser, L. F.; Leu, M.-T.; Molina, M. J.; Hanson, D. R.; Ravishankara, A. R.; Williams, L. R.; Tolbert, M. A., in *Progress and Problems in Atmospheric Chemistry*, Barker, J. R. (ed.), World Scientific, Singapore, 1995, pp. 771–875.
6. Molina, M. J.; Kolb, C. E., *Annu. Rev. Phys. Chem.*, 1996, **47**, 327–67.
7. Hanson, D. R.; Ravishankara, A. R.; Solomon, S., *J. Geophys. Res.*, 1994, **99**, 3625–9.
8. Danckwerts, P. V., *Trans. Faraday Soc.*, 1951, **47**, 1014–23.
9. Golden, D. M.; Spokes, G. N.; Benson, S. W., *Angew. Chem., Int. Ed. Engl.*, 1976, **12**, 534.
10. Schwartz, S. E., in *Chemistry of Multiphase Atmospheric Systems*, Jaeschke, W. (ed.), Proceedings of the NATO Advanced Study Institute on Chemistry of Multiphase Atmospheric Systems, Corfu, Greece, Springer, Berlin, 1983, pp. 415–71.
11. Shi, Q.; Davidovits, P.; Jayne, J. T.; Worsnop, D. R.; Kolb, C. E., *J. Phys. Chem. A*, 1999, **103**, 8812–23.
12. Taylor, R. S.; Ray, D.; Garrett, B. C., *J. Phys. Chem. B*, 1997, **101**, 5473–6.
13. Aikin, A. C.; Pesnell, W. D., *Geophys. Res. Lett.*, 1998, **25**, 1309–12.
14. Müller, F., www.mi.uni-hamburg.de/technische_meteorologie/Meso/homepages/fmueller/mass_acc.html.
15. Abbatt, J. P. D., *J. Geophys. Res.*, 1995, **100**, 14009–17.
16. Shorter, J. A.; De Bruyn, W. J.; Hu, J. H.: Swartz, E.; Davidovits, P.; Worsnop, D. R.; Zahniser, M. S.; Kolb, C. E., *Environ. Sci. Technol.*, 1995, **29**, 1171–8.
17. Kosak-Channing, L. F.; Helz, G. R., *Environ. Sci. Technol.*, 1983, **17**, 154.
18. Hoigne, J.; Bader, H.; Haag, W. R.; Staehelin, J., *Water Res.*, 1985, **19**, 993.
19. De Bruyn, W. J., *Thesis*, Boston College, 1994.
20. Caloz, F.; Fenter, F. F.; Tabor, K. D.; Rossi, M. J., *Rev. Sci. Instrum.*, 1997, **68**, 3172.
21. Howard, C. J., *J. Phys. Chem.*, 1979, **83**, 3.
22. Brown, R. L., *J. Res. Natl. Bur. Stand. U.S.*, 1978, **83**, 1.
23. Rudich, Y.; Talukdar, R. K.; Ravishankara, A. R.; Fox, R. W., *J. Geophys. Res.*, 1996, **101**, 21023–31.
24. Worsnop, D. R.; Zahniser, M. S.; Kolb, C. E.; Gardner, J. A.; Watson, L. R.; Van Doren, J. M.; Jayne, J. T.; Davidovits, P., *J. Phys. Chem.*, 1989, **93**, 1159–72.
25. Kolb, C. E.; Jayne, J. T.; Worsnop, D. R.; Davidovits, P., *Pure Appl. Chem.*, 1997, **69**, 959–68.
26. Robbins, R. C.; Cadle, R. D., *J. Phys. Chem.*, 1958, **62**, 469–71.
27. Huntzicker, J. J.; Cary, R. A.; Ling, C.-S., *Environ. Sci. Technol.*, 1980, **14**, 819.
28. McMurry, P. H.; Tahano, H.; Anderson, G. R., *Environ. Sci. Technol.*, 1983, **17**, 347.

29. Hu, J. H.; Abbatt, J. P. D., *J. Phys. Chem. A*, 1997, **101**, 871–8.
30. Mozurkewich, M.; McMurry, P. H.; Gupta, A.; Calvert, J. G., *J. Geophys. Res.*, 1987, **92**, 4163–4170.
31. Mozurkewich, M.; Calvert, J. G., *J. Geophys. Res.*, 1988, **93**, 15889.
32. Fried, A.; Henry, B. E.; Calvert, J. G.; Mozurkewich, M., *J. Geophys. Res.*, 1994, **99**, 3517–32.
33. Behnke, W.; George, C.; Scheer, V.; Zetzsch, C., *J.Geophys. Res.*, 1997, **102D**, 3795–804.
34. Behnke, W.; Krüger, H.-V.; Scheer, V.; Zetzsch, C., *J. Aerosol Sci.*, 1991, **S22**, 609–12.
35. Carstens, T.; Schurath, U., *Report by AIDA Information Service on Research Related to Aerosols and Heterogeneous Atmospheric Chemistry*, http://imk-aida.fzk.de/abstract/abs_etma.html, 1997.
36. Shimono, A.; Koda, S., *J. Chem. Eng. Jpn.*, 1995, **28**, 779.
37. Shimono, A.; Koda, S., *J. Phys. Chem.*, 1996, **100**, 10269.
38. Takami, A.; Kato, S.; Shimono, A.; Koda, S., *Chem. Phys.*, 1998, **231**, 215–227.
39. Reid, J., www.che.bham.ac.uk/publicsite/staff/staff%20details/reid/reid.htm, 2001.
40. Li, Y. Q.; Davidovits, P.; Shi, Q.; Jayne, J. T.; Kolb, C. E.; Worsnop, D. R., *J. Phys. Chem.*, 2001, **47**, 10627–34.
41. Fukuta, N.; Walter, L. A., *J. Atmos. Sci.*, 1970, **27**, 1160.
42. Davidovits, P.; Jayne, J. T., Duan, S. X.; Worsnop, D. R.; Zahniser, M. S.; Kolb, C. E., *J. Phys. Chem.*, 1991, **95**, 6337.
43. Gershenzon, M.; Davidovits, P.; Williams, L. R.; Shi, Q.; Jayne, J. T.; Kolb, C. E.; Worsnop, D. R., presented at the XXVII General Assembly of the European Geophysical Society, 21–26 April, 2002; www.cosis.net/abstracts/EGS02/00764/EGS02-A-00764–4.pdf.
44. Sander, S. P.; Friedl, R. R.; DeMore, W. B.; Ravishankara, A. R.; Golden, D. M.; Kolb, C. E.; Kurylo, M. J.; Hampson, R. F.; Huie, R. E.; Molina, M. J.; Moortgat, G. K., *JPL Publication 00-003: Chemical Kinetics and Photochemical Data for Use in Stratospheric Modeling. Evaluation Number 13. Supplement to Evaluation 12: Update of Key Reactions*, National Aeronautics and Space Administration, Jet Propulsion Laboratory, California Institute of Technology, Pasadena, CA, 2000, pp. 57–62.

CHAPTER 9

Accommodation Coefficients, Uptake Coefficients and Henry's Law Constants of Gases which React with Water or are Unstable

Peter Fogg
University of North London, UK (retired)

Gases in the atmosphere may undergo various changes in contact with the surface of pure water or aqueous solutions. In this chapter, such changes of significance to the study of atmospheric chemistry are summarized. Parameters from experimental measurements of absorption are tabulated. Chemical Abstracts Service (CAS) registry numbers are given.

1 OXYGEN, OZONE, HOx COMPOUNDS AND CARBON DIOXIDE
Oxygen, O_2, [CAS 7782-44-7]

Solubility data for oxygen in water are available (see Section 11.10). No measurements of the accommodation coefficient are available to the author. Oxygen gas, water vapour and carbon dioxide are involved in processes which lead to the fractionation of oxygen isotopes. These processes have been extensively studied because of the relationship between this fractionation and paleoclimatic changes (see Appendix II).

Ozone, O_3, [CAS 10028-15-6]

Utter *et al.* [1] investigated the uptake of ozone by water at 276 K. A wetted wall flow reactor with a flowing film of water of 0.2 mm thickness was used. The ozone concentration was 1011 molecules cm^{-3}. No reactive uptake was observed when pure deionized water was used. However, reactive uptake was measurable when

Chemicals in the Atmosphere – Solubility, Sources and Reactivity.   Edited by P. G. T. Fogg and J. Sangster
© 2003 IUPAC   ISBN: 0-471-98651-8

the water contained a reducing agent such as sodium sulfite or was acidified. The accommodation coefficient was reported to be greater than 2×10^{-2}. Magi et al. [2] were also unable to detect reactive uptake by pure water using a droplet tube apparatus.

A detailed evaluation of the measurements of the solubility of ozone in water by Peter Warneck is appended at the end of this chapter (Addendum II).

Hydroxyl Radicals, OH, [CAS 3352-57-6]

The uptake coefficient of OH radicals on water surface was measured by Takami et al. [3] using an impinging flow method with laser-induced fluorescence for measurement of concentrations of hydroxyl radicals. In this method, a stream of helium containing hydroxyl radicals and water vapour impinged coaxially on to a stream of water. The hydroxyl radicals were initially generated by the action of microwaves on the water vapour in the helium stream. The concentration of hydroxyl radicals at various distances from the water surface was found by measuring the fluorescence induced by a pulsed-dye laser. The uptake coefficient was experimentally determined to be $(4.2 \pm 2.8) \times 10^{-3}$ under the conditions of a contact time of 115 ms, pH 5.6 and a temperature of 293 K. The uptake coefficient was found to be larger in acidic and alkaline regions than in the neutral region. Negative temperature dependence was also observed. The measured uptake coefficient was fairly well reproduced by numerical simulation taking account of the known reactions of OH radicals in water and the Henry's law constant. From the simulation, the accommodation coefficient was estimated to be > 0.01 and was possibly close to unity.

A detailed discussion of estimations of the solubility of OH and of HO_2 by Yin-Nan Lee is appended at the end of this chapter (Addendum I).

Hydrogen Peroxide, H₂O₂, [CAS 7722-84-1]

Lind and Kok [4] measured partial pressures of H_2O_2 in an airstream which was in equilibrium with a solution of known concentration. Data fitted the equation

$$\ln k_{\mathrm{H}}/(\mathrm{mol\,dm^{-3}\,bar^{-1}}) = 6338/(T/\mathrm{K}) - 9.75$$

where k_{H} is a Henry's law type constant. Values of k_{H} were stated to be accurate to $\pm 9.74\,\%$.

A similar method was used by O'Sullivan et al. [5] A stream of air was passed through a solution of hydrogen peroxide and then through collection coils through which a stream of water was flowing. The solution flowing from the collection coils was analysed by high-performance liquid chromatography. Measurements by O'Sullivan et al. are shown in Table 9.1.

A summary of data from various sources is given in Table 9.2.

Curves fitting the various sets of data are shown in Figure 9.1.

Staffelbach and Kok [6] reported that the solubility of hydrogen peroxide depended on pH. This was not substantiated by the work of O'Sullivan et al. [7],

Table 9.1 Values of a Henry's law constant published by O'Sullivan et al. [5]

T/K	k_H/mol dm^{-3} bar^{-1}	Std deviation	No. of determinations
278.15	4.92×10^5	1.48×10^4	35
291.15	1.46×10^5	5.02×10^3	32
297.15	9.08×10^4	1.52×10^3	22

Table 9.2 Solubility data for hydrogen peroxide in water from various sources as quoted by O'Sullivan et al. [5][a]

Author	T/K	$c_{H_2O_2}$/ 10^{-3} mol dm^{-3}	A	B	$\Delta H°$/ kJmol^{-1}	$-\Delta G°$/ kJmol^{-1}	Ln(k_H/ mol dm^{-3} atm^{-1}) (25 °C)	Ln(k_H/ mol dm^{-3} bar^{-1}) (25 °C)
Schumb [7]			6990	12.28	58.1	27.7	11.17	11.16
Hwang and Dasgupta [8]			7920	15.45	65.8	27.6	11.11	11.11
Lind and Kok [4]			6338	9.75	52.7	28.6	11.52	11.51
Staffelbach and Kok [6]	278.2– 293.2	6–9	7514	13.64	62.5	28.7	11.57	11.56
NBS					54.9	28.4	11.46	11.45
O'Sullivan [5]	278.2– 301.2		7379	13.43	61.3	28.1	11.33	11.32

[a] $\Delta H°$ is the standard enthalpy change and $\Delta G°$ the standard Gibbs free energy change for the dissolution of 1 mol of H_2O_2. A and B are constants in equations of the type

$$\ln k_H/\text{mol dm}^{-3}\,\text{bar}^{-1} = A/(T/K) - B$$

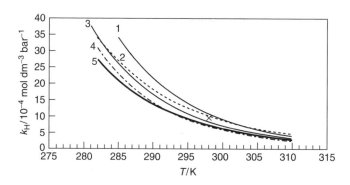

Figure 9.1 Solubility of hydrogen peroxide in water. 1, Staffelbach and Kok [6]; 2, Lind and Kok [4]; 3, O'Sullivan et al. [5]; 4, Schumb et al. [7]; 5, Hwang and Dasgupta [8]. A calculated value from National Bureau of Standards data quoted by O'Sullivan et al. is marked with a cross

who explained changes in solubility with concentration of sulfuric acid as an effect due to change in the activity of water. Staffelbach and Kok also reported an increase in solubility with increase in concentration of ammonium sulfate. This was attributed to an association between ammonia or ammonium ions and hydrogen peroxide.

Carbon Dioxide, CO_2, [CAS 124-38-9]

When carbon dioxide dissolves in water it exists in ionic equilibrium with hydrogen ions and bicarbonate ions:

$$CO_2 + H_2O \rightleftharpoons H^+ + HCO_3^-$$

$pK_a = 6.352$ at 298.15 K [8a], where K_a is the equilibrium constant for the above reaction.

Only about 0.35 % of CO_2 dissolved in water exists in the form of carbonic acid, H_2CO_3, at 298.2 K and a partial pressure of CO_2 of 1.013×10^5 Pa. Carbonic acid is a stronger acid than is indicated by the above pK_a value. The equilibrium constant for the reaction

$$H_2CO_3 \rightleftharpoons H^+ + HCO_3^-$$

is 1.32×10^{-4} at 298.2 K and the true pK_a for carbonic acid is therefore 3.88 [9].

Carbon dioxide undergoes oxygen isotope exchange when dissolved in water (see Appendix II).

2 C, H, O COMPOUNDS–FORMALDEHYDE, ACETALDEHYDE AND ACETONE

In the case of all aldehydes, distinction must be made between the Henry's law relationship between aldehyde in the gas phase and unhydrated aldehyde in the liquid phase on the one hand and the relationship between aldehyde in the gas phase and total hydrated and unhydrated aldehyde in the liquid phase on the other. In the case of dilute solution the following equilibrium constants need to be considered:

$$k_H = [RCHO]_{aq}/[RCHO]_g$$

$$\text{(intrinsic Henry's law relationship)} \tag{9.1}$$

$$k_H{}^* = ([RCHO]_{aq} + [RCH(OH)_2])/[RCHO]_g$$

$$\text{(effective Henry's law relationship)} \tag{9.2}$$

$$RCHO + H_2O \rightleftharpoons RCH(OH)_2$$

$$K_{hyd} = [RCH(OH)_2]/[RCHO]_{aq} \tag{9.3}$$

$$[RCHO]_t = [RCHO]_{aq} + [RCH(OH)_2]$$

From Equations (9.1)–(9.3):

$$k_H{}^* = k_H(1 + K_{hyd})$$

If $K_{hyd} \gg 1$ then $k_H{}^* \simeq k_H K_{hyd}$.

Methanal, (Formaldehyde), CH_2O, [CAS 50-00-0]

Formaldehyde reacts reversibly with water to form methylene glycol and poly(oxymethylene) glycols. Chemical analysis of a solution of formaldehyde gives the quantity of formaldehyde equivalent to all the reaction products.

$$HCHO + H_2O \rightleftharpoons HOCH_2OH$$

$$HO(CH_2O)_{i-1}H + HOCH_2OH \rightleftharpoons HO(CH_2O)_iH + H_2O, \text{ etc.}$$

Betterton and Hoffmann [10] used a bubble column apparatus to determine the solubility of formaldehyde in water and in $0.5 \, mol \, dm^{-3}$ sulfuric acid (Table 9.3).

Swartz *et al.* [11] used a horizontal bubble train apparatus to measure values of $k_H k^{1/2}$ at various pH values, where k is the pseudo-first-order rate constant for hydrolysis of formaldehyde. The hydrolysis is acid–base catalysed. Data for 293 K were consistent with the following equation based on the equilibrium constant given by Valenta [12] and the equation for the reverse reaction published by Bell and Evans [13]:

$$k/s^{-1} = 2.09 \times 10^{-1}[H_2O] + 6.14 \times 10^3[H^+] + 3.64 \times 10^6[OH^-]$$

Swartz *et al.* [11] also published a value of the Setchenov coefficient for NaCl solution, K_s, corresponding to the equation

$$\log_{10} k_H{}^0 / k_H = K_s M$$

where $K_s = 0.072 \pm 0.004 \, mol^{-1} \, dm^3$ at 293 K.

Jayne *et al.* [14] showed that the uptake of formaldehyde at pH 0–2 was larger than could be explained just by formation of the diols as indicated by the equilibrium above. They postulated the formation of a protonated species at high acidity according to the equilibrium

$$CH_2O + H^+ \rightleftharpoons CH_3O^+$$

They also suggested that at lower acidities in this pH range and less than 20 mass% of acid the formation of a surface complex affected the uptake, i.e.

$$CH_2O(g) \rightleftharpoons CH_2O^*(\text{surf})$$

Table 9.3 Solubility of formaldehyde in water and dilute sulfuric acid reported by Betterton and Hoffmann [10]

$t/°C$	T/K	$k_H*/mol \, dm^{-3} \, bar^{-1}$	$\Delta H_{298}/kJ \, mol^{-1a}$	$[HCHO]_t/mol \, dm^{-3}$
15	288.2	$(7.21 \pm 0.37) \times 10^{-3}$	-59.8 ± 1.8	1.1×10^{-3}
15	288.2	9.70×10^{-3b}		1.1×10^{-3}
25	298.2	$(2.93 \pm 0.69) \times 10^{-3}$		$(0.1-1.1) \times 10^{-2}$
35	308.2	$(1.48 \pm 0.07) \times 10^{-3}$		1.1×10^{-3}
45	318.2	$(6.64 \pm 1.89) \times 10^{-4c}$		1.1×10^{-3}

$^a \Delta H_{298}$ is the enthalpy change for the dissolution of formaldehyde in water at 298 K.
^{b} In $0.5 \, mol \, dm^{-3}$ H_2SO_4.
^{c} Given erroneously in the original paper as $(6.73 \pm 1.89) \times 10^{-3}$.

These authors published uptake coefficients as a function of gas–droplet contact time in graphical form. They also published a table of diffusion coefficients in the liquid phase and uptake coefficients at a contact time of 8 ms for various concentrations of sulfuric and/or nitric acid.

Albert *et al.* [15] measured equilibria between liquid and vapour phases in the system consisting of water, formaldehyde and the products of hydrolysis of formaldehyde. Measurements were made over a very wide concentration range. The measurements involved analysis of co-existing gas and liquid phase at various temperatures using the sodium sulfite method developed by Walker. They also were able to identify ^{13}C NMR peaks due to methylene glycol and poly(oxymethylene) glycols containing up to four formaldehyde units. Estimation of concentrations of each of these compounds could be made from measurements of peak areas. They also developed a new model to fit experimental variation of solubility with changes in temperature and pressure. Further, earlier models developed by different workers were reviewed by Albert *et al.* The curve corresponding to this latest model for variation with temperature of the partition of formaldehyde as HCHO between gas and liquid phases at infinite dilution is shown in Figure 9.2. Experimental data for the partition based on various sources [15–19] are also plotted.

Ethanal, (Acetaldehyde), C$_2$H$_4$O, [CAS 75-07-0]

Acetaldehyde is partially converted to the *gem*-diol in aqueous solution. The hydration constant, $[CH_3CH(OH)_2]/[(CH_3)CHO]$, is about 2×10^{-3} at

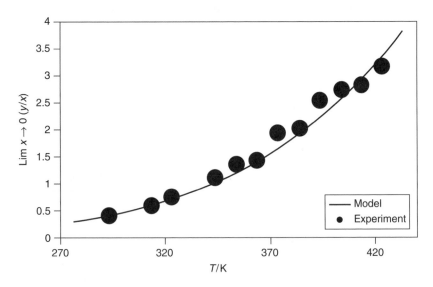

Figure 9.2 Variation with temperature of the limiting value of the partition coefficient of formaldehyde. x = mole fraction of formaldehyde in the liquid phase; y = mole fraction of formaldehyde in the vapour phase

298 K [20]. Equilibrium measurements of the Henry's law constant give the effective Henry's law constant. This was measured by Benkelberg *et al.* [21] by analysing the composition of the vapour over solutions of acetaldehyde in deionized water over the temperature range 283–313 K (Table 9.4). Mole fractions of acetaldehyde in the gas phase ranged from 3×10^{-6} to 5×10^{-3}. Replacement of the deionized water by rain water caused a negligible change in the solubility. Data obtained in this work are substantiated by earlier measurements by Snider and Dawson [22] and by Betterton and Hoffmann [10]. Measurements by Zhou and Mopper [23] are not consistent with measurements by the other authors (see Figure 9.3).

The effective Henry's law constant, k_H^*, is defined by the relationship

$$k_H^* = p/x$$

where x is the mole fraction of acetaldehyde plus diol in solution and p is the partial vapour pressure of acetaldehyde.

The experimental data from Benkelberg *et al.*, from Snider and Dawson and from Betterton and Hoffman fit the relationship published by Benkelberg *et al.*:

$$\ln(k_H^*/\text{bar}) = -(5671 \pm 22)/(T/\text{K}) + (20.4 \pm 0.1)$$

Data published by Zhou and Mopper are inconsistent with data from the other sources.

Benkelberg *et al.* also measured the solubility in synthetic seawater (see Table 9.4).

2-Propanone (Acetone), C$_3$H$_6$O, [CAS 67-64-1]

2-Propanone is believed to be a source of HOx radicals in the upper atmosphere. It is soluble in aqueous solutions and undergoes various reactions in the presence of sulfuric acid. Measurement of the uptake coefficient in sulfuric acid was

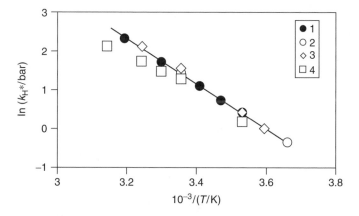

Figure 9.3 Variation of the solubility of acetaldehyde with temperature 1, Benkelberg *et al.* [21]; 2, Snider and Dawson [22]; 3, Betterton and Hoffmann [10]; 4, Zhou and Mopper [23]

Table 9.4 Solubility of acetaldehyde reported by Benkelberg et al. [21][a]

Nature of liquid phase	T/K	k_H^*/bar	No. of runs	y(acetaldehyde) $\times 10^6$
Deionized water	283.15	1.53 ± 0.05	3	900
	288.15	2.19 ± 0.05	3	76–900
	293.15	3.04 ± 0.18	4	76–900
	303.15	5.70 ± 0.22	4	76–900
	313.15	10.54 ± 0.31	3	900
Artificial seawater	289.15	2.82 ± 0.31	3	180
	298.15	4.78 ± 0.52	3	180
	303.15	6.49 ± 0.80	3	180
	313.15	10.95 ± 1.11	3	180

[a]The Henry's law constant, k_H, was defined as $k_H = p/x$, where p is the partial pressure of acetaldehyde and x the mole fraction in solution. y(acetaldehyde) = mole fraction of acetaldehyde in the vapor phase.

performed at low temperatures by Williams et al. [24] using a Knudsen cell. The initial uptake coefficient for 60 % H_2SO_4 at 233 K is 0.007. This is lower than the uptake coefficient on pure water. The effective Henry's law coefficient, k_H^*, is 2.2×10^4 mol dm^{-3} bar^{-1}.

Acetone is partially converted to the *gem*-diol in aqueous solution. The hydration constant, $[(CH_3)CH(OH)_2]/[(CH_3)_2CO]$, is about 2×10^{-3} at 298 K [20]. Equilibrium measurements of the Henry's law constant give the effective Henry's law constant. This was measured by Benkelberg et al. [21] by analysing the composition of the vapour over solutions of acetone in deionized water over the temperature range 278–313 K. Mole fractions of acetone in the gas phase ranged from 3×10^{-6} to 5×10^{-3}. Replacement of the deionized water by rain water caused negligible changes in the solubility. Data obtained in this work are substantiated by earlier measurements by Snider and Dawson [22] and Betterton [25]. Measurements by Zhou and Mopper [23] are not consistent with measurements by the other authors (see Figure 9.4).

The solubility in synthetic seawater of 3.5 % salinity was also measured. The solubility was about 15 % higher than in deionized water (see Table 9.5).

The effective Henry's law constant, k_H^*, is defined by the relationship

$$k_H^* = p/x$$

where x is the mole fraction of acetone plus diol in solution and p is the partial vapour pressure of acetone.

The experimental data from Benkelberg et al., from Snider and Dawson and from Betterton fit the relationship published by Benkelberg et al.:

$$\ln(k_H^*/\text{bar}) = -(5286 \pm 100)/(T/K) + (18.4 \pm 0.3)$$

The straight line in Figure 9.4 corresponds to this equation. Data published by Zhou and Mopper are not consistent with this equation (see Figure 9.4).

Since the fraction of acetone existing as the diol is about 0.002, the effective Henry's law constant may be equated with the Henry's law constant k_H giving

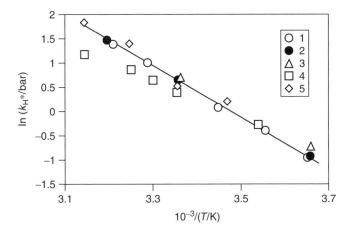

Figure 9.4 Variation of the solubility of 2-propanone (acetone) with temperature. 1, Deionized water [21]; 2, rain water [21]; 3, Snider and Dawson [22]; 4, Zhou and Mopper [23]; 5, Betterton [25]. The straight line is the linear regression line for data published by Benkelberg *et al.* [21]

Table 9.5 Average values of Henry's law constants published by Benkelberg *et al.* [21][a]

Nature of liquid phase	T/K	k_H*/bar	No. of runs	y(acetone) $\times 10^6$
Deionized water	273.66	0.401 ± 0.054	5	3–1430
	282.15	0.700 ± 0.077	4	13–1200
	289.76	1.136 ± 0.080	7	3–4800
	304.15	2.952 ± 0.170	5	15–3000
	311.66	4.174 ± 0.160	5	2–1200
Rain water	243.15	0.412 ± 0.014	5	36–2100
	298.15	2.075 ± 0.250	5	36–2100
	312.66	4.626 ± 0.170	5	120–3700
Artificial seawater	273.15	0.424 ± 0.03	5	33–4400
	287.66	1.22 ± 0.23	5	35–2100
	303.15	3.69 ± 0.52	4	33–2100
	312.66	5.72 ± 1.32	4	34–900

[a]The Henry's law constant, k_H, was defined as $k_H = p/x$, where p is the partial pressure of acetone and x the mole fraction in solution. y(acetone) = mole fraction of acetone in the vapor phase.

the ratio of the concentration of $(CH_3)_2CO$ molecules to the partial pressure of acetone (see Section 11.11).

Solubility of Carbonyl Compounds in the Presence of Sulfur Dioxide

The solubility of carbonyl compounds in the presence of sulfur dioxide has been discussed in detail by Benkelberg *et al.* [21]. Sulfur dioxide in the atmosphere dissolves in water to form bisulfite ions in solution. These ions exist in equilibrium with pyrosufite ions:

$$SO_2 + H_2O \rightleftharpoons HSO_3^- + H^+$$

$$2HSO_3^- \rightleftharpoons S_2O_5^{2-} + H_2O$$

The equilibrium constant of the second reaction is about 7×10^{-2} mol^{-1} dm^3 [26a]. At the concentration likely to be present in clouds, $[HSO_3^-] \gg [S_2O_5^{2-}]$.

Bisulfite ions form adducts with unhydrated forms of carbonyl compounds which can be generally represented as X(CO)Y:

$$X(CO)Y + HSO_3^- \rightleftharpoons adduct$$

$$K_1^* = [adduct]/[X(CO)Y][HSO_3^-]$$

Carbonyl compounds are partially hydrolysed in solution to form *gem*-diols.

A constant can be written for equilibria between the total hydrolysed + unhydrolysed carbonyl compound and the adduct, i.e.

$$K_1 = [adduct]/[HSO_3^-][total\ carbonyl\ compound]$$

It can be shown that

$$K_1^* = K_1(1 + K_h)$$

where K_h is the hydration constant, i.e. the ratio of the concentration of *gem*-diol to that of the unhydrated form.

Benkelberg *et al.* [21] have published enthalpy and entropy changes and equilibrium constants for dissociation of the adducts formed by acetone, acetaldehyde and formaldehyde with bisulfite ions based on solubility measurements over various temperature ranges (see Table 9.6). The equilibrium constants for acetone and acetaldehyde are in good agreement with earlier measurements by Betterton [25] and Kerp [26b].

The presence of dissolved sulfur dioxide makes a significant difference to the quantity of acetaldehyde and of formaldehyde absorbed by cloud droplets but has little effect on the quantity of acetone absorbed.

Table 9.6 Enthalpy and entropy changes for dissociation and equilibrium constants for formation of the adducts of acetone, acetaldehyde and formaldehyde with bisulfite ions[a]

Carbonyl compound	ΔH_d/kJ mol^{-1}	ΔS_d/J mol^{-1} K^{-1}	K_1^*/mol^{-1} dm^3 (298.15 K)
Acetone	41.3 ± 2.6	-93.1 ± 9.1	225
Acetaldehyde	63.3 ± 3.1	-98.6 ± 10.3	8.6×10^5
Formaldehyde	81.4 ± 1.1	-82.6 ± 3.7	8.9×10^9

[a] ΔH_d = enthalpy change for dissociation of adduct; ΔS_d = entropy change of dissociation of adduct; K_1^* = equilibrium constant for formation of the adduct with bisulfite ions.

3 OTHER ENVIRONMENTALLY IMPORTANT ALDEHYDES

Betterton and Hoffmann [10] measured Henry's law constants for several important aldehydes in addition to formaldehyde and acetaldehyde. They used a bubble column apparatus for all measurements except those on chloral, in which this case headspace analysis was used (Table 9.7). Data measured by Allen *et al.* [27], using an optical method, are marked with an asterisk in Table 9.7.

Table 9.7 Solubilities of various aldehydes reported by Betterton and Hoffman [10] and by Allen *et al.* [27]

Aldehyde	T/K	k_H^* /mol dm^{-3} bar^{-1}	$\Delta H_{298}/$ kJ mol^{-1a}	$[RCHO]_t/$ mol dm^{-3b}
Chloral, CCl_3CHO,	288.15	4.92×10^5	-29.1 ± 2.3	1.0×10^{-3}
[CAS 75-87-6]	298.15	3.40×10^5		$(0.1-1) \times 10^{-3}$
	308.15	2.42×10^5		1.0×10^{-3}
	318.15	1.54×10^5		1.0×10^{-3}
Glyoxal, $HCOHCO$,	288.15	$\geq 2.96 \times 10^5$		1.0×10^{-2}
[CAS 107-22-2]	298.15	$\geq 2.96 \times 10^5$		1.0×10^{-2}
	308.15	$\geq 2.96 \times 10^5$		1.0×10^{-2}
	318.15	$\geq 2.96 \times 10^5$		1.0×10^{-2}
Methylglyoxal,	288.15	8.36×10^3	-62.7 ± 5.0	1.4×10^{-3}
CH_3COCHO,	298.15	$(3.66 \pm 0.32) \times 10^3$		1.4×10^{-2}
[CAS 78-98-8]	308.15	1.27×10^3		1.4×10^{-3}
	318.15	7.69×10^2		1.4×10^{-3}
Formaldehyde,	288.15	$(7.21 \pm 0.37) \times 10^3$	-59.8 ± 1.8	1.1×10^{-3}
$HCHO$,	298.15	9.7×10^{3c}		1.1×10^{-3}
[CAS 50-00-0]				
	298.15	$(2.93 \pm 0.69) \times 10^3$		$(0.1-1.1) \times 10^{-2}$
	308.15	$(1.50 \pm 0.07) \times 10^3$		1.1×10^{-3}
	318.15	$(6.64 \pm 1.89) \times 10^3$		1.1×10^{-3}
Benzaldehyde	288.15	6.64×10	-42.2 ± 0.4	5.0×10^{-4}
C_6H_5CHO,	298.15	$(3.69 \pm 0.44) \times 10$		$(0.1-3.5) \times 10^{-3}$
[CAS 100-52-7]	308.15	2.15×10		5.0×10^{-4}
	318.15	(1.26×10)		5.0×10^{-4}
	278*	$(18.2 \pm 3.0) \times 10$		
	283*	$(12.1 \pm 1.8) \times 10$		
	288*	$(8.35 \pm 1.2) \times 10$		
	293*	$(5.55 \pm 0.84) \times 10$		
	298*	$(3.23 \pm 0.37) \times 10$		
Hydroxyacetaldehyde	298.15	4.09×10^4	-38.5	1.1×10^{-2}
$HOCH_2CHO$,	318.15	$(1.54 \pm 0.36) \times 10^4$		1.1×10^{-3}
[CAS 141-46-8]				
Acetaldehyde,	278.15	5.47×10	-52.1 ± 2.2	1.0×10^{-3}
CH_3CHO,	283.15	3.89×10		1.0×10^{-3}
[CAS 75-07-0]				
	298.15	$(1.13 \pm 0.04) \times 10$		2.0×10^{-3}
	308.15	6.47		1.0×10^{-3}

[a] ΔH_{298} = enthalpy change on solution.
[b] $[RCHO]_t = [RCHO]_{aq} + [RCH(OH)_2]$
[c] Measurement using 0.5 mol dm^{-3} H_2SO_4.

Table 9.8 Comparison of effective and intrinsic values of Henry's constants at 298 K[a] [10]

RCHO	$k_H*/\mathrm{mol\,dm^{-3}\,bar^{-1}}$	$K_H/\mathrm{mol\,dm^{-3}\,bar^{-1}}$	k_h	Ref. (K_h)
CCl$_3$CHO	3.40×10^5	1.4×10	2.5×10^4	20
HCOCHO	$\geq 2.96 \times 10^5$	≥ 1.4	2.2×10^5	28
CH$_3$COCHO	3.66×10^3	1.4	2.7×10^3	28
HOCH$_2$CHO	4.09×10^4	4.1×10^3	1.0×10	29
CH$_3$CHO	1.13×10	4.8	1.4	20
	1.25×10			30[b]

[a]Note that $k_H* = k_H \times K_h$ (allowing for rounding errors).
[b]Ref. to k_H.

Betterton and Hoffmann also published hydration constants K_h and intrinsic Henry's constants K_h. These were defined as

$$K_h = [\mathrm{RCH(OH)_2}]/[\mathrm{RCHO}]_\mathrm{aq}$$

$$k_H = [\mathrm{RCHO}]_\mathrm{g}/[\mathrm{RCHO}]_\mathrm{aq}$$

See Table 9.8.

Higher Aldehydes

Amoore and Buttery [31] reported values of the partition coefficient for air–water distribution, K_AW, for aldehydes from n-propanal to isobutyraldehyde at 298.15 K. These were defined as the ratio $[\mathrm{RCHO}]_\mathrm{g}$ to the sum of $[\mathrm{RCHO}]_\mathrm{aq} + [\mathrm{RC(OH)_2}]$. Values were derived from solubilities, S, and vapor pressures of pure aldehydes, p. This was based on the assumption that the partial pressure of a slightly soluble substance in a saturated solution of molar concentration S is equal to the vapour pressure p of pure substance. The molar concentration in the gas phase can then be estimated by assuming that the vapour approximates to an ideal gas. The equation for the behaviour of an ideal gas can then be used, i.e.

$$pV = nRT$$

where $n = $ No. of moles of gas and $R = $ gas constant.

Molar concentration in the gas phase, $c_g = $ No. of moles per unit volume

at temperature T

$$= n/V$$

$$= p/RT$$

Hence

$$K_\mathrm{AW}(\text{calculated}) = m_g/S$$

Amoore and Buttery compared these values of K_AW with experimental values of K_AW previously measured by Buttery et al. [32] using gas–liquid chromatography (Table 9.9). Effective Henry's law constants, $k_H{}^*$, relating $[RCHO]_{aq} + [RC(OH)_2]$ with the partial pressure of the aldehyde have been calculated from the vapour pressures and solubilities published by Amoore and Buttery (Table 9.9).

Olson [33] measured the solubility of 1,5-pentanediol (glutaraldehyde, $C_5H_8O_2$ [111-30-8]) in the solution concentration range 50–500 ppm He used a modified Gillespie still and a diluted sample of Union Carbide 505 solution. Effective Henry's law constants, $k_H{}^*$, were reported for the temperature range 38.22–84.76 °C (311.27–357.91 K). These data fit the equation $\ln(k_H{}^*/\text{bar}) = 24.5301 - 9187.99/(T/\text{K})$ with $R^2 = 0.999$ ($R = $ Pearson's correlation coefficient).

Unsaturated Aldehydes and Methyl Vinyl Ketone

Solubility data for methacrolein, acrolein and methyl vinyl ketone are show in Table 9.10.

Data for 2-methyl-2-propenal and 3-buten-2-one published by Allen et al. [27] fit a van't Hoff-type equation:

$$\ln k_H{}^* = A/(T/\text{K}) - B$$

(see Figure 9.5). Values of A and B are shown in Table 9.11 together with the heat of solution corresponding to the value of A.

Table 9.9 Solubilities of higher aldehydes at 298.15 K reported by Amoore and Buttery [31][a]

Aldehyde	vp/ mmHg	S/ g dm^{-3}	MW	K_{AW}(calc)	K_{AW}(expt)	$k_H{}^*$/mol dm^{-3} bar^{-1} (calc)
Propanal, C_3H_6O, [CAS 123-38-6]	318	306	58.1	0.0026	0.0030	12.43
Butanal, C_4H_8O, [CAS 123-72-8]	114	69.9	72.1	0.0060	0.0047	6.38
Pentanal, $C_5H_{10}O$, [CAS 110-62-3]	16*	11.7	86.1	0.0063	0.0060	6.37
1-Hexanal, $C_6H_{12}O$, [CAS 66-25-1]	10.6	5	100.2	0.0114	0.0087	3.53
1-Heptanal, $C_7H_{14}O$, [CAS 111-71-1]	2.93	2.39	114.2	0.0075	0.0110	5.36
1-Octanal, $C_8H_{16}O$, [CAS 124-13-0]	1.1	0.56	128.2	0.0140	0.0210	2.98
1-Nonanal, $C_9H_{18}O$, [CAS 124-19-6]	0.39	0.096	142.2	0.0310	0.0300	1.30
Isobutyraldehyde 2-Methylpropanal C_4H_8O, [CAS 78-84-2]	150	89	72.1	0.0060	0.0080	6.17

[a]$S = $ solubility; $K_{AW} = $ air–water distribution coefficient.

Table 9.10 Solubilities of propenal (acrolein), C_3H_4O, [CAS 107-02-8], 2-methyl-2-propenal (methacrolein), C_4H_6O, [CAS 78-85-3] and 3-buten-2-one (methyl vinyl ketone) C_4H_6O, [CAS 78-94-4]

Compound	Conc/mol l^{-1}	T/K	k_H*/mol dm^{-3} bar^{-1}	Method	Ref.
Propenal	~3.6×10^{-4}	273	35.2 ± 4.0	Headspace analysis	30
		298	7.3 ± 4.1		
2-Methyl-2-propenal	4.6×10^{-3}	298.5	6.4 ± 0.7	Headspace analysis	34
	1.0×10^{-3}	278.15	15.4 ± 2.2	Optical absorption	27
		283.15	10.9 ± 1.4		
		288.15	7.7 ± 1.0		
		293.15	5.9 ± 0.8		
		298.15	4.2 ± 0.6		
3-Buten-2-one		298.15	43.4		25
	4.8×10^{-5}	298.5	40.5 ± 7.0	Headspace analysis	34
	1.0×10^{-3}	278.15	149 ± 5	Optical absorption	27
		283.15	75.9 ± 3		
		288.15	52.2 ± 2		
		293.15	34.2 ± 1.4		
		298.15	21.2 ± 0.9		

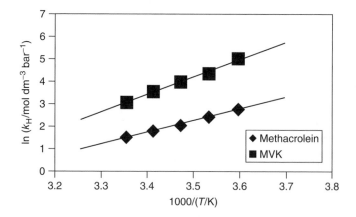

Figure 9.5 Variation with temperature of the solubility in water of 2-methyl-2-propenal (methacrolein) and of 3-buten-2-one (methyl vinyl ketone)

The data presented by Allen *et al.* are self-consistent but are significantly different from data published by Iraci *et al.* [34] and Betterton [25].

Allen *et al.* [27] developed a novel method of measuring Henry's law constants. This involved comparing the optical absorption of the liquid and the gas

Table 9.11 Solubility data for 2-methyl-2-propenal (methacrolein) and 3-buten-2-one (methyl vinyl ketone)[a]

Solute	A	B	R^2	$\Delta H^\circ/kJ\,mol^{-1}$
2-Methyl-2-propenal	5.1502×10^3	15.793	0.9979	-42.8
3-Buten-2-one	7.7929×10^3	23.086	0.9921	-64.8

[a]R = correlation coefficient for the linear regression line (see Figure 9.5); ΔH° = enthalpy change on solution.

phase when the system had reached equilibrium. The method is limited to solutes which show significant absorption in the visible or ultraviolet region.

4 METHYL HYDROPEROXIDE, CH_3O_2H, [CAS 3031-73-0], ETHYL HYDROPEROXIDE, $C_2H_5O_2H$ [CAS 3031-74-1] AND OTHER ORGANIC PEROXIDES

O'Sullivan et al. [5] measured solubility in water of methyl hydroperoxide, hydroxymethyl hydroperoxide, ethyl hydroperoxide and peroxyacetic acid. The technique used involved passing air over a moving, continuously replenished, thin film of a solution of an organic peroxide compound. Henry's law constants were calculated from the compositions of co-existing gas and liquid phases. The data obtained show satisfactory agreement with earlier measurements. Measurements of the solubility of bis(hydroxymethyl) peroxide, $HOCH_2OOCH_2OH$, have been reported by Staffelbach and Kok [6] and Zhou and Lee [35] (Table 9.12).

5 CARBOXYLIC ACIDS

The dissolution of a weak acid, HA, may be considered to be a two-stage process [Equations (9.4) and (9.5)]:

$$HA(g) \rightleftharpoons HA(aq) \tag{9.4}$$

$$k_H = [HA]/pHA$$

where k_H is the Henry's law constant for equilibrium between undissociated acid in the liquid and gas phase, [HA] the concentration of HA in the aqueous phase and p_{HA} the partial pressure of HA.

$$HA(aq) \rightleftharpoons H^+ + A^- \tag{9.5}$$

$$K_a = [H^+][A^-]/[HA]$$

where K_a is the acid dissociation constant

Total concentration of HA in solution = concentration of (undissociated

HA and dissociated HA)

$$= [HA] + [A^-]$$

Table 9.12　Solubilities of organic peroxides^a

Compound	Ref.	T/K	c_{XO_2} / mol dm^{-3}	A	B	ΔH° / kJ mol^{-1}	$-\Delta G^\circ$ / kJ mol^{-1}	Ln(k_H / mol dm^{-3} bar^{-1}) (25 °C)	k_H / mol dm^{-3} bar^{-1} (25 °C)	Method
Methyl hydroperoxide, CH$_3$OOH, [CAS 3031-73-0]	4			5322	12.15	44.2	14.2	5.70	298 ± 36	Analysis of gas stream in equilibrium with solution
	5	277–301	10^{-6}–10^{-2}	5241 ± 133	11.85 ± 0.18	43.6	14.2	5.74	307 ± 14	Analysis of gas stream in equilibrium with solution
Ethyl hydroperoxide, C$_2$H$_5$OOH, [CAS 3031-74-1]	5	277–301	10^{-6}–10^{-2}	5994	14.29	49.8	14.4	5.81	332 ± 20	Analysis of gas stream in equilibrium with solution
Hydroxymethyl hydroperoxide, HOCH$_2$OOH, [CAS 15932-89-5]	6	278–293	4.5–9 × 10^{-2}	10240	20.04	85.1	35.5	14.31	2.35 × 10^6 (295 K)	
	5	277–301	10^{-6}–10^{-2}	9652	18.05	80.2	35.5	14.32	1.65 × 10^6 ±0.35 × 10^6	Analysis of gas stream in equilibrium with solution
	35								4.9 × 10^5 (295 K)	
Bis(hydroxymethyl) peroxide, HOCH$_2$OOCH$_2$OH, [CAS 17088-73-2]	6	283	6–45 × 10^{-3}						≥1 × 10^7 (283 K)	
	35	283, 295							≥2 × 10^6 (283 K) ≥0.6 × 10^6 (295 K)	
Peroxyacetic acid, CH$_3$C(O)O$_2$H, [CAS 79-21-0]	4			5896	13.29	49.0	16.1	6.49	656 ± 117	
	5	277–301	10^{-6}–10^{-2}	5308	11.08	44.1	16.7	6.72	826 ± 175	Analysis of gas stream in equilibrium with solution

^{a} ΔH° = enthalpy change on solution; ΔG° = Gibbs free energy change on solution; k_H = Henry's law constant; c_{XO_2} = concentration of the peroxide; A, B = constants in the equation $\ln k_H /$mol dm^{-3} bar^{-1} = $A/(T/K) - B$.

Table 9.13 Solubility of organic acids at 298 K and very low partial pressures[a]

Compound	K_a at 298 K	T/K	k_H^*/mol kg^{-1} bar^{-1}	A	B	Ref.
Methanoic acid (formic acid), CH$_2$O$_2$, [CAS 64-18-6]	1.77×10^{-4} (293 K)	298.15	5458 ± 265	-10.32	5634	37,39
Ethanoic acid (acetic acid), C$_2$H$_4$O$_2$, [CAS 64-19-7]	1.76×10^{-5}	298.15	5430 ± 290	-25.68	8322	37,39
Propanoic acid (propionic acid), C$_3$H$_6$O$_2$, [CAS 79-09-4]	1.34×10^{-5}	298.15	5638 ± 340			37,39
Butanoic acid (n-butyric acid), C$_4$H$_8$O$_2$, [CAS 107-92-6]	1.54×10^{-5} (293 K)	298.15	4665 ± 178			39
2-Methylpropanoic acid (isobutyric acid), C$_4$H$_8$O$_2$, [CAS 79-31-2]	1.44×10^{-5} (291 K)	298.15	1112 ± 118			39
Pentanoic acid (n-valeric acid), C$_5$H$_{10}$O$_2$, [CAS 109-52-4]	1.51×10^{-5} (291 K)	278.15 288.15 298.15 308.15	11988 ± 500 5471 ± 229 2203 ± 76 1126 ± 69	-14.3503	6582.96	37,39
3-Methylbutanoic acid (isovaleric acid), C$_5$H$_{10}$O$_2$, [CAS 503-74-2]	1.7×10^{-5}	298.15	1179 ± 111			39
2,2-Dimethylpropanoic acid (neovaleric acid), C$_5$H$_{10}$O$_2$, [CAS 598-98-1]		298.15	348 ± 40			39
Hexanoic acid (n-caproic acid), C$_6$H$_{12}$O$_2$, [CAS 142-62-1]	1.43×10^{-5} (291 K)	298.15 308.15	1370 ± 54 657 ± 40	-13.9556	6303.73	39
2-Oxopropanoic acid (pyruvic acid), C$_3$H$_4$O$_3$, [CAS 127-17-3]		278.15 288.15 298.15 308.15	$(11.33 \pm 0.3) \times 10^5$ $(5.20 \pm 0.3) \times 10^5$ $(3.07 \pm 0.2) \times 10^5$ $(1.77 \pm 0.05) \times 10^5$	-4.43022	5087.92	39

$^a K_a$ = acid dissociation constant; k_H^* = effective Henry's law constant; A, B = constants in the equation $\ln k_H^* = A + B/(T/K)$.

The effective Henry's law constant, k_H^*, is the ratio between the total HA in solution and the partial pressure, i.e.

$$k_H^* = \frac{[HA] + [A^-]}{p_{HA}} = \frac{[HA] + K_a[HA]/[H^+]}{p_{HA}}$$

$$= \frac{[HA](1 + K_a/[H^+])}{p_{HA}}$$

Hence

$$k_H^* = k_H(1 + K_a/[H^+])$$

If concentrations are expressed in molalities, this equation becomes

$$k_H^* = k_H(1 + K_a/m_{H^+})$$

where m_{H^+} is the molal concentration of hydrogen ions.

In solutions of high ionic strength, concentrations of species X in the above relationships should be replaced by activities, a_X. The above equation consequently becomes

$$k_H^* = k_H(1 + K_a/a_{H^+})$$

If the dissociation of a weak acid is repressed by addition of sufficient strong acid, then

$$k_H \approx k_H^*$$

Data for the calculation of the solubility of formic and acetic acids at low partial pressures over a temperature range were published by Clegg and Brimblecombe [36] and Khan and Brimblecombe [37]. Data for pyruvic and methacrylic acid were published by Khan et al. [38]. Khan et al. [39] measured the equilibrium partial pressure of carboxylic acids from C_1 to C_6 over aqueous solutions at 298 K using a dynamic method developed by Scarano et al. [40]. In each case dissociation of the acids in solution was inhibited by addition of hydrochloric acid to give a pH of 5. This allowed the calculation of Henry's law constants corresponding to the ratio of undissociated acid in solution to the partial pressure of the organic acid. These authors also published equations for the variation of k_H^* with temperature. Solubility data from these sources for organic acids are shown in Table 9.13. Partition coefficients published by Amoore et al. [31] are given in Table 9.14.

Table 9.14 Air–water partition coefficients, K_{AW}, at 298 K for pentanoic acid and isovaleric acid published by Amoore et al. [31][a]

vp/mmHg		S/g dm^{-3}	MW	K_{AW}(calc)[b]	K_{AW}(expt)	k_H^*/mol dm^{-3} bar^{-1}
Pentanoic acid	0.29	50	102.1	0.000031		1267
Isovaleric acid	0.05	48	102.1	0.000055	0.000034	7052

[a]K_{AW} = concentration in air/concentration in water; S = solubility.
[b]Values of K_{AW} calculated from vp and the solubility of the acids. The experimental value for isovaleric acid was determined by a flow method [31].

6 HALOGENS AND COMPOUNDS CONTAINING HALOGENS

Chlorine, Cl$_2$, [CAS 7782-50-5]

The solubility of chlorine in water has been discussed by Fogg and Gerrard [41]. When chlorine dissolves in water it exists in equilibrium with chloride ions and hypochlorous acid:

$$Cl_2 + H_2O \rightleftharpoons H^+ + Cl^- + HClO$$

The equilibrium constant defined by

$$K = \frac{[H^+][Cl^-][HClO]}{[Cl_2]}$$

is 1.56×10^{-4} mol dm^{-3} at 273.15 K and 10.0×10^{-4} mol dm^{-3} at 343.15 K [42]. The interpolated value for 298.15 K is about 3.0×10^{-4} mol dm^{-3}. Water, saturated at 298.15 K with chlorine at a partial pressure of 1.013 bar, contains about half of the chlorine as chloride ions or hypochlorous acid. The bulk solubility of chlorine corresponds to the quantity of Cl$_2$ which is equivalent to the total Cl$_2$, Cl$^-$ and HClO in solution.

Hypochlorous acid dissociates:

$$HClO \rightleftharpoons H^+ + ClO^-$$

The dissociation constant is 2.95×10^{-5} mol dm^{-3} at 293.2 K in the approximate acid concentration range 0.1–0.01 mol dm^{-3} [43].

Chlorine forms a solid hydrate which may separate from aqueous solution. The formula of this hydrate is Cl$_2 \cdot$ 8H$_2$O and is in equilibrium with chlorine at a partial pressure of 1.013×10^5 Pa when the temperature is 282.8 K [44].

The solubility of chlorine in water has been evaluated by Battino [45], who recommended values of the mole fraction bulk solubility $x(Cl_2)$ for a partial pressure of 1.013 bar in the range 283.15–383.15 K (Table 9.15).

Table 9.15 Values of mole fraction bulk solubilities of chlorine in water recommended by Battino [45]

T/K	$x(Cl_2)$
283.15	0.00248
293.15	0.00188
303.15	0.0015
313.15	0.00123
323.15	0.00106
333.15	0.000939
343.15	0.000849
353.15	0.000784
363.15	0.000737
373.15	0.000697
383.15	0.000668

The variation of mole fraction solubility with pressure for 298.2 K from measurements by Whitney and Vivian [46] is shown in Figure 9.6. The lower the pressure the greater, in proportion, is the contribution of hydrolysis to the overall solubility. Linear extrapolations of mole fraction solubilities from 1.013×10^5 Pa to low pressures, based upon an assumption that Henry's law is applicable, cannot give reliable estimates of solubility.

Adding hydrochloric acid to water lowers the solubility of chlorine at low concentrations of acid. This is probably due to the reduction of the degree of hydrolysis of the chlorine. High concentrations of hydrochloric acid enhance the solubility of chlorine. The variation of solubility with concentration of acid at 1.013 bar and 298.15 K, based upon measurements by Sherrill and Izard [47], is shown in Figure 9.7.

The effect on the solubility of chlorine in water of adding barium, sodium and potassium chlorides has been studied by Sherrill and Izard [47] (see Figure 9.8).

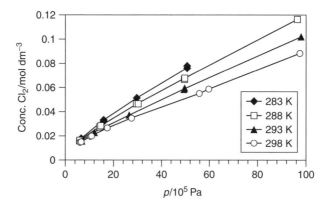

Figure 9.6 Variation with partial pressure of chlorine, p, of the concentration of chlorine in water at 283–298 K [47]

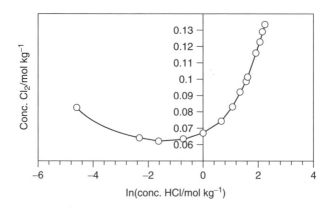

Figure 9.7 Variation of the solubility of chlorine at 298.15 K and a partial pressure of 1.013 bar in aqueous solutions of hydrogen chloride

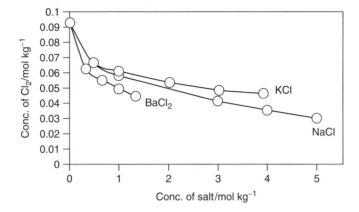

Figure 9.8 Effect of the concentration of salts on the solubility of chlorine in aqueous solutions under a barometric pressure at 298.15 K

At high concentrations of salt the variation of solubility with concentration of salt approximates to a Sechenov-type relationship:

$$\ln(s/\text{mol kg}^{-1}) = A - B(m/\text{mol kg}^{-1})$$

where s is the solubility of the gas, m is the concentration of salt and A and B are constants [41]. Adding chlorides reduces the degree of hydrolysis of chlorine. This is reflected in the marked change in solubility with addition of chlorides at low concentration of chlorides.

Bromine, Br$_2$ [CAS 7726-95-6] and Iodine, I$_2$, [CAS 7553-56-2]

Elemental bromine and iodine are believed to be formed during processes which occur in the marine boundary layer [48,49]. The solubility of the liquid, vapour pressure and ratio of solubility to vapour pressure are given in Table 9.16.

Table 9.16 Solubility and vapour pressure of bromine and iodine

Compound	T/K	Solubility/ g dm^{-3}	Vapor pressure/ 10^5 Pa[a]	Ratio of solubility to V.P./ mol dm^{-3} bar^{-1}[b]
Bromine	273.15	41.7	0.0865	3.02
	293.15	35.8	0.225	1
	323.15	35.2	0.7556	0.29
Iodine	293.15	0.29	0.000302	3.78
	298.15	0.3	0.000457	2.59
	323.15	0.78	0.00299	1.03

[a]From published data for the range 265–300 K.
[b]Identification of this ratio with a Henry's law constant applicable to infinite dilution may lead to errors because of hydrolysis as in the case of chlorine.

Hydrogen Chloride, HCl, [CAS 7647-01-0]

Carslaw *et al.* [50] defined a Henry's law constant based on the equilibrium

$$HY(g) \rightleftharpoons H^+(aq) + Y^-(aq)$$

on the assumption that there was an insignificant amount of undissociated acid in the liquid phase. A constant analogous to a Henry's law constant was defined as

$$k_H{}^\dagger = x_H f_H{}^* x_Y f_Y{}^* / p_{HY}$$

where x_i is the mole fraction of either ionic species i defined as a fraction of the total number of species which includes both ions and neutral molecules and f_i is the activity coefficient of species i on the basis that $f_i{}^* \to 1$ as $x_i \to 0$. This approach is useful if the partial pressure of hydrochloric acid is low.

These authors measured activity coefficients and vapour pressure of HCl solutions at 298.15 K. The corresponding value of k_H is 65.3 bar^{-1}. This is equivalent to a value of 2.013×10^6 mol^2 kg^{-2} bar^{-1} when concentrations of ions are expressed as molalities. They used heat capacity data and the value of the enthalpy of solution of HCl to derive the following expression for the value of k_H over a temperature range:

$$\ln(k_H{}^\dagger / \text{bar}^{-1}) = 6.4822 - 90027$$
$$\times 10^3 [1/(T_r/\text{K}) - 1/(T/\text{K})] - 65.346[T_r/T - 1 + \ln(T/T_r)] +$$
$$0.078178[T_r(T_r/T - 1) + T - T_r]/\text{K}$$

where T_r is equal to 298.15 K

Total pressures over solutions containing 9.45 and 10.51 mol kg^{-1} HCl were measured over the temperature range 199.5–289.4 K using a capacitance manometer. Equilibrium partial pressures were also measured by mass spectrometry from 234.6 to 274.8 K. The partial pressure of HCl as a fraction of the total pressure ranged from about 0.09 to 0.38. The total pressures estimated from the equation given above plus the values of the vapour pressure of water are up to about 3 % higher than experimental values from 289.4 to 250 K. At lower temperatures the calculated values are up to 10 % higher than experimental values.

Van Doren *et al.* [51] measured the diffusion coefficients in the gas phase and the uptake coefficient γ of HCl, HNO$_3$ and N$_2$O$_5$. A monodisperse train of water droplets of diameter 200 μm generated at a vibrating orifice passed through an atmosphere of helium and water vapour containing traces of the substance under test. Sample of droplets were trapped and analysed. Diffusion coefficients, D, of the trace gases in a mixture of He and H$_2$O were calculated from the relationship

$$1/D = (p_{H_2O}/D_{X-H_2O}) + (p_{He}/D_{X-He})$$

where D_{X-H_2O} and D_{X-He} are the binary diffusion coefficients for the trace gas, X, with water and helium, respectively. Values of D_{HCl-H_2O} and D_{HCl-He} were calculated as a function of temperature with CHEMKIN [52]. At 298 K,

$D_{\text{HCl–H}_2\text{O}} = 0.168 \, \text{bar cm}^2 \, \text{s}^{-1}$ and $D_{\text{HCl–He}} = 0.713 \, \text{bar cm}^2 \, \text{s}^{-1}$. Values of γ_{meas}, the measured uptake coefficient, were corrected to give values for zero pressure, γ_0 (see Table 9.17). The high value of the uptake coefficients indicated that they should be identified with accommodation coefficients [51].

The uptake of HCl by sulfuric acid solutions has been investigated by various workers. Abbatt [53] used a low-temperature flow tube coupled to a mass spectrometer. Values of $k_H^* \sqrt{D_l}$ were obtained, where k_H^* is the effective Henry law constant expressing the ratio of partial pressure to total HCl absorbed and D_l is the liquid-phase diffusion coefficient of HCl. These data were published in graphical form (see Figure 9.9).

Similar data were published by Hanson and Ravishankara [54] (see Table 9.18). These authors used a horizontally mounted cylindrical flow tube in conjunction with a chemical ionization mass spectrometer. A glass 'boat' containing sulfuric

Table 9.17 Uptake of hydrogen chloride by water[a]

No. of data	T/K	$P_{\text{tot}}/\text{bar}$	$p_{\text{H}_2\text{O}}/\text{bar}$	Orifice diam./μm	γ_{obs}	D/bar $\text{cm}^2 \, \text{s}^{-1}$	γ_0
12	274	7.2×10^{-3}	6.7×10^{-3}	62	0.131 ± 0.006	0.161	$0.18 +0.024/-0.017$
19	283	12.8×10^{-3}	12.4×10^{-3}	62	0.09 ± 0.003	0.171	$0.14 +0.024/-0.014$
25	294	25.2×10^{-3}	24.4×10^{-3}	62	0.046 ± 0.004	0.177	$0.06 +0.012/-0.008$

[a]No. of data = No. of experimental measurements at each temperature; P_{tot} = total pressure of water vapour and helium; $p_{\text{H}_2\text{O}}$ = partial pressure of water vapour; D = diffusion coefficient of HCl in the mixture of gases present.

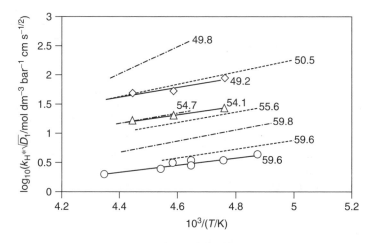

Figure 9.9 Values of k_H^*/D_l for uptake of hydrogen chloride by sulfuric acid of various concentrations as represented by Abbatt [53]. The solid lines and data points correspond to measurements by Abbatt. Best fit straight lines corresponding to data published by Hanson and Ravishankara [54] (dotted lines) and to data published by Williams and Golden [55] (dashed-dotted lines) are also shown. The numbers adjacent to the lines give the mass% of sulfuric acid in the solution. k_H^* = effective Henry's law constant; D_l = diffusion coefficient of HCl in the liquid phase

Table 9.18 Solution data for hydrogen chloride published by Hanson and Ravishankara [54]

$c_{H_2SO_4}/$ mass%	T/K	$k_H^* \sqrt{D_l}/$ $\mathrm{mol\,dm^{-3}\,bar^{-1}}$ $(\mathrm{cm^2\,s^{-1}})^{\frac{1}{2}}$	$c_{H_2SO_4}/$ mass%	T/K	$k_H^* \sqrt{D_l}/$ $\mathrm{mol\,dm^{-3}\,bar^{-1}}$ $(\mathrm{cm^2\,s^{-1}})^{\frac{1}{2}}$
59.6	200	7.0	55.6	225	10.9
	205	5.87		215	15.2
	205	6.3		205	27.7
	210	4.59	50.5	230	34.6
	220	3.0		215	76.5
	220	3.7		215	74.5
	200	8.0		207	144
55.6	225	11.3		200	154

Table 9.19 Variation of k_H^* for HCl with temperature and concentration of H_2SO_4 [54]

$c_{H_2SO_4}/$ mass%	$c_{HCl}/$ $\mathrm{mol\,dm^{-3}}$	T/K	$k_H^*/$ $\mathrm{mol\,dm^{-3}}$ $\mathrm{bar^{-1}}$	$c_{H_2SO_4}/$ mass%	$c_{HCl}/$ $\mathrm{mol\,dm^{-3}}$	T/K	$k_H^*/$ $\mathrm{mol\,dm^{-3}}$ $\mathrm{bar^{-1}}$
45	1×10^{-3}	218	1.7×10^6	50	9.2×10^{-3}	200	3.1×10^6
		203.2	7.1×10^6			200	3.6×10^6
		195	2.5×10^7			206	1.7×10^6
45.3	1.55×10^{-2}	200	1.3×10^7			212.5	8.4×10^5
		210	3.6×10^6	51.1	2.1×10^{-3}	205.3	1.1×10^6
		220	1.2×10^6			199	2.4×10^6
		200	1.3×10^7			209.8	5.9×10^5
		195	2.4×10^7			219.5	2.3×10^5
		195	2.2×10^7			199.7	2.8×10^6

acid was exposed to a flow of helium gas with or without traces of hydrogen chloride in the gas phase. For some experiments the inside of the tube was coated with sulfuric acid. The rate of absorption was found from the decrease in hydrogen chloride in the gas phase. The vapour pressure of hydrogen chloride was measured by passing helium over sulfuric acid containing hydrogen chloride.

Hanson and Ravishankara [54] also published values of k_H^* from direct measurements of vapour pressure over solutions of hydrogen chloride in sulfuric acid (Table 9.19).

Hydrogen Bromide, HBr, [CAS 10035-10-6]

Abbatt [53] measured the time-dependent uptake of HBr by supercooled solutions of sulfuric acid using a low-temperature flow tube. He obtained values of $k_H \sqrt{D_l}$, where k_H is the effective Henry's law constant relating total HBr absorbed to partial pressure of HBr. The diffusion coefficient of HBr in the liquid phase, D_l, may be estimated from the relationship $D_l = cT/\eta$, where η is

Table 9.20 Solubility of hydrogen bromide in sulfuric acid [57]

H_2SO_4/mass%	T/K	k_H*/ $mol\,dm^{-3}\,bar^{-1}$	H_2SO_4/mass%	T/K	k_H*/ $mol\,dm^{-3}\,bar^{-1}$
Cold flow tube measurements (HBr concentration in the liquid phase 0.10–0.30 M)					
40.3	248	5.2×10^7	48.8	258	8.5×10^5
40.3	258	2.8×10^7	60.5	218	1.4×10^6
40.3	268	7.0×10^6	60.5	228	5.0×10^5
48.8	218	8.4×10^7	60.5	238	1.5×10^5
48.8	228	2.7×10^7	60.5	248	5.4×10^4
48.8	238	9.5×10^6	60.5	258	3.2×10^4
48.8	248	2.6×10^6			
Room temperature experiments (HBr concentration 0.34–0.50 M)					
49.5	298	5.5×10^4	64.4	298	3.8×10^2
57.3	298	3.9×10^3	67.6	298	1.2×10^2

the measured viscosity of the sulfuric acid solution and c is a proportionality constant determined from aqueous solution at room temperature [56]. The data were published in graphical form.

Abbatt and Nowak [57] later published further work on the system in numerical form. Further measurements using a low-temperature flow tube were reported. However, a different technique was used for measurements of the vapour pressure of HBr at 298 K. In this case a stream of dry nitrogen was passed through a solution and the HBr flushed out, trapped in deionized water and estimated by electrical conductivity measurements on the resulting solution (Table 9.20).

The variation with temperature and acid concentration of the Henry's law constant, k_H*, published by Abbatt [53] and Abbatt and Novak [57] is shown in Figure 9.10. The figure also shows the variation of the constant from measurements by Williams et al. [58] as reported by Abbatt. Figure 9.11 shows a linear variation of log k_H* with concentration of acid at a fixed temperature of 230 K. This figure is based on the data presented in Figure 9.10. The consistency of the data from the different sources is a good indication of their reliability. Additional information about the behaviour of HBr is given in Section 7.3.

Chlorine Dioxide, ClO_2, [CAS 10049-04-4]

The uptake of chlorine dioxide by sulfuric acid solutions has been studied by Grothe et al. [59]. A Knudsen cell connected to a mass spectrometer and pumping system was used. The Knudsen cell consisted of two chambers connected by a valve. Chlorine dioxide diffused through the valve into the lower chamber and was absorbed by cold sulfuric acid covering the bottom of the cell. Uptake coefficients at 213 and 193 K were calculated from the rate of decrease of concentration of chlorine dioxide in the upper part of the cell, as monitored by the mass spectrometer. Addition of hydrogen peroxide lowered the uptake coefficient (see Table 9.21). No reaction products were detected in the gas phase.

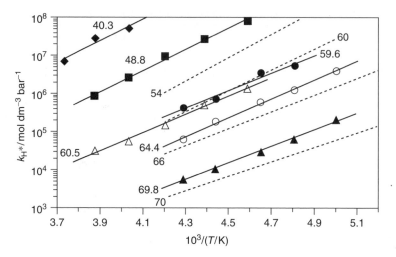

Figure 9.10 Dependence on temperature and concentration of acid of the solubility of hydrogen bromide in sulfuric acid. The data points and full lines correspond to measurements by Abbatt *et al.* [53,57] and the dotted line to measurements by Williams *et al.* [58]. Numbers adjacent to the lines show the mass% of sulfuric acid

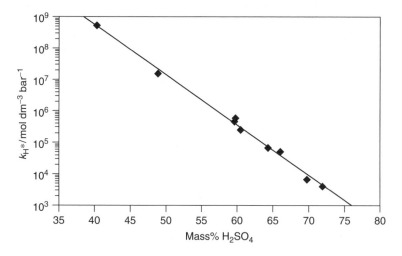

Figure 9.11 Variation of interpolated values of k_H^* for dissolution HBr in H_2SO_4 with concentration of sulfuric acid at an arbitrary temperature of 230 K (see Figure 9.10)

Hypochlorous Acid, HOCl, [CAS 7790-92-3]

Solubility data based on vapour pressure measurements have been published by Ourisson and Kastner [60], Holzwarth *et al.* [61] and Blatchley *et al.* [62] and reviewed by Huthwelker *et al.* [63]. Data from these sources as reported by Huthwelker *et al.* are shown in Figure 9.12.

Table 9.21 Measured uptake coefficients for dissolution of chlorine dioxide in sulfuric acid solutions

Conc. $H_2SO_4\%$	Conc. $H_2O_2\%$	T/K 193	T/K 213
50			5.37×10^{-5}
60		1.84×10^{-5}	1.30×10^{-4}
70			9.10×10^{-6}
80			1.10×10^{-5}
60	7		5.06×10^{-6}
60	7	1.81×10^{-6}	

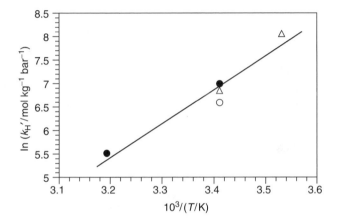

Figure 9.12 Variation of the solubility of HOCl in pure water with change of temperature [63]. ●, Ourisson and Kastner [60]; ○, Blatchley *et al.* [62]; △ Holzwarth *et al.* [61]

Abbatt and Nowak [57] used a low-temperature coated-wall flow tube coupled with an electron impact mass spectrometer to measure the time-dependent uptake of HOCl by sulfuric acid. This is related to $k_H^* \sqrt{D_1}$ by the relation

$$\text{time-dependent uptake coefficient} = 4RT k_H^* \sqrt{D_1} / c(\pi t)^{\frac{1}{2}}$$

where c is the mean molecular velocity, t is the time during which the gas is in contact with the surface, D_1 is the diffusion coefficient of HOCl gas, k_H^* is the effective Henry's law constant, R is the gas constant and T is the absolute temperature.

Hanson and Ravishankara [54] also obtained values of $k_H \sqrt{D_1}$ for HOCl dissolving in sulfuric acid. They used a similar apparatus consisting of a horizontally mounted cylindrical flow tube in conjunction with a chemical ionization mass spectrometer (see Section 2.3).

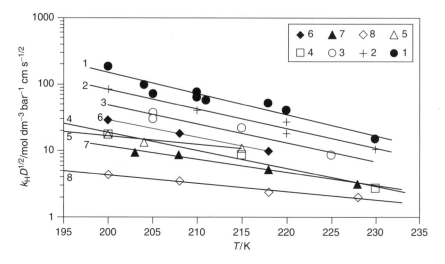

Figure 9.13 Solubility data for dissolution of HOCl in H_2SO_4 published by Hanson and Ravishankara [54] and by Abbatt and Nowak [57]. Mass% of H_2SO_4: [48] 1, 46; 2, 50.5; 3, 55.6; 4, 58; 5, 59.6; [49] 6, 59.7; 7, 64.6; 8, 69.3 k_H^* = effective henry's law constant; D_l = diffusion coefficient of HOCl gas in the liquid phase

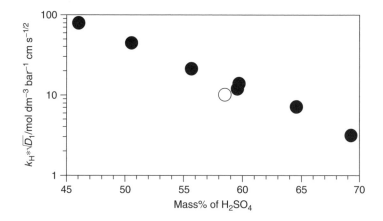

Figure 9.14 Apparent variation of k_H^*/D_l for HOCl with mass% of H_2SO_4 from Hanson and Ravishankara [54] and Abbatt and Nowak [57]. The data point marked with an open circle for 58.5 mass% H_2SO_4 from [54] appears to be inconsistent with other data points. k_H^* = effective Henry's law constant; D_l = diffusion coefficient of HOCl gas in the liquid phase

Data from the two sources are plotted in Figure 9.13. Figure 9.14 shows interpolated values of $k_H\sqrt{D_l}$ at 310 K from the two sources. This figure shows that, in general, the data from the two sources are consistent. However, the data reported by Hanson and Ravishankara for a concentration of sulfuric acid of 58.5 mass% are inconsistent with other data.

Huthwelker *et al.* [63] reviewed and assessed the available data on the solubility of HOCl in water and aqueous sulfuric acid (see Chapter 7). They derived the equation below for the effective Henry's law constant, based on the data from each source which had been weighted to reflect the estimated accuracy:

$$\ln(k_H^*/\text{mol kg}^{-1}\,\text{bar}^{-1}) = \ln(k'_{HT_0}/\text{mol kg}^{-1}\,\text{bar}^{-1}) - m_{(H_2SO_4)}k_1'/\text{mol kg}^{-1}$$
$$+ (\Delta H^\circ/R)[1/(T_0/\text{K}) - 1/(T/\text{K})]$$

where k'_{HT_0} is the thermodynamic Henry's law constant at T_0, i.e. the ratio of activity of undissociated HOCl in solution to the partial pressure of HOCl. ΔH° is the molar enthalpy change associated with dissolution of HOCl at T_0, k_1 is a fitting parameter and $m_{(H_2SO_4)}$ is the concentration of sulfuric acid in moles per kilogram of water.

$$T_0 = 298.15\,\text{K}$$

$$k_{HT_0} = k_H \text{ at } 298.15\,\text{K} = 653.6 \pm 6\,\text{mol kg}^{-1}\,\text{bar}^{-1}$$

$$\Delta H^\circ = -48.74 \pm 1.2\,\text{kJ mol}^{-1}$$

$$k_1 = -0.04107 + 54.56/(T/\text{K})$$

Of special importance for stratospheric ozone chemistry are the following reactions which take place in polar stratospheric clouds which consist of droplets of sulfuric acid:

$$ClONO_2 + HCl \longrightarrow Cl_2 + HNO_3$$

$$N_2O_5 + H_2O \longrightarrow 2HNO_3$$

$$HOCl + HCl \longrightarrow Cl_2 + H_2O$$

$$N_2O_5 + HCl \longrightarrow ClNO_2 + HNO_3$$

$$BrONO_2 + H_2O \longrightarrow HOBr + HNO_3$$

A kinetic model of the uptake of HOCl and also $ClONO_2$ by sulfuric acid in the presence of hydrogen chloride has been published by Shi *et al.* [64]. Details are also given in JPL Publication 00–003 [65]. This model is based on experimental data published by Robinson *et al.* [66], Donaldson *et al.* [67] and Hanson [68].

The model is valid from 180 to 260 K. Allowance is made for the mechanism of the reaction of HCl with both $ClONO_2$ and HOCl to include protonation to form $HClON_2^+$ and H_2OCl^+. Hydrolysis of $ClONO_2$ is assumed to take place partly via the protonated species. The model includes the calculation of solubilities based on Henry's law coefficients. Allowance is made for variation of reaction rate constants with variation of temperature and composition. Variations of viscosity and diffusion coefficients are based on work by Klassen *et al.* [69]. Variation of the activity of hydrogen ions is based on work by Michelsen [70] and Carslaw *et al.* [50].

The expression used for the Henry's law coefficient for $ClONO_2$ is

$$k_H/\text{mol bar}^{-1} = 1.6 \times 10^{-6}\exp[4710/(T/\text{K})]\exp[-S_{ClONO_2}(M/\text{mol dm}^{-3})]$$

where S_{ClONO_2} is a Sechenov-type coefficient given by

$$S_{ClONO_2}/\text{mol}^{-1} = 0.306 + 24/(T/\text{K})$$

and M is the molar concentration of sulfuric acid.

The expression for the Henry's law coefficient of HCl is

$$k_H/\text{mol bar}^{-1} = (0.093 - 0.60x + 1.2x^2)$$
$$\times \exp[-8.68 + (8515 - 10718x^{0.7})/(T/\text{K})]$$

where x is the mole fraction of H_2SO_4.

Shi *et al.* prepared a computer code for calculations from the model. The user inputs the temperature, the radius of the aerosol particles and the partial pressures of H_2O, HCl and $ClONO_2$. Reactive uptake coefficients of $ClONO_2$ with H_2O, $ClONO_2$ with HCl and HOCl with HCl are output. Figure 9.15 shows reactive uptake coefficients for a typical set of conditions.

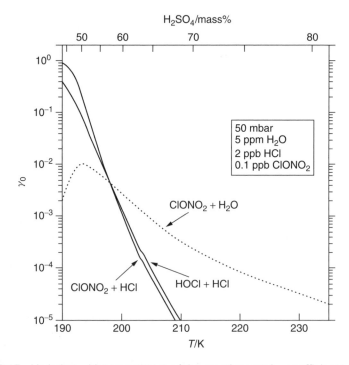

Figure 9.15 Variation with temperature of the reactive uptake coefficients for reaction of $ClONO_2$ and HOCl in sulfuric acid solutions for typical concentrations of HCl and $ClONO_2$ in the stratosphere. Curves are based on the model published by Shi *et al.* [64] and published in an addendum [71] to [65]

Hypobromous Acid, HOBr [CAS 13517-11-8]

Reactive bromine species play a significant role in the depletion of ozone in the atmosphere by the reaction

$$Br + O_3 \longrightarrow BrO + O_2$$

BrO can react in various ways to produce Br atoms and continue the chain reaction leading to destruction of further ozone molecules, e.g.

$$2BrO \longrightarrow 2\,Br + O_2$$

and

$$HO_2 + BrO \longrightarrow HOBr + O_2$$

HOBr is photolysed in the gas phase to produce bromine atoms:

$$HOBr + h\upsilon \longrightarrow OH + Br$$

Bromine atoms are very efficient ozone removers because they do not readily form stable species in the atmosphere.

Modelling studies have indicated the important role played by HOBr in producing bromine atoms. Not only is it photolysed in the gas phase as indicated above but also it reacts in the aqueous phase in the presence of bromide ions to produce bromine molecules, which are photolysed in the gas phase to give bromine atoms. In the presence of chloride ions HOBr reacts in the liquid phase to give ClBr, which is photolysed in the gas phase to give Cl and Br atoms.

Fickert *et al.* [72] studied the uptake of HOBr by aqueous solutions containing Cl^- and Br^-. A wetted wall flow tube reactor was used with helium or nitrogen as carrier gas. Under the conditions of the measurements, the rate of uptake was found to be limited by the rate of gas-phase diffusion of HOBr. The diffusion coefficient of HOBr is helium was found to be $319 \pm 48\,\mathrm{Torr\,cm^2\,s^{-1}}$ and that in nitrogen $84 \pm 7\,\mathrm{Torr\,cm^2\,s^{-1}}$ (errors correspond to two standard deviations). The accommodation coefficient, α, was found to be greater than 1×10^{-2}.

At a concentration of chloride ions of $1\,\mathrm{mol\,dm^{-3}}$ and bromide ions of $1 \times 10^{-3}\,\mathrm{mol\,dm^{-3}}$ in acidified solutions, more than 90 % of the HOBr which was taken up was followed by release of Br_2 to the atmosphere. As the concentration of bromide ions was lowered, BrCl became the dominant bromine compound which was released.

The following equilibria can explain the behaviour of HOBr in an aqueous solution such as seawater containing chloride and bromide ions. Data for rate constants (k_f, k_b) and equilibrium constants (K_3, K_4) have been published by Eigen and Kustin [73] and Wang *et al.* [74].

$$HOBr + Br^- + H^+ \rightleftharpoons Br_2 + H_2O \qquad (9.6)$$

$$k_f/\mathrm{mol^{-1}\,dm^3\,s^{-1}} = 1.6 \times 10^{10};\ k_b/\mathrm{s^{-1}} = 110\ (298\,\mathrm{K})$$

$$HOBr + Cl^- + H^+ \rightleftharpoons BrCl + H_2O \qquad (9.7)$$

$$k_f/\mathrm{mol^{-1}\,dm^3\,s^{-1}} = 5.6 \times 10^9;\ k_b/\mathrm{s^{-1}} > 1 \times 10^5\ (298\,\mathrm{K})$$

$$BrCl + Br^- \rightleftharpoons Br_2Cl^- \tag{9.8}$$

$$K_3 = [Br_2Cl^-]/[BrCl][Br^-] = (1.8 \pm 0.2) \times 10^4\,mol^{-1}\,dm^3 \; (298\,K)$$

$$Br_2 + Cl^- \rightleftharpoons Br_2Cl^- \tag{9.9}$$

$$K_4 = [Br_2Cl^-]/[Br_2][Cl^-] = (1.3 \pm 0.3)\,mol^{-1}\,dm^3 \; (298\,K)$$

$$BrCl + Cl^- \rightleftharpoons BrCl_2^-$$

$$Br_2 + Br^- \rightleftharpoons Br_3^-$$

$$BrCl(aq) \rightleftharpoons BrCl(g)$$

$$Br_2(aq) \rightleftharpoons Br_2(g)$$

It follows that

$$\frac{K_4}{K_3} = \frac{[Br_2Cl^-]/[Br_2][Cl^-]}{[Br_2Cl^-]/[BrCl][Br^-]} = \frac{[BrCl][Br^-]}{[Br_2][Cl^-]}$$

$$\frac{[BrCl]}{[Br_2]} = \frac{K_4[Cl^-]}{K_3[Br^-]}$$

The ratio K_4/K_3 is equal to $(7.22 \pm 1.8) \times 10^{-5}$.

Figure 9.16 shows experimental measurements of the BrCl/Br$_2$ ratio in the gas phase when a trace concentration of HOBr ($\sim 10^{12}$ molecules cm^{-3}) passed over a film of an aqueous solution of chloride ions (1 mol dm^{-3}) with various

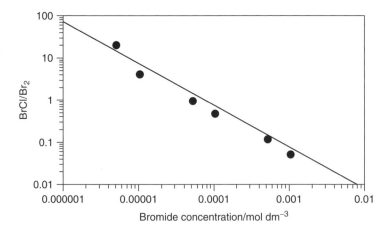

Figure 9.16 Ratio of BrCl to Br$_2$ in the gas phase produced by reaction of HOBr with a solution of NaCl (1 mol dm^{-3}) with various concentrations of NaBr in a wetted-wall flow tube reactor at 274 K measured by Fickert et al. [72]. The solid line shows the ratio predicted by the mode discussed in the text using equilibria data at 298 K published by Wang et al. [74]

concentrations of bromide ions [72]. The pH was adjusted to 5.5. The experimental ratios are in satisfactory agreement with those predicted by the model described above.

As indicated above, the initial reaction HOBr with Br^- and Cl^- to give Br_2 or BrCl in solution depends on the concentration of hydrogen ions. Fickert *et al.* [72] also investigated the effect of variation of pH on the concentration of bromine in the gas phase. The concentrations of chloride and bromide ions in the aqueous phase were maintained at 1 and 10^{-3} mol dm^{-3}, respectively. There was a negligible change in the formation of bromine in the pH range 4–5.5. At higher pH the concentration was reduced. At pH greater than 8.5 the yield was zero.

Other Halogen Compounds

De Bruyn *et al.* [75] carried out uptake studies on CCl_2O, CF_2O, CCl_3CClO, CF_3CFO and CF_3CClO. These compounds are of special interest because they are formed as intermediate products during the gas-phase oxidation in the atmosphere of volatile halogen compounds. It is important to be able to estimate the tropospheric lifetime of these compounds to determine whether they have a significant role in ozone depletion.

Values of the expression $k_H k_1^{1/2}$ were measured by time-resolved gas–liquid experiments with a bubble column apparatus as described earlier, where k_H is the Henry's law constant and k_1 the pseudo-first-order rate constant for hydrolysis of the carbonyl halide. Studies were carried out at 278 K and pH 1–13.

As can be seen from Figure 9.17, the limiting values of k_H at low values of k are about 0.15 and 1.0 mol dm^{-3} bar^{-1}, respectively. These limiting values

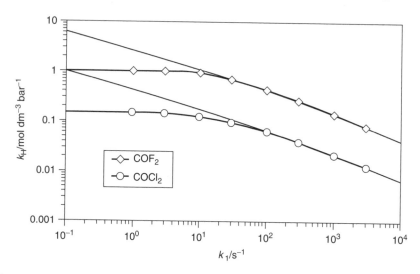

Figure 9.17 Best fit values of the Henry's law type constant, k_H, of CCl_2O and of CF_2O at pH 6 and 278 K versus eight different assumed values of the first-order rate constant for reaction with water, k_1. This graph is based on the model for absorption and the uptake coefficients discussed earlier

Table 9.22 Kinetic data for various halogen compounds

Compound	T/K	k_H/mol dm^{-3} bar^{-1}	k_1/s^{-1}	$k_H k_1^{\frac{1}{2}}$	$k_H k_1$/mol dm^{-3} bar^{-1} s^{-1}	k_{OH^-}/mol dm^{-3} s^{-1}	Method	Ref.
Trifluoroacetyl fluoride, CF$_3$CFO, [CAS 354-34-7]	278	$\leq$0.95		$\leq$3.8	4.2–95	$\geq$1.5 × 10^6	Bubble column	75
	284	3	150	37,60	444		Droplet train	76
Trifluoroacetyl chloride, CF$_3$CClO, [CAS 354-32-5]	278	$\leq$0.27		$\leq$1.2	1.2–27	$\geq$8.9 × 10^5	Bubble column	75
	284	2	220	30,60	434		Droplet train	76
Carbonyl chloride (phosgene), CCl$_2$O, [CAS 75-44-5]	278	$\leq$0.15		$\leq$0.66	0.67–15	$\geq$4.6 × 10^3	Bubble column	75
	298	$\leq$0.06		$\leq$0.29			Bubble column	75
	298	0.7	6	0.17	0.4		Pulsed jet	77
	253		~100					77
	296			0.17			Wetted-wall reactor	79,80
	296				3.8		Saturated NaNO$_3$ aerosol	79,80
Carbonyl fluoride, CF$_2$O, [CAS 353-50-4]	298		5.3				Pulsed radiolysis	81
	278	$\leq$1.0		$\leq$4.3	4.6–96	$\geq$2.9 × 10^4	Bubble column	75
	284			350			Droplet train	76
Trichloroacetyl chloride, CCl$_3$CClO, [CAS 76-02-8]	278	$\leq$2.0		$\leq$6.9	9.0–148	$\geq$1.4 × 10^5	Bubble column	75
	274–294	2	500	45	987		Droplet train	82
	288		~150				Stopped-flow	82

represent maximum values of k_H, i.e. the real values of k_H are equal to or less than these maximum values.

In the case of CCl_3CClO and CCl_2O the temperature varied from 278 to 298 K. Limits of the value of $k_H k_1$ were obtained. Data were compared with those of other workers (Table 9.22).

The hydrolysis of CCl_3CClO, CF_3CClO and CCl_2O is consistent with a mechanism which involves catalysis by OH ions at high pH. The observed first-order rate constant, k_1, can be written as

$$k_1 = k_{H_2O} + k_{OH^-}[OH^-]$$

At low pH, the OH^--catalysed reaction is negligible and $k_{OH^-}[OH^-]$ is negligible compared with k_{H_2O}. At pH above about 11, k_{H_2O} is negligible compared with $k_{OH^-}[OH^-]$ and $k_1 = k_{OH}[OH]$. By combining the $k_H k_1$ value with the upper limiting value of k_H a lower limit to the second-order rate constant k_{OH} can be obtained. These values are given in Table 9.22.

There are large discrepancies between data reported by George et al. [76,82] from droplet train experiments and those reported by De Bruyn et al. [75] The latter indicated that the bubble column apparatus was calibrated and tested by making measurements on 11 systems having know values of k_H and k. However, they could not explain the large difference between the two sets of measurements.

The data for CCl_2O from De Bruyn are consistent with data reported by Behnke et al. [79,80] and by Manogue and Pigford [77] and also by Ugi and Beck [78].

Evaluated tables of uptake coefficients for HOCl, HOBr, HOI, BrO, CCl_3CClO, CF_3CFO and CF_3CClO have been prepared by members of IUPAC Subcommittee for Gas Kinetic Data Evaluation. These tables are available from a Web site [83].

7 NITROGEN COMPOUNDS

Ammonia, NH₃, [CAS 7664-41-7]

Ammonia in aqueous solution exists in equilibrium with ammonium ions:

$$NH_3(g) \rightleftharpoons NH_3(aq)$$

$$NH_3(aq) + H_2O \rightleftharpoons NH_4^+(aq) + OH^-(aq)$$

The effect of ionic strength and pH has been studied by Clegg and co-workers [84,85].

The relationship between solubility and partial vapour pressure has been studied over the whole of the composition range by various workers [86–89]. Available data have been discussed by Fogg and Gerrard [90].

The system was investigated by Clifford and Hunter [91] at temperatures from 333 to 420 K and total pressures from about 1 to 16 bar. They also made use of earlier published work to present smoothed data for lower temperatures and

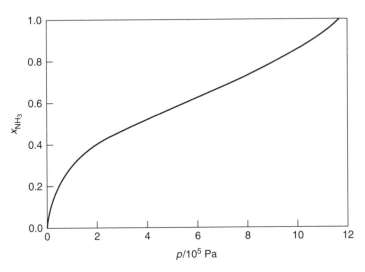

Figure 9.18 Variation of the mole fraction solubility of ammonia in water with partial pressure of ammonia, p, at 303.15 K from data published by Clifford and Hunter [91] (based on figure published by Fogg and Gerrard [90])

pressures. A plot of mole fraction solubility against pressure at 303.2 K, from data in this paper, is shown in Figure 9.18.

Various workers have measured mass accommodation coefficients for ammonia. This is defined as the fraction of ammonia molecules colliding with a liquid surface which are absorbed by the liquid. Ponche *et al.* [92] published an experimental value of 0.024 ± 0.01 determined by absorption of ammonia into a stream of monodisperse droplets generated by a vibrating orifice. This value was corrected for possible chemical interference to gave a value of 0.097 ± 0.009. This may be compared with the value of 0.04 $(+0.03; -0.005)$ given by Bongartz *et al.* [93] determined from the rate of absorption of 50–200 ppm of ammonia at 299 ± 1 K in a liquid jet as a function of the jet length.

Other less accurate estimations of the mass accommodation coefficients have been published by Larson and Taylor [94], Richardson and Hightower [95] and Harrison *et al.* [96].

Values of diffusion coefficients at 290 K in the gas phase, D_g, and the aqueous phase, D_a, have been published: $D_g = 0.230 \, \text{cm}^2 \, \text{s}^{-1}$ [97] and $D_a = 1.9 \times 10^{-5} \, \text{cm}^2 \, \text{s}^{-1}$ [98].

Swartz *et al.* [94] have published measurements of the uptake of ammonia by buffered aqueous solutions in the temperature range 263–293 K. A droplet train apparatus was used. At pH > 10 the uptake was limited by the physical solubility. The uptake increases with lower pH. At pH < 2 uptake is limited by the mass accommodation. This was stated to be ~0.08 at 283 K. At pH 13 the value of Henry's constant, k_H, was given by the equation

$$\log(k_H/\text{mol dm}^{-3} \, \text{bar}^{-1}) = 3.225 + 1396/(T/K)$$

Water at pH 4 was exposed to a mixture of NH_3 and SO_2. The uptake of NH_3 was enhanced owing to change of pH at the surface.

Uptake of ammonia by sulfuric acid at concentrations of 10–70 mass% was also studied in the temperature range 248–293 K. The uptake was found to be linearly proportional to the liquid surface area and to the gas–liquid contact time. The uptake on concentrated sulfuric acid is controlled by surface reaction with H^+ and the uptake coefficient is close to unity and is independent of temperature.

Carstens and Schurath [100] measured the mass accommodation coefficient of NH_3 on acidified water using a liquid jet technique with helium as carrier gas. The amount of NH_3 absorbed was calculated from the change in conductivity of the water due to replacement of H_3O^+ by NH_4^+. The temperature range was 275–326 K. Mass accommodation was assumed to be a two-step process with physical absorption followed by hydration. It was assumed that the following equation held:

$$\alpha/(1-\alpha) = \Delta G^*/RT$$

where ΔG^* is the Gibbs free energy of the transition state on going from gas phase to solvated species in solution and α is the mass accommodation coefficient.

The corresponding enthalpy and entropy changes were calculated from

$$\Delta G^* = \Delta H^* - T\Delta S^*$$

giving $\Delta H^* = -29.9 \pm 4.5 \, \text{kJ mol}^{-1}$ and $\Delta S^* = -126 \, \text{J mol}^{-1} \, \text{K}^{-1}$.

Nitrosyl Chloride, NOCl, [CAS 2696-92-6]

Scheer et al. [101] measured uptake coefficients of nitrosyl chloride by pure water and by aqueous solutions of electrolytes in the temperature range 273–293 K. They used wetted wall flow tubes operating at atmospheric and at reduced pressure (90 hPa) and also a droplet train flow tube. The data from measurements using the flow tube operating at atmospheric pressure are not reproduced here because they were much less reliable than other measurements. Uptake coefficients, γ, tended to increase with temperature as indicated in Figure 9.19. They also increased with increase in pH. Lower limits for $k_H k_1^{1/2}$ were calculated on the assumption that the reactive uptake of the NOCl was the rate-determining step in the overall uptake. On the basis of this assumption it follows that

$$\gamma_{\text{meas}} \simeq \frac{4 k_H RT}{\bar{c}} (D_1 k_1)^{\frac{1}{2}}$$

where γ_{meas} is the experimentally measured uptake coefficient, k_H is Henry's law constant, R is the gas constant, T is the absolute temperature, k_1 is the first-order rate constant for reaction with water and $\bar{c}$ is the mean molecular velocity (see Section 8.1). The value of the liquid-phase diffusion coefficient, D_1, was estimated by a method developed by Wilke and Chang as described by Reid et al. [102].

It was also estimated that the value of a Henry's law constant, k_H, was $>0.05 \, \text{mol dm}^{-3} \, \text{bar}^{-1}$ in the temperature range studied. A summary of data published by Scheer et al. [101] is given in Table 9.23.

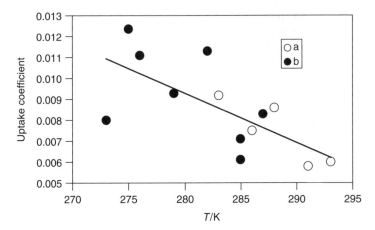

Figure 9.19 Plot of experimental values of the uptake coefficient for absorption of NOCl by water measured over a temperature range by Scheer *et al.* [101]. The trendline for the data is shown. a, Flow tube; b, droplet train

Table 9.23 Steady-state reactive uptake coefficient for dissolution of nitrosyl chloride in water and solutions of electrolytes [101]

Solvent	T/K	γ_{ss}	$\Delta\gamma^a$	$k_H k_1^{\frac{1}{2}}/$ $mol\,dm^{-3}$ $bar^{-1}\,s^{-\frac{1}{2b}}$	Method^c
Water	283	0.0092	+0.0035/−0.0021	894	a
	286	0.0075	+0.0006/−0.0005	690	a
	288	0.0086	+0.0072/−0.0030	792	a
	291	0.0058	+0.0045/−0.0021	497	a
	293	0.006	+0.0007/−0.0006	491	a
	273	0.008	+0.0017/−0.0016	953	b
	275	0.0124	+0.0016/−0.0015	1420	b
	276	0.0111	+0.0006/−0.0006	1245	b
	279	0.0093	+0.0019/−0.0017	990	b
	282	0.0113	+0.0019/−0.0018	1132	b
	285	0.0071	+0.0011/−0.0010	673	b
	285	0.0061	+0.0016/−0.0015	582	b
	287	0.0083	+0.0012/−0.0012	764	b
1 M NaCl	293	0.0134	+0.9866/−0.0107	1142	a
0.1 M HCl	293	0.0057	+0.0009/−0.0007	472	a
1 M HCl	293	0.0042	+0.0029/−0.0014	355	a
1 M NaOH	276	0.0294	+0.0017/−0.0016	3649	b
1 M NaCl	276	0.0063	+0.0004/−0.0005	726	b
1 M HCl	276	0.0021	+0.0003/−0.0002	242	b

^{a}Errors at 95 % confidence level.
^{b}The authors stated that these values should be regarded as lower limits.
^{c}a, Low-pressure flow tube; b, droplet chain.

Nitryl Chloride, ClNO$_2$, [CAS 13444-90-1]

Behnke *et al.* [103] measured the uptake coefficients at 291 K of nitryl chloride, ClNO$_2$, by pure water and solutions of sodium chloride using a wetted-wall flow tube. The ClNO$_2$ was generated by passing N$_2$O$_5$ over sodium chloride solution in a wetted-wall flow tube. The gas then passed over water or sodium chloride solution in a second wetted wall flow tube. Uptake coefficients are shown in Table 9.24.

The experimental points appear to be consistent and lie close to a smooth curve based on a model which takes into account reversible hydrolysis of ClNO$_2$ in the presence of excess chloride ions (Figure 9.20).

Table 9.24 Steady-state reactive uptake coefficient of nitryl chloride by pure water and solutions of sodium chloride at 291 K[a]

Conc. of NaCl/ mol dm^{-3}	Uptake coefficient, $\gamma_{ss}/10^{-6}$	Conc. of NaCl/ mol dm^{-3}	Uptake coefficient, $\gamma/10^{-6}$
0.0[b]	4.84 ± 0.13	1	1.27 ± 0.18[c]
0	4.54 ± 0.17	2	0.76 ± 0.05
0.1	3.10 ± 0.27	2.9	0.61 ± 0.06
0.2	2.46 ± 0.21	4	0.40 ± 0.05
0.3	2.30 ± 0.16	4.6	0.27 ± 0.02
0.6	1.61 ± 0.11		

[a]Experimental points appear to be consistent and lie close to a smooth curve based on a model which takes into account reversible hydrolysis of ClNO$_2$ in the presence of excess chloride ions (Figure 9.20).
[b]Concentration of ClNO$_2$ = 60 ppm. In all other runs the concentration was 50 ppm.
[c]The errors correspond to twice the standard deviation.

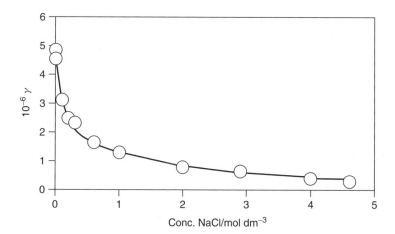

Figure 9.20 Variation of the uptake coefficient for ClNO$_2$ with variation of the concentration of NaCl at 291 K. The curve corresponds to a model presented by the authors [103]

Bromine Nitrate, BrNO₃, [CAS 40423-14-1]

Hanson *et al.* [121] investigated the uptake of $BrONO_2$ by submicron sulfuric acid aerosol. Reaction with the water present produced HOBr and HNO_3. The apparatus used consisted of a cylindrical flow reactor coupled with a chemical ionization mass spectrometer. The concentration of the acid was calculated from the partial pressure of water vapour and the temperature. Particle size and surface area was calculated from the UV extinction of the aerosol and the particle number density. Particle numbers, found by using an expansion counter, were $(0.4–7) \times 10^5$ cm^{-3} (see Table 9.25).

The uptake coefficient was almost unaffected by changes in particle radius. The data are scattered with high errors. There was no change with change in temperature which was outside error limits. Changes with change in concentration of acid from 47 to 73 mass% were also within error limits. However, there was an approximately linear decrease with increase in acid concentration from 73 to 83 mass%.

The authors showed that data fitted the following equation, within experimental error:

$$\frac{1}{\gamma_{meas}} = \frac{1}{\alpha} + \frac{1}{\Gamma_{rxn}}$$

where α is the mass accommodation coefficient. The rate-limiting effects of diffusion in the gas and liquid phases can be neglected in this case (see Section 8.1).

$$\Gamma_{rxn} = a \times a(H_2O)^b$$

where $a(H_2O)$ is the activity of the water. The best fit was given with the values $\alpha = 0.84 \pm 0.08$, $a = 211(\times 2.4, \div 2.4)$ and $b = 1.37 \pm 0.18$.

Table 9.25 Steady-state reactive uptake coefficient of BrNO₃ on sulfuric acid solutions at various partial pressures and temperatures[a]

T/K	Total pressure/ bar	$p(H_2O)$/Pa	Conc. H_2SO_4/ mass%	r_s/μm	Diffusion correction/%	γ_{ss}	Error
298	0.287	390	62.5	0.09	3	0.7	±0.15
297	0.264	160	69	0.09	5	0.91	±0.24
297	0.251	27	78	0.09	2	0.27	±0.06
298	0.251	27	78	0.17	2	0.24	±0.06
298	0.253	27	78	0.05	1	0.22	±0.06
294	0.235	29	76.5	0.09	1	0.26	±0.06
294	0.229	96	70.5	0.10	3	0.76	0.24
296	0.827	6.4	83	0.18	3	0.05	±0.02
296	0.827	72	73	0.18	20	0.53	±0.2
272	0.232	35	66	0.10	5	0.75	±0.25
249	0.225	32	47	0.10	6	0.81	±0.25
229	0.211	1.2	60[b]	0.1	3	0.9	±0.2
229	0.211	3.6	48[c]	0.11	9	1	±0.2

[a] r_s is the surface area-weighted mean radius; the uncertainty in the concentration of H_2SO_4 is ±1 %.
[b] HCl added ($\sim 10^{-3}$ mol dm^{-3}) to scavenge HOBr formed by hydrogenolysis.
[c] HCl added (~ 0.3 mol dm^{-3}). This may have affected the uptake of $BrONO_2$.

The value of γ for this compound is about 1000 times larger than that for $ClONO_2$. Reaction takes place over a very short reactive diffusive length and is therefore unaffected by particle size in the range studied. $BrONO_2$ is much more readily hydrolysed than $ClONO_2$. Recommended reactive uptake coefficients from JPL Publication 00–003 [65] are shown graphically in Figure 9.22.

Nitrogen Oxides, NO_2, NO_3, N_2O_5

Nitrogen Pentoxide, N_2O_5, [CAS 10102-03-1]

The uptake of N_2O_5 by water was measured by van Doren *et al.* [51] using a droplet train apparatus in an atmosphere of water vapour and helium. Data given by van Doren *et al.* are shown in Table 9.26.

The diffusion coefficient, D, of N_2O_5 in a mixture of He and H_2O was calculated in a similar way to values for hydrogen chloride as described above. The binary diffusion coefficients for N_2O_5 with water vapour and N_2O_5 with helium were taken to be equal to the values for SO_2 multiplied by a correction for the reduced mass difference. Values of the observed uptake coefficient, γ_{meas}, were corrected to give values for zero pressure, γ_0. The authors identified these high uptake coefficients with the accommodation coefficient.

Behnke *et al.* [103,122] investigated the uptake of N_2O_5 by NaCl solution. The method involved the interaction of N_2O_5 gas of known concentration with an aerosol of sodium chloride solution. The unreacted N_2O_5 was measured by photolysing the mixture to give NO_2 and NO. The NO_2 concentration was measured with a luminol analyser. Reaction produced nitric acid and nitryl chloride, $ClNO_2$. Uptake coefficients were measured at 291 K for relative humidities of 77.3–93.9 % but variation with humidity may not be significant. The average value obtained was $(3.2 \pm 0.2) \times 10^{-2}$.

Sander *et al.* [123] quoted the uptake coefficient for N_2O_5 on sulfuric acid as 0.1 at an unspecified temperature. This value was taken from a report by DeMore *et al.* [124].

Hu and Abbatt [125] measured the uptake of N_2O_5 on sulfuric acid solutions and on ammonium sulfate solutions. Reactions took place in a flow tube between N_2O_5 gas and the solutions in the form of an aerosol. Comparison by these authors of measurements of the uptake coefficients reported by various authors are shown in Figure 9.21. Uptake of N_2O_5 on ammonium sulfate solutions at 297 ± 1 K is

Table 9.26 Uptake of N_2O_5 by water [51][a]

No. of data	T/K	P_{tot}/bar	p_{H_2O}/bar	Orifice diam./ μm	γ_{meas}	D/bar $cm^2\,s^{-1}$	γ_0
7	271	6.69×10^{-3}	5.36×10^{-3}	60	0.057 ± 0.003	0.125	0.061 ± 0.004
20	282	12.35×10^{-3}	11.07×10^{-3}	61	0.036 ± 0.004	0.120	0.040 ± 0.005

[a]No. of data = No. of experimental measurements at each temperature. P_{tot} = total pressure; p_{H_2O} = partial pressure of water; D = diffusion coefficient of N_2O_5 in the mixture of water vapour and helium; γ_0 = corrected uptake coefficient for zero pressure.

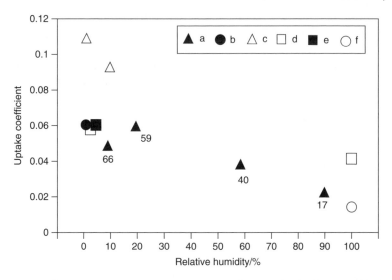

Figure 9.21 Experimental values of uptake coefficients for N_2O_5 on sulfuric acid solutions in the temperature range 277–298 K as a function of relative humidity as published by Hu and Abbatt [125]. a, Hu and Abbatt, 297 K [125]; b, Fried *et al.*, 293 K [127]; c, Mozurkewich and Calvert, 293 K [126]; d, Van Doren *et al.*, 282 K [51,128]; e, Lovejoy and Hanson, 298 K [129]; f, George *et al.*, 277 K [130]. The numbers next to the experimental points measured by Hu and Abbatt [125] correspond to the mass% of H_2SO_4 in the aerosol

Table 9.27 Variation with relative humidity of the steady-state reactive uptake coefficient at 297 K of N_2O_5 on $(NH_4)_2SO_4$ solution [125]

RH/%[a]	$(NH_4)_2SO_4$/mass%	Uptake coefficient, γ_{ss}
93.5	20	0.017 ± 0.002
83	40	0.023 ± 0.004
68.5	60	0.053 ± 0.006
50	80	0.044 ± 0.008

[a]RH = relative humidity.

shown in Table 9.27. Hu and Abbatt [125]stated that the rate-determining steps in each case were likely to be either the accommodation of the N_2O_5 on the surface or the dissociation of N_2O_5 into NO_2^+ and NO_3^-.

Mozurkewich and Calvert reported that dry ammonium sulfate was unreactive to N_2O_5 [126].

Recommended reactive uptake coefficients published in JPL Publication 00–003 [65] are shown graphically in Figure 9.22.

Nitrogen trioxide, NO₃, [CAS 12033-49-7]

Imamura *et al.* [104] measured the reactive uptake coefficient, γ, of the NO_3 radical on to aqueous solutions containing ions $X^- = HSO_3^-$, SO_3^{2-}, $HCOO^-$,

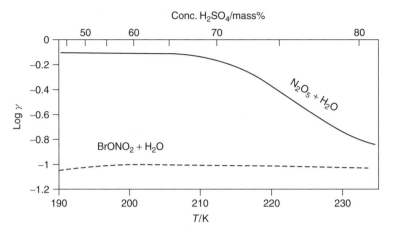

Figure 9.22 Recommended reactive uptake coefficients for BrONO$_2$ and N$_2$O$_5$ [65]

CH_3COO^- and OH^-. The authors used a wetted-wall flow tube reactor at 273 K. The values of the second-order rate constant, k, and also $k_H^2 D/k$ were found by measuring γ as a function of the composition of the solution, (k_H = Henry's law coefficient, D = diffusion coefficient in the liquid phase). The authors suggested that the reaction between NO_3 and the individual anions proceeded by electron transfer:

$$NO_3 + X^- \longrightarrow NO_3^- + X$$

The same group measured the uptake of NO_3 on potassium chloride, bromide and iodide solutions. Measurements indicated that reduction of iodide ions also proceeded by an electron transfer reaction [105,106].

By measuring the uptake coefficient as a function of salt concentration, the authors determined the value of $k_H D^{0.5}$ for NO_3 in pure water at 273 K as $(1.9 \pm 0.4) \times 10^{-3}$ mol dm^{-3} bar^{-1} s$^{-0.5}$. This was based on the assumption that the rate constant for the reaction of NO_3 with Cl^- is 2.76×10^6 mol^{-1} s^{-1} at 273 K as given by Exner *et al.* [107]. The Henry's law coefficient for NO_3 in water was estimated to be 0.6 ± 0.3 mol dm^{-3} bar^{-1} on the assumption that the diffusion coefficient of NO_3 in water is $(1.0 \pm 0.5) \times 10^{-5}$ cm^2 s^{-1}. This value of D was estimated by the following equation [108]:

$$D = kT/6\pi\eta R^*$$

where η is the viscosity of the liquid, k is the Boltzmann constant and R^* the effective radius of the solvated molecule.

Nitrogen dioxide, NO$_2$, [CAS 10102-44-0]

The kinetics of reactive dissolution of NO_2 in water have been studied by Schwartz and White [109]. The theory relating to low pressures has been summarized by Cheung *et al.* [110]. The behaviour of NO_2 when in interacts with water is complicated by the fact that NO_2 exists in equilibrium with N_2O_4 in both the gas phase and aqueous phase. It follows that

$$K_g = p_{N_2O_4}/(p_{NO_2})^2$$

$$K_{aq} = [N_2O_4(aq)]/[NO_2(aq)]^2$$

where K_g and K_{aq} are the equilibrium constants for dimerization in the gas and aqueous phase, respectively.

As shown by Schwartz and White [109], the total nitrogen in the oxidation state IV, N(IV), in the gas phase and aqueous phase is given by

$$[N(IV)aq] = [NO_2(aq)] + 2[N_2O_4(aq)]$$

$$= [NO_2(aq)] + 2K_{aq}[NO_2(aq)]^2$$

$$p_{N(IV)} = p_{NO_2} + 2p_{N_2O_4} = p_{NO_2} + 2K_g(p_{NO_2})^2$$

The solubility of the NO_2 species in the gas phase to give NO_2 in the aqueous phase is given by

$$[NO_2(aq)] = k_H p_{NO_2}$$

or

$$k_H = [NO_2(aq)]/p_{NO_2}$$

where k_H is the Henry's law constant for NO_2.

The effective Henry's law constant, k_H^*, is given by

$$k_H^* = \frac{[N(IV)aq]}{p_{N(IV)}} = \frac{k_H(1 + 2K_{aq}k_H p_{NO_2})}{(1 + 2K_g p_{NO_2})}$$

If p_{NO_2} is low enough, then $2K_g p_{NO_2} \ll 1$, and

$$k_H^* \simeq k_H(1 + 2K_{aq}k_H p_{NO_2})$$

As the pressure is decreased then k_H^* approaches k_H, which allows the measurement of k_H.

NO_2 reacts with water to form $HNO_2(aq)$ and $HNO_3(aq)$ by two different pathways:

$$2NO_2(aq) + H_2O(l) \xrightarrow{\ k'\ } HNO_2(aq) + HNO_3(aq)$$

$$2NO_2(aq) + H_2O(l) \underset{k_{back}}{\overset{k_{forw}}{\rightleftharpoons}} N_2O_4(aq) + H_2O(l) \xrightarrow{\ k''\ } HNO_2(aq) + HNO_3(aq)$$

Both of these pathways involve reactions which are second order with respect to NO_2 concentration. The rate of disappearance of NO_2 can be written as

$$-d[NO_2(aq)]/dt = 2k'[NO_2(aq)]^2 + 2k''[N_2O_4(aq)]$$

$$= 2k'[NO_2(aq)]^2 + 2k''[NO_2(aq)]^2 K_{aq}$$

$$= 2k_2[NO_2(aq)]^2$$

where $k_2 = k' + k'' K_{aq}$. The overall rate of disappearance of N(IV) is, however, mixed order in N(IV) concentration because the reaction of $N_2O_4(aq)$ to give $2NO_2(aq)$ is first order with respect to $N_2O_4(aq)$. The overall removal of N(IV) can therefore be represented as

$$-d[N(IV)aq]/dt = k_F[N(IV)aq]^n$$

At low concentrations, when the quantity of $N_2O_4(aq)$ is negligible, n equals 1. The value of n increases to 2 as concentrations tend to become high enough for $N_2O_4(aq)$ to be the dominant form of N(IV).

Cheung *et al.* [110] measured the uptake of NO_2 by water using a droplet train apparatus and also a horizontal bubble train apparatus. They showed how the uptake measurements could be interpreted to allow for a mixed-order kinetics rather than pseudo-first-order kinetics shown by other systems.

Measurements using the droplet train apparatus indicated that the uptake coefficient at 273 K was less than 5×10^{-4} at 273 K. This was consistent with a rate dominated by a bulk-phase process and did not provide evidence for a surface reaction postulated by Mertes and Wahner [111a].

Measurements using a horizontal bubble train apparatus provided values of the Henry's law constant for NO_2 (k_H) and the rate constant k_2. The authors published a table of values of these quantities measured by themselves and other workers (Table 9.28).

The measurements by Cheung *et al.* were consistent with a value of the dimerization constant K_{aq} less than about 6×10^4 mol dm^{-3}. This is in accord with a value of $(6.5 \pm 0.3) \times 10^4$ published by Grätzel *et al.* [116].

Kleffmann *et al.* [118] studied the heterogeneous conversion of NO_2 to HONO on sulfuric acid surfaces in a quartz reactor and bubbler system. The conversion

Table 9.28 Values of the Henry's law constant and the second-order rate constant for absorption and reaction of nitrogen dioxide with water[a]

Ref.	Method	T/K	$k_H k_2^{\frac{1}{2}}/$ $mol^{\frac{1}{2}} bar^{-1} s^{-\frac{1}{2}}$	$k_H \times 10^2/$ $mol\,dm^{-3}\,bar^{-1}$	$k_2 \times 10^{-7}/$ $mol\,dm^{-3}\,s^{-1}$
110	Uptake	293	70 ± 10	1.4 ± 0.2	2.7 ± 0.7
111b	Uptake	293		2.0 ± 0.3^b	
112	Uptake	295	69 ± 9		2.5 ± 0.6
113	Uptake	295	91 ± 20		4.3 ± 1.9
114	Uptake	293	105 ± 20^c		5.7 ± 2.2
115	Decay	293			2.6^d
116	Decay	293			6.5 ± 0.7
117	Decay	298			4.7 ± 1.0
110		293		1.4 ± 0.2	3.0 ± 0.9
		276		$2.3\,(+0.3, -0.9)$	2.2 ± 0.6

[a] k_H = Henry's law constant for NO_2; k_2 = rate constant for hydrolysis of NO_2; Uptake = gas uptake measurements; Decay = aqueous phase concentration decay measurements.
[b] Adjusted from 288 to 293 K.
[c] Adjusted from 283 to 293 K.
[d] Revised as indicated by Schwartz and White [109].

is first order with respect to concentration of NO_2 and is limited by an uptake coefficient of NO_2 of about 10^{-7}–10^{-6} under the conditions of the experiment. On surfaces of moderate acidity conversion proceeds by the reaction

$$2NO_2 + H_2O \longrightarrow HONO + HNO_3$$

with minor involvement of NO. The authors assumed that mass accommodation was the limiting factor so that the accommodation coefficient approximated to the uptake coefficient. The authors pointed out that the value of 10^{-7}–10^{-6} was comparable to the value of $<5 \times 10^{-6}$ for the uptake on 70 mass% sulfuric acid reported by Saastad et al. [119] but differed significantly from the value of 1.5×10^{-3} reported by Ponche and also from values published by Lee and Tang [119a] of $(6.3 \pm 0.7) \times 10^{-4}$ and by Mertes and Wahner [111] of $>2 \times 10^{-4}$.

Nitrous Acid, HONO, [CAS 7782-77-6]

Nitrous acid ionizes according to the equilibrium

$$HONO \rightleftharpoons H^+ + NO_2^-$$

The effective Henry's law constant (total concentration of HONO in solution/p_{HONO}), $k_H{}^*$, is related to the ratio of nonionized HONO in solution to the partial pressure of HONO, k_H, by the equation

$$k_H = k_H{}^*/(1 + K_a/[H])$$

where $k_H{}^*$ is the measured value and K_a the ionization constant of HONO. The ionization constant was reported by Park and Lee [113] to be 5.3×10^{-4} mol dm^{-3} at 298.15 K. At pH less than about 1.3 there is less than a 1 % difference between k_H and $k_H{}^*$ and therefore negligible and the correction for ionization may be neglected in strongly acidic solutions.

Becker et al. [120] measured the solubility of nitrous acid in sulfuric acid of various concentrations and temperatures (Table 9.29). The apparatus consisted of a Pyrex glass reactor connected to a temperature-controlled bubbler producing a stream of nitrogen gas. Sodium nitrite was added to sulfuric acid in the reactor so as to produce HONO in situ. Nitrogen gas passed through the apparatus for 2 h to allow equilibrium between the gas and liquid phases. Concentrations in the gas phase were then measured by spectrometry and in the liquid phase by ion chromatography.

Extrapolation of data for 298.15 K to zero concentration of acid (see Table 9.30) gave a value for $k_H{}^*$ of 47.2 ± 2.8 mol kg^{-1} bar^{-1}. This is close to the value of 48.5 ± 3 mol kg^{-1} bar^{-1} from data reported by Park and Lee [113]. This is also close to the value of 49.3 ± 3 mol kg^{-1} bar^{-1} estimated by Régimbal and Mozurkewich [131] from available thermodynamic data and based on earlier calculations by Schwartz and White [132].

Table 9.29 Effective Henry's law constant for the solution of HONO in H_2SO_4 at different temperatures[a]

$c_{H_2SO_4}/$ mass%	T/K	$k_H^*/$ mol kg^{-1} bar^{-1}	$c_{H_2SO_4}/$ mass%	T/K	$k_H^*/$ mol kg^{-1} bar^{-1}	$c_{H_2SO_4}/$ mass%	T/K	$k_H^*/$ mol kg^{-1} bar^{-1}
0.0	273	220 ± 11	31.5	298	10.9 ± 1.4	58.1	298	12.3 ± 1.7
	283	119.5 ± 5.9	39.7	249	253 ± 33	61.2	250	2720 ± 320
	295	59.4 ± 3.0		267	59.1 ± 7.5		268	398 ± 51
	303	37.7 ± 2		283	18 ± 3.0		283	112 ± 16
8.7	269	180 ± 23		298	7.8 ± 1.4		298	41.6 ± 6.5
	278	108.5 ± 18.6	51.2	249	211 ± 27	63.3	250	5300 ± 3150
	285	67.8 ± 9.3		267	44 ± 5.5		265	1155 ± 180
	291	46.5 ± 7.5		283	11.8 ± 1.4		283	283 ± 38
	298	32.1 ± 4.7		298	5.1 ± 0.7		298	120 ± 16
22.6	258	246 ± 29	56.4	246	226 ± 26	65.6	254	9350 ± 3700
	268	115 ± 17		259	92.1 ± 13.2		268	3190 ± 620
	279	56.7 ± 8.4		272	38 ± 5.3		283	897 ± 126
	288	28.9 ± 4.2		286	12.9 ± 2.0		298	366 ± 51
	298	17 ± 2.6		298	7.1 ± 1.4	67.3	251	26600 ± 15300
31.5	250	330 ± 40	58.1	251	440 ± 65		268	7500 ± 3080
	267	87.2 ± 11.9		266	97.1 ± 13.1		283	2520 ± 621
	283	24.4 ± 3.2		283	27 ± 4.0		298	986 ± 100

[a] $c_{H_2SO_4}$ = concentration of sulfuric acid; k_H^* = effective Henry's law constant.

Table 9.30 Effective Henry's law constant for nitrous acid in solutions of sulfuric acid at 298.15 K[a]

$c_{H_2SO_4}/$ mass%	$k_H^*/$mol kg^{-1} bar^{-1}	$c_{H_2SO_4}/$ mass%	$k_H^*/$mol kg^{-1} bar^{-1}	$c_{H_2SO_4}/$ mass%	$k_H^*/$mol kg^{-1} bar^{-1}
0.3	47.2 ± 6.9	31.5	10.9 ± 1.6	61.2	41.6 ± 6.5
4.6	38.3 ± 5.7	39.7	7.8 ± 1.4	63.3	120 ± 16
8.7	32.1 ± 4.7	51.2	5.1 ± 0.7	65.6	366 ± 51
14.6	24.5 ± 3.6	52.3	4.7 ± 1.4	67.3	986 ± 100
22.6	17 ± 2.6	56.4	7.1 ± 1.4	73.6	$22290(+40000, -10000)$
30	14 ± 2.4	58.1	12.3 ± 1.7		

[a] $c_{H_2SO_4}$ = concentration of sulfuric acid; k_H^* = effective Henry's law constant.

The variation of k_H^* with concentration of sulfuric acid is consistent with a Setchenov-type equation, i.e.

$$\ln(k_H^0/k_H^c) = k_s c$$

where k_H^0 is the value of k_H^* for pure water, k_H^c is the value of k_H^* for concentration of sulfuric acid c and k_s is a Setchenov constant. The value of k_s for 298 K is 0.044 ± 0.002 mass%$^{-1}$.

At sulfuric acid concentrations above 53 mass%, formation of nitrosylsulfuric acid by the following reaction is significant:

$$HONO + H_2SO_4 \rightleftharpoons H_2O + HSO_4^- + NO^+$$

Under these conditions, the effective value of Henry's law constant, k_H^*, increases with increase in acid concentration, but this value no longer corresponds to a ratio of molecular HONO in the liquid and gas phases. In solutions of high acidity the following equilibrium exists:

$$HONO + H^+ \rightleftharpoons H_2O + NO^+$$

The value of the equilibrium constant was estimated by the authors to be $1.2(+1.0, -0.5) \times 10^{-8}$ at 298 K. This is consistent with the value of 1.4×10^{-8} reported by Seel and Winkler [133].

At fixed concentrations of acid the value of k_H^* decreases with increase in temperature and follows a van't Hoff-type equation which may be written as

$$\ln k_H^* = -\Delta H/RT + \Delta S/R$$

At concentrations below 53 mass% the thermodynamic quantities may be equated with heat of solution and entropy of solution. Above this concentration chemical changes are involved and the quantities must be considered as parameters describing the variation of the effective value of the Henry's law constant.

Becker et al. [134] derived the following equation to take account of the variation in solubility of HONO with both temperature and concentration of sulfuric acid:

$$k_H^*/\text{mol kg}^{-1}\,\text{bar}^{-1} = A_T \exp(B_T c) + D_t \exp(E_T c/\text{mass\%})$$

where

$$c = \text{concentration of } H_2SO_4 \text{ mass\%}$$

$$A_T/\text{mol kg}^{-1}\,\text{bar}^{-1} = 3.910 \times 10^{-6} \exp[4873/(T/\text{K})]$$

$$B_T = [12.86/(T/\text{K}) - 0.0946]/\text{mass\%}$$

$$D_T/\text{mol kg}^{-1}\,\text{bar}^{-1} = 1.771 \times 10^{-24} \exp[7902/(T/\text{K})]$$

$$E_T = [-35.69/(T/\text{K}) + 0.6398]/\text{mass\%}$$

Curves based on these equations are depicted in Figure 9.23.

Becker et al. [134] also measured the solubility of nitrous acid in ammonium sulfate solutions using a similar apparatus. k_H was again calculated from the measured k_H^* from the relationship

$$k_H = k_H^*/(1 + K_a/[H^+])$$

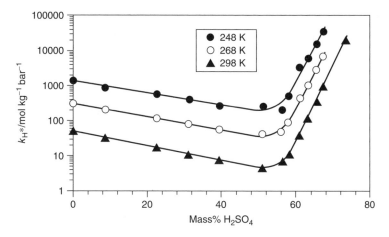

Figure 9.23 Variation of the effective Henry's law constant for dissolution of HONO with variation of temperature and concentration of sulfuric acid. The data points are based on data given in Tables 9.30 and 9.31. The solid curves are based on an equation derived by Becker *et al.* which gave the best fit to the experimental data [120]

Table 9.31 Variation with temperature and concentration of the limiting solubility of nitrous acid in solutions of ammonium sulfate[a]

$c_{(NH_4)_2SO_4}/$ mass%	T/K	$k_H*/$ mol kg^{-1} bar^{-1}	$c_{(NH_4)_2SO_4}/$ mass%	T/K	$k_H*/$ mol kg^{-1} bar^{-1}
0.0[b]	273.4	220 ± 11	20	267.9	203 ± 31
	283	119.5 ± 5.9		278.6	97 ± 15
	295	59.4 ± 3.0		288.2	54 ± 8
	303	37.7 ± 2.0		298.1	28 ± 4
5	273	204 ± 31	29.9	263.3	262 ± 39
	280.7	112 ± 17		273.4	134 ± 20
	288.2	76 ± 11		283.6	58 ± 9
	298.2	41 ± 6		293.4	31 ± 5
10	271	187 ± 28		299.9	21 ± 3
	277.8	119 ± 18	39.2	268.2	157 ± 26
	287.8	61 ± 9		278.3	67 ± 10
	298.3	35 ± 5		288.1	34 ± 5
				298.3	19 ± 3

[a]$c_{H_2SO_4}$ = concentration of sulfuric acid; k_H* = effective Henry's law constant.
[b]Values for pure water were taken from Park and Lee [113].

Experimental data are shown in Table 9.31 These data satisfactorily fit the equation

$$k_H*/\text{mol kg}^{-1}\,\text{bar}^{-1} = A \exp[Bc_{(NH_4)_2SO_4}/\text{mass\%}]$$

where

$$A = 3.06 \times 10^{-6} \exp[4932/(T/K)]$$

$$B = 17.6/(T/K) - 0.083$$

Nitric Acid, HNO₃, [CAS 7697-37-2]

Clegg and Brimblecombe [135] defined a constant H' relating the product of activities of H^+ and NO_3^- to the partial pressure of nitric acid over its aqueous solution as

$$H' = x(H^+)x(NO_3^-)f_{\pm}^2/p(HNO_3)$$

where $f_{\pm}$ is the mean rational activity coefficient taking the standard state to correspond to infinite dilution. Complete dissociation of acid is assumed and mole fractions, x, are calculated on the basis of the total number of components, i.e.

$$x(H^+) = x(NO_3^-) = n(H^+)/[n(H^+) + n(NO_3^-) + n(H_2O)]$$

where n is the number of moles.

Clegg and Brimblecombe [135] derived the following equation for H':

$$\ln(H'/\text{atm}^{-1}) = 385.972199 - 3020.3522/(T/K) - 71.001998\ln(T/K)$$
$$+ 0.131442311(T/K) - (0.420928363 \times 10^{-4})(T/K)^2$$

This may be written as

$$\ln(H'/\text{Pa}^{-1}) = 374.4461105 - 3020.3522/(T/K) - 71.001998\ln(T/K)$$
$$+ 0.131442311(T/K) - (0.420928363 \times 10^{-4})(T/K)^2$$

This equation generates values of H' shown in Table 9.32.

Table 9.32 Product of the activities of hydrogen and nitrate ion in an aqueous solution of nitric acid as a function of partial pressure of nitric acid as predicted by Clegg and Brimblecombe [135]

T/K	H'/Pa^{-1}
198	3261
208	644
218	139
228	32.6
238	8.27
248	2.25
258	6.56×10^{-1}
268	2.04×10^{-1}
278	6.70×10^{-2}
288	2.33×10^{-2}
298	8.54×10^{-3}
308	3.29×10^{-3}
318	1.32×10^{-3}
328	5.57×10^{-4}
338	2.44×10^{-4}

This relationship may be used to estimate the vapour pressures of solutions of nitric acid and gives good agreement with experimental values compiled by Holeci [136]. The use of values to H' to estimate the solubility of nitric acid requires a knowledge of $f_{\pm}$, the mean activity coefficient. The estimation of this coefficient has been discussed by Clegg and Brimblecombe [135].

The partial pressure of nitric acid in the atmosphere is often about 10^{-4} Pa [42]. At 298 K the value of H' is 8.42×10^{-3} Pa^{-1} (Table 9.32). The value of $x(H^+)x(NO_3^-)f_{\pm}^2$ is therefore $8.54 \times 10^{-3} \times 10^{-4} = 8.42 \times 10^{-7}$. The concentration of nitric acid in an aqueous solution in equilibrium with nitric acid vapour in the atmosphere at 298 K is therefore very low. It may be assumed that the mean activity coefficient is close to unity if no other ionic species are present. Hence

$$x(H^+)x(NO_3^-) = H'p(HNO_3)$$

At a partial pressure of 10^{-4} Pa at 298.15 K:

$$x(H^+)x(NO_3^-) = 8.54 \times 10^{-7}$$

Hence for solution in pure water $x(H^+) = x(NO_3^-) = 9.24 \times 10^{-4}$. The acid may be assumed to be completely ionized and hence

$$n(HNO_3) = n(H^+) = n(NO_3^-)$$

where n is the number of moles of H^+, NO_3^- and ionized HNO_3. It follows that

$$x(HNO_3) = n(HNO_3)/[2n(HNO_3) + n(H_2O)]$$

$$\simeq n(HNO_3)/n(H_2O)$$

The approximate solubility of nitric acid at a partial pressure of 10^{-4} Pa and 298 K

$$= 9.24 \times 10^{-4} \times 1000/18.02 \text{ mol kg}^{-1}$$

$$= 0.05 \text{ mol kg}^{-1}$$

Van Doren et al. [51] reported uptake coefficients of nitric acid which they identified as accommodation coefficients (Table 9.33). A monodisperse train of water droplets of diameter 200 μm generated at a vibrating orifice passed through an atmosphere of helium and water vapour containing traces of substance under test. Samples of droplets were trapped and analysed. Ponche et al. [92] reported values of uptake coefficients which he also identified as mass accommodation coefficients, (Table 9.33). A similar droplet train flow tube apparatus was used.

Peroxynitric Acid, HO$_2$NO$_2$, [CAS 26404-66-0]

Régimbal and Mozurkewich [131] published an estimated value of the Henry's law constant of peroxynitric acid in pure water using the thermochemical cycle

Table 9.33 Uptake coefficients for dissolution of nitric acid in water[a]

No. of data	T/K	P_{tot}/bar	p_{H_2O}/bar	Orifice diam./μm	γ_{obs}	D/bar cm^2 s^{-1}	γ_0
Data from Van Doren et al. [51]:							
7	268	6.39×10^{-3}	4.21×10^{-3}	60	0.159 ± 0.005	0.160	$0.193 +0.021/-0.015$
22	273	7.25×10^{-3}	6.24×10^{-3}	60	0.125 ± 0.006	0.135	$0.158 +0.020/-0.014$
11	282	11.95×10^{-3}	11.33×10^{-3}	60	0.107 ± 0.005	0.129	$0.167 +0.033/-0.020$
14	289	21.20×10^{-3}	18.41×10^{-3}	60	0.064 ± 0.005	0.141	$0.102 +0.022/-0.014$
7	293	25.65×10^{-3}	22.97×10^{-3}	60	0.050 ± 0.010	0.140	$0.071 +0.020/-0.017$
Data from Ponche et al. [92]:							
52	297				0.050 ± 1.0	0.168[b]	0.11 ± 0.01 at 298 K

[a]No. of data = No. of experimental measurements at each temperature; P_{tot} = total pressure; p_{H_2O} = partial pressure of water; γ_{obs} = observed value of the uptake coefficient at the experimental pressure; γ_0 = value of the uptake coefficient corrected to zero pressure; D = diffusion coefficient of HNO_3 in the mixture of water vapour and helium.
[b]Value from Reid and Sherwood [137] quoted by Ponche et al. [92].

based on the equilibrium with HO_2 and NO_2 which exists in both the gas and liquid phases, i.e.

$$\text{Gas phase:} \qquad HO_2 + NO_2 \rightleftharpoons HOONO_2$$

$$\downarrow \qquad \downarrow \qquad\qquad \downarrow$$

$$\uparrow \qquad \uparrow \qquad\qquad \uparrow$$

$$\text{Liquid phase :} \qquad HO_2 + NO_2 \rightleftharpoons HOONO_2$$

The estimated value was $(12.4 \pm 4.7) \times 10^3\,\text{mol dm}^{-3}\,\text{bar}^{-1}$ at 298 K. Park and Lee [138] reported a value of $\sim 19.6 \times 10^3\,\text{mol dm}^{-3}\,\text{bar}^{-1}$ at 293 K.

Zhang *et al.* [89] measured the solubility in aqueous sulfuric acid at concentrations of acid from 11.46 to 29.04 mol kg^{-1} of water. They used a fast flow horizontal flow tube apparatus, the inner walls of which were covered with a thin layer of the sulfuric acid solution.

Leu and Zhang [139] published an equation to give the best fit to the experimental data. The present author considers that the best form of the equation is

$$\ln(k_H{}^*/\text{mol dm}^{-3}\,\text{bar}^{-1}) = 3.68 - m H_2SO_4 \times [-0.25 + 65/(T/\text{K})] - 8400$$

$$\times\, [1/(T_0/\text{K}) - 1/(T/\text{K})]$$

where $k_H{}^*$ is the Henry's law constant for dissolution of $HOONO_2$ in the aqueous solution, $m H_2SO_4$ is the molal concentration of sulfuric acid and T_0 is 298.15 K.

In the equation as published by Leu and Zhang, the units of $k_H{}^*$ were given as $\text{mol kg}^{-1}\,\text{atm}^{-1}$. This appears to be incorrect because the equation fits the experimental values of $k_H{}^*$ published by Zhang *et al.* [89] when these data are expressed in the original units of $\text{mol dm}^{-3}\,\text{atm}^{-1}$. The experimental data fit this equation with a standard deviation in values of $\ln (k_H{}^*/\text{mol dm}^{-3}\,\text{bar}^{-1})$ of 0.153. This deviation is equivalent to a factor of 1.165.

Experimental measurements and smoothed values from this equation at the temperatures of measurement are shown in Table 9.34.

This equation is not compatible with available data for dissolution of HO_2NO_2 in pure water discussed above. The equation indicates a value for the Henry's law constant of about $40\,\text{mol dm}^{-3}\,\text{bar}^{-1}$ at 298.15 K. This is very much smaller that the experimental values of about 13 000 at 298 K and 20 000 at 293 K given above. Leu and Zhang suggest that their smoothing equation may not be valid outside the range of the experimental data on which it is based. They also indicate the possibility that the processes involving the uptake of HO_2NO_2 may differ at pH 7 from those at lower pH.

Peroxyacyl Nitrates

Peroxyacyl nitrates were extensively studied by Kames and Schurath [140] by a dynamic method using a bubble column apparatus. These peroxy compounds were generated by photolysis of acetone or 3-pentanone in the presence of NO_2 to

Table 9.34 Experimental values of the Henry's law constants for dissolution of peroxynitric acid in aqueous solutions of sulfuric acid measured by Zhang et al. [89] together with smoothed values published by Leu and Zhang [139][a]

H_2SO_4 content/ mol dm^{-3} bar^{-1}	T/K	$k_H^* \sqrt{D_l}/$ mol dm^3 bar^{-1} cm s$^{-\frac{1}{2}}$	$k_H^*/$mol dm^{-3} bar^{-1} (experimental values)	$k_H^*/$mol dm^{-3} bar^{-1} (smoothed)
11.46	208.9	608.5	3.8×10^6	3.226×10^6
	208.9	624.2	3.9×10^6	3.226×10^6
	214.4	333.5	1.6×10^6	1.260×10^6
	224.1	101.9	3.2×10^5	2.702×10^5
	229.1	47.7	1.7×10^5	1.276×10^5
14.27	207.9	326.4	2.9×10^6	3.226×10^6
	207.9	320.7	2.9×10^6	3.226×10^6
	213.5	200.4	1.4×10^6	1.260×10^6
	218.9	113.3	5.4×10^5	5.280×10^5
	223.5	65.1	2.5×10^5	2.622×10^5
	226.8	41.8	1.5×10^5	1.607×10^5
14.74	210.9	257.0	1.9×10^6	1.880×10^6
	215.6	138.0	7.8×10^5	8.705×10^5
	217.4	109.1	4.9×10^5	6.514×10^5
	222	72.2	3.0×10^5	3.203×10^5
20.17	201.4	176.5	4.9×10^6	6.628×10^6
	202.6	179.2	4.4×10^6	5.373×10^6
	205.4	129.8	2.5×10^6	3.325×10^6
	210.4	93.6	1.2×10^6	1.465×10^6
	215.1	64.8	5.9×10^5	6.986×10^5
	215.1	63.8	5.8×10^5	6.986×10^5
28.74	204.1	35.6	2.6×10^6	2.251×10^6
	204.2	34.7	2.5×10^6	2.228×10^6
	208	29.6	1.3×10^6	1.236×10^6
	213.6	26.4	6.3×10^5	5.440×10^5
	219	17.0	2.5×10^5	2.570×10^5
	223.6	15.1	1.6×10^5	1.383×10^5
	223.6	15.6	1.6×10^5	1.383×10^5
	223.6	14.6	1.5×10^5	1.383×10^5
29.04	204.9	30.8	2.1×10^6	1.957×10^6
	215.1	21.2	4.4×10^6	4.322×10^5
	221.2	15.9	2.0×10^6	1.885×10^5

[a]D_l = diffusion coefficient for PAN in the liquid phase; k_H^* = effective Henry's law constant.

produce peroxyacetyl nitrate (PAN) and peroxypropionyl nitrate (PPN). Addition of aldehydes to acetone resulted in the formation of higher peroxynitrates. Rate constants for hydrolysis, k_h, were obtained and compared with data from other authors. The authors reported the ratio of the concentration in the aqueous layer to that in the gas phase and also the ratio of concentration in the aqueous layer to the pressure of the gas. A summary of the data for pure water measured by Kames and Schurath is given in Table 9.35. Data from other sources are also included in the table. Thermodynamic quantities for calculation of the variation

Table 9.35 Henry's law constants and hydrolysis rate constants for peroxyacyl nitrates [140]

Compound	k_H (293.3 K) (dimensionless concentration ratio)	k_H/mol dm^{-3} bar^{-1} (293.2 K)	k_h/(10^{-4} s^{-1}) at 293.2 K	Ref.[a]
Peroxyacetyl nitrate (PAN), [CAS 2278-22-0]	97.9 ± 2.0	4.05 ± 0.08	3.4 ± 0.14	
	98.3 ± 4.8	4.05 ± 0.20	4.0 ± 1.2 (295 K)	141
	97.4 ± 3.6	4.05 ± 0.15	6.4 ± 1.9	142
			3.9 ± 0.9	143
Peroxypropionyl nitrate (PPN), [CAS 5796-89-4]	70.8 ± 1.5	2.86 ± 0.06	3.1 ± 0.22	
Peroxy-n-butyryl nitrate (PnBN), [CAS 138779-12-1]	54.8 ± 1.5	2.27 ± 0.06	2.4 ± 0.4	
Peroxymethacryloyl nitrate (MPAN), [CAS 88181-75-3]	41 ± 3.8	1.68 ± 0.16	5.6 ± 1.2	
Peroxyisobutyryl nitrate (PiBN), [CAS 65424-60-4]	24 ± 0.7	0.99 ± 0.03	7.4 ± 0.5	

[a] Sources of k_H.

Table 9.36 Data for calculation of variation of k_H and of k_h for peroxyacyl nitrates with change of temperature[a]

Compound	ΔH°_{solv}/kJ mol^{-1}	ΔS°_{solv}/J K^{-1} mol^{-1}	E_a/J mol^{-1}	Log$_{10}A$
PAN	-47.4 ± 0.8	-150 ± 2.5	54.5 ± 0.5	6.2 ± 0.1
	-49.2 ± 4.7^a	-157 ± 16^a	60.7 ± 11.7^b	$7.4 \pm 2.2^{a,b}$
	-54.2 ± 1.0^c	-173 ± 4^c	55.5 ± 5.5^c	6.6 ± 1.0^c
PPN	-49.4 ± 0.9	-160 ± 3	54.4 ± 3.1	6.2 ± 0.6

[a] ΔH°_{solv} = standard solvation enthalpy; ΔS°_{solv} = standard solvation entropy; E_a = activation energy for hydrolysis; A = pre-exponential factor in the equation $k_H = A\,e^{E_a/RT}$; ^{a}[141]; ^{b}[143]; ^{c}[142].

of k_H and k_h with change of temperature are given in Table 9.36. These can be used to calculate k_H and k_h by the equations

$$\ln(k_H/\text{mol dm}^{-3}\,\text{bar}^{-1}) = -(\Delta H^\circ_{sol} - T\Delta S^\circ_{sol})/RT$$

where ΔH_{sol}° is the standard enthalpy change of solution and ΔS_{sol}° is the corresponding entropy change, and

$$k_h = A\exp(-E_a/RT)$$

where E_a is the activation energy for the hydrolysis.

Kames and Schurath also investigated the effect of electrolytes on the solubility and hydrolysis of peroxyacyl nitrates. The results are shown in Table 9.37. The authors reported that there was negligible increase in the rate of hydrolysis of PAN and PPN by the presence of electrolytes in the synthetic seawater.

Table 9.37 Effect of electrolytes on the solubility and hydrolysis of peroxyacyl nitrates

Compound	k_H (293.2 K) (dimensionless concentration ratio)	k_H/mol dm^{-3} bar^{-1} (293 K)	k_h/10^{-4} s^{-1} (293.2 K)	Liquid[a]
PAN	97.9 ± 2.0	4.05 ± 0.08	3.4 ± 0.14	a
	85.3 ± 3.7	3.6 ± 0.2	3.7 ± 0.4	b
			15	c
			33	d
			$\sim$44 $\pm$ 7	e
PPN	70.8 ± 1.5	2.86 ± 0.06	3.1 ± 0.22	a
	59 ± 9	2.5 ± 0.5	3.9 ± 0.2	b

[a]a, Distilled water, pH 6.8; b, synthetic seawater, ionic strength $I = 0.69$ M made up of 0.48 M NaCl, 0.03 M CaCl$_2$ and 0.03 M MgSO$_4$; c, 11 mM phosphate buffer solution, pH 7.8; d, 33 mM phosphate buffer solution, pH 7.2 [144]; e, filtered North Sea water, pH 7.8.

Table 9.38 Experimental values of the Henry's law constants for dissolution of peroxyacetyl nitrate (PAN) in aqueous solutions of sulfuric acid measured by Zhang and Leu [145] together with smoothed values published by Leu and Zhang [139][a]

H$_2$SO$_4$ content/ mol dm^{-3} bar^{-1}	T/K	$k_H^* \sqrt{D_l}$/ mol dm^{-3} bar^{-1} cm s$^{-\frac{1}{2}}$	k_H^*/ mol dm^{-3} bar^{-1} (experimental values)	k_H^*/ mol dm^{-3} bar^{-1} (smoothed)
46	199.1	7.11	6.38×10^4	8.815×10^4
46	199.1	7.30	6.56×10^4	8.815×10^4
46	202.3	5.72	4.12×10^4	5.035×10^4
46	207.2	4.22	2.25×10^4	2.174×10^4
46	211.2	2.75	1.18×10^4	1.135×10^4
46	216.5	1.64	5.47×10^3	4.948×10^3
54	207.8	3.06	1.97×10^4	2.263×10^4
54	208.2	2.96	1.87×10^4	2.110×10^4
54	213.8	2.22	1.04×10^4	8.078×10^3
54	214.9	1.98	8.73×10^3	6.747×10^3
54	216.5	1.04	4.21×10^3	5.151×10^3
54	216.8	1.11	4.44×10^3	4.948×10^3
54	219.4	0.64	2.29×10^3	3.251×10^3
54	226	0.32	8.64×10^2	1.172×10^3
59	207.8	3.08	2.39×10^4	2.526×10^4
59	208.9	3.08	2.23×10^4	2.068×10^4
59	209.6	2.96	2.05×10^4	1.816×10^4
59	211.5	2.67	1.66×10^4	1.279×10^4
59	212.8	2.12	1.22×10^4	1.016×10^4
59	214.7	1.55	8.06×10^3	7.309×10^3
59	216.6	0.84	3.96×10^3	5.254×10^3
72	207.8	2.19	5.51×10^4	4.082×10^4
72	208	2.08	5.07×10^4	3.922×10^4
72	208.1	2.09	5.15×10^4	3.806×10^4
72	211.7	1.98	3.30×10^4	1.710×10^4
72	212.6	1.78	2.84×10^4	1.414×10^4
72	213.4	1.34	1.97×10^4	1.181×10^4
72	214.8	1.04	1.38×10^4	8.750×10^3
72	217	0.81	9.12×10^3	5.524×10^3
72	219.2	0.59	5.69×10^3	3.522×10^3
72	221.6	0.38	3.09×10^3	2.157×10^3

[a]D_l = diffusion coefficient for PAN in the liquid phase; k_H^* = effective Henry's law constant.

The solubility of PAN in concentrated sulfuric acid was investigated by Zhang and Leu [145] using a horizontal flow tube reactor. Data are shown in Table 9.38.

Leu and Zhang [139] published an equation which gave the best fit to this experimental data and also to the data available for solubility in pure water. The present author considers that the best form of the equation is

$$\ln(k_H^*/\text{mol}\,\text{dm}^{-3}\,\text{bar}^{-1}) = 1.06 - m\text{H}_2\text{SO}_4 \times [0.69 - 152/(T/\text{K})] - 5810$$
$$\times [1/(T_0/\text{K}) - 1/(T/\text{K})]$$

where k_H^* is the effective Henry's law constant for dissolution of PAN in aqueous solution, $m\text{H}_2\text{SO}_4$ is the molal concentration of sulfuric acid, T is the temperature of the solution and T_0 is 298.15 K. In the equation as published by Leu and Zhang the units of k_H^* were $\text{mol}\,\text{kg}^{-1}\,\text{atm}^{-1}$. This appears to be incorrect because the equation fits the experimental values of k_H published by Zhang and Leu when these data are expressed in the original units of $\text{mol}\,\text{dm}^{-3}\,\text{atm}^{-1}$.

The experimental data fit the equation above with a standard deviation in values of $\ln(k_H^*/\text{mol}\,\text{dm}^{-3}\,\text{bar}^{-1})$ of 0.317. This deviation is equivalent to a factor of 1.372.

Smoothed values from this equation at the temperatures of measurement are shown in Table 9.38.

8 COMPOUNDS CONTAINING SULFUR

Sulfur Dioxide, SO$_2$, [CAS 7446–09–5]

The solubility of sulfur dioxide in water has been discussed by Fogg and Gerrard [153]. Sulfur dioxide, dissolved in water, is in equilibrium with other species in solution (see Section 2.3). Literature values of solubilities usually refer to bulk solubilities which include the sulfur dioxide equivalent to the sulfur, which is in all species derived from the dissolved sulfur dioxide.

Battino [45] evaluated the available measurements of the solubility of sulfur dioxide in water. He derived a smoothing equation for the bulk mole fraction solubility at a partial pressure of 1.013 bar which was based upon measurements by Hudson [154], Beuschlein and Simenson [155], Rabe and Harris [156] and Lavrova and Tudorovskaya [157]. The corresponding equation for a partial pressure of 1 bar may be written as

$$\ln x(\text{SO}_2) = -51.458 + 4575.5/(T/\text{K}) + 5.6854\ln(T/\text{K})$$

The bulk mole fraction solubility, $x(\text{SO}_2)$ is defined as

$$x(\text{SO}_2) = n(\text{total SO}_2)/[n(\text{total SO}_2) + n(\text{H}_2\text{O})]$$

where $n(\text{total SO}_2)$ is the number of moles of SO$_2$ in solution plus the sulfur dioxide equivalent to species derived from the dissolved SO$_2$ and $n(\text{H}_2\text{O})$ is the number of moles of H$_2$O present plus the water molecules which have reacted with the SO$_2$. The standard deviation in values of $x(\text{SO}_2)$ is 0.00017 (temperature range 283–380 K).

The variation of solubility of sulfur dioxide with pressure does not approximate to Henry's law, even at very low pressures, as indicated by the measurements of Johnstone and Leppla [158] (see Figure 9.24). The lower the concentration of the sulfur dioxide, the larger is the proportion that is ionized. As a consequence, the lower the partial pressure of sulfur dioxide, the greater is the relative contribution of the ionization process to the overall dissolution of sulfur dioxide. At 298.2 K under a partial pressure of 1.0 bar, about 10 % of the dissolved sulfur dioxide is ionized. The proportion is about 75 % under a partial pressure of 0.10 bar.

Various workers have studied the dissolution of sulfur dioxide in solutions of electrolytes. For comprehensive information, the reader should refer to the compilation and evaluation of solubilities of sulfur dioxide edited by Young [159].

The solubility of sulfur dioxide in aqueous solutions of sulfuric acid of various concentrations under a total pressure of 101.3 bar was measured by Miles and Fenton [160] at 293 K and by Cupr [161] at 314.2 and 335.2 K. The mole fraction of sulfur dioxide in solution may be calculated without taking ionization or chemical change of the sulfur dioxide into account. It may be defined as

$$\text{bulk mole fraction solubility of } SO_2 = \text{moles of}$$

$$SO_2/(\text{total moles of } H_2SO_4 + SO_2 + H_2O)$$

The number of moles of SO_2 in solution includes the number of moles of SO_2 which have undergone chemical change. Cupr's [161] measurements at 314.2 and 335.2 K show that the mole fraction solubility passes through a minimum with increase in acid concentration (Figure 9.25). Measurements by Miles and Fenton [160] confirm that, at high acid concentrations, the mole fraction solubility of gas increases with concentration of acid.

Boniface et al. [162] used a droplet train flow reactor and also a bubble train reactor to measure the uptake of SO_2 on water at various pH values in the temperature range 263–293 K. At low pH there is evidence of the formation

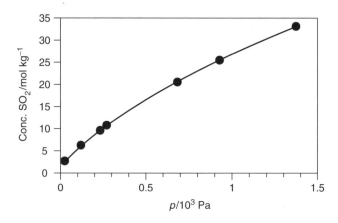

Figure 9.24   Variation of the solubility in water of SO_2 with partial pressure, p, at 298.2 K [158]

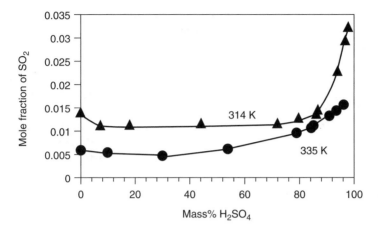

Figure 9.25 Variation of the mole fraction solubility of SO_2 with concentration of H_2SO_4 in aqueous solution at a total pressure of 1.013×10^5 Pa [161]

of a surface complex. With increase in pH the uptake rises towards the mass accommodation coefficient, probably because of the reaction of SO_2 with OH^-:

$$SO_2 + OH^- \longrightarrow HSO_3^-$$

Values of the mass accommodation coefficient, α, were fitted to the equation

$$\alpha/(1 - \alpha) = \exp(-\Delta G/RT)$$

where $\Delta H = -21.3 \, kJ \, mol^{-1}$
$\Delta S = 84.9 \, J \, mol^{-1} \, K^{-1}$
$\alpha = 0.36$ at 283.15 K

The solubility of gases such as SO_2, H_2S and of CO_2 is enhanced because of their ability to form ionic species in solution. An effective Henry's law coefficient, may be defined as

k_H^* = total concentration of ionized and nonionized species in solution/partial pressure of the gas

Another constant of the Henry's law may be defined as

k_H = concentration of nonionized gas in solution/partial pressure of gas

In the case of these acids with first and second dissociation constants K_{a_1} and K_{a_2}:

$$k_H^* = k_H \left(1 + \frac{K_{a_1}}{[H^+]} + \frac{K_{a_1} K_{a_2}}{[H^+]^2} \right)$$

However, under the experimental conditions the term relating to the second dissociation constant may be neglected to give

$$k_H^* = k_H \left(1 + \frac{K_{a_1}}{[H^+]} \right)$$

This work by Boniface *et al.* also gave pseudo-first-order rate constants k_1 for reaction of the gases with water and second-order rate constants k_2 for reaction with hydroxide ion. These constants and parameters characterizing the reactions from other sources are shown in Table 9.39.

Uptake coefficients as a function of pH for SO_2 and also for H_2S and for CO_2 are shown in Figure 9.26. The curves displayed correspond to models for the three systems based on the parameters given in Table 9.39. At low pH the model for SO_2 predicts the dotted curve and is not consistent with the experimental data. Modification of the model to take account of the formation of a surface complex at low pH as shown by the full curve gives satisfactory agreement.

Table 9.39 Data characterizing the dissolution of SO_2, H_2S and CO_2 in water and aqueous solutions

Gas	SO_2	H_2S	CO_2
k_1/s^{-1}	3.4×10^{6a}	$<1 \times 10^4$	1.4×10^{-2b}
$k_2/mol\,dm^{-3}\,s^{-1}$	$(1.1 \pm 0.2) \times 10^{10}$	$(1.7 \pm 0.2) \times 10^9$	$(4.0 \pm 0.7) \times 10^3$
$K_{a_1}/mol\,dm^{-3}$	1.55×10^{-2c}	8.35×10^{-8d}	4.0×10^{-7e}
$K_{a_2}/mol\,dm^{-3}$	1.02×10^{-7f}	6.49×10^{-15e}	3.2×10^{-11e}
$k_H/mol\,dm^{-3}\,bar^{-1}$	1.6^c	0.10^g	0.04^f

[a] Eigen *et al.* [146].
[b] Pinsent *et al.* [147].
[c] Maahs [148].
[d] Millero [149].
[e] Perrin [150].
[f] Lide [151].
[g] De Bruyn *et al.* [152].

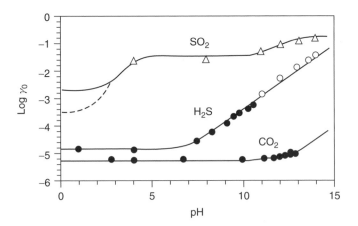

Figure 9.26 Uptake of SO_2, H_2S and CO_2 as a function of pH at 291 K and a gas–liquid contact time of 5 ms. The points marked with open symbols correspond to measurements with a droplet train flow reactor and those marked with solid symbols to measurements with a horizontal bubble train. Points marked with an open square correspond to measurements by Jayne *et al.* [163], which were adjusted to the same temperature an gas–liquid interaction time

Table 9.40   Selected values of accommodation coefficients [124] from JPL publication 97-4

Gas	Liquid	Composition	T/K	α	Ref.
O_3	Water		292	$>2 \times 10^{-3}$	1
OH	Water		275	$>4 \times 10^{-3}$	165
	Sulfuric acid	28 mass% H_2SO_4	275	>0.07	165
		97 mass% H_2SO_4	298	$>5 \times 10^{-4}$	166
HO_2	Water		275	>0.02	165
	Salt solns	$NH_4HSO_4 + H_2O$	293	>0.2	167
		$LiNO_3 + H_2O$	295	>0.2	168
H_2O	Nitric acid	$HNO_3 + H_2O$	278	>0.3	169
	Sulfuric acid	96 mass% H_2SO_4	298	$>2 \times 10^{-3}$	170
	NaCl soln		$\sim$298	>0.5	171
H_2O_2	Water		273	0.18	172
	Sulfuric acid	96 mass% H_2SO_4	298	$>8 \times 10^{-4}$	170
HNO_3	Water		268	0.2	128
			293	0.07	37
	Nitric acid	$HNO_3 + nH_2O$	278	0.6	169
	Sulfuric acid	73 mass% H_2SO_4	283	0.1	128
		75 mass% H_2SO_4	230	$>2 \times 10^{-3}$	173
		97 mass% H_2SO_4	295	$>2.4 \times 10^{-3}$	170
NH_3	Water		$\sim$295	0.06	92,93
CH_3OH	Water		260–291	0.12–0.02	174
C_2H_5OH	Water		260–291	0.13–0.02	174
$CH_3CH_2CH_2OH$	Water		260–291	0.08–0.02	174
$CH_3CH_2(OH)CH_3$	Water		260–291	0.10–0.02	174
$HOCH_2CH_2OH$	Water		26–291	0.13–0.04	174
CH_2O	Water		26–270	0.04	14
	Sulfuric acid	$H_2SO_4 + nH_2O$	235–300	0.04	14
CH_3CHO	Water		267	>0.03	175
CH_3COCH_3	Water		260–285	0.07–0.01	176
HCOOH	Water		260–291	0.10–0.02	174
CH_3COOH	Water		260–291	0.15–0.03	174
HCl	Water		274	0.2	51
	Sulfuric acid	$\leq$40 mass% H_2SO_4	283	0.15	177
HOBr	Sulfuric acid	58 mass% H_2SO_4	$\sim$228	>0.005	178
$CHBr_3$	Sulfuric acid	97 mass% H_2SO_4	220	$>3 \times 10^{-3}$	179
CF_3COOH	Water		263–288	0.2–0.1	180
SO_2	Water		260–292	0.11	172
$CH_3S(O)CH_3$	Water		262–281	0.16–0.08	181
$CH_3S(O_2)CH_3$	Water		262–281	0.27–0.08	181
$CH_3S(O_2)OH$	Water		264–278	0.17–0.11	181

Table 9.41 Selected values of evaluated reactive uptake coefficients from JPL publications 00–003 [65]; 97-04 [124]

Gas	Surface	Composition	T/K	γ	Ref.
$OH + surface \longrightarrow$ Products					
OH	Hydrochloric acid	$HCl \cdot nH_2O(l)$	220	>0.2	182
$HO_2 + Surface \longrightarrow$ Products					
	Sulfuric acid	$H_2SO_4 \cdot nH_2O(l)$ (28 wt%)	275	0.07	165,182,183
		(55 wt%)	223	>0.05	
		(80–96 wt%)	243	>0.2	
$NO_3 + H_2O(l) \longrightarrow HNO_3 + OH$					
NO_3	Liquid water	$H_2O(l)$	273	2×10^{-4}	105,106
$N_2O_5 + H_2O \longrightarrow 2HNO_3$					
N_2O_5	Liquid water	$H_2O(l)$	260–295	0.05	51,130
	Sulfuric acid	$H_2SO_4 \cdot nH_2O(l)$	180–300	a	b
$N_2O_5 + NaCl \longrightarrow ClNO_2 + NaNO_3$					
N_2O_5	Sodium chloride	NaCl(aq)		~0.03	130,184–186
$HONO + H_2O \longrightarrow$ Products					
HONO	Liquid water	$H_2O(l)$	247–297	0.04	187
$NH_3 + H_2SO_4 \longrightarrow NH_4HSO_4$					
NH_3	Sulfuric acid	$H_2SO_4 \cdot nH_2O$	288–300	0.4	188–191
$CH_3C(O)O_2 + H_2O \longrightarrow CH_3C(O)OH + HO_2$					
$CH_3C(O)O_2$	Liquid water	$H_2O(l)$	274	4×10^{-3}	192
	Sulfuric acid	34 wt% H_2SO_4	246	3×10^{-3}	192
		51 wt% H_2SO_4	273	1×10^{-3}	
		71 wt% H_2SO_4	298	1×10^{-3}	

	Surface	Products			
Cl + Surface ⟶ Products					
Cl	Sulfuric acid	$H_2SO_4 \cdot nH_2O(l)$	221–296	2×10^{-4}	193
HOBr + HCl(s) ⟶ BrCl + H₂O					
HOBr	Sulfuric acid	$H_2SO_4 \cdot nH_2O$ (60–69 mass% H_2SO_4)	228	0.2	194
ClONO₂ + H₂O ⟶ HOCl + HNO₃					
ClNO₃	Sulfuric acid	$H_2SO_4 \cdot nH_2O(l)$ (30–80 mass% H_2SO_4)	200–265	b	66
BrONO₂ + H₂O ⟶ HOBr + HNO₃					
BrONO₂	Sulfuric acid	$H_2SO_4 \cdot nH_2O$	210–298	0.8	194,195
BrONO₂ + HCl ⟶ BrCl + HNO₃					
BrONO₂/HCl	Sulfuric acid	$H_2SO_4 \cdot nH_2O$	229	0.9	194,195
CF₃OH + H₂O ⟶ Products					
CF₃OH	Sulfuric acid	$H_2SO_4 \cdot nH_2O$ (40 wt% H_2SO_4)	210–250	0.07	196
		45 wt% H_2SO_4	210–250	0.04	196
		50 wt% H_2SO_4	210–250	0.01	196
		50 wt% H_2SO_4	210–250	0.001	196

[a]Experimental data for absorption on surfaces of sulfuric acid (40–80 mass%) considered to be free of saturation effects together with data for uptake on pure water and on ammonium sulfate aerosol at high humidity were used to find a best fit equation for the experimental values of the uptake coefficient. The form of this equation is

$$\gamma_0 = \exp[k_0 + k_1/(T/K) + k_2/(T/K)^2].$$

The parameters k_0, k_1 and k_2 giving the best fit are

$$k_0 = -25.5265 - 0.133188\,ms + 0.00930846\,ms^2 - 9.0194 \times 10^{-5}\,ms^3$$

$$k_1 = 9283.76 + 115.345\,ms - 5.19258\,ms^2 + 0.0483464\,ms^3$$

$$k_2 = -851801 - 22191.2\,ms + 766.916\,ms^2 - 6.85427\,ms^3$$

Where ms is the mass percentage of H_2SO_4. The error in the value of γ_0 from the equations has been estimated to be ±15 %.
[b]Refs. 127, 129, 197–201.

Table 9.42 Evaluated Henry's law constants for important reactions from JPL publication 97-4 [124]

Compound	T/K	Mass% H_2SO_4	k_H or k_H^*/mol cm^{-3} bar^{-1}	Ref.
Nitrous acid, HONO, [CAS 7782-77-6] in $H_2SO_4 \cdot nH_2O(l)$	248–298	0.3–53	k_H^*/mol cm^{-3} bar^{-1} $= (46.6 \pm 2.8) \exp(-0.044 \pm 0.002)X$ (X = mass% H_2SO_4)	120
Hypochlorous acid, HOCl, [CAS 7790-92-3] in $H_2SO_4 \cdot nH_2O(l)$	200–230	46–80	$\ln(k_H^*$/mol kg^{-1} bar$^{-1})$ $= -6.5078 - m[-0.04107 + 54.56/(T/K)]$ $- 5862[1/(T_0/K) - 1/(T/K)]$ where $T_0 = 298$ K and m is the concentration of H_2SO_4/mol kg^{-1}	63
Formic acid, HCOOH, [CAS 64-18-6] in $H_2O(l)$	298	0	1.1×10^4	202
Acetic acid, CH_3COOH, [CAS 64-19-7] in $H_2O(l)$	298	0	7×10^3	202
Peroxyacetic acid, $CH_3C(O)O_2H$, [CAS 79-21-0] in $H_2O(l)$	275	0	≤ 0.1	192
Peroxyacetyl nitrate, $CH_3C(O)OONO_2$ [CAS 2278-22-0] in $H_2O(l)$	293	0	4.0	140
Peroxypropionyl nitrate (PPN) $CH_3CH_2C(O)OONO_2$ [CAS 5796-89-4] in $H_2O(l)$	293	0	2.9	140
Carbonyl chloride, CCl_2O, [CAS 75-44-5] in $H_2O(l)$	278	0	<0.2	75
	298	0	<0.1	75

Substance	T/K			Ref.
Trichloroacetyl chloride, CCl_3CClO, [CAS 76-02-8] in $H_2O(l)$	278	0	≤ 2	76
Carbonyl fluoride, CF_2O [CAS 353-50-4] in $H_2O(l)$	278	0	<1	203
Carbonyl fluoride, CF_2O, [CAS 353-50-4] in $H_2SO_4 \cdot nH_2O$	215–230	60	<5	204
Trifluoroacetyl fluoride, CF_3CFO, [354-34-7] in $H_2O(l)$	278	0	<1	203
Trifluoroacetyl chloride, CF_3CClO, [354-32-5] in $H_2O(l)$	278	0	≤ 0.5	204
Trifluoromethanol, CF_3OH, [1493-11-4] in $H_2SO_4 \cdot nH_2O(l)$	250	40	>237	196
Trifluoroacetic acid, $CF_3C(O)OH$, [76-05-1], in $H_2O(l)$	278–308	50	0.6 ± 0.3	207
		0	$k_H/\text{mol kg}^{-1}\,\text{bar}^{-1} = 8.891 - 9.206 \times 10^3$ $[1/(T_0/K) - 1/(T/K)]$ where $T_0 = 298.15$ K[a]	205
Nitrogen trioxide, NO_3, [CAS 12033-49-7] in $H_2O(l)$	273	0		105

[a]Suggested equation based on vapour pressure measurements over water at 278.15, 298.15 and 308.15 K [205].

Sulfuric Acid, H₂SO₄, [CAS 7664-93-9]

Pöschl *et al.* [104] have reported measurements of the mass accommodation coefficient of H_2SO_4 vapour on aqueous sulfuric acid surfaces. They used a coated cylindrical fast flow reactor at 303 K to measure the loss of H_2SO_4 vapour from the walls in an atmosphere of N_2 and H_2O. The concentration of sulfuric acid in the surface layer was varied from 70 to 95 mass% by varying the partial pressure of the water vapour. The mass accommodation coefficient was calculated from the dependence on pressure of the loss of sulfuric acid from the walls. The mean observed value of the mass accommodation coefficient was 0.65 with a lower limit of 0.43 within one standard deviation. The experiments also yielded a value of the diffusion coefficient of H_2SO_4 in N_2 of 66.8 $\pm$ 1.1 Torr cm^2 s^{-1} or 0.891 $\pm$ 0.015 Pa m^2 s^{-1}.

9   RECENT CRITICAL EVALUATIONS OF SOLUBILITY DATA FOR IMPORTANT REACTIONS

A selection of recent critical evaluations of accommodation coefficients tabulated in JPL Publication 97-4 [124] are shown in Table 9.40.

A selection of reactive uptake coefficients tabulated in JPL Publication 00–3 [65] are shown in Table 9.41.

A selection of evaluated values of Henry's law constants tabulated in JPL Publication 97-4 [124] are shown in Table 9.42.

REFERENCES

1. Utter, R. G.; Burkholder, J. B.; Howard, C. J.; Ravishankara, A. R., *J. Phys. Chem.*, 1992, **96**, 4973–8.
2. Magi, L.; Schweitzer, F.; Pallares, C.; Cherif, S.; Mirabel, Ph. George, Ch., *J. Phys. Chem. A*, 1997, **101**, 4943.
3. Takami, A.; Kato, S.; Shimono, A.; Koda, S., *Chem. Phys.*, 1998, **231**, 215–27.
4. Lind, J. A.; Kok, G. L., *J. Geophys. Res.*, 1986, **91**, 7889–7895; 1994, **99**(D10), 21119.
5. O'Sullivan, D. W. O.; Lee, M.; Noone, B. C.; Heikes, B. G., *J. Phys. Chem.*, 1996, **100**, 3241–7.
6. Staffelbach, T. A.; Kok, G. L., *J. Geophys. Res.*, 1993, **98**, 12713–7.
7. Schumb, W. C.; Satterfield, C. N.; Wentworth, R. L., *ACS Monogr. Ser.*, 1955, **128**, 221–47.
8. Hwang, H.; Dasgupta, P. K., *Environ. Sci. Technol.*, 1985, **19**, 255–8.
8a. Harned, H. S.; Davies, R., *J. Am. Chem. Soc.*, 1943, **65**, 2030.
9. Berg, D.; Patterson, A., *J. Am. Chem. Soc.*, 1953, **75**, 5197.
10. Betterton, E. A.; Hoffmann, M. R., *Environ. Sci. Technol.*, 1988, **22**, 1415–8.
11. Swartz, E.; Boniface, J.; Tchertkov, I.; Rattigan, O. V.; Robinson, D. V.; Davidovits, P.; Worsnop, D. R.; Jayne, J. T.; Kolb, C. E., *Environ. Sci. Technol.*, 1997, **31**, 2634–41.
12. Valenta, P., *Collect. Czech Chem. Commun.*, 1960, **25**, 853.
13. Bell, R. P.; Evans, P. G., *Proc. R. Soc., Ser. A*, 1966, **291**, 297.
14. Jayne, J. T.; Worsnop, D. R.; Kolb, C. E.; Swartz, E.; Davidovits, P., *J. Phys. Chem.*, 1996, **100**, 8015–22.
15. Albert, M.; Garcia, B. C.; Kreiter, C.; Maurer, G., *AIChE J.*, 1999, **45**, 2024–33.

16. Albert, M.; Hahnenstein, I.; Maurer, G., *AIChE J.*, 1996, **42**, 1741.
17. Kogan, L. V.; Blazhin, Yu. M.; Ogorodikov, S. K.; Kafarov, V. V., *Zh. Prikl. Khim.*, 1977, **50**, 2682.
18. Credali, L.; Mortillaro, L.; Galiazzo, G.; Russo, M.; De Checci, C., *Chim. Ind. (Milan)*, 1965, **47**, 732.
19. Maurer, G., *AIChE J.*, 1986, **32**, 932.
20. Bell, R. P., *Adv. Phys. Org. Chem.*, 1966, **4**, 1–29.
21. Benkelberg, H.-J., Hamm, S.; Warneck, P. J., *Atmos. Chem.*, 1995, **20**, 17–34.
22. Snider, J. R.; Dawson, G. A., *J. Geophys. Res.*, 1985, **90**, 3797–805.
23. Zhou, X.; Mopper, K., *Environ. Sci. Technol.*, 1990, **24**, 1864–9.
24. Williams, L. R.; Klassen, J. K.; Golden, D. M., presented at the 1997 Conference on the Atmospheric Effects of Aviation; http://hyperion.gsfc.nasa.gov/AEAP/98 williamspos.html.
25. Betterton, E. A., *Atmos. Environ.*, 1991, **25A**, 1473–7.
26a. Golding, R. M., *J. Chem. Soc.*, 1960, 3711–6.
26b. Kerp, W., *Chem. Zentralbl.*, 1904, **75**, 57–9.
27. Allen, J. M.; Balcavage, W. X.; Ramachandran, B. R.; Shrout, A. L., *Environ. Toxicol. Chem.*, 1998, **17**, 1216–21.
28. Wasa, T.; Musha, S., *Univ. Osaka Prefect. Ser. A*, 1970, **19**, 169–80.
29. Sorensen, P. E., *Acta Chem. Scand.*, 1972, **26**, 3357–65.
30. Snider, J. R.; Dawson, G. A., *J. Geophys. Res.*, 1985, **90**, 3797–805.
31. Amoore, J. E.; Buttery, R. G., *Chem. Senses Flavour*, 1978, **3**, 57.
32. Buttery, R. G., Ling, L. C.; Guadagni, D. G., *J. Agric. Food Chem.*, 1969, **17**, 385–9.
33. Olson, J. D., *Fluid Phase Equilib.*, 1998, **150–151**, 713–20.
34. Iraci, L. T.; Baker, B. M.; Tyndall, G. S.; Orlando, J. J., *J. Atmos. Chem.*, 1999, **33**, 321–30.
35. Zhou, X., Lee, Y., *J. Phys. Chem.*, 1992, **96**, 265–72.
36. Clegg, S. L.; Brimblecombe, P., in Melchior, D. C.; Bassett, R. L. (eds), *Chemical Modelling of Aqueous Systems II*, American Chemical Society, Washington, DC, 1990, pp. 58–73.
37. Khan, I.; Brimblecombe, P., *J.Aerosol Sci.*, 1992, **23**, Suppl. 1, S897–S900.
38. Khan, I.; Brimblecombe, P.; Clegg, S. L., *Environ. Technol.*, 1992, **13**, 587–93.
39. Khan, I.; Brimblecombe, P.; Clegg, S. L., *J. Atmos. Chem.*, 1995, **22**, 285–302.
40. Scarano, E.; Gay, G.; Forina, M., *Anal. Chem.*, 1971, **43**, 206–11.
41. Fogg, P. G. T.; Gerrard, W., *Solubility of Gases in Liquids*, Wiley, Chichester, 1990, pp. 174–7.
42. Jakowlin, A. A., *Z. Phys. Chem.*, 1899, **29**, 613.
43. Lide, D. R. (ed.), *CRC Handbook of Chemistry and Physics*, 73rd edn, CRC Press, Boca Raton, FL, 1992.
44. Harris, I., *Nature*, 1943, **151**, 309.
45. Battino, R., *Sulfur Dioxide, Chlorine, Fluorine, and Chlorine Oxides*, IUPAC Solubility Data Series, Vol. 12, Young, C. L. (ed.), Pergamon Press, Oxford, 1983.
46. Whitney, R. P.; Vivian, J. E., *Ind. Eng. Chem.*, 1941, **33**, 741–4.
47. Sherrill, M. S.; Izard, E. F., *J. Am. Chem. Soc.*, 1928, **50**, 1665; 1931, **53**, 1667.
48. Vogt, R.; Sander, R.; Von Glasow, R.; Crutzen, P. J., *J. Atmos. Chem.*, 1999, **32**, 375–95.
49. McFiggans, G.; Plane, J. M. C.; Allan, B. J.; Carpenter, L. J., *J. Geophys. Res.*, 2000, **105**(D11), 14371–85.
50. Carslaw, K. S.; Clegg, S. L.; Brimblecombe, P., *J. Phys. Chem.*, 1995, **99**, 11557–74.
51. Van Doren, J. M.; Watson, L. R.; Davidovits, P.; Worsnop, D. R.; Zahniser, M. S.; Kolb, C. E., *J. Phys. Chem.*, 1990, **94**, 3265–9.

52. Kee, R. J.; Dixon-Lewis, G.; Warnatz, J.; Coltrin, M. E.; Miller, J. A., *SAN086-8246*, Sandia National Laboratory, Albuquerque, NM, 1986.
53. Abbatt, J. P. D., *J. Geophys. Res.*, 1995, **100**(D7), 14009–17.
54. Hanson, D. R.; Ravishankara, A. R., *J. Phys. Chem.*, 1993, **97**, 12309–19.
55. Williams, L. R.; Golden, D. M., *Geophys. Res. Lett.*, 1993, **20**, 2227–30.
56. Tyndall, G. S.; Winker, D.; Anderson, T.; Eisele, F., *Advances in Measurement Methods for Atmospheric Chemistry*, Chapt. 8, http://medias.obs-mip.fr:8000/igac/html/book/chap8/chap8.html, 2000.
57. Abbatt, J. P. D.; Nowak, J. B., *J. Phys. Chem.*, 1997, **101**, 2131–7.
58. Williams, L. R.; Golden, D. M.; Huestis, D. L., *J. Geophys. Res.*, 1995, **100**, 7329–35.
59. Grothe, H.; Knözinger, E.; Myhre, C. L.; Nielsen, C., *Air Pollution Research Report 66, Polar Stratospheric Ozone 1997. Proceedings of the 4th European Symposium, 22–26 September 1997, Schliersee*, Germany, Harris, N. R. P. (ed.), European Commission, Brussels, 1998.
60. Ourisson, J.; Kastner, M., *Bull. Soc. Chim. Fr., Ser. 5*, 1939, **6**, 1307–11.
61. Holzwarth, G.; Balmer, R. G.; Soni, L., *Water Res.*, 1984, **18**, 1421–7.
62. Bletchley, E. R.; Johnson, R. W.; Alleman, J. E.; McCoy, W. F., *Water Res.*, 1992, **26**, 99–106.
63. Huthwelker, T.; Peter, Th.; Luo, B. P.; Clegg, S. L.; Carslaw, K. S.; Brimblecombe, P., *J. Atmos. Chem.*, 1995, **21**(1), 81–95.
64. Shi, Q.; Davidovits, P.; Jayne, D. R.; Kolb, C. E.; Worsnop, D. R., *J. Geophys. Res.*, 2001, **106**, 24259–74.
65. Sander, S. P.; Friedl, R. R.; DeMore, W. B.; Ravishankara, A. R.; Golden, D. M.; Kolb, C. E.; Kurylo, M. J.; Hampson, R. F.; Huie, R. E.; Molina, M. J.; Moortgat, G. K., *JPL Publication 00–003: Chemical Kinetics and Photochemical Data for Use in Stratospheric Modeling. Evaluation Number 13. Supplement to Evaluation 12: Update of Key Reactions*. National Aeronautics and Space Administration, Jet Propulsion Laboratory, California Institute of Technology, Pasadena, CA, 2000, pp. 57–62.
66. Robinson, G. N.; Worsnop, D. R.; Jayne, J. T.; Kolb, C. E. *J. Geophys. Res.*, 1997, **102**, 3583–601.
67. Donaldson, D. J.; Ravishankara, A. R.; Hanson, D. R., *J. Phys. Chem. A*, 1997, **101**, 4717–25.
68. Hanson, D. R., *J. Phys. Chem. A*, 1998, **102**, 4794–807.
69. Klassen, J. K.; Hu, Z.; William, L. R., *J. Geophys. Res.*, 1998, **103**, 16197–202.
70. Michelsen, H. A., *Geophys. Res. Lett.*, 1998, **25**, 3571–3.
71. *JPL-003: Heterogeneous Chemistry Section Addenda and Errata for JPL Publication 00–003; Chemical Kinetics and Photochemical Data for Use in Stratospheric Modeling. Evaluation Number 13*, National Aeronautics and Space Administration, Washington, DC, http://jpldataeval.jpl.nasa.gov/thirteenerrata.htm.
72. Fickert, S.; Adams, J. W.; Crowley, J. N., *J. Geophys. Res.*, 1999, **104**(D19), 23719–27.
73. Eigen, M.; Kustin, K., *J. Am. Chem. Soc.*, 1962, **84**, 1355–61.
74. Wang, T. X.; Kelley, M. D.; Cooper, J. N.; Beckwith, C.; Margerum, D. W., *Inorg. Chem.*, 1994, **33**, 5872–8.
75. De Bruyn, W. J.; Shorter, J. A.; Davidovits, P.; Worsnop, D. R.; Zahniser, M. S.; Kolb, C. E., *Environ. Sci. Technol.*, 1995, **29**, 1179–85.
76. George, Ch.; Lagrange, J.; Lagrange, Ph.; Mirabel, Ph.; Pallares, C.; Ponche, J. L., *J. Geophys. Res.*, 1994, **99**, 1255–62.
77. Manogue, W. H.; Pigford, R., *AIChE J.*, 1960, **6**, 494.
78. Ugi, I.; Beck, F., *Chem. Ber.*, 1961, **94**, 1839.
79. Behnke, W.; Elend, M.; Krüger, H.-V.; Zetzsch, C., in *Proceedings of the AFEAS Workshop*, Brussels, Belgium, 1992, p. 68.

80. Behnke, W.; Elend, M.; Krüger, H.-V.; Zetzsch, C., in *Proceedings of the STEP–HALOCSIDE/AFEAS Workshop*, University College, Dublin, Ireland, 1993.
81. Mertens R. von Sonntag, C.; Lind, J.; Merenyi, G., *Angew. Chem., Int. Ed. Engl.*, 1994, **33**, 1259.
82. George, Ch.; Saison, J. Y.; Ponche, J. L.; Mirabel, Ph., *J. Phys. Chem.*, 1994, **98**, 10857–62.
83. IUPAC Subcommittee for Gas Kinetic Data Evaluation, *Evaluated Kinetic Data*, http://www.iupac-kinetic.ch.cam.ac.uk/, 2001.
84. Clegg, S. L.; Brimblecombe, P., *J. Phys. Chem.*, 1989, **93**, 7237–48.
85. Clegg, S. L.; Brimblecombe, P.; Wexler, A. S., *J. Phys. Chem. A.*, 1998, **102**, 2137–54.
86. Neuhausen, B. S.; Patrick, W. A., *J. Phys. Chem.*, 1921, **25**, 693.
87. Perman, J., *J. Chem. Soc.*, 1903, **83**, 1168.
88. Sherwood, T. K., *Ind. Eng. Chem.*, 1925, **17**, 745.
89. Zhang, R.; Leu, M.-T.; Keyser, L. F.; *J. Phys. Chem. A*, 1997, **101**, 3324–30.
90. Fogg, P. G. T.; Gerrard, W., *Solubility of Gases in Liquids*, Wiley, Chichester, 1990, pp. 88–89.
91. Clifford, I. L.; Hunter, E., *J. Phys. Chem.*, 1932, **36**, 101.
92. Ponche, J. L.; George, C.; Mirabel, P., *J. Atmos. Chem.*, 1993, **16**, 1–21.
93. Bongartz, A.; Schweighoefer, S.; Roose, C.; Schurath, U., *J. Atmos. Chem.*, 1995, **20**, 35–58.
94. Larson, T. V.; Taylor, G. S., *Atmos. Environ.*, 1983, **12**, 2489–95.
95. Richardson, C. B.; Hightower, R. L., *Atmos. Environ.*, 1987, **21**, 971–5.
96. Harrison, R. M.; Sturges, W. T.; Kitto, N.; Li, Y.-Q., *Atmos. Environ.*, 1990, **24**, 1883–8.
97. Reid, C. R.; Sherwood, T. K., *The Properties of Gases and Liquids*, McGraw-Hill, New York, 1986, pp. 520–65.
98. Leaist, D. G., *J. Phys. Chem.*, 1987, **91**, 4635–8.
99. Swartz, E.; Shi, Q.; Davidovits, P.; Jayne, D. R.; Worsnop, D. R.; Kolb, C. E., *J. Phys. Chem. A*, 1999, **103**, 8812–23, 8824–33.
100. Carstens, T.; Schurath, U., presented at the EUROTRAC Symposium, 25–29 March 1996, Garmisch-Partenkirchen, Germany; http://imk-aida.fzk.de/abstract/abs_etma.html.
101. Scheer, V.; Frenzel, A.; Behnke, W.; Zetzsch, C.; Magi, L.; George, Ch.; Mirabel, Ph., *J. Phys. Chem. A*, 1997, **101**, 9359–66.
102. Reid, R. C.; Prausnitz, J. M.; Poling, B. E., *The Properties of Gases and Liquids*, 4th edn., McGraw-Hill, New York, 1987.
103. Behnke, W.; George, C.; Scheer, V.; Zetzsch, C., *J. Geophys. Res.*, 1997, **102D**, 3795–804.
104. Imamura, T.; Rudich, Y.; Talukdar, R. K.; Fox, R. W.; Ravishankara, A. R., *J. Phys. Chem. A*, 1997, **101**, 2316–22.
105. Rudich, Y.; Talukdar, R. K.; Ravishankara, A. R.; Fox, R. W., *J. Geophys. Res.*, 1996, **101**, 21023–31.
106. Rudich, Y.; Talukdar, R. K.; Imamura, T.; Fox, R. W.; Ravishankara, A. R., *Chem. Phys. Lett.*, 1996, **261**, 467–73.
107. Exner, M.; Herrman, H.; Zellner, R., *Ber. Bunsen-Ges. Phys. Chem.*, 1992, **96**, 470–7.
108. Atkins, P. W., *Physical Chemistry*, Oxford University Press, New York, 1990.
109. Schwartz, S. E.; White, W. H., Kinetics of reactive dissolution of nitrogen oxides into aqueous solution, in *Advances in Environmental Science and Technology*, Schwartz, S. E. (ed.), Wiley, New York, 1983, Vol. 12, pp. 1–116.
110. Cheung, J. L.; Li, Y. Q.; Boniface, J.; Shi, Q.; Davidovits, P.; Worsnop, D. R.; Jayne, J. T.; Kolb, C. E., *J. Phys. Chem. A*, 2000, **104**, 2655–62.
111a. Mertes, S.; Wahner, A., *J. Phys. Chem.*, 1995, **99**, 14000–6.

111b. Komiyama, H.; Inoue, M., *Chem. Eng. Sci.*, 1980, **35**, 154–61.
112. Lee, Y.-N.; Schwartz, S. E., *J. Phys. Chem.*, 1981, **85**, 840–8.
113. Park, J.-Y.; Lee, Y.-N., *J. Phys. Chem.*, 1988, **92**, 6294–302.
114. Cape, J. N.; Storeton-West, R. L.; Devine, S. F.; Beatty, R. N.; Murdoch, A., *Atmos. Environ.*, 1993, **27A**, 2613–21.
115. Moll, A. J., The rate of hydrolysis of nitrogen tetroxide, *PhD Thesis*, University of Washington, 1966.
116. Grätzel, M.; Henglein, A.; Little, J.; Beck, G., *Ber. Bunsen-Ges. Phys. Chem.*, 1969, **73**, 646–53.
117. Treinin, A.; Haydon, E., *J. Am. Chem. Soc.*, 1970, **92**, 5821–8.
118. Kleffman, J.; Becker, K. H.; Wiesen, P., *Atmos. Environ.*, 1998, **32**, 2721–9.
119. Saastad, O. W.; Ellermann, T.; Nielsen, C. J. *Geophys. Res. Lett.*, 1993, **20**, 1191–3.
119a. Lee, J. H.; Tang, I. N., *Atmos. Environ.* 1988, **22**, 1147–51.
120. Becker, K. H.; Kleffmann, J.; Kurtenbach, R.; Wiesen, P., *J. Phys. Chem.*, 1996, **100**, 14984–90.
121. Hanson, D. R.; Ravishankara, A. R.; Lovejoy, E. R., *J. Geophys. Res.*, 1996, **101D**, 9063–9.
122. Behnke, W.; Krüger, H.-V.; Scheer, V.; Zetzsch, C., *J. Aerosol Sci.*, 1991, **S22**, 609–12.
123. Sander, R.; Rudich, Y.; von Glasow, R.; Crutzen, P. J., *Geophys. Res. Lett.*, 1999, **26**, 2857–60.
124. DeMore, W. B.; Sander, S. P.; Golden, D. M.; Hampson, R. F.; Kurylo, M. J.; Howard, C. J.; Ravishankara, A. R.; Kolb, C. E.; Molina, M. J., *JPL Publication 97-4: Chemical Kinetics and Photochemical Data for Use in Stratospheric Modeling,* Jet Propulsion Laboratory, Pasadena, CA, 1997.
125. Hu, J. H.; Abbatt, J. P. D., *J. Phys. Chem. A*, 1997, **101**, 871–8.
126. Mozurkewich, M.; Calvert, J. G., *J. Geophys. Res.*, 1988, **93**, 15889.
127. Fried, A.; Henry, B. E.; Calvert, J. G.; Mozurkewich, M., *J. Geophys. Res.*, 1994, **99**, 3517–32.
128. Van Doren, J. M.; Watson, L. R.; Davidovits, P.; Worsnop, D. R.; Zahniser, M. S.; Kolb, C. E., *J. Phys. Chem.*, 1991, **95**, 1684–9.
129. Lovejoy, E. R.; Hanson, D. R., *J. Phys. Chem.*, 1995, **99**, 2080–7.
130. George, Ch.; Ponche, J. L.; Mirabel, Ph.; Behnke, W.; Scheer, V.; Zetzsch, C., *J. Phys. Chem.*, 1994, **98**, 8780–4.
131. Régimbal, J.-M.; Mozurkewich, M., *J. Phys. Chem. A*, 1997, **101**, 8822–9.
132. Schwartz, S. E.; White, W. H., *Adv. Environ. Sci. Eng.*, 1981, **4**, 1.
133. Seel, F.; Winkler, R., *Z. Phys. Chem., NF*, 1960, **25**, 217.
134. Becker, K. H.; Kleffmann, J.; Negri, R. M.; Wiesen, P. *J. Chem. Soc., Faraday Trans.*, 1998, **94**, 1583–6.
135. Clegg, S. L.; Brimblecombe, P., *J. Phys. Chem.*, 1990, **94**, 5369–80; 1992, **96**, 6854.
136. Holeci, I., *Chem. Prum.*, 1968, **16**, 267–70.
137. Reid, C. R.; Sherwood, T. K., *The Properties of Gases and Liquids*, McGraw-Hill, New York, 1986, pp. 520–65.
138. Park, J. Y.; Lee, Y. N., Aqueous solubility and hydrolysis kinetics of peroxynitric acid, paper presented at the 193rd American Chemical Society Meeting, Denver, CO, April 5–10, 1987.
139. Leu, M.-T.; Zhang, R., *Geophys. Res. Lett.*, 1999, **26**, 1129–32.
140. Kames, J.; Schurath, U., *J. Atmos. Chem.*, 1995, **21**, 151–64.
141. Lee, Y. N.; in *Gas–Liquid Chemistry of Natural Waters*, Vol. 1, Brookhaven National Laboratories, Brookhaven, NY, 1984, BNL 51757 21/1–21/7.
142. Kames, J.; Schweighoefer, S.; Schurath, U., *J. Atmos. Chem.*, 1991, **12**, 169–80.
143. Holdren, M. W.; Spice, C. W.; Hales, J. M., *Atmos. Environ.*, 1984, **18**, 1171–3.

144. Mudd, J. B., *J. Biol. Chem.*, 1966, **241**, 4077–80.
145. Zhang, R. Leu, M.-T., *J. Geophys. Res.*, 1997, **102**(D/7), 8837–43.
146. Eigen, M.; Kustin, K.; Maas, G. Z., *Z. Phys. Chem. (Munich)*, 1961, **30**, 130.
147. Pinsent, B. R. W.; Pearson, L.; Roughton, F. J. W., *Trans. Faraday Soc.*, 1956, **52**, 1512.
148. Maahs, H. G., in *Heterogeneous Atmospheric Chemistry*, Schryer, D. R. (ed.), Geophysical Monograph No. 26; American Geophysical Union, Washington, DC, 1982, p. 187.
149. Millero, F. J., *Mar. Chem.*, 1986, **18**, 121.
150. Perrin, D. D., *Ionization Constants of Inorganic Acids and Bases in Aqueous Solution*, IUPAC Chemical Data Series No. 29, Pergamon Press, Oxford, 1982, p. 20.
151. Lide, D. R. (ed.), *Handbook of Chemistry and Physics*, 73rd edn, CRC Press, Boca Raton, FL, 1992, pp. 6–3 and 6–10.
152. De Bruyn, W. J.; Swartz, E.; Hu, J. H.; Shorter, J. A.; Davidovits, P.; Worsnop, D. R.; Zahniser, M. S.; Kolb, C. E., *J. Geophys. Res.*, 1995, **100**, 7245.
153. Fogg, P. G. T.; Gerrard, W., *Solubility of Gases in Liquids*, Wiley, Chichester, 1990, pp. 41–4.
154. Hudson, J. C., *J. Chem. Soc.*, 1925, 1332.
155. Beuschlein, W. L.; Simenson, L. O., *J. Am. Chem. Soc.*, 1940, **62**, 610.
156. Rabe, A. E.; Harris, J. F., *J. Chem. Eng. Data*, 1963, **8**, 333.
157. Lavrova, E. M.; Tudorovskaya, L. L., *Zh. Prikl. Khim.*, 1977, **50**, 1146; *J. Appl. Chem. USSR*, 1977, **50**, 1102.
158. Johnstone, H. F.; Leppla, P. W., *J. Am. Chem. Soc.*, 1934, **56**, 2233–8.
159. Young, C. L. (ed.), *Sulfur Dioxide, Chlorine, Fluorine, and Chlorine Oxides*, IUPAC Solubility Data Series, Vol. 12, Pergamon Press, Oxford, 1983.
160. Miles, F. D.; Fenton, J., *J. Chem. Soc.*, 1920, **117**, 59.
161. Cupr, V. *Recl. Trav. Chim. Pays-Bas*, 1928, **47**, 55.
162. Boniface, J.; Shi, Q.; Li, Y. Q.; Cheung, J. L.; Rattigan, O. V.; Davidovits, P.; Worsnop, D. R.; Jayne, D. R.; Kolb, C. E., *J. Phys. Chem. A*, 2000, **104**, 7502–10.
163. Jayne, J. T.; Davidovits, P.; Worsnop, D. R.; Zahniser, M. S.; Kolb, C. E., *J. Phys. Chem.*, 1990, **94**, 6041.
164. Pöschl, U.; Canagaratna, M.; Jayne, J. T.; Molina, L. T.; Worsnop, D. R.; Kolb, C. E.; Molina, M. J., *J. Phys. Chem. A*, 1998, **102**, 10082.
165. Hanson, D. R.; Burkholder, J. B.; Howard, C. J.; Ravishankara, A. R. *J. Phys. Chem.*, 1992, **96**, 4979–85.
166. Baldwin, A. C.; Golden, D. M., *J. Geophys. Res.*, 1980, **85**, 2888–9.
167. Taylor, R. S.; Ray, D.; Garrett, B. C., *J. Phys. Chem. B*, 1997, **101**, 5473–6.
168. Mozurkewich, M.; McMurry, P. H.; Gupta, A.; Calvert, J. G., *J. Geophys. Res.*, 1987, **92**, 4163–70.
169. Rudolf, R.; Wagner, P. E., *J. Aerosol Sci.*, 1994, **25**, 597–8.
170. Baldwin, A. C.; Golden, D. M., *Science*, 1979, **206**, 562.
171. Fung, K. N.; Tang, I. N.; Munkelwitz, H. R., *Appl. Opt.*, 1987, **26**, 1282–7.
172. Worsnop, D. R.; Zahniser, M. S.; Kolb, C. E.; Gardner, J. A.; Watson, L. R.; Van Doren, J. M.; Jayne, J. T.; Davidovits, P., *J. Phys. Chem.*, 1989, **93**, 1159–72.
173. Tolbert, M. A.; Rossi, M. J.; Golden, D. M., *Geophys. Res. Lett.*, 1988, **15**, 847–50.
174. Jayne, J. T.; Duan, S. X.; Davidovits, P.; Worsnop, D. R.; Zahniser, M. S.; Kolb, C. E., *J. Phys. Chem.*, 1991, **95**, 6329–36.
175. Jayne, J. T., Duan, S. X.; Davidovits, P.; Worsnop, D. R.; Zahniser, M. S.; Kolb, C. E. *J. Phys. Chem.*, 1992, **96**, 5452–60.
176. Duan, S. X.; Jayne, J. T.; Davidovits, P.; Worsnop, D. R.; Zahniser, M. S.; Kolb, C. E., *J. Phys. Chem.*, 1993, **97**, 2284–8.
177. Watson, L. R.; Doren, J. M. V.; Davidovits, P.; Worsnop, D. R.; Zahniser, M. S.; Kolb, C. E. *J. Geophys. Res.*, 1990, **95**, 5631–8.

178. Abbatt, J. P. D., *Geophys. Res. Lett.*, 1994, **21**, 665–8.
179. Hanson, D. R.; Ravishankara, A. R., in *The Tropospheric Chemistry of Ozone in the Polar Regions*, Niki, H.; Becker K. H. (eds), NATO ASI Series, Series 1, Vol. 7, Springer, Berlin, 1993, pp. 17281–90.
180. Hu, J. H.; Shorter, J. A.; Davidovits, P.; Worsnop, D. R.; Zahniser, M. S.; Kolb, C. E. *J. Phys. Chem.*, 1993, **97**, 11037–42.
181. De Bruyn, W. J.; Shorter, J. A.; Davidovits, P.; Worsnop, D. R.; Zahniser, M. S.; Kolb, C. E., *J. Geophys. Res.*, 1994, **99**, 16927–32.
182. Cooper, P. L.; Abbatt, J. P. D., *J. Phys. Chem.*, 1996, **100**, 2249–54.
183. Gersherzon, V. M.; Grigorieva, V. M.; Ivanov, A. V.; Remorov, R. G., *Faraday Discuss.*, 1995, **100**, 83–100.
184. Behnke, W.; Kruger, H.-U.; Scheer, V.; Zetzsch, C., *J. Aerosol Sci.*, 1992, **23**, S923–S936.
185. Behnke, W.; Scheer, V.; Zetzsch, C., *J. Aerosol Sci.*, 1993, **24**, S115–S116.
186. Zetzsch, C.; Behnke, W., *Ber. Bunsen-Ges. Phys. Chem.*, 1992, **96**, 488–93.
187. Bongartz, A.; Kames, J.; Schurath, U.; George, C.; Mirabel, P.; Ponche, J. L., *J. Atmos. Chem.*, 1994, **18**, 149–60.
188. Robbins, R. C.; Cadle, R. D., *J. Phys. Chem.*, 1958, **62**, 469–71.
189. Huntzicker, J. J.; Cary, R. A.; Ling, C.-S. *Environ. Sci. Technol.*, 1980, **14**, 819–24.
190. McMurry, P. H.; Takano, H.; Anderson, G. R., *Environ. Sci. Technol.*, 1983, **17**, 347–57.
191. Daumer, B.; Niessner, R.; Klockow, D., *J. Aerosol Sci.*, 1992, **23**, 315–25.
192. Villalta, P. W.; Lovejoy, E. R.; Hanson, D. R., *Geophys. Res. Lett.*, 1996, **23**, 1765–8.
193. Martin, L. R.; Judeikis, H. S.; Wun, M., *J. Geophys. Res.*, 1980, **85**, 5511–8.
194. Hanson, D. R.; Ravishankara, A. R., *Geophys. Res. Lett.*, 1995, **22**, 385–8.
195. Hanson, D. R.; Lovejoy, E. R., *Science*, 1995, **267**, 1326–9.
196. Lovejoy, E. R.; Huey, L. G.; Hanson, D. R., *J. Geophys. Res.*, 1995, **100**, 18775–80.
197. Hanson, D. R.; Lovejoy, E. R., *Geophys. Res. Lett.*, 1994, **21**, 2401–4.
198. Hanson, D. R.; Ravishankara, A. R., *J. Phys. Chem.*, 1992, **96**, 2682–91.
199. Zhang, R.; Leu, M.-T.; Keyser, L. F., *J. Geophys. Res.*, 1995, **100**, 18845–54.
200. Williams, L. R.; Manion, J. A.; Golden, D. M.; Tolbert, M. A., *J. Appl. Meteorol.*, 1994, **33**, 785–90.
201. Manion, J. A.; Fittschen, C. M.; Golden, D. M.; Williams, L. R.; Tolbert, M. A., *Isr. J. Chem.*, 1994, **34**, 355–63.
202. Johnson, B. J.; Betterton, E. A.; Craig, D., *J. Atmos. Chem.*, 1996, **24**, 113–9.
203. De Bruyn, W. J.; Duan, S. X. Shi, X. Q.; Davidovits, P.; Worsnop, D. R.; Zahniser, M. S.; Kolb, C. E., *Geophys. Res. Lett.*, 1992, **19**, 1939–42.
204. Hanson, D. R.; Ravishankara, A. R., *Geophys. Res. Lett.*, 1991, **18**, 1699–701.
205. Bowder, D. J.; Clegg, S. L.; Brimblecombe, P., *Chemosphere*, 1996, **32**, 405–20, 785–90.

Addendum I: Henry's Law Constants of OH and HO$_2$

Yin-Nan Lee
Atmospheric Sciences Division, Brookhaven National Laboratory, Upton, NY, USA

The Henry's law type constants of OH and HO$_2$ have not been experimentally determined for obvious reasons: it is extremely difficult to measure the concentrations of these reactive species in either the gas phase or the aqueous phase, let alone simultaneously in both phases. At a more fundamental level, because these radicals react rapidly in both phases, e.g. [1–3], compared with mass-transfer rates characterizing typical laboratory multi-phase systems, the gas–liquid equilibrium which is necessary for such measurements to be feasible is typically not attainable. Consequently, the Henry's law constants of these radicals are traditionally evaluated from the free energy of solution, $\Delta_{sol}G°(X)$ accompanying the process of transferring a molecule X from the gas phase, denoted g, to the aqueous phase, a, i.e.

$$X_g \rightleftharpoons X_a \qquad (9.10)$$

using the equation

$$\Delta_{sol}G°(X) = -RT \ln k_H \qquad (9.11)$$

$\Delta_{sol}G°(X)$ is defined as

$$\Delta_{sol}G°(X) = \Delta_f G°(X)_a - \Delta_f G°(X)_g \qquad (9.12)$$

where the free energies of formation of X in the gas phase and in the aqueous phase are typically evaluated using thermochemical cycles.

It should be pointed out that, because the Henry's law constant determined in the way described above is a function of the difference between two comparable numbers in the exponent, the uncertainty is therefore generally sizable. A small uncertainty of 0.8 kJ mol^{-1} in the individual quantity would correspond to ~50 % uncertainty in the value of k_H. Since a typical combined uncertainty is rarely

Chemicals in the Atmosphere – Solubility, Sources and Reactivity. Edited by P. G. T. Fogg and J. Sangster
© 2003 IUPAC ISBN: 0-471-98651-8

smaller than $2.1\,kJ\,mol^{-1}$ the Henry's law constant evaluated in this fashion cannot be expected to have an uncertainty smaller than a factor of 2.

Finally, it is noted that both of these radicals undergo acid–base dissociation in aqueous solution. Consequently, the effective Henry's law solubilities, k_H^*, of these species depend on their pK_a and pH of the solution, i.e.

$$k_H^* = k_H(1 + K_a/[H^+]) \qquad (9.13)$$

The pK_a values of HO$_2$ and OH have been determined as 4.9 [4] and 11.9 [5], respectively. The heats of ionization of HO$_2$ and OH were determined as 0 and $42 \pm 8\,kJ\,mol^{-1}$ [6]. Later, Bielski [7] recommended a value of 4.69 for the pK_a of HO$_2$. For liquid water typically encountered in the environment, such as cloud, rain and seawater, the pH falls in the range between $\sim$3 and $\sim$8.5, encompassing the pK_a of HO$_2$. Consequently, the effective Henry's law constant of HO$_2$ must be considered for HO$_2$. However, because of the large pK_a of OH compared with the aforementioned pH range, its effective Henry's law solubility is not expected to be affected by the solution pH of these aqueous media.

1   OH [CAS 3352-57-6]

Values of thermodynamic properties of OH in aqueous solution are listed in Table 9.43. Values for OH in the gas phase are listed in Table 9.44. The Henry's law constant of OH has been estimated by various investigators and the values range from 25 to $1 \times 10^5\,mol\,dm^{-3}\,bar^{-1}$ [18]. While the high values were simply assumed by Chameides [19], Jacob [20] adhered to a much smaller value of $25\,mol\,dm^{-3}\,bar^{-1}$ (298 K), as did Schwartz [21], based on Schwarz and Dodson [11]. However, it is recognized that the large discrepancies in the value of Henry's law constant of OH have little substantive impact on a model's ability to describe the atmospheric chemistry and distribution of this important species, as stated for example by Lelieveld and Crutzen [18].

Table 9.43 Thermodynamic properties of aqueous OH

Quantity	Value	Ref.
$\Delta_f G_a^\circ/kJ\,mol^{-1}$	13.39	8,9
	19.00	10
	25.10	11
	26.82	12
	35.69	13
$\Delta_f H_a^\circ/kJ\,mol^{-1}$	7.1	14
	−4.2	10
$S^\circ/J\,mol^{-1}K^{-1}$	96	14
$\Delta_{sol}G^\circ/kJ\,mol^{-1}$	−10.0	11
	−20.9	8
pK_a	11.9	15

Table 9.44 Thermodynamic properties of gaseous OH

Quantity	Value	Ref.
$\Delta_f G_g^\circ$/kJ mol^{-1}	34.31	16
$\Delta_f H_g^\circ$/kJ mol^{-1}	38.99	16
	44.4	17
	37.28	17
	39.12	17
	41.8	17

Table 9.45 Gibbs energy of solution, $\Delta_{sol}G^\circ$

Species	Value/kJ mol^{-1}	Ref.
HO	−25	22
	−15.5	10
	−5.0	23
	−8.4	24

The value of $\Delta_{sol}G^\circ$ for OH was estimated by Koppenol and Liebman [8] to be -21 kJ mol^{-1}. This corresponds to a Henry's law constant of 4.6×10^3 mol dm^{-3} bar^{-1}. The values of $\Delta_{sol}G^\circ$ which have appeared in the literature range from 17.6 to -25 kJ mol^{-1} (Table 9.45), and would result in Henry's law constant values from 8 to 2.5×10^4 mol dm^{-3} bar^{-1}. However, using the average value of $\Delta_f G_g^\circ = 34.64 \pm 3.51$ kJ mol^{-1} and $\Delta_f G_a^\circ = 21.09 \pm 6.11$ kJ mol^{-1} results in a Henry's law constant of 237 mol dm^{-3} bar^{-1} for HO, with a large uncertainty of 15 to 4×10^4. This is of little consequence in the development of models of the atmospheric chemistry of HO.

2   HO$_2$ [CAS 3170-83-0]

Values of thermodynamic properties of HO$_2$ in aqueous solution are listed in Table 9.46. The first explicit evaluation of the Henry's law constant of HO$_2$

Table 9.46 Thermodynamic properties of aqueous HO$_2$

Quantity	Value	Ref.
$\Delta_f G_a^\circ$/kJ mol^{-1}	5.06 ± 0.84	25
$\Delta_f H_a^\circ$/kJ mol^{-1}	-36.0 ± 4.2	14
S°/J mol^{-1}K^{-1}	138 ± 4	14
$\Delta_{sol}G_a^\circ$/kJ mol^{-1}	-356 (O$_2^-$)	26
pK_a	4.8 ± 0.1	3

Table 9.47   Thermodynamic properties of HO$_2$(g)

Quantity	Value	Ref.
$\Delta_f G_g^\circ$/kJ mol^{-1}	22.6 ± 2.5	29
	14.43	16
	26.8	30
$\Delta_f H_g^\circ$/kJ mol^{-1}	10.5 ± 2.5	29
	2.1 ± 8.4	16
	$14.6 + 4.2/-2.1$	27
	14.6	31
	16.82 ± 2.51	32
	15.90 ± 5.02	28
	22.2	33
	24.06	34
$T\Delta_f S_g^\circ$/kJ mol^{-1}	-12.30	29
S°/J mol^{-1} K^{-1}	228.99	29

using a thermochemical cycle was given by Schwartz [25] as shown below.

$$HO_2(g) \longrightarrow O_2(g) + \tfrac{1}{2}H_2(g) \qquad (9.14)$$

$$HO_2(aq) \longrightarrow O_2^-(aq) + H^+(aq) \qquad (9.15)$$

A value of 1.2×10^3 mol dm^{-3} bar^{-1} was recommended based on an estimated value of 5.06 ± 0.84 kJ mol^{-1} for $\Delta_f G^\circ(HO_2)_a$ in combination with a $\Delta_f G^\circ(HO_2)_g = 22.6$ kJ mol^{-1}. It was also pointed out that, because of the uncertainty in the value of the enthalpy of formation of HO$_2$, $\Delta_f H_g^\circ$, which was used to deduce $\Delta G^\circ(HO_2)_g$, the value of $k_H(HO_2)$ may be as large as 6.7×10^3 mol dm^{-3} bar^{-1} if $\Delta_f H_g^\circ = 14.6$ kJ mol^{-1} were used. The recommended value by Shum and Benson [27] for $\Delta_f H_g^\circ$ is $14.6(+4.2, -2.1)$ kJ mol^{-1}, contrasting with a JANAF value of 2.1 ± 8.4 kJ mol^{-1} [16]. Although Chameides [19] adopted a high value for $k_H(HO_2)$, i.e. 9×10^3 mol dm^{-3} bar^{-1} (291 K), Jacob [20] adhered to an intermediate value of 4×10^3 mol dm^{-3} bar^{-1} in a review of atmospheric heterogeneous chemistry.

Since the study of Shum and Benson [27], several additional values of $\Delta_f H_g^\circ$ have appeared in the literature. These values are compiled in Table 9.47. Fisher and Armentrout [28] compared their experimentally determined value, 15.9 ± 5.0 kJ mol^{-1}, with the available data and found good agreement with both the value recommended by Shum and Benson [9] and that measured by Howard [29] (Table 9.47). In the case of the experimentally determined appearance energy of ion radicals, Holmes *et al.* [29] reported a value of 14.6 kJ mol^{-1} for $\Delta_f H^\circ(HO_2)$, in excellent agreement with the recommended value by Shum and Benson [27]. Later, Espinosa-Garcia [32] reported a theoretical value of 16.82 ± 2.51 kJ mol^{-1} and suggested that earlier calculations such as that of Sana *et al.* [34] may represent overestimates. Judging from the fairly consistent experimentally determined results and a downward trend in the calculated values which asymptotically approach the experimental values, the value recommended for $\Delta_f H^\circ (HO_2)_g$ by Shum and Benson [27] is the most appropriate to use. With the use of the value $14.6 \pm 4.2 \, kJ \, mol^{-1}$ for $\Delta_f H_g^\circ$ which leads to $\Delta_f G_g^\circ = 26.94 \pm 4.18 \, kJ \, mol^{-1}$ and $\Delta_f G_a^\circ = 5.06 \pm 0.84 \, kJ \, mol^{-1}$ evaluated by Schwartz [25], the Henry's law constant of HO$_2$ is estimated to be $6.8 \times 10^3 \, mol \, dm^{-3} \, bar^{-1}$. However, in view of the fairly large uncertainties of the thermodynamic values involved in the calculation, 1.3×10^3 to $3.7 \times 10^4 \, mol \, dm^{-3} \, bar^{-1}$ may be used as bounds.

REFERENCES

1. Atkinson, R.; Baulch, D. L.; Cox, R. A.; Hampson, R. F., Jr; Kerr, J. A.; Troe, J., *J. Phys. Chem. Ref. Data, suppl. IV*, 1992, **21**, 1125–1568.
2. Buxton, G. V.; Greenstock, C. L.; Helman, W. P.; Ross, A. B., *J. Phys. Chem. Ref. Data*, 1988, **17**, 513–886.
3. Bielski, B. H. J.; Cabelli, D. E.; Arudi, R.; Ross, A. B., *J. Phys. Chem. Ref. Data*, 1985, **14**, 1041–100.
4. Behar, D.; Czapski, G.; Rabani, J.; Dorfman, L. M.; Schwarz, H. A., *J. Phys. Chem.*, 1970, **74**, 3209.
5. Weeks, J. L.; Rabani, J., *J. Phys. Chem.*, 1966, **70**, 2100.
6. Baxendale, J. H.; Ward, M. D.; Wardman, P., *Trans. Faraday Soc.*, 1971, **67**, 2532–7.
7. Bielski, B. H. J., *Photochem. Photobiol.*, 1978, **28**, 645–9.
8. Koppenol, W. H.; Liebman, J. F., *J. Phys. Chem.*, 1984, **88**, 99.
9. Koppenol, W. H.; Butler, J., *Adv. Free Rad. Biol. Med.*, 1985, **1**, 91.
10. Schwarz, H. A., *J. Chem. Educ.*, 1981, **58**, 101–5.
11. Schwarz, H. A.; Dodson, R. W., *J. Phys. Chem.*, 1984, **88**, 3643.
12. Kläning, U. K.; Sehested, K.; Holcman, J., *J. Phys. Chem.*, 1985, **89**, 760–3.
13. Henglein, A., *Radiat. Phys. Chem.*, 1980, **15**, 151.
14. Benson, S. W.; Nangia, P. S., *J. Am. Chem. Soc.*, 1980, **102**, 2843–4.
15. Rabani, J.; Matheson, M. S., *J. Am. Chem. Soc.*, 1964, **86**, 3175.
16. Chase, M. W., Jr; Davies, C. A.; Downey, J. R., Jr; Frurip, D. J.; McDonald, R. A.; Syverud, A. N. JANAF Thermochemical Tables, 3rd edn, *J. Phys. Chem. Ref. Data*, 1985, **14**, suppl. 1.
17. *Gmelin Handbook of Inorganic Chemistry*, Springer, Berlin, 1998.
18. Lelieveld, J.; Crutzen, P. J., *J. Atmos. Chem.*, 1991, **12**, 229–67.
19. Chameides, W. L., *J. Geophys. Res.*, 1984, **89**, 4739–55.
20. Jacob, D. J., Heterogeneous chemistry and tropospheric ozone. NARSTO Critical Review Paper, *Atmos. Environ.*, 2000, **34**, 2103.
21. Schwartz, S. E., Chemical conversions in clouds, in *Aerosols: Research, Risk Assessment and Control Strategies, Proceedings of the Second US–Dutch International Symposium, Williamsburg, Virginia, May 19–25, 1985*, Lewis, Chelsea, MI, 1986.
22. Stein, G., *J. Phys. Chem.*, 1965, **42**, 2986.
23. George, P., in *Oxidases and Related Redox Systems*, King, T. E.; Mason, S.; Morrison, M. (eds), Wiley, New York, 1965, pp. 3–36.
24. Berdnikov, V. M.; Bazhin, N. M., *Russ. J. Phys. Chem.*, 1970, **44**, 395–8.
25. Schwartz, S. E., *J. Geophys. Res.*, 1984, **89**, 11589–98.
26. Koppenol, W. H., in *Oxy Radicals and Their Scavenger Systems*, Vol. 1, Cohen, G.; Greenwald, R. A. (eds), Elsevier Biomedical, New York, 1983, pp. 274–7.
27. Shum, L. G. S.; Benson, S. W., *J. Phys. Chem.*, 1983, **87**, 3479–82.
28. Fisher, E. R.; Armentrout, P. B. *J. Phys. Chem.*, 1990, **94**, 4396.
29. Howard, C. J., *J. Am. Chem. Soc.*, 1980, **102**, 6937–41.

30. Koppenol, W. H., *Bioelectrochem. Bioenerg.*, 1987, **18**, 3–11.
31. Holmes, J. L.; Lossing, F. P.; Mayer, P. M., *J. Am. Chem. Soc.*, 1991, **113**, 9723–8.
32. Espinosa-Garcia, J., *Mol. Phys.*, 1993, **79**, 445–7.
33. Boutalib, A.; Cardy, H.; Chevaldonnet, C.; Chaillet, M., *Chem. Phys.*, 1986, **110**, 295.
34. Sana, M.; Leroy, G.; Peeters, D.; Younang, E., *J. Mol. Struct. (THEOCHEM)*, 1987, **151**, 325.

Addendum II: The Solubility of Ozone in Water

Peter Warneck

Max-Planck-Institut für Chemie, Abteilung Biogeochemie, Mainz, Germany

Data for the system are shown in Table 9.48 and Figure 9.27.

1   COMMENTS

Data reported during the period 1872–1967 have been reviewed by Roth and Sullivan [4]. They considered these data unreliable and did not recommend their use. The main problem is that ozone undergoes decomposition in water, and that the decay rate rises with increasing pH and with temperature. Even at pH 7 the rate of' decomposition is too high to allow reliable data for k_H to be obtained.

This suggests that only the absorption of ozone in acidified solutions be considered. However, the degree of absorption also depends on the salt content/ionic strength of the solution, and in strongly acidified media the uptake may be significantly reduced. The experimental method commonly applied when studying a substance undergoing decomposition in solution is the continuous flow technique, i.e. a gas stream containing ozone is passed through a water column until a steady state is reached. Under conditions where the decomposition is slow, the steady state will correspond closely to gas–liquid equilibrium, otherwise corrections must he made.

The most careful experimental study to be designed to overcome these problems is that of Kosak-Channing and Helz [5], who worked at pH 3.4 ± 0.1. For comparison, a number of data from earlier measurements were selected on the basis of the criterion that the solvent was mildly acidic, and if the pH was given, that it fell in the range 2–4. Data for temperatures exceeding 50 °C were not used to guard against the increasing rate of ozone decomposition. As the solubility of ozone in water is fairly low and experimentally attainable concentrations also are small, the observed gas–liquid partition coefficient is expected to be always in

Chemicals in the Atmosphere – Solubility, Sources and Reactivity.   Edited by P. G. T. Fogg and J. Sangster
© 2003 IUPAC   ISBN: 0-471-98651-8

Table 9.48   Solubility data for ozone, O_3, [CAS 10028-15-6], in water[a]

T/K	L	Z	$k_H{}^\dagger/\text{mol dm}^{-3}\,\text{bar}^{-1}$	Ln $(k_H/\text{mol dm}^{-3}\,\text{bar}^{-1})$	Conditions	Ref.
303.15	0.24		0.00952	−4.654	Weak H_2SO_4 soln	1
306.15	0.224		0.00880	−4.733		
315.9	0.174		0.00663	−5.017		
322.15	0.156		0.00582	−5.146		
273.15	0.487		0.02145	−3.847	Weakly acidic soln	2
288.35			0.01505	−4.196	0.01 mol dm^{-3} HClO$_4$	
293.15			0.01215	−4.41	Original unit of pressure mmHg at 0 °C.	
298.15			0.01106	−4.505		
303.15			0.01017	−4.588		
276.15			0.02185	−3.824	pH 2.7	4
293.15			0.01289	−4.351	pH 2.9, 1.95 average	
313.15			0.00824	−4.798	pH 3.2, 2.3 average	
278.15		2.2	0.01965	−3.929	pH ~3.4	5
283.15		2.71	0.01567	−4.156		
288.15		2.86	0.01459	−4.227		
293.15		3.3	0.01243	−4.387		
298.15		3.68	0.01096	−4.513		
303.15		4.15	0.00956	−4.65		

[a]L = Ostwald coefficient; Z = gas–liquid molar concentration ratio.

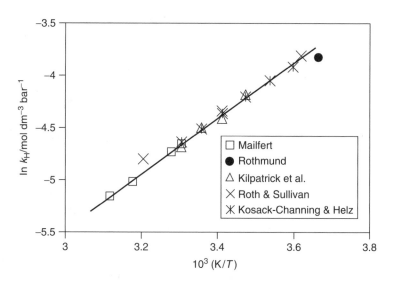

Figure 9.27   Measurements of the solubility of ozone by various authors

the Henry's law regime. Several authors have tested this assumption and found k_H (or the equivalent quantity) to be independent of concentration. Iodometric techniques generally were used to analyse for ozone, both in the gas phase and in solution.

2   DATA EVALUATION

The Ostwald coefficient L is defined as the volume of gas at system temperature T and partial pressure p dissolved per unit volume of solvent. The Henry's law coefficient the Ostwald coefficient, the concentration in the aqueous phase c_{aq}, and the gas–liquid concentration ratio Z are related by

$$k_H = c_{aq}/p = L/R_g T = 1/Z R_g T$$

where $R_g = 8.3143 \times 10^{-2} \, \text{bar dm}^3 \, \text{mol}^{-1} \, \text{K}^{-1}$ is the gas constant.

It was assumed that over the limited temperature range 0–50 °C (273–323 K) the density of water is $1 \, \text{kg dm}^{-3}$ to a good approximation. The approximate relation

$$c_{aq}/x \approx 10^3/M_w = 55.494$$

was used to convert mole fraction x of a dilute solute in water to molar concentrations c_{aq}, where 'M_w is the molar mass of water in g mol^{-1}.

Figure 9.27 shows a semi-logarithmic plot of the k_H values against $1/T$. The data scatter reasonably well around a straight line. All the points shown except those of Roth and Sullivan [4] are individual values for the temperatures given. Roth and Sullivan [4] reported their data graphically as a function of pH, although there is no obvious reason for the assumption that the Henry's law coefficient for ozone should depend on the pH. The values were read off the graphs, but the data exhibit appreciable scatter. At the temperatures 293. 15 and 313.15 K, where two data points were given for the pH range 2–4, an average was taken and entered in Table 9.48 and in Figure 9.27. The 293.15 K point agrees well with the other data, whereas the 313.15 K value lies markedly below the least-squares line. The point was nevertheless retained in the total data set.

The straight line shown in Figure 9.27 results from a least-squares treatment of the 18 data points leading to

$$\ln k_H = -(12.44 \pm 0.24) + (2363 \pm 70)/T$$

for the temperature range 273–322 K. This may be compared with the evaluation of Kosak-Channing and Helz [5]:

$$\ln k_H = -(12.20 \pm 0.30) + (2297 \pm 88)/T$$

from the six data points determined in the temperature range 278–303 K (corrected from atm to bar).

The values derived from these equations for the Henry's law coefficient at 298.15 K are 1.095×10^{-2} and $1.116 \times 10^{-2} \, \text{mol}^{-1} \, \text{dm}^3 \, \text{bar}^{-1}$, respectively, with about 35 % uncertainty. The average of the two direct determinations at 298.15 K [3,5] is $0.01101 \pm 0.0007 \, \text{mol}^{-1} \, \text{dm}^{-3} \, \text{bar}^{-1}$.

REFERENCES

1. Mailfert, M., *C. R. Acad. Sci.*, 1894, **119**, 951–3.
2. Rothmund, V., *Nernst Festschr.*, 1912, 391–4.
3. Kilpatrick, M. L.; Herrick, C. C.; Kilpatrick, M., *J. Am. Chem. Soc.*, 1956, **78**, 1784–9.
4. Roth, J. A.; Sullivan, D. P., *Ind. Eng. Chem. Fundam.*, 1981, **20**, 137–40.
5. Kosak-Channing, L. F.; Helz, G. R., *Environ. Sci. Technol.*, 1983, **17**, 145–9.

CHAPTER 10

Solid Particulates and Surface Reactions in the Atmosphere

Peter Fogg
University of North London, UK (retired)

Organic and inorganic material exists in the atmosphere as solid particulates from various sources. Soot is formed from combustion processes. Mineral dust is blown from deserts and can travel large distances. A 23-year study of dust over Miami indicates that during some summers up to 50 % of the fine particulates come from Africa [1]. Marine salts are taken up into the atmosphere in sea spray which can evaporate to leave the solid salts. Ammonium sulfate is formed in solution in the cloud droplets which may evaporate to leave the solid. Cirrus clouds consist of ice crystals. Solid particles of organic substances may condense from hot vapors emitted from industrial processes.

In the UK, it has been estimated [2] that 25 % of particulate matter is from road transport, 24 % from non-combustion sources, 17 % from industrial processes involving combustion, 16 % from residential combustion and 15 % from public power generation. The biggest natural sources are volcanoes but, in the UK, these are less significant than man made sources. In urban areas annual mean values concentrations of particulates are usually about $10-40\,\mu g\,m^{-3}$. In times of heavy pollution levels may rise to several hundred $\mu g\,m^{-3}$. In rural areas levels are normally $0-10\,\mu g\,m^{-3}$.

Pollutant particulate matter ranges in diameter from 0.1 to $50\,\mu m$. Larger particles do not remain suspended for any significant time. Particles smaller than $10\,\mu m$ (PM_{10}) can be inhaled and represent a potential danger to health. Particles smaller than $2.5\,\mu m$ are the most hazardous because they are not blocked by the nose and mucous membranes before reaching the lungs.

Gases in the atmosphere are adsorbed on the surface of solid particulates and, in some cases, undergo reactions on the surface.

Chemicals in the Atmosphere – Solubility, Sources and Reactivity. Edited by P. G. T. Fogg and J. Sangster
© 2003 IUPAC   ISBN: 0-471-98651-8

1 PRIMARY SOLID PARTICULATES

Soil Dust

Soil dust can be an important component of the particulates in the atmosphere. The principal sources are deserts, dry lake beds and semi-arid fringes of deserts. It can also arise from other dry areas where vegetation has been removed or the soil surface disturbed by agricultural activities. It has been estimated that up to 50 % of dust in the atmosphere arises from disturbed land surfaces. The uptake of dust from a land surface depends on the wind speed, grain size and soil moisture.

Large particles of dust quickly settle from the atmosphere. Particles of diameter less than 1 μm can remain in the atmosphere for several weeks [11].

Sea Salt

Particles of salt result from evaporation of aerosol droplets of seawater taken up from the sea. The bursting of air bubbles in the white head of a wave is the principal way in which the aerosol is generated and generation is strongly dependent on wind speed. Sea salt particles are very efficient cloud condensation nuclei. They also cause significant light scattering.

Sea salt particles have a range of diameters from about 0.05 to 10 μm. They therefore have a wide range of lifetimes and vertical distribution [11].

Industrial Dust

Various industrial and human activities produce dust. Such activities include cement manufacture, waste incineration and metallurgy. Much of the dust from such sources has a diameter greater than 1 μm and soon precipitates from the atmosphere [11].

Carbonaceous Aerosols

Carbonaceous material forms a large part of the aerosols in the atmosphere. Soot and tarry material in the atmosphere is often called *black carbon* to make a distinction from organic compounds which cause insignificant direct light absorption. Some solid compounds in the atmosphere are soluble in water and can cause cloud droplet nucleation.

Much of the carbonaceous material arises from biomass and carbonaceous fuel burning. The particles are predominately less that 1 μm in diameter.

Some of the primary carbonaceous aerosol particles in the atmosphere are of biogenic origin. These include particles originating from leaf fragments and other plant debris, pollen, spores, fungi, viruses, bacteria and humic matter [11].

The estimated world-wide primary particle emissions from sources other than volcanoes during 2000 are shown in Table 10.1.

Table 10.1 Estimated rate of emissions of primary solid particles during 2000/Tg yr^{-1} reported in *Climate Change 2001* [11]; reproduced by permission of the Intergovernmental Panel on Climate Change

Material	NH[a]	SH[a]	Global	Error range[b]	Ref.
Carbonaceous aerosols					
Organic matter (0–2 μm)					
Biomass burning	28	26	54	45–80	3,4
Fossil fuel	28	0.4	28	10–30	5,6
Biogenic			56	0–90	5
Black carbon (soot, etc.)					
Biomass burning	2.9	2.7	5.7	5–9	3,4
Fossil fuel	6.5	0.1	6.6	6–8	5,6
Aircraft	0.005	0.0004	0.006		5,6
Industrial dust (>1 μm)			100	40–100	7,8
Sea salt					9
diameter <1 μm	23	31	54	18–100	
diameter 1–16 μm	1420	1870	3290	1000–6000	
Total	1440	1900	3340	1000–6000	
Mineral dust (soil, sand etc.)					10
diameter <1 μm	90	17	110		
diameter 1–2 μm	240	50	290		
diameter 2–20 μm	1470	282	1750		
Total	1800	349	2150	1000–3000	

[a]NH = Northern Hemisphere; SH = Southern Hemisphere.
[b]Lowest and highest estimates reported in the literature.

Ice and Hydrates of Sulfuric Acid and Nitric Acid

Ice is formed in the atmosphere and precipitates as snow and hail. At high altitudes solid hydrates of nitric acid and of sulfuric acid may be present. Type I polar stratospheric clouds consist mainly of nitric acid trihydrate. Type II consist mainly of relatively pure water ice.

2 SECONDARY SOLID PARTICULATES

Organic Aerosols from Atmospheric Oxidation of Hydrocarbons

Volatile organic compounds in the atmosphere can be oxidized. Some of these oxidation products are solid. The conditions of formation leads to particles less than 1 μm in diameter. Most of this formation of this solid material is by oxidation of terpenes and other volatile substances of plant origin. Volatile aromatic organic compounds of anthropogenic origin can also be oxidized in the atmosphere to form solid particles. (see Table 10.3).

Laboratory experiments indicate that the yield of solid material from volatile organic compounds appears to depend upon the oxidizing agent involved. Oxidation by O_3 or NO_3 yields more solid material than oxidation by OH [12,13].

Emission of NO_x in urban areas leads to a greater concentration of O_3 and of NO_3 and hence a greater tendency for the formation of solid particles from volatile organic compounds [11].

Sulfates

Sulfates occur in sea salt and some gypsum dust is taken up into the atmosphere. Most of the sulfates in the atmosphere are due to sulfur dioxide and dimethyl sulfide entering the atmosphere (see Tables 10.2 and 10.3). Hydrogen sulfide entering the atmosphere is readily oxidized to sulfur dioxide.

Estimates of the global input of sulfur into the atmosphere range from 80 to $130\,Tg\,S\,yr^{-1}$ Much of the sulfur dioxide in the atmosphere results from the

Table 10.2 Recent estimation of emissions of precursors of secondary aerosols reported in *Climate Change 2001* [11]; reproduced by permission of the Intergovernmental Panel on Climate Change

Substance	NH[a]	SH[a]	Global	Error range[b]	Ref.
NO_x (as $Tg\,N\,yr^{-1}$)	32	9	41		
Fossil fuel	20	1.1	21		17
Aircraft	0.54	0.04	0.58	0.4 – 0.9	18,19
Biomass burning	3.3	3.1	6.4	2 – 12	20,21
Soils	3.5	2.0	5.5	3 – 12	22,23
Agricultural soils			2.2	0 – 4	
Natural soils			3.2	3 – 8	
Lightning	4.4	2.6	7.0	2 – 12	24
NH_3 (as $Tg\,N\,yr^{-1}$)	41	13	54	40 – 70	25
Domestic animals	18	4.1	21.6	10 – 30	
Agriculture	12	1.1	12.6	6 – 18	
Human	2.3	0.3	2.6	1.3 – 3.9	
Biomass burning	3.5	2.2	5.7	3 – 8	
Fossil fuel and industry	0.29	0.01	0.3	0.1 – 0.5	
Natural soils	1.4	1.1	2.4	1 – 10	
Wild animals	0.10	0.02	0.1	0 – 1	
Oceans	3.6	4.5	8.2	3 – 16	
SO_2 (as $Tg\,S\,yr^{-1}$)	76	12	88	67 – 130	
Fossil fuel and industry	68	8	76	60 – 100	17
Aircraft	0.06	0.004	0.06	0.03 – 1.0	18,26,27
Biomass burning	1.2	1.0	2.2	1 – 6	28
Volcanoes	6.3	3.0	9.3	6 – 20	29
DMS and H_2S (as $Tg\,S\,yr^{-1}$)	11.6	13.4	25.0	12 – 42	
Oceans	11	13	24	13 – 36	30
Land biota and soils	0.6	0.4	1.0	0.4 – 5.6	31,32
VOC (as $Tg\,C\,yr^{-1}$)	171	65	236	100 – 560	
Anthropogenic	104	5	109	60 – 160	33
Terpenes	67	60	127	40 – 400	34

[a,b]See Table 10.1.

Table 10.3 Estimated quantities of substances forming secondary aerosols/(Tg substance/yr) reported in *Climate Change 2001* [11]; reproduced by permission of the Intergovernmental Panel on Climate Change

Material	NH[a]	SH[a]	Global	Error range[b]	Ref.
Sulfate (as NH$_4$HSO$_4$)	145	55	200	107–374	11
Anthropogenic	106	15	122	69–214	
Biogenic	25	32	57	28–118	
Volcanic	14	7	21	9–48	
Nitrate (as NO$_3{}^-$)					
Anthropogenic	12.4	1.8	14.2	9.6–19.2	
Natural	2.2	1.7	3.9	1.9–7.6	
Organic compounds					
Anthropogenic VOCs[c]	0.15	0.45	0.6	0.3–1.8	
Biogenic VOCs	8.2	7.4	16	8–40	35,36

[a,b]See Table 10.1.
[c]Volatile organic compounds.

burning of fossil fuel. Some is due to volcanic activity. Dimethyl sulfide originates from biogenic sources and is mainly produced by marine plankton. It is readily oxidized to sulfur dioxide [11].

Some of the sulfur dioxide is oxidized and forms sulfates. Most of this oxidation is thought to occur after dissolution in cloud droplets with a smaller proportion occurring in the gas phase [11]. Recent estimates of the total proportion oxidized vary from 46 to 82 % [11]. The remainder is washed out from the atmosphere without oxidation. In the presence of ammonia the oxidized sulfur dioxide forms ammonium sulfate or ammonium hydrogensulfate. Evaporation of solution leads to solid sulfate particles, mostly less than 1 μm in diameter. These particles are efficient scatterers of light. In the absence of ammonia, sulfuric acid is formed. Sulfuric acid can interact with any sea salt present to form sodium and other sulfates. Sulfates in the atmosphere can nucleate cloud droplets and hence increase cloud cover and density [11].

It has been estimated that the average global average $NH_4{}^+/SO_4{}^{2-}$ molar ratio is ~1 [14].

Nitrates

If excess ammonia is present, beyond that required to convert sulfuric acid to ammonium sulfate, then nitric acid can be converted to ammonium nitrate. Evaporation of solution leads to solid particles of the salt which can scatter light. The global burden of ammonium nitrate aerosol has been estimated to be 0.24–0.4 Tg [11].

Volcanoes

The most significant materials in volcanic emissions are dust particles and sulfur compounds [11]. Dust particles tend to be coarse and usually settle from the atmosphere in about 1–2 months. Most of the sulfur is in the form of sulfur dioxide, usually with less than 1 % of H_2S and SO_4^{2-} The sulfur dioxide from volcanoes can reach the upper troposphere. It can therefore remain in the atmosphere longer and be more effective in influencing the climate than gas from anthropogenic sources which is more readily deposited [15,16].

A few very violent volcanic eruptions transfer sulfur into the stratosphere. In the absence of recent volcanic activity, the sulfur load in the stratosphere is about 0.15 Tg S. This is about 15 % of the sulfur in the whole of the atmosphere. However, a massive eruption can increase the load in the stratosphere by about 100 Tg S. Such events caused global cooling of -0.14 to $-0.31\,°C$ during periods in the 19th and 20th centuries [11]. Volcanoes can also transfer solid sodium chloride to the stratosphere.

3   REACTIONS ON THE SURFACE OF SOLID PARTICULATES

Reactions which take place on the surface of ice crystals, solid nitric acid or sulfuric acid hydrates, salt crystals or on the surface of soot or other impure forms of carbon play an important part in the chemistry of the atmosphere. In addition, there is evidence that inorganic particulates such as metal oxides or silica can have photocatalytic activity [37].

Models of Surface Reactions

Models of reactions between gases in the atmosphere which take place at a gas–solid interface are less satisfactory than models of reactions at a gas–liquid interface. One possibility is that a reaction at the gas–solid interface takes place between adsorbed gaseous species. Another possibility is that reaction takes place in the first few layers of a disordered crystal lattice. Another possibility is that reaction takes places in a liquid-like surface layer. It is also possible that the solid is porous due to the presence of microcrystallites. The problems have been reviewed by Carslaw and Peter [38] with special reference to the reactions of $ClONO_2$ and of HOCl with HCl on nitric acid trihydrate and sulfuric acid tetrahydrate.

Models which have been put forward for surface reactions in the atmosphere have been reviewed by Carslaw and Peter [39].

Various models have been used to interpret reactions on ice and on nitric acid trihydrate (NAT) and on sulfuric acid tetrahydrate (SAT). Tabazadeh and Turco [40] and also Mozurkewich [41] assumed that adsorption would follow a Langmuir-type isotherm and that the reaction rate would depend on the concentrations of reactants on the surface. Keyser et al. [42] and Chu et al. [43] made allowances for the porosity of surfaces.

Carslaw and Peter [38] derived the following equation for the reactive uptake coefficient γ_X of a species X reacting with a species Y on a non-porous surface:

$$\frac{1}{\gamma_X} = \frac{1}{\alpha_X} + \frac{\bar{c}\sigma^2}{4K_X{}^*k_BTk_r{}^s\theta_Y}$$

where $K_X{}^* = K_X/(1 + K_X)$, where K_X is the Langmuir equilibrium constant

$\quad k_r{}^s$ = second-order surface reaction rate coefficient with units $cm^2\,s^{-1}$

$\quad \alpha_X$ = mass accommodation coefficient of X

$\quad \bar{c}$ = mean molecular velocity

$\quad \sigma$ = surface area of an adsorption site

$\quad \theta_Y$ = fraction of the surface covered by Y

$\quad k_B$ = Boltzmann constant.

Henson et al. [44] showed that the Langmuir constant for adsorption of HCl on NAT and on SAT by HCl depended on the relative humidity. Carslaw and Peter made allowance for this in estimating θ_{HCl}.

The model was used to interpret experimental data obtained for the following reactions:

1. $ClONO_2 + HCl$ on NAT [45,46];
2. $ClONO_2 + HCl$ on SAT [46,47];
3. $HOCl + HCl$ on NAT [48,49].

This model cannot distinguish between a surface and a multilayer reaction. It was, however, able to indicate apparent discrepancies between data from different sources.

Clegg and Abbatt [50] investigated the uptake of SO_2 and also H_2O_2 on ice surfaces at temperatures of 213–238 K and partial pressures of 10^{-5}–10^{-2} bar. In the case of SO_2, the uptake appears to be proportional to the square root of the partial pressure within the experimental error. This suggests that adsorption is consistent with dissociation of the hydrated SO_2 on the surface of the ice. In the case of H_2O_2 uptake is proportional to pressure. The uptake of SO_2 increases with pH whereas that of H_2O_2 shows little effect of pH change.

Clegg and Abbatt [51] also studied the oxidation of sulfur dioxide by hydrogen peroxide which can take place on the surface of ice. This may be a significant route for oxidation of SO_2 in the atmosphere. The reaction can also take place in aqueous solution in clouds containing liquid droplets. The following mechanism for the surface reaction is consistent with the dependence of the reaction rate upon the partial pressure of SO_2 and H_2O_2 found under laboratory conditions:

$$H_2O_2(g) \longrightarrow H_2O_2(ads)$$

$$SO_2(g) \longrightarrow SO_2(ads)$$

$$SO_2(ads) + H_2O \longrightarrow H^+ + HSO_3{}^-(ads)$$

$$HSO_3{}^-(ads) + H_2O_2(ads) \longrightarrow HOOSO_2{}^-(ads) + H_2O$$

$$HOOSO_2{}^-(ads) + H^+(ads) \longrightarrow 2H^+(ads) + SO_4{}^{2-}(ads)$$

where (g) = gas and (ads) = adsorbed.

Table 10.4 Additional reactions on the surfaces of sodium and potassium halides reported in JPL publication 97-4 [52] and elsewhere[a]

Reaction	T/K	Reactive uptake coefficient, γ	Ref.
$NO_2 + NaCl(s) \rightarrow ClNO + NaNO_3$	298	$(1.3 \pm 0.6) \times 10^{-4}$	53,54
	298	$(1.3 \pm 0.3) \times 10^{-6}$	55
$N_2O_5 + NaCl(s) \rightarrow ClNO_2 + NaNO_3$	≈ 300	5×10^{-4}	56
	298	$>2.5 \times 10^{-3}$	57
$ClNO + NaCl(s) \rightarrow$ products	298	$>1 \times 10^{-5}$	58
$ClNO_2 + NaCl(s) \rightarrow$ products	298	$<1 \times 10^{-5}$	58
$ClONO_2 + NaCl(s) \rightarrow Cl_2 + NaNO_3$	~ 298		59
	225–296	4–7×10^{-3}	60
	~ 298	0.23 ± 0.06	61
$ClONO_2 + NaBr(s) \rightarrow BrCl + NaNO_3$	~ 295		62
$ClONO_2 + KBr(s) \rightarrow BrCl + KNO_3$	~ 295	0.35 ± 0.06	61
$BrCl + KBr(s) \rightarrow Br_2 + KCl$	~ 295	>0.1	61
$OH \rightarrow$ products	245–339	$1.2 \times 10^{-5} \exp 1750/(T/K)$	63
$HOBr + NaCl/NaBr(s) \rightarrow Br_2$	243	0.02	64
$HOBr + NaCl(s) \rightarrow BrCl$	233	0.06 (initial)	64
$HOI + NaCl/NaBr(s) \rightarrow IBr + ICl$	298	$>10^{-2}$	65
$HOI + NaCl(s) \rightarrow ICl$	253		65
$IONO_2 + NaCl/NaBr(s) \rightarrow IBr + ICl$	298	$>10^{-2}$	65
$IONO_2 + NaCl(s) \rightarrow ICl$	253		65
$HOI + NaCl(s) \rightarrow I_2$	298		66
$HOI + NaBr(s) \rightarrow I_2 + IBr$	298	$(6 \pm 2) \times 10^{-2}$	66

[a]Note that gaps in the table indicate that no quantitative measurements were reported. Reactive uptake coefficients correspond to the first reactant in each equation.

Reactions on the Surface of Salts

Details of measurements of uptake coefficients of reactions on sodium and potassium halides are shown in Table 10.4.

Mozurkewich and Calvert found that dry ammonium sulfate was highly unreactive to N_2O_5 [67]. Jefferson *et al.* [68] used a flow tube apparatus to measure the rate of uptake of sulfuric acid vapour by solid aerosols (20–400 nm diameter) of ammonium sulfate and also of sodium chloride. The mass accommodation coefficients were 0.73 ± 0.21 for loss on ammonium sulfate and 0.79 ± 0.23 for loss on sodium chloride. When the NaCl aerosol particles were coated with stearic acid, the mass accommodation coefficient was reduced to 0.31 and 0.19 for high and low stearic acid coverages, respectively.

Davies and Cox [69] studied the heterogeneous reaction of HNO_3 with NaCl (solid):

$$HNO_3 + NaCl(solid) \rightleftharpoons HCl + NaNO_3$$

This reaction is believed to take place in the atmosphere. Uptake coefficients were measured at room temperature over a range of partial pressures of water from 0.002 to 10 mbar. The concentration of HNO_3 varied from 4 to 700 × 10^{11} molecules cm^{-3}. Uptake coefficients of HNO_3 were measured using a tubular flow reactor with the wall coated with salt films or crystallite grains. Values increased with increasing partial pressure of water and constant concentration of HNO_3. Values decreased with increasing concentration of HNO_3 at constant partial pressure of water.

The authors proposed that the mechanism involved ionization of adsorbed HNO_3 at defect sites where multilayer adsorption of water took place. This was assumed to be followed by reaction of H_3O^+ with Cl^- and desorption of HCl molecules. Good evidence of this is the similarity between the variation of uptake coefficient and the variation in adsorption of water when the partial pressure of water is varied as reported by Barraclough and Hall [70] (Figure 10.1). The reaction had previously been studied by Vogt and Finlayson-Pitts [53,54], who showed that there was adsorption of HNO_3 but no reaction if the NaCl were absolutely dry. Fenter et al. [56] showed that NaBr, KCl and KBr behaved in a similar manner to NaCl when these salts were exposed to HNO_3.

Koch and Rossi [71] developed a method of measuring the average residence time of both reactive and unreactive adsorbants on solid substrates. The method involved sending a pulse of molecules of the gas under investigation through a tube coated with the solid under test. The arrival time of the pulse at the end of the tube was recorded by a mass spectrometer in a high-vacuum chamber. Measurements on the average residence time, τ, of nitryl chloride molecules on the surface of sodium chloride and that of chlorine nitrate on potassium bromide were reported:

$$ClNO_2/KBr \ \tau = <0.02 \, ms$$

$$ClONO_2/NaCl \ \tau = 0.7 \, ms$$

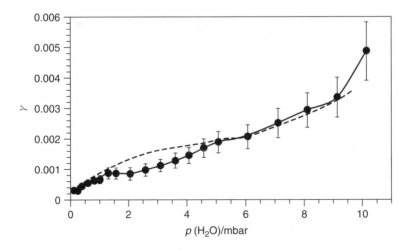

Figure 10.1 Variation of the uptake coefficient of HNO_3 at $(6.0 \pm 1.0) \times 10^{12}$ molecules cm^{-3} with variation of partial pressure of water vapour reported by Davies and Cox [64]. The broken line shows the scaled isotherm for adsorption of water vapour as reported by Barraclough and Hall [70]

The authors assumed the following uptake coefficients measured by Caloz et al. [61,72]:

$$ClNO_2/KBr \ \gamma = 2 \times 10^{-4}$$
$$ClONO_2/NaCl \ \gamma = 2 \times 10^{-1}$$

Reactions on solid sodium and potassium halides by various other species have been investigated.

Soot

Soot is formed by incomplete combustion of compounds containing carbon. It is widely dispersed in the atmosphere and contains traces of organic compounds. Samples from different sources may differ in physical and chemical properties, and laboratory samples may differ from material in the atmosphere. Measurements of uptake coefficients for reactions on carbon/soot are given in Table 10.5.

Ammann et al. [79,80] have found evidence for the formation of nitrous acid by reaction of nitrogen dioxide on the surface of soot particles from a laboratory soot source and from a diesel engine. Data were consistent with adsorption of NO_2 followed by its reaction with reduced organic species on the surface of the soot. Experiments [79,81] have shown that the formation of nitrous acid by this surface reaction in diesel fumes is much less than the formation by homogeneous mechanisms. Kirchner et al. [82] showed by means of Fourier transfer infrared spectroscopy that freshly emitted diesel soot particles are hydrophobic whereas partially oxidized soot is hydroscopic.

Alcala-Jornod and Rossi [83] used the molecular diffusion tube technique to investigate the interaction of NO_2 and water with samples of soot from various sources at ambient temperature. This technique consisted in injecting a pulse of molecules of the gas under test into a tube with an internal coating of the surface under test. The time of arrival at the other end of the tube was found by recording the signal from a mass spectrometer. In cases where gas molecules stuck to the

Table 10.5 Values of uptake coefficients for gases on samples of carbon/soot

Reaction	Kind of carbon	T/K	Reactive uptake coefficient, γ	Ref.
$O_3 \rightarrow$ products		~300	3.3×10^{-3} to ~2×10^{-4}	73,74
		~300	3×10^{-3}	75
			10^{-3} to 10^{-5}	76
		~298	$(1 \pm 0.5) \times 10^{-3}$	77
$HNO_3 \rightarrow$ products	Degussa FW2	190–440	$(3.8 \pm 0.8) \times 10^{-2}$	77
	Degussa FW2	220 ± 1	$(13 \pm 1) \times 10^{-2}$	85
	Degussa FW2	295 ± 1	$(6.7 \pm 0.5) \times 10^{-2}$	85
	Hexane soot	295 ± 1	$(2.3 \pm 0.4) \times 10^{-2}$	85
	Kerosene soot	220 ± 1	$(9.3 \pm 0.2) \times 10^{-2}$	85
	Kerosene soot	295 ± 1	$(6.0 \pm 0.5) \times 10^{-2}$	85

surface but did not react, the surface residence time of such molecules could be found by comparison with the arrival time of molecules which did not interact with the surface. In practice, a second sample of the gas under test was also injected into a tube of identical size which has been coated with Teflon. This provided the reference. If the gas under test was not only retarded by the surface of the first tube but also reacted with the surface, it was also necessary to monitor the quantity of gas under test which emerged. from the reaction tube. Values of the initial uptake coefficient, γ_0, were calculated.

The reaction tube was first coated with soot by placing it over a flame fuelled by the hydrocarbon under test. Three sources of soot were investigated: methylbenzene, ethyne and decane. Pulses of NO_2 were injected and the gas emerging was monitored for HONO. In the case of soot from ethyne there was no signal at m/z 47, indicating that there was no overall formation of HONO. The yield of HONO was 80–90 % in the case of the decane and methylbenzene soot. The initial uptake coefficient for ethyne soot was 0.26 and that for decane and methylbenzene soot was 0.1 on average. The maximum residence time for NO_2 was 70 ms on ethyne soot and 50 ms in the case of soot from decane and methylbenzene.

In the case of the interaction of H_2O with soot, the initial uptake coefficient was $<2 \times 10^{-3}$ for all samples. The residence times were 0.1 ms (ethyne soot), 2–3 ms (methylbenzene soot and diesel soot) and 5 ms (decane soot).

Thilibi and Petit [84] and Rogaski *et al.* [77] reported the decomposition of HNO_3 on amorphous carbon. These workers found decomposition to give NO and NO_2. The interaction of HNO_3 with various forms of amorphous carbon was studied more thoroughly by Choi and Leu [85]. These workers used a fast flow tube reactor coupled to a quadrupole mass spectrometer. The behaviour of HNO_3 on the following samples of carbon was investigated:

1. FW2 channel-type black supplied by Degussa (Ridgefield Park, NJ, USA). This contained about 18 % of volatile material and the BET surface area was $368\ m^2\ g^{-1}$.
2. Graphite powder; BET surface area $15\ m^2\ g^{-1}$.
3. Hexane; soot BET surface area $46\ m^2\ g^{-1}$.
4. Kerosene soot; BET surface area $91\ m^2\ g^{-1}$.

At pressures of HNO_3 of 10^{-2} bar or above and 295 K, decomposition to give mostly NO_x and H_2O occurred on black carbon FW2. A similar but less extensive decomposition took place on graphite under these conditions. Negligible decomposition took place on hexane and kerosene soot. No decomposition was observed on any of the samples at 220 K and 10^{-2} bar or at 295 K and 7×10^{-5} bar. Physical adsorption took place in each case. It was suggested that decomposition involved two adjacent HNO_3 molecules adsorbed on the surface of the carbon. At very low pressures there was low surface coverage and hence negligible pairs of adjacent molecules.

The authors concluded that the decomposition of HNO_3 on the surface of soot was unlikely in the upper troposphere and lower stratosphere because its partial pressure was too low.

Stephens *et al.* [76] and Smith and Chughtai [86] have reported that O_3 reactively decomposes on soot particles to give O_2, CO and CO_2 as products. Novakov *et al.* [87] and Baldwin [88] have reported that SO_2 is oxidized to sulfate on soot particles.

Reactions on Water ice

Details of measurements of reactions on the surface of ice are shown in Table 10.6.

Polar stratospheric clouds form in polar winters. They have a greater tendency to form over the Antarctic than over the Arctic because of different wind flow patterns. The absorption of chlorine and bromine nitrates, $ClONO_2$ and $BrONO_2$, on to water ice and nitric acid ice in polar stratospheric clouds is believed to play an important role in the formation of the ozone hole. Chlorine and bromine nitrates are formed from photochemical breakdown products of halogen compounds in the stratosphere and nitrogen oxides. They are stable in the gaseous state and, together with HCl, serve as chlorine and bromine reservoirs. They react readily when adsorbed on the surface of water ice or nitric acid ice. In the case of $ClONO_2$ the following reactions leading to the formation of chlorine molecules are believed to take place [103]:

$$HCl + ClONO_2 \longrightarrow HNO_3 + Cl_2$$

$$ClONO_2 + H_2O \longrightarrow HNO_3 + HOCl$$

$$HCl + HOCl \longrightarrow H_2O + Cl_2$$

$BrONO_2$ undergoes similar reactions to form molecular bromine.

Molecular chlorine and bromine do not, themselves, initiate chain reactions leading to the decomposition of ozone. However, they are photochemically decomposed to atoms when subject to solar radiation. In polar regions this cannot happen until the end of the dark polar winter.

The periodicity of the destruction of the ozone layer can therefore be explained. During the polar winter, the presence of polar stratospheric clouds allows the build-up of chlorine and bromine molecules. With the return of the sun in the polar spring these molecules are photochemically decomposed to atoms which can initiate gas-phase chain reactions such as

$$ClO + ClO + M \longrightarrow Cl_2O_2 + M$$

$$Cl_2O_2 + hv \longrightarrow Cl + ClO_2$$

$$ClO_2 + M \longrightarrow Cl + O_2 + M$$

$$Cl + O_3 \longrightarrow ClO + O_2$$

Table 10.6 Reactive uptake coefficients for reactions on water ice reported in JPL publication 97-4 [52] and elsewhere[a]

Reaction	T/K	Reactive uptake, coefficient, γ	Ref.
OH → products	205–230	>0.01	89
HO_2 → products	223	0.025 ± 0.005	89
$N_2O_5 + H_2O \rightarrow 2\ HNO_3$	195–200	0.28 ± 0.11	90
	195–200	0.24 ± 0.07	91
	188	>0.03	92
$N_2O_5 + HCl \rightarrow ClNO_2 + HNO_3$	190–220	0.028 ± 0.11	90
		$>1 \times 10^{-3}$	93
$HONO + HCl \rightarrow ClNO + H_2O$	180–200	0.02–0.13	94
$Cl_2 + HBr \rightarrow BrCl + HCl$	200	>0.2	95
ClO → products	183	$(8 \pm 2) \times 10^{-5}$	96
	183	$>1 \times 10^{-3}$	90
	213	$<1 \times 10^{-5}$	89
$HOCl + HCl \rightarrow Cl_2 + H_2O$	195–200	High	97
		High	98
		High	45
$ClONO_2 + H_2O \rightarrow HOCl + HNO_3$	200	See text	
$ClONO_2 + HCl \rightarrow Cl_2 + HNO_3$	200	0.27 (+0.73, −0.13)	99
	191	0.3 (+0.7, −0.1)	97
	188	$(2.7 \pm 1.9) \times 10^{-1}$	45
	180	$(6.4 \pm 0.7) \times 10^{-1}$	100
	200	$(2.7 \pm 0.7) \times 10^{-1}$	100
$ClONO_2 + HBr \rightarrow BrCl + HNO_3$	200	>0.3	95
$2\ BrO \rightarrow Br_2 + O_2$	253	$(1.0 \pm 0.4) \times 10^{-3}$	89
$HOBr + HCl \rightarrow BrCl + H_2O$	228	0.25 (+0.10, −0.05)	101
$HOBr + HBr \rightarrow Br_2 + H_2O$	228	0.12 ± 0.03	101
$BrONO_2 + H_2O \rightarrow HOBr + HNO_3$	200	≥ 0.3	102
$BrONO_2 + HCl \rightarrow BrCl + HNO_3$	200	~0.25	102
	190	0.32 ± 0.04	100
	200	0.27 ± 0.05	100
HOI[b]	243	$>10^{-2}$	65
$IONO_2$[b]	243	$>10^{-2}$	65
$HOI \rightarrow I_2$	180	$(4.3-8) \times 10^{-2}$	66
	190	$(3.6-6.4) \times 10^{-2}$	66
	200	$(2-4.8) \times 10^{-2}$	66

[a]Reactive uptake coefficients correspond to the first reactant in each equation.
[b]Partially reversible with slight traces of I_2 formed.

where M = any other gas molecule. Other reaction sequences are probably involved. It has been estimated that about 70 % of the ozone loss over the Antarctic is due to the above sequence.

Various workers have obtained uptake coefficients for $ClONO_2$ in the surface reaction with excess HCl. The most recent measurement is by Oppliger *et al.* [100], who found a decrease in the uptake coefficients when the temperature was increased from 180 to 200 K. This is consistent with the decrease in the surface concentration of hydrogen chloride with increase in temperature.

The reaction of $ClONO_2$ with water which takes place on the surface of water ice is an important source of HOCl, i.e.

$$ClONO_2 + H_2O \longrightarrow HNO_3 + HOCl$$

HOCl is adsorbed on the ice surface and can subsequently react with HCl to form chlorine molecules. The reaction has been studied by various workers [43,47,91,97,99,100,102,104,105]. Measurements by Oppliger et al. [100] indicated that the uptake coefficient in the temperature range 180–200 K decreased from $(2.2 \pm 0.3) \times 10^{-1}$ to $(3.1 \pm 0.5) \times 10^{-2}$. Adsorbed HOCl resulting from the reaction has been directly observed on an ice surface by Geiger et al. [106] by using second-order nonlinear optical harmonic generation techniques.

Field measurements by Valdez et al. [78] showed that sulfur dioxide is converted to S(VI) on the surface of snow. Hydrogen peroxide has been detected on the surface of fresh snow and it was assumed that SO_2 was oxidized by H_2O_2. This was confirmed by Mitra et al. [107] and Conklin et al. [108]. They showed that SO_2 is oxidized by H_2O_2 in the presence of ice. In addition, it was shown by Chu et al. [109] using a flow tube that an increase in concentration of H_2O_2 in the ice increased the oxidation of SO_2 and that sulfate is the main product. Clegg and Abbatt [110,111] investigated the effect of the partial pressure of each gas and of the temperature. The reactive uptake of SO_2 increased from about 10^{-3} to 10^{-2} when the partial pressure of H_2O_2 increased from 2.7×10^{-3} to about $(2-7) \times 10^{-2}$ Pa at a partial pressure of SO_2 of 1.5×10^{-4} Pa at 228 K. This is because the surface is unsaturated with H_2O_2 under these conditions. The reactive uptake coefficient of SO_2 decreased from about 10^{-2} to about 10^{-3} when the partial pressure of SO_2 was increased from 2.7×10^{-5} to about 1.3×10^{-3} at a partial pressure of H_2O_2 of 8.7×10^{-3} Pa at 228 K.

Surface Reactions on Frozen Salt Solutions

Adams et al. [64,65] investigated the uptake of halogen compounds on frozen solutions of NaCl/NaBr. Results of these measurements are summarized in Table 10.7.

Surface Reactions on Nitric Acid ice ($HNO_3 \cdot 3H_2O$)

Hanson and Ravishankara [91,97,102] measured the reactive uptake of $ClONO_2$ on nitric acid trihydrate in the reaction

$$ClONO_2 + H_2O \longrightarrow HNO_3 + HOCl$$

The value given in the latest paper [102] is $(1.0 \pm 0.3) \times 10^{-3}$. They also found [46] that the value decreased markedly with decrease in relative humidity with the reactive uptake coefficient 2.0×10^{-3} at 90 % relative humidity (191 K), 5.0×10^{-4} at 50 % (194 K) and 3.0×10^{-4} at 25 % (198 K). The apparent effect of relative humidity is consistent with a value of $(2.0 \pm 0.3) \times 10^{-3}$ obtained by Zhang et al. [112] at 195 K (see evaluations of these data in [113–116]).

Table 10.7 Reactive uptake coefficients for reactions of various halogen compounds on the surface of frozen solutions of NaCl and NaBr[a]

Reaction	T/K	Uptake coefficient, γ	Ref.
BrCl + NaCl/NaBr (frozen soln) → Br_2	233	>3.4 × 10^{-2} (initial)	64
Br_2 + NaCl/NaBr (frozen soln) → Br_2[b]	233	>2.5 × 10^{-2} (initial)	64
Br_2 + NaBr (frozen soln) → Br_2[b]	233	>1.0 × 10^{-2} (initial)	64
Cl_2 + NaCl/NaBr (frozen soln) → Br_2	233	>2.8 × 10^{-2} (initial)	64
HOI + NaCl/NaBr (frozen soln) → ICl + IBr	243	>5 × 10^{-2}	65
$IONO_2$ + NaCl/NaBr (frozen soln) → ICl + IBr	243		65
ICl + NaCl/NaBr (frozen soln) → IBr	243	>10^{-2} (initial)	65
IBr + NaCl/NaBr (frozen soln) → IBr[b]	243		65

[a] The NaCl/NaBr surfaces were made from a solution containing NaCl (2 mol dm^{-3}) and NaBr (0.003 mol dm^{-3}). The ratio of Cl$^-$ to Br$^-$ was typical of frozen sea spray. The NaBr surface was from a solution of NaBr (0.003 mol dm^{-3}).
[b] Reversible adsorption.

The uptake of $ClONO_2$ is much greater in the presence of HCl when the following reaction takes place:

$$ClONO_2 + HCl \longrightarrow HNO_3 + Cl_2$$

Hanson and Ravishankara [91] and also Leu et al. [104] obtained values of the uptake coefficient of about 0.3 in the range 191–200 K. Abbatt and Molina [45] showed that the value dropped to $(3.0 \pm 1.0) \times 10^{-3}$ over HNO_3-rich nitric acid trihydrate. Hanson and Ravishankara [97] showed that there was little change in the uptake coefficient when the partial pressure of HCl was reduced from $(4.0–5.3) \times 10^{-7}$ to 6.7×10^{-8} mbar at 90 % relative humidity but there was about a sevenfold decrease if the relative humidity was reduced from 90 to 30 %, i.e.

	$p(HCl) = 6.7 \times 10^{-8}$ mbar	$p(HCl) = (4.0–5.3) \times 10^{-7}$ mbar
90 % relative humidity	$\gamma = 0.23$ (190 K)	$\gamma = 0.20$ (190 K)
30 % relative humidity		$\gamma = 0.030$ (197 K)

Details of other reactions on the surface of nitric acid trihydrate are given in Table 10.8.

Inorganic Oxides

Catalytic decomposition of chloro compounds has been experimentally demonstrated in the case of alumina. Robinson et al. [118] measured the dissociative uptake of CF_2Cl_2 and CF_2Br_2 on α-alumina at 210 and 315 K. Values of dissociative uptake coefficients of about 2×10^{-5} were obtained. The same

Table 10.8 Reactive uptakes coefficients for additional reactions on nitric acid trihydrate reported in JPL publication 97-4 [52][a]

Reaction	T/K	Reactive uptake, coefficient, γ	Ref.
OH → products	200, 228	>0.2	89
$N_2O_5 + H_2O \rightarrow 2\,HNO_3$	200	$0.0006 \pm 30\,\%$	91, 102
		0.015 ± 0.006	92
$N_2O_5 + HCl \rightarrow ClNO_2 + HNO_3$	~200	$0.0032 \pm 30\,\%$	91
$N_2O_5 + HBr \rightarrow BrNO_3 + HNO_3$	~200	~0.005 (some values ~0.04)	95
ClO → products	183	$<(8 \pm 4) \times 10^{-5}$	96
$HOCl + HCl \rightarrow Cl_2 + H_2O$		Yield drops, relative to water ice as proportion of HNO_3 on the surface increases	97 117
$ClONO_2 + H_2O \rightarrow HOCl + HNO_3$		See text	
$ClONO_2 + HCl \rightarrow Cl_2 + HNO_3$		See text	
$ClONO_2 + HBr \rightarrow BrCl + HNO_3$		≥ 0.3	95

[a]Reactive uptake coefficients correspond to the first reactant in each equation.

group [119,120] made similar measurements of the dissociative uptake of CF_3Cl, CF_2Cl_2, $CFCl_3$ and CCl_4 on dehydroxylated γ-alumina powders in the temperature range 120–300 K. Dissociative uptake coefficients ranged from 0.4×10^{-5} to 1.0×10^{-5}.

Vogt and co-workers [121] have reported an investigation of the heterogeneous reactions of nitric acid and dinitrogen pentoxide on dry samples of γ-alumina, Arizona test dust, Saharan dust and Gobi dust at 300 K. The formation of surface products was monitored by diffuse reflectance infrared Fourier transform spectroscopy (DRIFTS) during adsorption. Nitrate and water were the reaction products together with the loss of surface OH groups. It was suggested that the first step in each case was the attachment of the nitrogen compound to a surface OH group, i.e.

$$\{OMOH \cdots HNO_3\} \longrightarrow \{(OM) + NO_3^-\} + H_2O$$

where M = Al or Si, etc. The authors reported uptake coefficients, surface areas and reaction orders for various inorganic oxides as shown in Table 10.9. Uptake coefficients were calculated using the geometric surface area of the sample. This may not give reliable results. Exposure of the Saharan dust to water vapour during the adsorption caused an increase in both the uptake coefficient and adsorption [121].

Similar data have been reported by Hanisch and co-workers [122], who investigated the uptake of HNO_3 on various mineral dust and mineral oxide surfaces at 298 K. These workers used a low-pressure Knudsen reactor coupled with a quadrupole mass spectrometer for the analysis of reactants and products. Values of the initial uptake coefficients were in the range $(6–18.6) \times 10^{-2}$ and were estimated to be accurate to $\pm 25\,\%$ (Table 10.10). Nitrate was formed on the

Table 10.9 Reactive uptake coefficients and reaction orders for various reactions on inorganic oxides at 300 K reported by Vogt and co-workers [121]

Reaction	Uptake coefficient, γ	Reaction order	BET surface area/$m^2\,g^{-1}$
$HNO_3 + Al_2O_3$	7.7×10^{-3}	1.01	10.2
+ Saharan dust	5.4×10^{-2}	1.01	49.7
+ Arizona dust	1.6×10^{-2}	0.85	6/8
+ Gobi dust	1.1×10^{-2}	0.94	10.9
$N_2O_5 + Al_2O_3$	6.4×10^{-3}	1.00	10.2
+ Saharan dust	9.1×10^{-3}	0.97	49.7

Table 10.10 Initial uptake coefficients for gaseous nitric acid on the surfaces of various solids reported by Hanisch and co-workers [122]

Solid	$\gamma_0/10^{-2}$	Solid	$\gamma_0/10^{-2}$
Saharan dust (Cape Verde)	13.5	Ca-montmorillonite	10.0
Chinese dust (Takla Makan)	12.1	Na-montmorillonite	8.2
Arizona dust	6	Palgorskite	18.6
Ripodolite (chlorite)	6.6	Dolomite	14.0
$CaCO_3$ (preheated)	10	Orthoclase	8.4
$CaCO_3$ (unheated)	18	Illite	6.7
Illite–smectite (70:30)	6.3	Kaolinite	5.5

surfaces. The authors considered that the high uptake of nitric acid by dust samples indicated that mineral dust had an important role in the removal of nitrate from the gaseous phase to the particulate phase in the upper atmosphere.

Börensen et al. [123] investigated the reaction of NO_2 on alumina dust by DRIFTS. Nitrate was formed on the alumina surface. NO_2 concentrations ranged from 2.5×10^{13} to 8.5×10^{14} molecules cm^{-3} at 299 K. The reaction order with respect to NO_2 was 1.86 ± 0.1. As a consequence, the reactive uptake coefficient varied linearly with concentration of NO_2 from 7.3×10^{-10} to 1.3×10^{-8}, as shown in Figure 10.2. Three measurements at 363 K were also reported. These indicated that the reactive uptake coefficient decreased with increase in temperature (see Figure 10.2). These workers concluded that the reactive uptake coefficient of NO_2 to form nitrate under normal environmental conditions would be about 10^{-9} and therefore such a reaction would be negligible.

Zamaraev et al. [124] have discussed the possibility that heterogeneous photocatalytic processes may be significant in atmospheric chemistry. When a semiconducting material absorbs a quantum of light with an energy that exceeds the width of the gap between the valence band and conduction band, an electron is excited and passes into the conduction band. This can lead to catalytic activity. The excited electron can cause the reduction of an adsorbed species. The hole in the valence band can act as an oxidant.

The energy of a light quantum is inversely proportional to the wavelength. The width of the gap therefore forces an upper limit to the wavelength of light

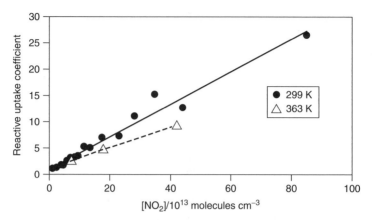

Figure 10.2 Variation of the reactive uptake of NO_2 on alumina with concentration of NO_2 [123]

which excites an electron. In the case of pure bulk samples of Fe_2O_3 this upper limit is 570 nm and in the case of pure TiO_2 it is 420 nm. Most of the radiation reaching the troposphere has a wavelength greater than about 300 nm. Particles in the upper atmosphere are subject to much radiation of shorter wavelength and hence higher energy. However, the proportion of metal oxides reaching the upper atmosphere is small because of high sedimentation rates.

Zamaraev *et al.* [124] were particularly interested in the behaviour of Fe_2O_3, TiO_2 and ZnO because of their relative abundance in dust particles. The following heterogeneous photochemical reactions involving N_2, CO_2 and H_2O have been studied under laboratory conditions using pure samples of the oxides:

$$N_2 + H_2O \xrightarrow[TiO_2]{h\upsilon} NH_3 + O_2 \qquad [125-127]$$

$$N_2 + H_2O \xrightarrow[TiO_2]{h\upsilon} N_2H_4 + O_2 \qquad [124-127]$$

These reactions have been studied in a nitrogen atmosphere. The TiO_2 powder carried an adsorbed layer of water. The first reaction can also occur in the presence of a ferric oxide catalyst:

$$N_2 + H_2O \xrightarrow[ZnO-Fe_2O_3]{h\upsilon} H_2 + HNO_3 \qquad [128]$$

$$H_2O \xrightarrow[TiO_2]{h\upsilon} H_2 + O_2 \qquad [37,125-127,129-133]$$

$$HO + O_2 \xrightarrow[TiO_2]{h\upsilon} H_2O_2 \qquad [37,126,130]$$

$$CO_2 + H_2O \xrightarrow[TiO_2]{h\upsilon} HCOOH, CH_2O, CH_3OH \qquad [134]$$

Various photocatalytic reactions catalysed by metal oxides and other types of semiconductors involve compounds which are also trace compounds in the atmosphere have been studied by other workers. Some reactions are of significance in waste water treatment and others in the development of new synthetic methods. Various examples were quoted by Zamaraev *et al.* [124]. These include the following:

$$C_2H_6 + O_2 \xrightarrow[\text{TiO}_2]{h\nu} HCOOH, CH_2O, CH_3OH \qquad [125,135]$$

$$C_6H_6 + O_2 \xrightarrow[\text{TiO}_2]{h\nu} CO_2 + H_2O \qquad [125,136]$$

Photocatalytic decomposition of CCl_4, $CHCl_3$, $CFCl_3$, CF_2Cl_2, $C_2F_3Cl_3$ and $C_2F_4Cl_2$ in the presence of O_2 over particles of ZnO, TiO_2, Fe_2O_3, volcanic ash, chalk and desert sands has been reported [129].

Decomposition of H_2S and NO_x has also been reported:

$$H_2S \xrightarrow[\text{CdS, ZnS}]{h\nu} H_2 + S \qquad [133]$$

$$NO_x \xrightarrow[\substack{\text{metal halides,}\\\text{aluminosilicates}}]{h\nu} N_2 + O \qquad [137-140]$$

Many of the laboratory studies of heterogeneous photocatalysed reactions which may be significant in the atmosphere have been carried out using suspensions of catalysts in water. However, these studies are relevant because oxide and other solid particles in the atmosphere are likely to be covered with an adsorbed layer of water. Zamaraev *et al.* [124] have shown that many such reactions could be significant in the atmosphere, especially in the presence of the large concentrations of dust during volcanic eruptions. However, precise estimation of their importance could not be carried out because of uncertainties in quantum yields under atmospheric conditions [141].

Chianelli *et al.* [142] investigated atmospheric aerosols over Mexico City, a heavily polluted area. They found that solid aerosols consisted of two intermixed components. One component was carbonaceous material with variable composition. Material with a fullerene-type structure was detected. The other component consisted mostly of clay minerals with nanoparticles of metal oxides and sulfides. They stated that some of these oxides and sulfides, such as Fe_2O_3, MnO_2 and FeS_2, have bandgaps which enable them to enhance the photocatalytic activity of aluminosilicates and titanate compounds which are also present in the aerosols. They suggested that this could be significant in the photocatalytic production of alkyl free radicals. A similar range of components was found when single particles collected over Atlanta were analysed by laser mass spectrometry during the Atlanta SuperSite Experiment in August 1999 [143,144].

Sulfuric Acid Hydrates

Zhang et al. [145] investigated the uptake of N_2O_5 on sulfuric acid monohydrate using a coated flow tube. Values tended to decrease with increase in temperature but increased with partial pressure of water vapour (see Table 10.11)

Zhang et al. [146] found no increase in the uptake of N_2O_5 on sulfuric acid monohydrate at 200–220 K when HCl was present. The uptake coefficient for reaction with HCl under these conditions appears to be less than 10^{-4}.

Hanson and Ravishankara [147] found that the uptake coefficient of N_2O_5 on sulfuric acid tetrahydrate was $(5.0–8.0) \times 10^{-3}$ at 195–205 K.

Measurements by Hanson and Ravishankara [147] of the uptake coefficient of $ClONO_2$ for the reaction

$$ClONO_2 + HCl \longrightarrow HNO_3 + Cl_2$$

appear to show a decrease from 1.25×10^{-1} at 192 K to 2.4×10^{-4} at 205 K. Measurements by Zhang et al. [47] showed a marked dependence on the relative humidity at 195 K. At 100 % relative humidity the value found was $(1.2 \pm 0.3) \times 10^{-1}$ and at 0.7 % relative humidity it was is $(7.0 \pm 2.0) \times 10^{-4}$.

Hanson and Ravishankara [147] measured the uptake coefficient on sulfuric acid tetrahydrate for the reaction

$$ClONO_2 + H_2O \longrightarrow HNO_3 + HOCl$$

Table 10.11 Uptake coefficients for reaction of N_2O_5 on sulfuric acid monohydrate [145]

T/K	p_{H_2O}/mbar			
	1.5	2.0	2.8	4.8
215	1.1×10^{-3}			
220		8.5×10^{-4}	9.1×10^{-4}	
230			5.2×10^{-4}	7.5×10^{-4}

Table 10.12 Reactive uptake coefficients of $ClONO_2$ for the reaction $ClONO_2 + H_2O \rightarrow HNO_3 + HOCl$ on sulfuric acid tetrahydrate at various relative humidities (R.H.)

T/K	R.H./%						Ref.
	0.8	7	16	30	90	100	
191.5					2.0×10^{-3}		147
195	(5.0 ± 1.3) $\times 10^{-4}$					(1.6 ± 0.4) $\times 10^{-2}$	47
196				2.0×10^{-3}			147
200			5.0×10^{-4}				147
205		1.0×10^{-4}					147

The values showed a marked decrease as the temperature was raised and the relative humidity decreased.

Measurements by Zhang *et al.* [47] indicate that change in relative humidity is probably the more important factor (see Table 10.12).

Zhang *et al.* [146] reported an initial value, γ_0, of $\leq 2 \times 10^{-4}$ for reaction on sulfuric acid monohydrate at $210\,K$ and $p_{H_2O} = 1.3 \times 10^{-4}$ mbar.

REFERENCES

1. Spotts, P. N., *Christian Sci. Monit.*, August 5, 1999.
2. Information from the Atmospheric, Climate and Environmental Information Programme maintained by Manchester Metropolitan University and supported by the UK Department of the Environment, Food and Rural Affairs (DEFRA); http://www.doc.mmu.ac.uk/aric/eae/Air_Quality/Older/Particulates.html.
3. Liousse, C.; Penner, J. E.; Chuang, C.; Walton, J. J.; Eddleman, H.; Cachier, H., *J. Geophys. Res. Atmos.*, 1996, **101**, 19411–32.
4. Scholes, M.; Andreae, M. O., *Ambio*, 2000, **29**, 23–9.
5. Penner, J. E.; Eddleman, H.; Novakov, T., *Atmos. Environ.*, 1993, **27A**, 1277–95.
6. Cooke, W. F.; Liousse, C.; Cachier, H.; Feichter, J., *J. Geophys. Res.*, 1999, **104**, 22137–62.
7. Wolf, M. E.; Hidy, G. M., *Geophys. Res. Atmos.*, 1997, **102**, 11113–21.
8. Andreae, M. O., in *World Survey of Climatology, Vol 16: Future Climates of the World*, Henderson-Sellers, A. (ed.), Elsevier, Amsterdam, 1995, pp. 341–92.
9. Gong, S. L.; Barrie, L. A.; Blanchet, J.-P.; Spacek, L., in *Air Pollution Modeling and Its Applications XII*, Gryning, S.-E.; Chaumerliac, N. (eds), Plenum Press, New York, 1998.
10. Ginoux, P.; Chin, M.; Tegen, I.; Prospero, J.; Holben, B.; Dubovik, O.; Lin, S.-J., *J. Geophys. Res.*, 2001, **106**, 20255–74.
11. Penner, J. E.; Andreae, M.; Annegarn, H.; Barrie, L.; Feichter, J.; Hegg, D.; Jayaraman, A.; Leaitch, R.; Murphy, D.; Nganga, J.; Pitari, G., Aerosols, their direct and indirect effects, in *Climate Change 2001: the Scientific Basis. Contribution of Working Group I to the Third Assessment Report of the Intergovernmental Panel on Climate Change*, Houghton, J. T.; Ding, Y.; Griggs, D. J.; Noguer, M.; van der Linden, P. J.; Dai, X.; Maskell, K.; Johnson, C. A. (eds), Cambridge University Press, Cambridge, 2001, pp. 881.
12. Griffin, R. J.; Cocker, D. R.; Flagan, R. C.; Seinfeld, J. H., *J. Geophys. Res.*, 1999, **104**, 3555–67.
13. Hoffman, T.; Odum, J. R.; Bowman, F.; Collins, D.; Klockow, D.; Flagan, R. C.; Seinfeld, J. H., *J. Atmos. Chem.*, 1997, **26**, 189–222.
14. Adams, P. J.; Seinfeld, J. H.; Koch, D. M., *J. Geophys. Res.*, 1999, **104**, 13791–823.
15. Graf, H. F.; Feichter, J.; Langmann, B., *J. Geophys. Res.*, 1997, **102**, 10727–38.
16. Benkovitz, C. M.; Berkovitz, C. M.; Easter, S.; Nemesure, S.; Wagner, R.; Schwartz, S. E., *J. Geophys. Res.*, 1994, **99**, 20725–56.
17. Benkovitz, C. M.; Scholtz, M. T.; Pacyna, J.; Tarrason, L.; Dignon, J.; Voldner, E. C.; Spiro, P. A.; Logan, J. A.; Graedel, T. E., *J. Geophys. Res.*, 1996, **101**, 29239–53.
18. Penner, J. E.; Bergmann, D.; Walton, J. J.; Kinnison, D.; Prather, M. J.; Rotman, D.; Price, C.; Pickering, K. E.; Baughcum, S. L., *J. Geophys. Res.*, 1999, **103**, 22097–114.
19. Daggett, D. L.; Sutkus, D. J., Jr; DuPois, D. P.; Baughcum, S. L., An Evaluation of Aircraft Emissions Inventory Methodology by Comparisons with Reported Airline data, NASA/CR-1999-209480, NASA, Washington, DC, 1999.

20. Liousse, C.; Penner, J. E.; Chuang, C.; Walton, J. J.; Eddleman, H.; Cachier, H., *J. Geophys. Res. Atmos.*, 1996, **101**, 19411–32.

21. Atherton, C. A. *Lawrence Livermore National Laboratory Report UCRL-ID-122583*, Lawrence Livermore National Laboratory, Livermore, CA, 1996.

22. Yienger, J. J.; Levy, H., *J. Geophys. Res.*, 1995, **100**, 11447–64.

23. Price, C.; Penner, J. E.; Prather, M. J., *J. Geophys. Res.*, 1997, **102**, 5929–41.

24. Lawrence, M. G.; Chameides, W. L.; Kasibhatla, P. S.; Levy, H., II; Moxim, W., Lightning and atmospheric chemistry: the rate of atmospheric NO production, in *Handbook of Atmospheric Electrodynamics*, Volland, I. H. (ed.), CRC Press, Boca Raton, FL, 1995, pp. 189–202.

25. Bouwman, A. F.; Lee, D. S.; Asman, W. A. H.; Dentener, F. J.; van der Hoek, K. W.; Olivier, J. G. J., *Global Biochem. Cycles*, 1997, **11**, 561–88.

26. Penner, J. E.; Bergmann, D.; Walton, J. J.; Kinnison, D.; Prather, M. J.; Rotman, D.; Price, C.; Pickering, K. E.; Baughcum, S. L., *J. Geophys. Res.*, 1998, **103**, 22097–114.

27. Fahey, D. W.; Schumann, U.; Ackerman, S.; Artaxo, P.; Boucher, O.; Danilin, M. Y.; Kärcher, B.; Minnis, P.; Nakajima, T.; Toon, O. B., in *Aviation and the Global Atmosphere*, Penner, J. E.; Lister, D. H.; Griggs, D. J.; Dokken, D. L.; McFarland, M. (eds), Cambridge University Press, Cambridge, 1999, pp. 65–120.

28. Spiro, P. A.; Jacob, D. J.; Logan, J. A., *J. Geophys. Res.*, 1992, **97**, 6023–36.

29. Andres, R. J.; Kasgnoc, A. D., *J. Geophys. Res. Atmos.*, 1998, **103**, 25251–61.

30. Kettle, A. J.; Andreae, M. O., *J. Geophys. Res.*, 2000, **105**, 26793–808.

31. Bates, T. S.; Lamb, B. K.; Guenther, A.; Dignon, J.; Stoiber, R. E., *J. Atmos. Chem.*, 1992, **14**, 315–37.

32. Andreae, M. O.; Jaeschke, W. A., *Wetlands, Terrestrial Ecosystems, and Associated Water Bodies, SCOPE 48*, Howarth, R. W.; Stewart, J. W. B.; Ivanov, M. V. (eds), Wiley, Chichester, 1992, pp. 27–61.

33. Picot, S. D.; Watson, J. J.; Jones, J. W., *J. Geophys. Res.*, 1992, **97**, 9897–912.

34. Guenther, A.; Hewitt, C.; Erickson, D.; Fall, R.; Geron, C.; Graedel, T.; Harley, P.; Klinger, L.; Lerdau, M.; McKay, W.; Pierce, T.; Scholes, B.; Steinbrecher, R.; Tallamraju, R.; Taylor, J.; Zimmermann, P., *J. Geophys. Res. Atmos.*, 1995, **100**, 8873–92.

35. Griffin, R. J.; Cocker, D. R., III; Seinfeld, J. H.; Dabdub, D., *Geophys. Res. Lett.*, 1999, **26**, 2721–34.

36. Penner, J. E.; Chuang, C. C.; Grant, K., Climate change and radiative forcing by anthropogenic aerosols. A review of research during the last five years, presented at the La Jolla International School of Science, Institute for Advanced Physics Studies, La Jolla, CA, 1999.

37. Zamaraev, K. I.; Parmon, V. N., *Catal. Rev. Sci. Eng.*, 1980, **22**, 261.

38. Carslaw, K. S.; Peter, T., *Geophys. Res. Lett.*, 1997, **24**, 1743–6.

39. Carslaw, K. S.; Peter, T., *Rev. Geophys.*, 1997, **35**, 125–54.

40. Tabazadeh, A.; Turco, R. P., *J. Geophys. Res.*, 1993, **12**, 12727–40.

41. Mozurkewich, M., *Geophys. Res. Lett.*, 1993, **20**, 355–8.

42. Keyser, L. F.; Moore, S. B.; Leu, M.-T., *J. Phys. Chem.*, 1993, **97**, 2800–1.

43. Chu, L. T.; Leu, M.-T.; Keyser, L. F., *J. Phys. Chem.*, 1993, **97**, 12798–804.

44. Henson, B. F.; Wilson, K. R.; Robinson, J. M., *Geophys. Res. Lett.*, 1996, **23**, 1021–4.

45. Abbatt, J. P. D.; Molina, M. J., *J. Phys. Chem.*, 1992, **96**, 7674–9.

46. Hanson, D. R.; Ravishankara, A. R., *J. Geophys. Res.*, 1993, **98**, 22931–6.

47. Zhang R.; Jayne, J. T.; Molina, M. J., *J. Phys. Chem.*, 1994, **98**, 867–74.

48. Abbatt, J. P. D.; Molina, M. J., *Geophys. Res. Lett.*, 1992, **19**, 461–4.

49. Hanson, D. R.; Ravishankara, A. R., *J. Phys. Chem.*, 1992, **96**, 2682–91.

50. Clegg, S. M.; Abbatt, J. P. D., *J. Phys. Chem. A*, 2001, **105**, 6630–6.

51. Clegg, S. M.; Abbatt, J. P. D., *Atmos. Chem. Phys. Discuss.*, 2001, **1**, 77–92.

52. DeMore, W. B.; Sander, S. P.; Golden, D. M.; Hampson, R. F.; Kurylo, M. J.; Howard, C. J.; Ravishankara, A. R.; Kolb, C. E.; Molina, M. J., *Chemical Kinetics and Photochemical Data for Use in Stratospheric Modeling*, JPL Publication 97-4, Jet Propulsion Laboratory, Pasadena, CA, 1997.

53. Vogt, R.; Finlayson-Pitts, B. F., *J. Phys. Chem.*, 1994, **98**, 3747–55.

54. Vogt, R.; Finlayson-Pitts, B. F., *Geophys. Res. Lett.*, 1994, **21**, 2291–4.

55. Peters, S. J., Ewing, G. E., *J. Phys. Chem.*, 1996, **100**, 14093–102.

56. Fenter, F. F.; Caloz, F.; Rossi, M. J., *J. Phys. Chem.*, 1994, **98**, 9801–10.

57. Livingston, F. E.; Finlayson-Pitts, B. J., *Geophys. Res. Lett.*, 1991, **18**, 17–20.

58. Beichert, P.; Finlayson-Pitts, B. J., *J. Phys. Chem.*, 1996, **100**, 15218–28.

59. Finlayson-Pitts, B. J.; Ezell, M. J.; Pitts, J. N., Jr; *Nature*, 1989, **337**, 241–4.

60. Timonen, R. S., Chu, L. T.; Leu, M.-T.; Keyser, L. F., *J. Phys. Chem.*, 1994, **98**, 9509–17.

61. Caloz, F., Fenter, F. F.; Rossi, M. J., *J. Phys. Chem.*, 1996, **100**, 7494–501.

62. Berko, H. N.; McCaslin, P. C.; Finlayson-Pitts, B. J., *J. Phys. Chem.*, 1991, **95**, 6951–8.

63. Ivanov, A. V.; Gersherzon, Y. M.; Gratpanche, F.; Devolder, P.; Saverysyn, J.-P., *Am. Geophys.*, 1996, **14**, 659–64.

64. Adams, J. W.; Holmes, N. S.; Crowley, J. N., *Atmos. Chem. Phys.*, 2002, **2**, 79–91.

65. Holmes, N. S.; Adams, J. W.; Crowley, J. N., *Phys. Chem. Chem. Phys.*, 2001, **3**, 1679–87.

66. Allanic, A.; Rossi, M. J., *J. Geophys. Res.*, 1999, **104**, 18689–96.

67. Mozurkewich, M.; Calvert, J. G., *J. Geophys. Res.*, 1988, **93**, 15889–96.

68. Jefferson, A.; Eisele, F. L.; Ziemann, P. J.; Weber, R. J.; Marti, J. J.; McMurry, P. H., *J. Geophys. Res.*, 1997, **102**(D15), 19021–8.

69. Davies, J. A.; Cox, R. A., *Environ. Sci. Pollut. Res.*, 1998, **5**, 134.

70. Barraclough, P. B.; Hall, P. G., *Surf. Sci.*, 1974, **46**, 393–417.

71. Koch, T. G.; Rossi, M. J., *J. Phys. Chem. A*, 1998, **102**, 9193–201.

72. Caloz, F.; Seisel, S.; Fenter, F.; Rossi, M. J., *J. Phys. Chem.*, 1998, **102**, 7470–9.

73. Fendel, W.; Matter, D.; Burtscher, H.; Schimdt-Ott, A., *Atmos. Environ.*, 1995, **29**, 967–73.

74. Fendel, W.; Ott, A. S., *J. Aerosol Sci.*, 1993, **24**, S317–S318.

75. Smith, D. M.; Welch, W. F.; Jassim, J. A.; Chughtai, A. R.; Stedman, D. H., *Appl. Spectrosc.*, 1988, **42**, 1473–82.

76. Stephens, S. ; Rossi, M. J.; Golden, D. M., *Int. J. Chem. Kinet.*, 1986, **18**, 1133–49.

77. Rogaski, C. A.; Golden, D. M.; Williams, L. R., *Geophys. Res. Lett.*, 1996, **24**, 381–4.

78. Valdez, M. P.; Bales, R. C.; Stanley, D. A.; Dawson, G. A., *J. Geophys. Res.*, 1987, **92**, 9779–87.

79. Ammann, M.; Arens, F.; Gutzwiller, L.; Baltensperger, U.; Gäggeler, H. W., 3rd *CMD Annual Report*, 2000, 140–3.

80. Ammann, M.; Kalberger, M.; Arens, F.; Lavanchy, V.; Gäggeler, H. W.; Baltensperger, U., *Environ. Sci. Pollut. Res.*, 1998, **5**, 134.

81. Kleffman, J.; Heland, J.; Kurtenbach, R.; Lörzer, J. C.; Wiesen, P.; Ammann, M.; Gutzwiller, L.; Rodenas Garcia, M.; Pons, M.; Wirtz, K.; Scheer, V.; Vogt, R., 3rd *CMD Annual Report*, 2000, 177–180; *J. Air Waste Manage. Assoc.*, in press.

82. Kirchner, U.; Scheer, V.; Börensen, C.; Vogt, R., *Environ. Sci. Pollut. Res.*, 1998, **5**, 186–7.

83. Alcala-Jornod, C.; Rossi, M. J., *Phys. Chem. Chem. Phys.*, 2000, **2**, 5584–93.

84. Thilibi, J.; Petit, J. C., in *Impact of Emissions from Aircraft and Spacecraft upon the Atmosphere*, Schumann, U.; Wurzel, D. (eds), DLR Mitteilung 94-06, Deutsches Zentrum für Luft- und Raum Fahrt, Cologne, 1994.

85. Choi, W.; Leu, M.-T., *J. Phys. Chem. A*, 1998, **102**, 7618–30.

86. Smith, D. M.; Chughtai, A. R., *J. Geophys. Res.*, 1996, **101**, 19607.

87. Novakov, T.; Chang, S. G.; Harker, A. B., *Science*, 1974, **186**, 259.

88. Baldwin, A. C., *Int. J. Chem. Kinet.*, 1986, **14**, 1133.
89. Cooper, P. L.; Abbatt, J. P. D., *J. Phys. Chem.*, 1996, **100**, 2249–54.
90. Leu, M. T., *Geophys. Res. Lett.*, 1988, **15**, 851–4.
91. Hanson, D. R.; Ravishankara, A. R., *J. Geophys. Res.*, 1991, **96**, 5081–90.
92. Quinlan, M. A., Reihs, C. M.; Golden, D. M.; Tolbert, M. A., *J. Phys. Chem.*, 1990, **94**, 3255–60.
93. Tolbert, M. A.; Rossi, M. J.; Golden, D. M., *Science*, 1988, **240**, 1018–21.
94. Fenter, F. F.; Rossi, M. J., *J. Phys. Chem.*, 1996, **100**, 13765–75.
95. Hanson, D. R.; Ravishankara, A. R., *J. Phys. Chem.*, 1992, **96**, 9441–6.
96. Kenner, R. D.; Plumb, I. C.; Ryan, K. R., *Geophys. Res. Lett.*, 1993, **20**, 193–6.
97. Hanson, D. R.; Ravishankara, A. R., *J. Phys. Chem.*, 1992, **96**, 2682–91.
98. Abbatt, J. P. D.; Molina, M. J., *Geophys. Res. Lett.*, 1992, **19**, 461–4.
99. Leu, M. T., *Geophys. Res. Lett.*, 1988, **15**, 17–20.
100. Oppliger, R.; Allanic, A.; Rossi, M. J., *J. Phys. Chem. A*, 1997, **101**, 1903–11.
101. Abbatt, J. P. D., *Geophys. Res. Lett.*, 1994, **21**, 665–8.
102. Hanson, D. R.; Ravishankara, A. R., *J. Phys. Chem.*, 1993, **97**, 2802–3.
103. World Meteorological Organization, *Scientific Assessment of Ozone Depletion, 1991, Report 25*, WMO, Geneva, 1992.
104. Leu, M.-T.; Moore, S. B.; Keyser, L. F., *J. Phys. Chem.*, 1991, **95**, 7763–71.
105. Tolbert, M. A.; Rossi, M. J.; Malhotra, R.; Golden, D. M., *Science*, 1987, **238**, 1258–60.
106. Geiger, F. M.; Tridico, A. C.; Hicks, J. M., *J. Phys. Chem. B*, 1999, **103**, 8205–15.
107. Mitra, S. K.; Barth, S.; Pruppacher, H. R., *Atmos. Environ.*, 1990, **24A**, 2307–12.
108. Conklin, M. H.; Sommerfield, R. A.; Laird, S. K.; Villinski, J. E., *Atmos. Environ.*, 1993, **27A**, 159–66.
109. Chu, L.; Diao, G.; Chu. L. T., *J. Phys. Chem. A*, 2000, **104**, 7565–73.
110. Clegg, S.; Abbatt, J. P. D., *J. Phys. Chem. A*, 2001, **105**, 6630–6.
111. Clegg, S.; Abbatt, J. P. D., *Atmos. Chem. Phys. Discuss.*, 2001, **1**, 77–92.
112. Zhang, R.; Jayne, J. T.; Molina, M. J., *J. Phys. Chem.*, 1994, **98**, 867.
113. *IUPAC Subcommittee on Gas Kinetic Data Evaluation – Data Sheet R10*, November 2000.
114. DeMore, W. B.; Howard, C. J.; Sander, S. P.; Ravishankara, A. R.; Golden, D. M.; Kolb, C. E.; Hampson, R. F.; Molina, M. J.; Kurylo, M. J., *Chemical Kinetics and Photochemical Data for Use in Stratospheric Modeling, Evaluation 12*, JPL Publication 97-4, NASA, Jet Propulsion Laboratory, California Institute of Technology, Pasadena, CA, 1997.
115. Sander, S. P.; Friedl, R. R.; DeMore, W. B.; Ravishankara, A. R.; Golden, D. M.; Kolb, C. E.; Kurylo, M. J. Hampson, R. F.; Huie, R. E.; Molina, M. J.; Moortgat, G. K., *Chemical Kinetics and Photochemical Data for Use in Stratospheric Modeling, Evaluation 13*, JPL Publication 97-4, NASA, Jet Propulsion Laboratory, California Institute of Technology, Pasadena, CA, 1997.
116. *IUPAC Subcommittee on Gas Kinetic Data Evaluation – Data Sheet R9*, November 2000.
117. Abbatt, J. P. D.; Molina, M. J., *Geophys. Res. Lett.*, 1992, **19**, 461–4.
118. Robinson, G. N.; Freedman, A.; Kolb, C. E.; Worsnop, D. R., *Geophys. Res. Lett.*, 1996, **23**, 317.
119. Dai, Q.; Robinson, G. N.; Freedman, A., *J. Phys. Chem. B*, 1996, **101**, 4940–6.
120. Robinson, G. N.; Dai, Q.; Freedman, A., *J. Phys. Chem. B*, 1996, **101**, 4947–53.
121. Vogt, R.; Börensen, C.; Seisel, S.; Zellner, R., 3rd *CMD Annual Report*, 2000, 173–175; Börensen, C.; Seisel, S.; Vogt, R.; Zellner, R., presented at CMD EC/EUROTRAC Joint Workshop, Lausanne, 11–13 September 2000.
122. Hanisch, F.; Crowley, J. N., *J. Phys. Chem.*, 2001, **105**, 3096–106; Crowley, J. N.; Hanisch, F.; Holmes, N.; Winkler, K., 3rd*CMD Annual Report*, 2000, 152–5.

123. Börensen, C.; Kirchner, U.; Scheer, V.; Vogt, R.; Zellner, R., *J. Phys. Chem. A*, 2000, **104**, 5036–45.
124. Zamaraev, K. I.; Khramov, M. I.; Parmon, V. N., *Catal. Rev. Sci. Eng.*, 1994, **36**, 617–44.
125. Ollis, D. F.; Pruden, A. L., in *Heterogeneous Atmospheric Chemistry*, Schryer, R. (ed.), American Geophysical Union, Washington, DC, 1982.
126. Pelizzetti, E.; Serpone, N. (eds), *Photocatalysis. Fundamentals and Application*, Wiley, New York, 1990.
127. Schrauzer, G. N.; Guth, T. D., *J. Am. Chem. Soc.*, 1977, **99**, 7189.
128. Tennakone, K.; Ilperuma, O. A.; Thaminimulla, C. T. K.; Bandara, J. M. S., *J. Photochem. Photobiol. A: Chem.*, 1992, **66**, 375.
129. Isidorov, V. A. *Organicheskaya Khimiya Atmosphery*, Khimiya, St. Petersburg, 1992.
130. Grätzel, M., *Energy Resources through Photochemistry and Catalysis*, Academic Press, New York, 1990.
131. Sciavello, M. (ed.), *Photocatalysis and Environment. Trends and Applications*, Kluwer, Dordrecht, 1988.
132. Bulatov, A. V.; Khidekel, M. L., *Izv. Akad. Nauk SSSR, Ser. Khim.*, 1976, (8), 1902.
133. Fujishima, A.; Honda, K., *Nature*, 1972, **238**, 37.
134. Halmann, M., in *Energy Resources Through Photochemistry and Catalysis*, Grätzel, M. (ed.), Academic Press, New York, 1983, p. 507.
135. Daroux, M.; Kivana, D.; Duran, M.; Bideau, M., *Can. J. Chem. Eng.*, 1985, **63**, 668.
136. Matthews, R. W., *Aust. J. Chem.*, 1987, **40**, 667.
137. Alyea, F. N.; Cunnold, D. M.; Prinn, R. G., *Atmos. Environ.*, 1978, **12**, 1009.
138. Dixon-Warren, St. J.; Jackson, R. C.; Polanyo, J. C.; Rieley, R.; Shapter, J. G.; Weiss, H., *J. Phys. Chem.*, 1992, **96**, 10983.
139. Kudo, A.; Sakata, T., *Chem. Lett.*, 1992, **12**, 2381.
140. Martin, L. P.; Van-Fogl, M., in *Heterogeneous Atmospheric Chemistry*, Schryer, D. R. (ed.), American Geophysical Union, Washington, DC, 1982.
141. Gruzdkov, Yu. A.; Savinov, E. N.; Makarshin, L. L.; Parmon, V. N., in *Fotokataliticheskoe Preobrazovanie Solnechnoi Energii*, Parmon, V. N.; Zamaraev, K. I. (eds), Nauka, Novosibirsk, 1991, p. 186.
142. Chianelli, R. R.; Yácaman, M. J.; Arenas, J.; Adape, F., *J. Hazard. Subst. Res.*, 1998, **1**, 1–17.
143. http://cires.colorado.edu/~shanhu/PALMS%20ATL%20PartI_new.
144. Lee, S.-H.; Murphy, D. M.; Thomson, D. S.; Middlebrook, A. M., *J. Geophys. Res.*, 2002, **107**, 101029.
145. Zhang, R.; Leu, M.-T.; Keyser, L. F., *Geophys. Res. Lett.*, 1995, **22**, 1493–6.
146. Zhang, R.; Leu, M.-T.; Keyser, L. F., *J. Geophys. Res.*, 1995, **100**, 18845–54.
147. Hanson, D. R.; Ravishankara, A. R., *J. Geophys. Res.*, 1993, **98**, 22931–6.

CHAPTER 11

Henry's Law Constants for Compounds Stable in Water

James Sangster
Sangster Research Laboratories, Montreal, Canada

This chapter deals with 'unreactive' compounds, whereas Chapter 9 dealt with 'reactive' compounds. The term 'reactive' means that some or all of the solute in solution is not the same chemical species as in the gas phase. Thus, substances such as N_2, benzene, methane and hexachlorobenzene are found in this chapter, whereas HCl, aldehydes, radicals, atoms and organic acids are in the next chapter. Although this distinction may be rather arbitrary in a few cases, it is a convenient way of organizing relevant material.

In this chapter, the results of a literature search for Henry's law constant data or solubility or infinite dilution activity coefficients are presented. As shown in Chapter 3, solubility and infinite dilution activity coefficients are legitimate sources of information for Henry's law constants. The list of compounds is not meant to be exhaustive, but examples were chosen which were thought to be of interest to atmospheric and environmental chemists and chemical engineers. As far as the data permit, a critical evaluation has been attempted for all compounds, and recommended data are indicated by the capitalized word RECOMMENDED (recommended data may be single values or tabulated data or a fitting equation).

It may be helpful here to explain briefly what is meant by 'critical evaluation' in this context. The purpose is not, of course, to choose certain data as the 'best' for all time. Rather, it is to distinguish the more reliable from the less reliable of the data available to the authors at the time of writing. To this end, a few notes on critical evaluation are given here, according to the nature (source) of the data for individual compounds.

1 SIMPLE GASES ABOVE THEIR CRITICAL TEMPERATURES

In this category are inert atmospheric gases, methane, etc. The relevant data are solubilities at 1 atm partial pressure of gas, published as critical evaluations, of

Chemicals in the Atmosphere – Solubility, Sources and Reactivity.   Edited by P. G. T. Fogg and J. Sangster
© 2003 IUPAC   ISBN: 0-471-98651-8

all previously available data, in the IUPAC Solubility Data Series (1979–) or in similar single articles in the scientific literature. Such data are taken as RECOMMENDED and the Henry's law constants derived from them (as described in Chapter 4) are also taken to be RECOMMENDED.

2 GASES, LIQUIDS AND SOLIDS FOR WHICH WATER SOLUBILITY IS THE ONLY OR MOST RELIABLE OR MOST EXTENSIVE TYPE OF DATA AVAILABLE

Examples are CH_2Cl_2, ethane, butane, and toluene. In these cases, the Henry's law constant data from solubility were compared with those from other methods (if any) in order to ensure that adopted data are reasonable. In general, the data from solubility were retained as RECOMMENDED data, with few exceptions. In the case of liquids and solids, the liquid (or supercooled liquid) vapour pressure was needed for the calculation (Chapter 4). Sources of vapour pressure are cited; they are believed to be reliable. In some instances, the source is not given because the original authors supplied the data.

3 COMPOUNDS FOR WHICH BOTH RELIABLE SOLUBILITY AND EXTENSIVE EXPERIMENTAL MEASUREMENTS ARE AVAILABLE

Examples are *cis*- and *trans*-1,2-dichloroethene, C_2Cl_4, trimethylbenzenes and dichlorobenzenes. Usually the data are combined in order to find a comprehensive fitting equation.

4 COMPOUNDS FOR WHICH ONLY EXPERIMENTAL MEASUREMENTS (APART FROM SOLUBILITY) ARE AVAILABLE

Examples are methanol, acetonitrile, acetone, methyl vinyl ketone and chlorobenzene.

5 GENERAL REMARKS

In all cases, data for temperatures other than $25\,°C$ were sought. For ease of reproducing data, a fitting equation of the type

$$\ln k_H = A - B/(T/K)$$

was always attempted if a fitting equation was not already available from original articles. The values of the parameters A and B are given, together with the standard error of estimate (goodness of fit) of k_H from (to) the equation.

In theory, the parameters A and B may be related to the enthalpy and entropy of solution of the solute at infinite dilution (Chapter 3). Here, they are presented as simple fitting parameters, and no attempt has been made to interpret them further.

In general, all Henry's law constant data from all chosen sources were combined to deduce a comprehensive fitting equation. The term 'chosen sources' is crucial here. The first step in the evaluation was to put all data on a common temperature plot. Obvious outliers were identified and excluded from further consideration; such outliers are not listed in the tables. The temperature plots of k_H from experimental measurements show variable scatter after elimination of outliers; no attempt was made to reduce this scatter by further data selection.

There are two universal assumptions which have been made in these evaluations:

- gases and vapour phases are assumed to be ideal;
- the solution is assumed always to be dilute enough that the solute concentration is within the Henry's law region, i.e. the vapour pressure is low enough for concentration to be proportional to pressure within the limits of experimental measurements.

6 PRESENTATION OF DATA

In this chapter, the inorganic gases are presented first. They are listed in alphabetical order according to their usual chemical formulae. The remainder (all organic) are then listed, in the Chemical Abstracts (Hill system) order. At the top of each evaluation, the solute is identified according to molecular formula, name and CAS Registry Number. The solvent is also identified.

The organic compounds include, among others:

- alkyl, alkenyl, alkynyl, aromatic and polycyclic aromatic hydrocarbons
- alcohols
- ketones
- chlorinated hydrocarbons
- phenol and substituted phenols
- sulfur compounds
- agrochemicals
- anesthetic gases
- esters
- terpenes
- polychlorinated biphenyls (PCBs)
- chlorinated dibenzodioxins.

7 SYMBOLS AND GLOSSARY

atm	atmosphere (pressure)
Bunsen	Bunsen coefficient
°C	degrees celsius
E	power of ten
γ^∞	infinite dilution activity coefficient
K	kelvin
k_H	limiting solubility (Henry's law constant)
k_{AW}	air–water partition coefficient

k_{WA}	water–air partition coefficient
limiting solubility ratio	generic name for k_H, k_{AW}, k_{WA}, γ^∞
M	molarity
m	molality
mmHg	millimetres of mercury (pressure)
Ostwald	Ostwald coefficient
Pa	pascal (pressure)
R	gas constant ($1.98722\,cal\,mol^{-1}\,K^{-1}$ or $8.3145\,J\,mol^{-1}\,K^{-1}$)
S	solubility (with units)
t	temperature (celsius)
T	temperature (kelvin)
w	mass fraction
x	mole fraction

8   METHOD DESCRIPTORS (CHAPTERS 3 AND 4)

- DIRECT
- EBULLIOMETRY
- EPICS (Equilibrium Partitioning In Closed Systems)
- GC (gas chromatography)
- FOG (fog chamber)
- HEADSPACE (headspace chromatography)
- INDIRECT
- PURGE (inert gas stripping)
- VP/AS (vapour pressure aqueous solubility)
- WET WALL (wetted-wall column technique)

9   CALCULATION OF k_H BY THE VP/AS METHOD FOR SOLID COMPOUNDS

In these few cases, the calculation was done according to the 'simplified VP/AS method,' involving Equations (4.13), (4.14) and (4.23) of Chapter 4. The source of the $p^{\circ,l}$ datum is mentioned, as well as what value of $\Delta_{fus}S^\circ$ is used.

10   HENRY'S LAW CONSTANTS FOR ELEMENTS AND INORGANIC COMPOUNDS

Solute		
Formula	Name	CAS Registry Number
Ar	Argon	7440-37-1
Solvent: Water		

Clever [1] presented critically evaluated smoothed data for the solubility of this gas at a partial pressure of 101.3 kPa in the range $273 < T/K < 348$. The data were represented by

$$\ln x = -57.6661 + 74.7627/(T/100\ \text{K}) + 20.1398 \ln(T/100\ \text{K})$$

The standard deviation of fit was 0.3 % [1]. The corresponding equation for k_H is

$$\ln(k_H/\text{mol kg}^{-1}\,\text{Pa}^{-1}) = -65.1756 + 74.7627/(T/100\ \text{K})$$
$$+ 20.1398 \ln(T/100\ \text{K})$$

$t/°C$	T/K	Solubility, x	$k_H/\text{mol kg}^{-1}\,\text{Pa}^{-1}$
0	273.2	4.284E−5	2.35E−8
5	278.2	3.775E−5	2.07E−8
10	283.2	3.362E−5	1.84E−8
15	288.2	3.025E−5	1.66E−8
20	293.2	2.748E−5	1.51E−8
25	298.2	2.519E−5	1.38E−8
30	303.2	2.328E−5	1.28E−8
35	308.2	2.169E−5	1.19E−8
40	313.2	2.036E−5	1.12E−8
45	318.2	1.925E−5	1.06E−8
50	323.2	1.832E−5	1.00E−8
55	328.2	1.754E−5	0.961E−8
60	333.2	1.690E−5	0.926E−8
65	338.2	1.637E−5	0.897E−8
70	343.2	1.594E−5	0.873E−8
75	348.2	1.560E−5	0.855E−8

In these calculations, it has been assumed that the concentration of the solute was within the region where Henry's law is valid. These are RECOMMENDED data in the temperature range 273–348 K.

Solute		
Formula	Name	CAS Registry Number
ClO_2	Chlorine dioxide	10049-04-4
Solvent: Water		

Young [2] presented critically evaluated smoothed data for the solubility of this gas at a partial pressure of 101.3 kPa in the range $283 < T/K < 333$. The data were represented by

$$\ln x = 7.9163 + 0.4791/(T/100\,\text{K}) - 11.0593\ln(T/100\,\text{K})$$

The standard deviation in x from this equation is 5 % [2]. The corresponding equation for k_H is

$$\ln(k_H/\text{mol}\,\text{kg}^{-1}\,\text{Pa}^{-1}) = 0.4070 + 0.4791/(T/100\,\text{K}) - 11.0593\ln(T/100\,\text{K})$$

$t/°C$	T/K	Solubility, x	$k_H/\text{mol}\,\text{kg}^{-1}\,\text{Pa}^{-1}$
10	283.2	3.255E−2	17.8E−6
15	288.2	2.674E−2	14.7E−6
20	293.2	2.204E−2	12.1E−6
25	298.2	1.823E−2	9.99E−6
30	303.2	1.513E−2	8.29E−6
35	308.2	1.259E−2	6.90E−6
40	313.2	1.051E−2	5.76E−6
45	318.2	0.880E−2	4.82E−6
50	323.2	0.739E−2	4.05E−6
55	328.2	0.622E−2	3.41E−6
60	333.2	0.525E−2	2.88E−6

In these calculations, it has been assumed that the concentration of the solute was within the region where Henry's law is valid. These are RECOMMENDED data in the range 283–333 K.

Solute		
Formula	Name	CAS Registry Number
CO	Carbon monoxide	630-08-0
Solvent: Water		

Cargill [3] presented critically evaluated smoothed data for the solubility of this gas at a partial pressure of 101.3 kPa in the range $273 < T/\text{K} < 328$. The data were represented by

$$\ln[(1/k_H)/\text{Pa}] = 126.753761 - 152.599953/(T/100\,\text{K})$$
$$- 67.8429542\ln(T/100\,\text{K}) + 7.04595356(T/100\,\text{K})$$

The standard deviation of fit in x was 0.04 % [3]. The corresponding equation for k_H in preferred units is

$$\ln(k_H/\text{mol}\,\text{kg}^{-1}\,\text{Pa}^{-1}) = -435.165523 + 1.52599953\text{E}4/(T/\text{K})$$
$$+ 67.8429542\ln(T/\text{K}) - 704.595356(T/\text{K})$$

$t/°C$	T/K	Solubility, x	$k_H/\text{mol kg}^{-1}\,\text{Pa}^{-1}$
0	273.2	2.939E−5	1.61E−8
5	278.2	2.591E−5	1.42E−8
10	283.2	2.316E−5	1.27E−8
15	288.2	2.096E−5	1.15E−8
20	293.2	1.918E−5	1.05E−8
25	298.2	1.774E−5	0.972E−8
30	303.2	1.658E−5	0.908E−8
35	308.2	1.562E−5	0.856E−8
40	313.2	1.484E−5	0.813E−8
45	318.2	1.421E−5	0.779E−8
50	323.2	1.379E−5	0.755E−8
55	328.2	1.330E−5	0.728E−8

In these calculations, it has been assumed that the concentration of the solute was within the region where Henry's law is valid. These are RECOMMENDED data in the temperature range 273–328 K.

Solute		
Formula	Name	CAS Registry Number
CO_2	Carbon dioxide	124-38-9
Solvent: Water		

Scharlin [4] presented critically evaluated smoothed data for the solubility of this gas in water in the range $273 < T/K < 433$ at a partial pressure of 101.3 kPa. This was a re-working of a previous critical evaluation of Carroll et al. [5] for the solubility at a partial pressure of less than 1 MPa in the same temperature range. The data were tabulated and a fitting equation given [4]; unfortunately, the fitting equation, as published, is erroneous and does not reproduce the tabular data. For present purposes, data from the original publication [5] were used.

In the earlier article, the solubility data were presented in the form of an equation for k_H:

$$\ln[(1/k_H)/\text{MPa}] = -6.8346 + 1.2817\text{E}4/(T/K) - 3.7688\text{E}6/(T/K)^2$$
$$+ 2.997\text{E}8/(T/K)^3$$

or

$$\ln(k_H/\text{mol kg}^{-1}\,\text{Pa}^{-1}) = -2.9644 - 1.2817\text{E}4/(T/K) + 3.7688\text{E}6/(T/K)^2$$
$$- 2.997\text{E}8/(T/K)^3$$

(The solubility, back-calculated from this equation, reproduced the data presented by Scharlin [4] within fitting error.)

$t/°C$	T/K	$k_H/mol\,kg^{-1}\,Pa^{-1}$
0	273.2	7.66E−7
10	283.2	5.43E−7
20	293.2	4.01E−7
30	303.2	3.08E−7
40	313.2	2.45E−7
50	323.2	2.01E−7
60	333.2	1.70E−7
70	343.2	1.48E−7
80	353.2	1.32E−7
90	363.2	1.20E−7
100	373.2	1.11E−7
110	383.2	1.05E−7
120	393.2	1.00E−7
130	403.2	0.976E−7
140	413.2	0.961E−7
150	423.2	0.958E−7
160	433.2	0.964E−7

In these calculations, it has been assumed that the concentration of the solute was within the region where Henry's law is valid. These are RECOMMENDED data in the range 273–433 K.

Solute		
Formula	Name	CAS Registry Number
COS	Carbon oxysulfide	463-58-1
Solvent: Water		

Wilhelm *et al.* [6] presented critically evaluated smoothed data for the solubility of this gas at a partial pressure of 101.3 kPa in the range $273 < T/K < 303$. The data were represented by

$$(R/cal\,mol^{-1})\ln x = -439.589 + 23896.1/(T/K) + 60.3429\ln(T/K)$$

The standard deviation of fit was 0.8 % [6]. The corresponding equation for k_H is

$$\ln(k_H/mol\,kg^{-1}\,Pa^{-1}) = -228.727 + 12025.4/(T/K) + 30.3699\ln(T/K)$$

$t/°C$	T/K	Solubility, x	$k_H/mol\,kg^{-1}\,Pa^{-1}$
0	273.2	10.76E−4	5.90E−7
5	278.2	8.463E−4	4.64E−7
10	283.2	6.775E−4	3.71E−7
15	288.2	5.517E−4	3.02E−7
20	293.2	4.565E−4	2.50E−7
25	298.2	3.835E−4	2.10E−7
30	303.2	3.267E−4	1.79E−7

It has been assumed in this table that (1) the concentration of the solute was within the region where Henry's law is valid and (2) COS is unchanged in aqueous solution, although it reacts slowly with water [7,8]:

$$COS + H_2O = H_2S + CO_2$$

These are RECOMMENDED data in the temperature range 273–303 K.

Solute		
Formula	Name	CAS Registry Number
H_2	Hydrogen	1333-74-0
Solvent: Water		

Young [9a] presented critically evaluated smoothed data for the solubility of this gas at a partial pressure of 101.3 kPa in the range $273 < T/K < 353$. The data were represented by

$$\ln x = -48.1611 + 55.2845/(T/100\,K) + 16.8893\ln(T/100\,K)$$

The standard deviation of fit was 0.5 % [9a]. The corresponding equation for k_H is

$$\ln(k_H/mol\,kg^{-1}\,Pa^{-1}) = -133.4487 + 5.52845E3/(T/K) + 16.8893\ln(T/K)$$

$t/°C$	T/K	Solubility, x	$k_H/mol\,kg^{-1}\,Pa^{-1}$
0	273.2	1.755E−5	9.62E−9
5	278.2	1.657E−5	9.08E−9
10	283.2	1.576E−5	8.63E−9
15	288.2	1.510E−5	8.27E−9
20	293.2	1.455E−5	7.97E−9
25	298.2	1.411E−5	7.73E−9

$t/°C$	T/K	Solubility, x	$k_H/\mathrm{mol\,kg^{-1}\,Pa^{-1}}$
30	303.2	1.377E−5	7.54E−9
35	308.2	1.350E−5	7.40E−9
40	313.2	1.330E−5	7.29E−9
45	318.2	1.317E−5	7.22E−9
50	323.2	1.310E−5	7.18E−9
55	328.2	1.308E−5	7.17E−9
60	333.2	1.312E−5	7.19E−9
65	338.2	1.320E−5	7.23E−9
70	343.2	1.333E−5	7.30E−9
75	348.2	1.350E−5	7.40E−9
80	353.2	1.371E−5	7.51E−9

In these calculations, it has been assumed that the concentration of the solute was within the region where Henry's law is valid. These are RECOMMENDED data in the temperature range 273–353 K.

Solute		
Formula	Name	CAS Registry Number
H_2O_2	Hydrogen peroxide	7722-84-1
Solvent: Water		

Selected experimental measurements of k_H for this liquid are as follows:

Ref.	T/K	Quantity	Unit	Value	$\mathrm{mol\,kg^{-1}\,Pa^{-1}}$	Method
9c	276.15	k_H	M atm^{-1}	5.08E5	5.01	DIRECT
10	278.15	k_H	M atm^{-1}	4.98E5	4.91	DIRECT
11	278.15	k_H	M atm^{-1}	3.82E5	3.77	VLE
10	281.15	k_H	M atm^{-1}	1.48E5	1.46	DIRECT
9c	283.15	k_H	M atm^{-1}	2.92E5	2.88	DIRECT
11	283.15	k_H	M atm^{-1}	2.45E5	2.42	VLE
12	283.15	k_H	M atm^{-1}	2.73E5	2.69	DIRECT
13	288.15	k_H	M atm^{-1}	2.10E5	2.07	DIRECT
14	288.15	k_H	M atm^{-1}	2.54E5	2.51	DIRECT
9c	293.15	k_H	M atm^{-1}	1.35E5	1.33	DIRECT
15	293.15	k_H	M atm^{-1}	1.42E5	1.40	DIRECT
14	293.15	k_H	M atm^{-1}	1.63E5	1.61	DIRECT
13	293.15	k_H	M atm^{-1}	1.44E5	1.42	DIRECT
12	295.15	k_H	M atm^{-1}	1.07E5	1.06	DIRECT

Ref.	T/K	Quantity	Unit	Value	mol kg^{-1} Pa^{-1}	Method
10	297.15	k_H	M atm^{-1}	9.2E4	0.908	DIRECT
11	298.15	k_H	M atm^{-1}	1.0E5	0.987	THERMO
14	298.15	k_H	M atm^{-1}	1.06E5	1.05	DIRECT

These data may be represented by the equation

$$\ln(k_H/\text{mol kg}^{-1}\,\text{Pa}^{-1}) = -20.449 + 6091.5/(T/\text{K})$$

The standard error of estimate of k_H by this equation is 12 %. This equation generates RECOMMENDED values.

Solute		
Formula	Name	CAS Registry Number
He	Helium	7440-59-7
Solvent: Water		

Clever [16] presented critically evaluated smoothed data for the solubility of this gas at a partial pressure of 101.3 kPa in the range $273 < T/\text{K} < 348$. The data were represented by

$$\ln x = -41.4611 + 42.5962/(T/100\,\text{K}) + 14.0094\ln(T/100\,\text{K})$$

The standard deviation of fit was 0.5 % [16]. The corresponding equation for k_H is

$$\ln(k_H/\text{mol kg}^{-1}\,\text{Pa}^{-1}) = -113.4867 + 4.25962\text{E}3/(T/100\,\text{K})$$
$$+ 14.0094\ln(T/\text{K})$$

$t/°C$	T/K	Solubility, x	k_H/mol kg^{-1} Pa^{-1}
0	273.2	7.585E−6	4.16E−9
5	278.2	7.389E−6	4.05E−9
10	283.2	7.237E−6	3.97E−9
15	288.2	7.123E−6	3.90E−9
20	293.2	7.044E−6	3.86E−9
25	298.2	6.997E−6	3.83E−9
30	303.2	6.978E−6	3.82E−9
35	308.2	6.987E−6	3.83E−9
40	313.2	7.020E−6	3.85E−9

$t/°C$	T/K	Solubility, x	$k_H/\text{mol kg}^{-1}\,\text{Pa}^{-1}$
45	318.2	7.077E−6	3.88E−9
50	323.2	7.158E−6	3.92E−9
55	328.2	7.261E−6	3.98E−9
60	333.2	7.385E−6	4.05E−9
65	338.2	7.532E−6	4.13E−9
70	343.2	7.700E−6	4.22E−9
75	348.2	7.890E−6	4.32E−9

In these calculations, it has been assumed that the concentration of the solute was within the region where Henry's law is valid. These are RECOMMENDED data in the temperature range 273–348 K.

Solute		
Formula	Name	CAS Registry Number
Kr	Krypton	7439-90-9
Solvent: Water		

Clever [17] presented critically evaluated smoothed data for the solubility of this gas at a partial pressure of 101.3 kPa in the range $273 < T/K < 353$. The data were represented by

$$\ln x = -66.9928 + 91.0166/(T/100\,\text{K}) + 24.2207\ln(T/100\,\text{K})$$

The corresponding equation for k_H is

$$\ln(k_H/\text{mol kg}^{-1}\,\text{Pa}^{-1}) = -74.5023 + 91.0166/(T/100\,\text{K})$$
$$+ 24.2207\ln(T/100\,\text{K})$$

$t/°C$	T/K	Solubility, x	$k_H/\text{mol kg}^{-1}\,\text{Pa}^{-1}$
0	273.2	8.841E−5	4.84E−8
5	278.2	7.537E−5	4.13E−8
10	283.2	6.511E−5	3.57E−8
15	288.2	5.696E−5	3.12E−8
20	293.2	5.041E−5	2.76E−8
25	298.2	4.511E−5	2.47E−8
30	303.2	4.079E−5	2.23E−8
35	308.2	3.725E−5	2.04E−8
40	313.2	3.432E−5	1.88E−8

$t/°C$	T/K	Solubility, x	$k_H/\text{mol kg}^{-1}\,\text{Pa}^{-1}$
45	318.2	3.190E−5	1.75E−8
50	323.2	2.990E−5	1.64E−8
55	328.2	2.823E−5	1.55E−8
60	333.2	2.686E−5	1.47E−8
65	338.2	2.572E−5	1.41E−8
70	343.2	2.480E−5	1.36E−8
75	348.2	2.405E−5	1.32E−8
80	353.2	2.346E−5	1.29E−8

In these calculations, it has been assumed that the concentration of the solute was within the region where Henry's law is valid. These are RECOMMENDED data in thee temperature range 273–353 K.

Solute		
Formula	Name	CAS Registry Number
N_2	Nitrogen	7727-37-9
Solvent: Water		

Battino *et al.* [18] presented critically evaluated smoothed data for the solubility of this gas at a partial pressure of 101.3 kPa in the range $273 < T/K < 348$. The data were represented by

$$\ln x = -67.38765 + 86.32129/(T/100\text{ K}) + 24.79808\ln(T/100\text{ K})$$

The standard deviation of fit was 0.72 % [18]. The corresponding equation for k_H is

$$\ln(k_H/\text{mol kg}^{-1}\,\text{Pa}^{-1}) = -190.25042 + 8.632129\text{E3}/(T/K)$$
$$+ 24.79808\ln(T/K)$$

$t/°C$	T/K	Solubility, x	$k_H/\text{mol kg}^{-1}\,\text{Pa}^{-1}$
0	273.2	1.908E−5	10.5E−9
5	278.2	1.695E−5	9.29E−9
10	283.2	1.524E−5	8.35E−9
15	288.2	1.386E−5	7.59E−9
20	293.2	1.274E−5	6.98E−9
25	298.2	1.183E−5	6.48E−9
30	303.2	1.108E−5	6.07E−9

$t/°C$	T/K	Solubility, x	$k_H/\text{mol kg}^{-1}\,\text{Pa}^{-1}$
35	308.2	1.047E−5	5.74E−9
40	313.2	0.9981E−5	5.47E−9
45	318.2	0.9585E−5	5.25E−9
50	323.2	0.9273E−5	5.08E−9
55	328.2	0.9033E−5	4.95E−9
60	333.2	0.8855E−5	4.85E−9
65	338.2	0.8735E−5	4.79E−9
70	343.2	0.8666E−5	4.75E−9
75	348.2	0.8644E−5	4.74E−9

In these calculations, it has been assumed that the concentration of the solute was within the region where Henry's law is valid. These are RECOMMENDED data in the temperature range 273–348 K.

Solute		
Formula	Name	CAS Registry Number
N_2O	Nitrous oxide	10024-97-2
Solvent: Water		

Young [9b] presented critically evaluated smoothed data for the solubility of this gas at a partial pressure of 101.3 kPa in the range $273 < T/K < 313$. The data were represented by

$$\ln x = -60.7467 + 88.8280/(T/100\,\text{K}) + 21.2531 \ln(T/100\,\text{K})$$

The standard deviation of fit was 1.2 % [9b]. The corresponding equation for k_H is

$$\ln(k_H/\text{mol kg}^{-1}\,\text{Pa}^{-1}) = -166.1303 + 8.88280\text{E}3/(T/K) + 21.2531 \ln(T/K)$$

$t/°C$	T/K	Solubility, x	$k_H/\text{mol kg}^{-1}\,\text{Pa}^{-1}$
0	273.2	10.38E−4	5.69E−7
5	278.2	8.505E−4	4.66E−7
10	283.2	7.067E−4	3.87E−7
15	288.2	5.948E−4	3.26E−7
20	293.2	5.068E−4	2.78E−7
25	298.2	4.367E−4	2.39E−7
30	303.2	3.805E−4	2.09E−7

$t/°C$	T/K	Solubility, x	$k_H/\text{mol kg}^{-1}\text{Pa}^{-1}$
35	308.2	3.348E−4	1.83E−7
40	313.2	2.975E−4	1.63E−7

In these calculations, it has been assumed that the concentration of the solute was within the region where Henry's law is valid. These are RECOMMENDED data in the temperature range 273–313 K.

Solute		
Formula	Name	CAS Registry Number
NO	Nitric oxide	10102-43-9
Solvent: Water		

Young [9b] presented critically evaluated smoothed data for the solubility of this gas at a partial pressure of 101.3 kPa in the range $273 < T/K < 358$. The data were represented by the equation

$$\ln x = -62.8086 + 82.3420/(T/100\,\text{K}) + 22.8155\ln(T/100\,\text{K})$$

The standard deviation of fit was 0.8 % [9b]. The corresponding equation for k_H is

$$\ln(k_H/\text{mol kg}^{-1}\text{Pa}^{-1}) = -74.923 + 8234.2/(T/\text{K}) + 22.816\ln(T/\text{K})$$

$t/°C$	T/K	Solubility, x	$k_H/\text{mol kg}^{-1}\text{Pa}^{-1}$
0	273.2	5.905E−5	3.24E−8
5	278.2	5.196E−5	2.85E−8
10	283.2	4.625E−5	2.53E−8
15	288.2	4.163E−5	2.28E−8
20	293.2	3.786E−5	2.07E−8
25	298.2	3.477E−5	1.91E−8
30	303.2	3.222E−5	1.77E−8
35	308.2	3.012E−5	1.65E−8
40	313.2	2.838E−5	1.56E−8
45	318.2	2.695E−5	1.48E−8
50	323.2	2.577E−5	1.41E−8
55	328.2	2.481E−5	1.36E−8
60	333.2	2.404E−5	1.32E−8
65	338.2	2.343E−5	1.28E−8

$t/°C$	T/K	Solubility, x	k_H/mol kg^{-1} Pa^{-1}
70	343.2	2.297E$-$5	1.26E$-$8
75	348.2	2.264E$-$5	1.24E$-$8
80	353.2	2.242E$-$5	1.23E$-$8
85	358.2	2.232E$-$5	1.22E$-$8

In these calculations, it has been assumed that the concentration of the solute was within the region where Henry's law is valid. These are RECOMMENDED data in the temperature range 273–358 K.

Solute		
Formula	Name	CAS Registry Number
O_2	Oxygen	7727-37-9
Solvent: Water		

Battino [19] presented critically evaluated smoothed data for the solubility of this gas at a partial pressure of 101.3 kPa in the range $273 < T/K < 348$. The data were represented by

$$\ln x = -64.21517 + 83.91236/(T/100\,K) + 23.24323 \ln(T/100\,K)$$

The standard deviation of fit was 0.17 % [19]. The corresponding equation for k_H is

$$\ln(k_H/\text{mol kg}^{-1}\,\text{Pa}^{-1}) = -187.07794 + 8.391236E3/(T/K) + 23.24323 \ln(T/K)$$

$t/°C$	T/K	Solubility, x	k_H/mol kg^{-1} Pa^{-1}
0	273.2	3.949E$-$5	21.63E$-$9
5	278.2	3.460E$-$5	18.96E$-$9
10	283.2	3.070E$-$5	16.82E$-$9
15	288.2	2.756E$-$5	15.10E$-$9
20	293.2	2.501E$-$5	13.70E$-$9
25	298.2	2.293E$-$5	12.56E$-$9
30	303.2	2.122E$-$5	11.63E$-$9
35	308.2	1.982E$-$5	10.86E$-$9
40	313.2	1.867E$-$5	10.23E$-$9
45	318.2	1.773E$-$5	9.71E$-$9
50	323.2	1.697E$-$5	9.30E$-$9
55	328.2	1.635E$-$5	8.96E$-$9

$t/°C$	T/K	Solubility, x	k_H/mol kg^{-1} Pa^{-1}
60	333.2	1.586E−5	8.69E−9
65	338.2	1.549E−5	8.49E−9
70	343.2	1.521E−5	8.33E−9
75	348.2	1.502E−5	8.23E−9

In these calculations, it has been assumed that the concentration of the solute was within the region where Henry's law is valid. These are RECOMMENDED data in the temperature range 273–348 K.

Solute		
Formula	Name	CAS Registry Number
$^{222}Rn_{86}$	Radon-222	14859-67-7
Solvent: Water		

Clever [17] presented critically evaluated smoothed data for the solubility of this gas at a partial pressure of 101.3 kPa in the range $273 < T/K < 373$.

$t/°C$	T/K	Solubility, x	k_H/mol kg^{-1} Pa^{-1}
0	273.2	4.217E−4	23.1E−8
5	278.2	3.382E−4	18.5E−8
10	283.2	2.764E−4	15.1E−8
15	288.2	2.299E−4	12.6E−8
20	293.2	1.445E−4	10.7E−8
25	298.2	1.671E−4	9.15E−8
30	303.2	1.457E−4	7.98E−8
35	308.2	1.288E−4	7.06E−8
40	313.2	1.153E−4	6.32E−8
45	318.2	1.046E−4	5.73E−8
50	323.2	0.959E−4	5.25E−8
55	328.2	0.889E−4	4.87E−8
60	333.2	0.833E−4	4.56E−8
65	338.2	0.788E−4	4.32E−8
70	343.2	0.752E−4	4.12E−8
75	348.2	0.724E−4	3.97E−8
80	353.2	0.703E−4	3.85E−8
85	358.2	0.688E−4	3.77E−8
90	363.2	0.678E−4	3.71E−8
95	368.2	0.673E−4	3.69E−8
100	373.2	0.672E−4	3.68E−8

These data may be represented by the equation

$$\ln(k_{\mathrm{H}}/\mathrm{mol\,kg^{-1}\,Pa^{-1}}) = 12.3530 - 0.162594(T/\mathrm{K}) + 2.2433\mathrm{E} - 4(T/\mathrm{K})^2$$

The standard error of estimate of k_{H} by this equation is 2 %. This equation generates RECOMMENDED values in the temperature range 273–373 K.

Solute		
Formula	Name	CAS Registry Number
SF_6	Sulfur hexafluoride	2551-62-4
Solvent: Water		

Wilhelm *et al.* [6] presented critically evaluated smoothed data for the solubility of this gas at a partial pressure of 101.3 kPa in the range $273 < T/\mathrm{K} < 328$. The data were represented by

$$(R/\mathrm{cal\,mol^{-1}})\ln x = -877.854 + 42051.0/(T/\mathrm{K}) + 125.018\ln(T/\mathrm{K})$$

This equation represents x within 0.5 % [6]. The corresponding expression for k_{H} is

$$\ln(k_{\mathrm{H}}/\mathrm{mol\,kg^{-1}\,Pa^{-1}}) = -449.260 + 21160.7/(T/\mathrm{K}) + 62.9110\ln(T/\mathrm{K})$$

This equation generates RECOMMENDED data in the range 273–328 K. Mroczek [20] determined k_{H} by direct measurement at 101.3 kPa partial pressure in the range $348 < T/\mathrm{K} < 503$ and included earlier measurements of Ashton *et al.* [21] and Cosgrove and Walkley [22] at lower temperatures. Mroczek's data are included in the table in parentheses:

$t/°C$	T/K	Solubility, x	$k_{\mathrm{H}}/\mathrm{mol\,kg^{-1}\,Pa^{-1}}$
0	273.2	11.78E−6	6.45E−9
5	278.2	9.164E−6	5.02E−9
10	283.2	7.335E−6	4.02E−9
15	288.2	6.032E−6	3.31E−9
20	293.2	5.088E−6	2.79E−9
25	298.2	4.394E−6	2.41(2.42)E−9
30	303.2	3.881E−6	2.13E−9
35	308.2	3.499E−6	1.92E−9
40	313.2	3.218E−6	1.76E−9
45	318.2	3.014E−6	1.65E−9
50	323.2	2.872E−6	1.57(1.59)E−9
55	328.2	2.782E−6	1.52E−9

$t/^\circ C$	T/K	Solubility, x	$k_H/\text{mol kg}^{-1}\,\text{Pa}^{-1}$
75	348.2	–	– (1.32E−9)
100	373.2	–	– (1.28E−9)
125	398.2	–	– (1.38E−9)
150	423.2	–	– (1.64E−9)
175	448.2	–	– (2.04E−9)
200	473.2	–	– (2.68E−9)
225	498.2	–	– (3.58E−9)
230	503.2	–	– (3.80E−9)

In these calculations, it has been assumed that the concentration of the solute was within the region where Henry's law is valid.

11 HENRY'S LAW CONSTANTS FOR ORGANIC COMPOUNDS (C_1)

Solute		
Formula	Name	CAS Registry Number
CBr_3F	Tribromofluoromethane	353-54-8
Solvent: Water		

The solubility of this liquid in water at 25 °C was reported by O'Connell [23]. The datum is 0.04 g solute per 100 g water ($x = 2.66E−5$). Vapour pressure data for this compound are not available; an estimated value using Predict II is 1000 Pa [24]. The calculated k_H is 1.48E−6 mol kg^{-1} Pa^{-1}. This is an estimate only.

Solute		
Formula	Name	CAS Registry Number
CBr_4	Tetrabromomethane	558-13-4
Solvent: Water		

There are two reports of the solubility of this solid in water at 30 °C. The vapour pressure of the supercooled liquid at this temperature was obtained by extrapolating the data of Kudchadker et al. [25]: 59.1 Pa.

Ref.	T/K	Solubility, m	Solubility, x	$10^5 \ln k_H/\text{mol kg}^{-1}\,\text{Pa}^{-1}$
26	303.15	7.0E−4	1.26E−5	1.18
27	303.15	7.2E−4	1.30E−5	1.22

Solute		
Formula	Name	CAS Registry Number
$CClF_3$	Chlorotrifluoromethane	75-72-9
Solvent: Water		

Wilhelm *et al.* [16] presented critically evaluated smoothed data for the solubility of this gas at a partial pressure of 101.3 kPa in water in the range $298 < T/K < 348$. The data were represented by the equation

$$(RT/\text{cal mol}^{-1})\ln x = 3204.55 - 32.5600(T/K)$$

or

$$\ln x = 1612.58/(T/K) - 16.3847$$

The standard deviation in x calculated by this equation is 1 %. The equation for k_H is

$$\ln(k_H/\text{mol kg}^{-1}\,\text{Pa}^{-1}) = -23.8942 + 1612.58/(T/K)$$

$t/°C$	T/K	Solubility, x	$k_H/\text{mol kg}^{-1}\,\text{Pa}^{-1}$
25	298.2	1.711E−5	9.37E−9
30	303.2	1.565E−5	8.57E−9
35	308.2	1.435E−5	7.86E−9
40	313.2	1.320E−5	7.23E−9
45	318.2	1.217E−5	6.67E−9
50	323.2	1.126E−5	6.17E−9
55	328.2	1.043E−5	5.72E−9
60	333.2	0.969E−5	5.31E−9
65	338.2	0.902E−5	4.94E−9
70	343.2	0.787E−5	4.31E−9

In these calculations, it has been assumed that the concentration of the solute was within the region where Henry's law is valid. These are RECOMMENDED values in the temperature range 298–343 K.

Solute		
Formula	Name	CAS Registry Number
CCl_2F_2	Dichlorodifluoromethane	75-71-8
Solvent: Water		

There are two reported estimates of k_H of this liquid:

			Original measurements		
Ref.	$t/°C$	T/K	Limiting solubility ratio	Method	$k_H/\mathrm{mol\,kg^{-1}\,Pa^{-1}}$
28	20	293.2	$k_{WA} = 0.06$	VP/AS	2.46E−8
29	–	–	$(1/k_H)/\mathrm{atm\,m^3\,mol^{-1}} = 1.5$	VP/AS	0.987E−8

In these calculations, it has been assumed that the concentration of the solute was within the region where Henry's law is valid.

Solute		
Formula	Name	CAS Registry Number
CCl_3F	Trichlorofluoromethane	75-69-4
Solvent: Water		

Horvath and Getzen [30] presented critically evaluated smoothed data for the solubility of this liquid in water in the range $273 < T/K < 313$. The data were represented by

$$w = 0.250094 - 1.6263\mathrm{E}{-}3(T/K) + 2.6547\mathrm{E}{-}6(T/K)^2$$

where w is mass fraction. The mole fraction solubility is

$$x = 3.27903\mathrm{E}{-}2 - 2.13227\mathrm{E}{-}4(T/K) + 3.48063\mathrm{E}{-}7(T/K)^2$$

The k_H data were calculated with this equation and the vapour pressure data of Yurttaş *et al.* [31]:

$t/°C$	T/K	Solubility, x	$k_H/\text{mol kg}^{-1}\,\text{Pa}^{-1}$
0	273.2	5.166E−4	7.25E−7
5	278.2	4.099E−4	4.68E−7
10	283.2	3.207E−4	3.00E−7
15	288.2	2.488E−4	1.92E−7
20	293.2	1.943E−4	1.25E−7
25	298.2	1.572E−4	0.844E−7
30	303.2	1.375E−4	0.621E−7
35	308.2	1.352E−4	0.517E−7
40	313.2	1.500E−4	0.487E−7

In these calculations, it has been assumed that the concentration of the solute was within the region where Henry's law is valid. These data may be represented by the equation

$$\ln(k_H/\text{mol kg}^{-1}\,\text{Pa}^{-1}) = -36.699 + 6135.4/(T/K)$$

The standard error of estimate of k_H by this equation is 10 %. These are RECOMMENDED data in the temperature range 273–313 K.

Solute		
Formula	Name	CAS Registry Number
CCl_3NO_2	Chloropicrin	76-06-2
Solvent: Water		

There are two reports of k_H for this solid:

			Original measurements		$k_H/\text{mol kg}^{-1}\,\text{Pa}^{-1}$
Ref.	$t/°C$	T/K	Limiting solubility ratio	Method	
32	20	293.2	$(1/k_H)/\text{Pa m}^3\,\text{mol}^{-1} = 197.27$	VP/AS	5.07E−6
33	25	298.2	$k_{AW} = 8.4\text{E}-2$	HEADSPACE	4.80E−6

In these calculations, it has been assumed that the concentration of the solute was within the region where Henry's law is valid.

An order-of-magnitude estimate for k_H of this compound is 4.9E−6 mol kg^{-1} Pa^{-1}.

Solute		
Formula	Name	CAS Registry Number
CCl_4	Tetrachloromethane	56-23-5
Solvent: Water		

Horvath and Getzen [30] presented critically evaluated smoothed data for the solubility of this liquid in water in the range $273 < T/K < 308$. The data were represented by

$$w = 3.4653E-2 - 2.30285E-4(T/K) + 3.91621E-7(T/K)^2$$

where w is mass fraction. The mole fraction solubility is given by

$$x = 4.05894E-3 - 2.69735E-5(T/K) + 4.5871E-8(T/K)^2$$

The k_H values were calculated with this equation, together with vapour pressure of tetrachloromethane from Gmehling et al. [34].

$t/°C$	T/K	Solubility, x	$k_H/\text{mol kg}^{-1}\,\text{Pa}^{-1}$
0	273.2	1.137E−4	14.1E−7
5	278.2	1.052E−4	10.0E−7
10	283.2	0.9916E−4	7.29E−7
15	288.2	0.9529E−4	5.49E−7
20	293.2	0.9376E−4	4.27E−7
25	298.2	0.9529E−4	3.44E−7
30	303.2	0.9752E−4	2.86E−7
35	308.2	1.029E−4	2.45E−7

In these calculations, it has been assumed that the concentration of the solute was within the region where Henry's law is valid. These data may be represented by the equation

$$\ln(k_H/\text{mol kg}^{-1}\,\text{Pa}^{-1}) = -29.033 + 4229.3/(T/K)$$

The standard error of estimate of k_H by this equation is 5 %. This equation generates RECOMMENDED k_H data in the temperature range 273–308 K.

Solute		
Formula	Name	CAS Registry Number
CF_4	Tetrafluoromethane	75-73-0
Solvent: Water		

Wilhelm *et al.* [6] presented critically evaluated smoothed data for the solubility of this gas in water at a partial pressure of 101.3 kPa. The data were represented by

$$(R/\text{cal mol}^{-1}\,\text{K}^{-1})\ln x = -644.690 + 30\,657.7T/\text{K} + 90.7528\ln(T/\text{K})$$

valid for the range $276 < T/\text{K} < 323$. The standard deviation in x given by the authors [6] is 0.52 %. The corresponding equation for k_H is

$$k_H/\text{mol kg}^{-1}\,\text{Pa}^{-1} = \exp[-324.418 + 15\,427.4\ T/\text{K} + 45.6682\ln(T/\text{K})]$$

$t/°C$	T/K	Solubility, x	$k_H/\text{mol kg}^{-1}\,\text{Pa}^{-1}$
5	278.2	6.617E−6	3.63E−9
10	283.2	5.066E−6	3.07E−9
15	288.2	4.844E−6	2.65E−9
20	293.2	4.265E−6	2.34E−9
25	298.2	3.819E−6	2.09E−9
30	303.2	3.477E−6	1.91E−9
35	308.2	3.214E−6	1.76E−9
40	313.2	3.014E−6	1.65E−9
45	318.2	2.864E−6	1.57E−9
50	323.2	2.757E−6	1.51E−9

In these calculations, it has been assumed that the concentration of the solute was within the region where Henry's law is valid. These k_H data are RECOMMENDED for the temperature range 278–323 K.

Solute		
Formula	Name	CAS Registry Number
CHBrCl$_2$	Bromodichloromethane	75-27-4
Solvent: Water		

Available experimental measurements of k_H of this liquid are as follows:

			Original measurements		
Ref.	$t/°C$	T/K	Limiting solubility ratio	Method	$k_H/\text{mol kg}^{-1}\,\text{Pa}^{-1}$
35	20	293.2	$(1/k_H)/\text{atm m}^3\,\text{mol}^{-1} = 1.6\text{E}{-}3$	PURGE	6.16E−6
36	20	293.2	$(1/k_H)/\text{atm m}^3\,\text{mol}^{-1} = 1.6\text{E}{-}3$	GC	6.16E−6

			Original measurements		
Ref.	$t/°C$	T/K	Limiting solubility ratio	Method	$k_H/\text{mol kg}^{-1}\,\text{Pa}^{-1}$
36	30	303.2	$(1/k_H)/\text{atm m}^3\,\text{mol}^{-1} = 2.6E{-}3$	GC	3.80E−6
37	35	308.2	$k_{AW} = 0.126$	EPICS	3.10E−6
36	40	313.2	$(1/k_H)/\text{atm m}^3\,\text{mol}^{-1} = 4.0E{-}3$	GC	2.47E−6
37	45	318.2	$k_{AW} = 0.171$	EPICS	2.21E−6
37	55	328.2	$k_{AW} = 0.242$	EPICS	1.52E−6
37	60	333.2	$k_{AW} = 0.260$	EPICS	1.39E−6

In these calculations, it has been assumed that the concentration of the solute was within the region where Henry's law is valid.

These data may be represented by the equation

$$\ln(k_H/\text{mol kg}^{-1}\,\text{Pa}^{-1}) = -24.690 + 3710.0/(T/K)$$

The standard error of estimate of k_H by this equation is 4 %. This equation generates RECOMMENDED k_H data in the range 293–333 K.

Solute		
Formula	Name	CAS Registry Number
CHBr$_2$Cl	Dibromochloromethane	124-48-1
Solvent: Water		

Available experimental measurements of k_H of this liquid are as follows:

			Original measurements		k_H/mol
Ref.	$t/°C$	T/K	Limiting solubility ratio	Method	$\text{kg}^{-1}\,\text{Pa}^{-1}$
35	20	293.2	$(1/k_H)/\text{atm m}^3\,\text{mol}^{-1} = 8.7E{-}4$	PURGE	11.3E−6
36	20	293.2	$(1/k_H)/\text{atm m}^3\,\text{mol}^{-1} = 8.0E{-}4$	GC	12.3E−6
36	30	303.2	$(1/k_H)/\text{atm m}^3\,\text{mol}^{-1} = 1.4E{-}3$	GC	7.05E−6
37	35	308.2	$k_{AW} = 0.087$	EPICS	4.49E−6
36	40	313.2	$(1/k_H)/\text{atm m}^3\,\text{mol}^{-1} = 2.2E{-}3$	GC	4.49E−6
37	45	318.2	$k_{AW} = 0.116$	EPICS	3.26E−6
37	55	328.2	$k_{AW} = 0.153$	EPICS	2.40E−6
37	60	333.2	$k_{AW} = 0.184$	EPICS	1.96E−6

In these calculations, it has been assumed that the concentration of the solute was within the region where Henry's law is valid. These data may be represented by the equation

$$\ln(k_H/\text{mol kg}^{-1}\,\text{Pa}^{-1}) = -26.348 + 4383.0/(T/K)$$

The standard error of estimate of k_H by this equation is 6%. This equation generates RECOMMENDED data in the temperature range 293–333 K.

Solute		
Formula	Name	CAS Registry Number
CHBr$_3$	Tribromomethane	75-25-2
Solvent: Water		

Horvath and Getzen [30] presented critically evaluated smoothed data for the solubility of this liquid in water in the range $283 < T/K < 303$. The data were represented by

$$w = 2.371\text{E}{-}2 - 1.517\text{E}{-}4(T/K) + 2.7808\text{E}{-}7(T/K)^2$$

where w is mass fraction. The mole fraction solubility is then

$$x = 1.690\text{E}{-}3 - 1.081\text{E}{-}5(T/K) + 1.982\text{E}{-}8(T/K)^2$$

The k_H values were calculated from this equation and vapour pressure from Gmehling *et al.* [34]:

$t/°\text{C}$	T/K	Solubility, x	$k_H/\text{mol kg}^{-1}\,\text{Pa}^{-1}$
10	283.2	2.182E−4	4.60E−5
15	288.2	2.208E−4	3.28E−5
20	293.2	2.243E−4	2.39E−5
25	298.2	2.289E−4	1.78E−5
30	303.2	2.344E−4	1.34E−5

In these calculations, it has been assumed that the concentration of the solute was within the region where Henry's law is valid. These data may be represented by the equation

$$\ln(k_H/\text{mol kg}^{-1}\,\text{Pa}^{-1}) = -28.667 + 5286.7/(T/K)$$

The standard error of estimate of k_H by this equation is 1%. This equation generates RECOMMENDED values of k_H in the temperature range 283–303 K.

Solute		
Formula	Name	CAS Registry Number
$CHClF_2$	Chlorodifluoromethane	75-45-6
Solvent: Water		

Wilhelm *et al.* [6] presented critically evaluated smoothed data for the solubility of this gas at a partial pressure of 101.3 kPa in water in the range $297 < T/K < 352$. The data were represented by the equation

$$(R/\text{cal mol}^{-1}\ \text{K}^{-1}) \ln x = -378.939 + 25\,999.6/(T/K) + 45.2647 \ln(T/K)$$
$$+ 0.0642996(T/K)$$

or

$$\ln x = -190.688 + 13.083.4/(T/K) + 22.7779 \ln(T/K)$$
$$+ 3.23566E{-}2(T/K)$$

The standard deviation in x calculated by this equation is 4 % [6]. The equation for k_H is

$$\ln(k_H/\text{mol kg}^{-1}\,\text{Pa}^{-1}) = -198.198 + 13083.4/(T/K) + 22.7779 \ln(T/K)$$
$$+ 3.23566E{-}2(T/K)$$

$t/°C$	T/K	Solubility, x	$k_H/\text{mol kg}^{-1}\,\text{Pa}^{-1}$
20	293.2	7.629E−4	4.19E−7
25	298.2	6.237E−4	3.43E−7
30	303.2	5.192E−4	2.85E−7
35	308.2	4.399E−4	2.42E−7
40	313.2	3.788E−4	2.08E−7
45	318.2	3.313E−4	1.82E−7
50	323.2	2.940E−4	1.62E−7
55	328.2	2.646E−4	1.45E−7

In these calculations, it has been assumed that the concentration of the solute was within the region where Henry's law is valid. These are RECOMMENDED values in the temperature range 293–353 K.

Solute		
Formula	Name	CAS Registry Number
$CHCl_3$	Trichloromethane	67-66-3
Solvent: Water		

Horvath and Getzen [30] presented critically evaluated smoothed data for the solubility of this liquid in water in the range $273 < T/K < 333$. The data were represented by

$$w = 0.125333 - 7.40557\text{E}{-4}(T/K) + 1.16374\text{E}{-6}(T/K)^2$$

where w is mass fraction. The mole fraction solubility is

$$x = 1.89101\text{E}{-2} - 1.11734\text{E}{-4}(T/K) + 1.75584\text{E}{-7}(T/K)^2$$

The k_H data were calculated from this equation with vapour pressure data from Gmehling et al. [34]:

$t/^\circ\text{C}$	T/K	Solubility, x	$k_H/\text{mol kg}^{-1}\,\text{Pa}^{-1}$
0	273.2	1.490E−3	10.33E−6
5	278.2	1.416	7.59E−6
10	283.2	1.350E−3	5.65E−6
15	288.2	1.293E−3	4.27E−6
20	293.2	1.244E−3	3.28E−6
25	298.2	1.205E−3	2.55E−6
30	303.2	1.174E−3	2.02E−6
35	308.2	1.152E−3	1.62E−6
40	313.2	1.139E−3	1.32E−6

In these calculations, it has been assumed that the concentration of the solute was within the region where Henry's law is valid. These data may be represented by the equation

$$\ln(k_H/\text{mol kg}^{-1}\,\text{Pa}^{-1}) = -27.672 + 4415.8/(T/K)$$

The standard error of estimate of k_H by this equation is 2 %. This equation generates RECOMMENDED data in the temperature range 273–313 K.

Solute		
Formula	Name	CAS Registry Number
CHF_3	Trifluoromethane	75-46-7
Solvent: Water		

Wilhelm et al. [6] presented critically evaluated smoothed data for the solubility of this gas at a partial pressure of 101.3 kPa in water in the range $298 < T/K < 348$. The data were represented by

$$(RT/\text{cal mol}^{-1}) \ln x = 6386.96 - 37.9627(T/K)$$

or

$$\ln x = -19.1034 + 3214.02/(T/K)$$

The expression for k_H is

$$\ln(k_H/\text{mol kg}^{-1}\,\text{Pa}^{-1}) = -26.615 + 3214.4/(T/K)$$

$t/°C$	T/K	Solubility, x	$k_H/\text{mol kg}^{-1}\,\text{Pa}^{-1}$
25	298.2	2.427E−4	1.33E−7
30	303.2	2.032E−4	1.11E−7
35	308.2	1.711E−4	0.936E−7
40	313.2	1.448E−4	0.793E−7
45	318.2	1.233E−4	0.675E−7
50	323.2	1.054E−4	0.577E−7
55	328.2	0.906E−4	0.496E−7
60	333.2	0.782E−4	0.428E−7
65	338.2	0.678E−4	0.371E−7
70	343.2	0.590E−4	0.323E−7
75	348.2	0.516E−4	0.282E−7

In these calculations, it has been assumed that the concentration of the solute was within the region where Henry's law is valid. These are RECOMMENDED data in the temperature range 298–348 K.

Solute		
Formula	Name	CAS Registry Number
CHI$_3$	Triiodomethane	75-47-8
Solvent: Water		

The solubility of this solid in water at 30 °C was reported by van Arkel and Vies [26]. The datum is 3.0E−4 m ($x = 5.4$E−6). Vapour pressure data for this solid are not available; the value for the supercooled liquid at 30 °C was estimated using Predict II to be 69.5 Pa [24]. A value for k_H was calculated by the simplified VP/AS method, using $\Delta_{fus}S° = 39\,\text{J mol}^{-1}\,\text{K}^{-1}$. Thus $k_H = 6.22$E−6 mol kg^{-1} Pa^{-1}.

A value for this quantity at 25 °C was published by Yaws et al. [38], who gave no information on its origin; it was probably calculated from water solubility and vapour pressure. The datum was reported as 161.7 atm, which is equivalent to 3.39E−6 mol kg^{-1} Pa^{-1}.

An order-of-magnitude estimate of k_H for this solid is 4.8E−6 mol kg^{-1} Pa^{-1}.

Solute		
Formula	Name	CAS Registry Number
CH_2Br_2	Dibromomethane	74-95-3
Solvent: Water		

Horvath and Getzen [30] presented critically evaluated smoothed data for the solubility of this liquid in water in the range $273 < T/\text{K} < 323$. The data were represented by

$$w = 0.101890 - 6.336789\text{E}{-}4(T/\text{K}) + 1.10906\text{E}{-}6(T/\text{K})^2$$

where w is mass fraction. The mole fraction solubility is thus

$$x = 1.05611\text{E}{-}2 - 6.56821\text{E}{-}5(T/\text{K}) + 1.14956\text{E}{-}7(T/\text{K})^2$$

The k_H were calculated from this equation and the vapour pressure from Gmehling *et al.* [34]:

$t/°C$	T/K	Solubility, x	$k_H/\text{mol kg}^{-1}\,\text{Pa}^{-1}$
0	273.2	1.197E−3	4.87E−5
5	278.2	1.185E−3	3.63E−5
10	283.2	1.180E−3	2.75E−5
15	288.2	1.180E−3	2.10E−5
20	293.2	1.185E−3	1.63E−5
25	298.2	1.197E−3	1.28E−5
30	303.2	1.214E−3	1.02E−5
35	308.2	1.237E−3	0.815E−5
40	313.2	1.266E−3	0.662E−5
45	318.2	1.300E−3	0.543E−5
50	323.2	1.340E−3	0.450E−5

In these calculations, it has been assumed that the concentration of the solute was within the region where Henry's law is valid. These data may be represented by the equation

$$\ln(k_H/\text{mol kg}^{-1}\,\text{Pa}^{-1}) = -25.363 + 4208.5/(T/\text{K})$$

The standard error of estimate of k_H by this equation is 1%. This equation generates RECOMMENDED data in the temperature range 273–323 K.

Solute		
Formula	Name	CAS Registry Number
CH$_2$BrCl	Bromochloromethane	74-97-5
Solvent: Water		

Horvath and Getzen [30] evaluated available data for the solubility of this liquid in water; they recommended two data. The k_H were calculated from the mole fraction solubilities and from vapour pressure data of Gmehling *et al.* [34]:

Original measurements				
Ref.	$t/°C$	T/K	Solubility, x	$k_H/\text{mol kg}^{-1}\,\text{Pa}^{-1}$
23	20	293.2	1.16E-2	61.0
39	25	298.2	2.3E-3	9.70R

In these calculations, it has been assumed that the concentration of the solute was within the region where Henry's law is valid. The datum at 25 °C is recommended because the saturated solution was prepared by the generator method.

Solute		
Formula	Name	CAS Registry Number
CH$_2$Cl$_2$	Dichloromethane	75-09-2
Solvent: Water		

Horvath and Getzen [30] presented critically evaluated smoothed data for the solubility of this liquid in the range $273 < T/K < 308$. The data were represented by

$$w = 0.58838 - 3.8224\text{E}{-}3(T/K) + 6.3928\text{E}{-}6(T/K)^2$$

where w is mass fraction. The corresponding mole fraction solubility is

$$x = 0.124847 - 8.11067\text{E}{-}4(T/K) + 1.35648\text{E}{-}6(T/K)^2$$

and the k_H were calculated with this equation and vapour pressure from Gmehling *et al.* [34]:

$t/°C$	T/K	Solubility, x	$k_H/\text{mol kg}^{-1}\,\text{Pa}^{-1}$
0	273.2	4.512E−3	1.30E−5
5	278.2	4.196E−3	0.952E−5
10	283.2	3.948E−3	0.712E−5
15	288.2	3.767E−3	0.545E−5
20	293.2	3.654E−3	0.428E−5
25	298.2	3.610E−3	0.345E−5
30	303.2	3.632E−3	0.285E−5
35	308.2	3.723E−3	0.242E−5

In these calculations, it has been assumed that the concentration of the solute was within the region where Henry's law is valid. These data may be represented by the equation

$$\ln(k_H/\text{mol kg}^{-1}\,\text{Pa}^{-1}) = -26.183 + 4064.2/(T/K)$$

The standard error of estimate of k_H by this equation is 3 % in the temperature range 273–308 K. This equation generates RECOMMENDED values.

Solute		
Formula	Name	CAS Registry Number
CH_2I_2	Diiodomethane	75-11-6
Solvent: Water		

There are two estimates for k_H of this liquid at 25 °C:

Original measurements					
Ref.	$t/°C$	T/K	Limiting solubility ratio	Method	$k_H/\text{mol kg}^{-1}\,\text{Pa}^{-1}$
40	25	298.2	$\gamma^\infty = 1.21E4$	VP/AS	2.54E−5
38	25	298.2	$(1/k_H)/\text{atm} = 19.68$	VP/AS	2.79E−5

In these calculations, it has been assumed that the concentration of the solute was within the region where Henry's law is valid. An order-of-magnitude estimate of $k_H/\text{mol kg}^{-1}\,\text{Pa}^{-1}$ is thus 2.67E−5.

Solute		
Formula	Name	CAS Registry Number
CH_3Br	Bromomethane	74-83-9
Solvent: Water		

Wilhelm *et al.* [6] presented critically evaluated smoothed data for the solubility of this gas at a partial pressure of 101.3 kPa in water in the range $278 < T/K < 353$. The data were represented by the equation

$$(R/\text{cal mol}^{-1}\ \text{K}^{-1}) \ln x = -325.392 + 19159.9/(T/\text{K}) + 43.797 \ln(T/\text{K})$$

or

$$\ln x = -163.742 + 9641.56/(T/\text{K}) + 22.0393 \ln(T/\text{K})$$

The standard deviation in x from this equation is 5 % [6]. The k_H were calculated from this equation:

$t/°C$	T/K	Solubility, x	$k_H/\text{mol kg}^{-1}\ \text{Pa}^-$
5	278.2	64.84E−4	3.55E−6
10	283.2	52.06E−4	2.85E−6
15	288.2	42.42E−4	2.32E−6
20	293.2	35.02E−4	1.92E−6
25	298.2	28.28E−4	1.60E−6
30	303.2	24.78E−4	1.36E−6
35	308.2	21.21E−4	1.16E−6
40	313.2	18.35E−4	1.01E−6
45	318.2	16.04E−4	0.879E−6
50	323.2	14.15E−4	0.775E−6
55	328.2	12.60E−4	0.690E−6

In these calculations, it has been assumed that the concentration of the solute was within the region where Henry's law is valid. These data may be represented by the equation

$$\ln(k_H/\text{mol kg}^{-1}\ \text{Pa}^{-1}) = -171.252 + 9641.6/(T/\text{K})$$

The standard error of estimate of k_H by this equation is 5 %. These are RECOMMENDED values in the temperature range 278–353 K.

Solute		
Formula	Name	CAS Registry Number
CH_3Cl	Chloromethane	74-87-3
Solvent: Water		

Wilhelm *et al.* [6] presented critically evaluated smoothed data for the solubility of this gas at a partial pressure of 101.3 kPa in water in the range $277 < T/K < 353$. The data were represented by the equation

$$(R/\text{cal mol}^{-1}\,\text{K}^{-1})\ln x = -342.796 + 19412.2/(T/\text{K}) + 46.5481\ln(T/\text{K})$$

or

$$\ln x = -172.519 + 9769.6/(T/\text{K}) + 23.4263\ln(T/\text{K})$$

The standard deviation in x calculated by this equation is 7 % [6]. The equation for k_H is

$$\ln(k_H/\text{mol kg}^{-1}\,\text{Pa}^{-1}) = -180.029 + 9769.6/(T/\text{K}) + 23.4263\ln(T/\text{K})$$

This equation generates RECOMMENDED data in the temperature range 278–353 K.

$t/°C$	T/K	Solubility, x	$k_H/\text{mol kg}^{-1}\,\text{Pa}^{-1}$
5	278.2	39.94E−4	21.3E−7
10	283.2	31.79E−4	17.4E−7
15	288.2	26.33E−4	14.4E−7
20	293.2	22.09E−4	12.1E−7
25	298.2	18.78E−4	10.3E−7
30	303.2	16.15E−4	8.84E−7
35	308.2	14.04E−4	7.69E−7
40	313.2	12.34E−4	6.76E−7
45	318.2	10.95E−4	6.00E−7
50	323.2	9.815E−4	5.38E−7
55	328.2	8.873E−4	4.86E−7
60	333.2	8.089E−4	4.43E−7
65	338.2	7.432E−4	4.07E−7
70	343.2	6.881E−4	3.77E−7
75	348.2	6.416E−4	3.51E−7
80	353.2	6.022E−4	3.30E−7

In these calculations, it has been assumed that the concentration of the solute was within the region where Henry's law is valid.

Solute		
Formula	Name	CAS Registry Number
CH_3F	Fluoromethane	593-53-3
Solvent: Water		

Wilhelm *et al.* [6] presented critically evaluated smoothed data for the solubility of this gas in water at a partial pressure of 101.3 kPa in the temperature range $273 < T/K < 353$. The data were represented by

$$(R/cal\,mol^{-1}\,K^{-1})\ln x = -270.079 + 15103.1/(T/K) + 36.1231\ln(T/K)$$

or

$$\ln x = -135.908 + 7600.11/(T/K) + 18.1777\ln(T/K)$$

The standard deviation in x given by this equation is 1 % [6] and the equation for k_H is

$$\ln(k_H/mol\,kg^{-1}\,Pa^{-1}) = -143.418 + 7600.11/(T/K) + 18.1777\ln(T/K)$$

$t/°C$	T/K	Solubility, x	$k_H/mol\,kg^{-1}\,Pa^{-1}$
0	273.2	22.28E−4	12.2E−7
5	278.2	18.79E−4	10.3E−7
10	283.2	16.03E−4	8.78E−7
15	288.2	13.83E−4	7.57E−7
20	293.2	12.06E−4	6.60E−7
25	298.2	10.62E−4	5.81E−7
30	303.2	9.433E−4	5.17E−7
35	308.2	8.455E−4	4.63E−7
40	313.5	7.642E−4	4.18E−7
45	318.2	6.960E−4	3.81E−7
50	323.2	6.385E−4	3.50E−7
55	328.2	5.899E−4	3.23E−7
60	333.2	5.485E−4	3.00E−7
65	338.3	5.132E−4	2.81E−7
70	343.2	4.830E−4	2.64E−7
75	348.2	4.571E−4	2.50E−7
80	353.2	4.348E−4	2.38E−7

In these calculations, it has been assumed that the concentration of the solute was within the region where Henry's law is valid. These are RECOMMENDED data for the temperature range 273–353 K.

Solute		
Formula	Name	CAS Registry Number
CH_3I	Iodomethane	74-88-4
Solvent: Water		

Horvath and Getzen [30] presented critically evaluated smoothed data for the solubility of this liquid in water in the range $273 < T/K < 313$. The data were represented by

$$w = 0.190036 - 1.1723E{-}3(T/K) + 1.95356E{-}6(T/K)^2$$

where w is mass fraction. The mole fraction solubility is given by

$$x = 2.41258E{-}2 - 1.48828E{-}4(T/K) + 2.48012E{-}7(T/K)^2$$

The k_H were calculated from this equation and vapour pressure from Gmehling et al. [34]:

$t/°C$	T/K	Solubility, x	$k_H/\mathrm{mol\,kg^{-1}\,Pa^{-1}}$
0	273.2	1.978E−3	5.91E−6
5	278.2	1.917E−3	4.55E−6
10	283.2	1.869E−3	3.55E−6
15	288.2	1.834E−3	2.82E−6
20	293.2	1.810E−3	2.27E−6
25	298.2	1.799E−3	1.85E−6
30	303.2	1.801E−3	1.53E−6
35	308.2	1.815E−3	1.29E−6
40	313.2	1.841E−3	1.10E−6

In these calculations, it has been assumed that the concentration of the solute was within the region where Henry's law is valid. These data may be represented by the equation

$$\ln(k_H/\mathrm{mol\,kg^{-1}\,Pa^{-1}}) = -25.273 + 3605.9/(T/K)$$

The standard error of estimate of k_H by this equation is 2 %. This equation generates RECOMMENDED values in the temperature range 273–313 K.

Solute		
Formula	Name	CAS Registry Number
CH_3NO_2	Methyl nitrate	598-58-3
Solvent: Water		

Kames and Schurath [41] reported measurements, by the direct method, of k_H of this liquid in water in the range $273 < T/K < 298$. The data were presented only as the equation

$$\ln(k_H/\text{M atm}^{-1}) = -0.1520 + 4.74/(T/K)$$

Original measurements			
$t/°C$	T/K	Limiting solubility ratio	$k_H/\text{mol kg}^{-1}\,\text{Pa}^{-1}$
0	273.2	$k_H/\text{M atm}^{-1} = 8.58$	8.47E−5
5	278.2	$k_H/\text{M atm}^{-1} = 6.28$	6.20E−5
10	283.2	$k_H/\text{M atm}^{-1} = 4.65$	4.59E−5
15	288.2	$k_H/\text{M atm}^{-1} = 3.78$	3.73E−5
20	293.2	$k_H/\text{M atm}^{-1} = 2.63$	2.60E−5
25	298.2	$k_H/\text{M atm}^{-1} = 2.00$	1.97E−5

In these calculations, it has been assumed that the concentration of the solute was within the region where Henry's law is valid. These data are RECOMMENDED in the temperature range 273–298 K.

Solute		
Formula	Name	CAS Registry Number
CH_4	Methane	74-82-8
Solvent: Water		

Clever and Young [42] presented critically evaluated smoothed data for the solubility of this gas in water at a gas partial pressure of 101.3 kPa in the range $273 < T/K < 328$. The data were represented by the equation

$$\ln x = -99.14188 + 132.821/(T/100\,\text{K}) + 51.91445 \ln(T/100\,\text{K})$$
$$- 4.25831\,(T/100\,\text{K})$$

The standard deviation in x given by this equation is 1.5 %. Thus

$$\ln(k_H/\text{mol kg}^{-1}\,\text{Pa}^{-1}) = -106.651 + 132.821/(T/100\,\text{K})$$
$$+ 51.91445\ln(T/100\,\text{K}) - 4.25831\,(T/100\,\text{K})$$

$t/°C$	T/K	Solubility, x	$k_H/\text{mol kg}^{-1}\,\text{Pa}^{-1}$
0	273.3	4.6666E−5	2.55E−8
5	278.2	4.0221E−5	2.20E−8
10	283.2	3.5192E−5	1.93E−8
15	288.2	3.1224E−5	1.71E−8
20	293.2	2.8062E−5	1.54E−8
25	298.2	2.5523E−5	1.40E−8
30	303.2	2.3469E−5	1.29E−8
35	308.2	2.1802E−5	1.19E−8
40	313.2	2.0445E−5	1.12E−8
45	318.2	1.9340E−5	1.06E−8
50	323.2	1.8442E−5	1.01E−8
55	328.2	1.7717E−5	0.964E−8

In these calculations, it has been assumed that the concentration of the solute was within the region where Henry's law is valid. These are RECOMMENDED data for the temperature range 273–328 K.

Solute		
Formula	Name	CAS Registry Number
CH_4O	Methanol	67-56-1
Solvent: Water		

Available experimental data of k_H for this liquid are given in the table.

Original measurements					$k_H/\text{mol kg}^{-1}\,\text{Pa}^{-1}$
Ref.	$t/°C$	T/K	Limiting solubility ratio	Method	
43	0	273.2	$k_{AW} = 4.00E−5$	DIRECT	11.3E−3
44	20	293.2	$\gamma^\infty = 2.69$	GC	1.59E−3
45	20	293.2	$\gamma^\infty = 1.68$	DIRECT	2.54E−3
46	24	297.2	$\gamma^\infty = 2.12$	GC	1.63E−3
47	25	298.2	$\gamma^\infty = 1.64$	DIRECT	2.01E−3

Ref.	$t/°C$	T/K	Limiting solubility ratio	Method	k_H/mol $kg^{-1} Pa^{-1}$
			Original measurements		
48	25	298.2	$\gamma^\infty = 1.46$	INDIRECT	2.26E−3
49	25	298.2	$k_{AW} = 3.33E−4$	PURGE	1.21E−3
50	25	298.2	$\gamma^\infty = 1.70$	DIRECT	1.94E−3
51	25	298.2	$\gamma^\infty = 1.64$	DIRECT	2.01E−3
52	25	298.2	$\gamma^\infty = 1.74$	GC	1.90E−3
53	25	298.2	$\gamma^\infty = 1.65$	GC	2.00E−3
54	25	298.2	$(1/k_H/mmHg) = 187$	HEADSPACE	2.23E−3
44	30	303.2	$\gamma^\infty = 2.53$	GC	1.00E−3
45	30	303.2	$\gamma^\infty = 1.77$	DIRECT	1.44E−3
44	40	313.2	$\gamma^\infty = 2.30$	GC	0.682E−3
45	40	313.2	$\gamma^\infty = 1.92$	DIRECT	0.817E−3
55	40	313.2	$k_{AW} = 4.7E−4$	HEADSPACE	0.818E−3
56	45	318.2	$\gamma^\infty = 1.46$	EBULLIOMETRY	0.855E−3
57	50	323.2	$\gamma^\infty = 1.814$	GC	0.551E−3
55	50	323.2	$k_{AW} = 6.5E−4$	EADSPACE	0.573E−3
56	65	338.2	$\gamma^\infty = 1.59$	EBULLIOMETRY	0.339E−3
55	70	343.2	$k_{AW} = 1.7E−3$	HEADSPACE	0.212E−3
56	75	348.2	$\gamma^\infty = 1.52$	EBULLIOMETRY	0.242E−3
55	80	353.2	$k_{AW} = 2.67E−3$	HEADSPACE	0.128E−3
56	85	358.2	$\gamma^\infty = 1.78$	EBULLIOMETRY	0.145E−3
58	100	373.2	$\gamma^\infty = 2.36$	EBULLIMETRY	0.666E−4

In these calculations, it has been assumed that the concentration of the solute was within the region where Henry's law is valid. These data may be represented by the equation

$$\ln(k_H/mol\,kg^{-1}\,Pa^{-1}) = -21.446 + 4500.3/(T/K)$$

The standard error of estimate of k_H by this equation is 18 % in the temperature range 273–373 K. This equation generates RECOMMENDED values.

Solute		
Formula	Name	CAS Registry Number
CH_4S	Methanethiol	74-93-1
Solvent: Water		

There are three reports for k_H of this liquid in water:

			Original measurements		
Ref.	$t/°C$	T/K	Limiting solubility ratio	Method	$k_H/mol\,kg^{-1}\,Pa^{-1}$
40	25	298.2	$\gamma^\infty = 110$	VP/AS	2.51E−6
59	25	298.2	$(k_H/atm^{-1}) = 3.66E{-}3$	DIRECT	2.01E−6
60	25	298.2	$k_{WA} = 9.88$	HEADSPACE	3.99E−6

In these calculations, it has been assumed that the concentration of the solute was within the region where Henry's law is valid.

12 HENRY'S LAW CONSTANTS FOR ORGANIC COMPOUNDS (C₂–C₉)

Solute		
Formula	Name	CAS Registry Number
C_2Cl_4	Tetrachloroethene	127-18-4
Solvent: Water		

There are 43 measured data in the range 2–80 °C for k_H of this liquid in water. When all the data are put on a common plot, five data sets indicate that the temperature coefficient of k_H is negative, and two sets indicate a positive dependence. There is no apparent correlation with method. The table below contains the data for the five sets showing a negative temperature dependence.

			Original measurements		k_H/mol
Ref.	$t/°C$	T/K	Limiting solubility ratio	Method	$kg^{-1}\,Pa^{-1}$
61	10	283.2	$(1/k_H)/atm\,m^3\,mol^{-1} = 6.82E{-}3$	EPICS	14.5E−7
62	15	288.2	$k_{AW} = 0.424$	HEADSPACE	9.85E−7
63	15	288.2	$(1/k_H)/kPa\,m^3\,mol^{-1} = 1.0335$	EPICS	9.68E−7
61	18	291.2	$(1/k_H)/atm\,m^3\,mol^{-1} = 1.17E{-}2$	EPICS	8.44E−7
62	20	293.2	$k_{AW} = 0.513$	HEADSPACE	8.00E−7
65	20	293.2	$(1/k_H)/MPa = 65.5$	PURGE	8.48E−7
65	22	295.2	$k_{AW} = 0.526$	DIRECT	7.75E−7
66	25	298.2	$(1/k_H)/atm\,m^3\,mol^{-1} = 1.7E{-}2$	HEADSPACE	5.81E−7
62	25	298.2	$k_{AW} = 0.690$	HEADSPACE	5.81E−7
61	25	298.2	$(1/k_H)/atm\,m^3\,mol^{-1} = 1.77E{-}2$	EPICS	5.58E−7
66	30	303.2	$(1/k_H)/atm\,m^3\,mol^{-1} = 2.31E{-}2$	HEADSPACE	4.27E−7

Ref.	$t/°C$	T/K	Original measurements		k_H/mol
			Limiting solubility ratio	Method	$\text{kg}^{-1}\,\text{Pa}^{-1}$
62	30	303.2	$k_{AW} = 0.857$	HEADSPACE	4.63E−7
62	30	303.2	$k_{AW} = 1.059$	HEADSPACE	3.69E−7
61	35	308.2	$(1/k_H)/\text{atm m}^3\,\text{mol}^{-1} = 2.82\text{E}{-}2$	EPICS	3.50E−7
63	35	308.2	$(1/k_H)/\text{kPa m}^3\,\text{mol}^{-1} = 2.653$	EPICS	3.77E−7
66	40	313.2	$(1/k_H)/\text{atm m}^3\,\text{mol}^{-1} = 3.03\text{E}{-}2$	HEADSPACE	3.26E−7
67	40	313.2	$k_{AW} = 1.303$	EPICS	2.95E−7
62	40	313.2	$k_{AW} = 1.261$	HEADSPACE	3.05E−7
66	45	318.2	$(1/k_H)/\text{atm m}^3\,\text{mol}^{-1} = 3.81\text{E}{-}2$	HEADSPACE	2.59E−7
62	45	318.2	$k_{AW} = 1.553$	HEADSPACE	2.44E−7
37	45	318.2	$k_{AW} = 1.30$	EPICS	2.91E−7
67	50	323.2	$k_{AW} = 1.773$	EPICS	2.10E−7
67	60	333.2	$k_{AW} = 2.491$	EPICS	1.45E−7

In these calculations, it has been assumed that the concentration of the solute was within the region where Henry's law is valid. These data may be represented by the equation

$$\ln(k_H/\text{mol kg}^{-1}\,\text{Pa}^{-1}) = -28.395 + 4196.5/(T/K)$$

The standard error of estimate of k_H by this equation is 6 %. This equation generates RECOMMENDED data in the range 283–333 K.

Horvath et al. [68] presented critically evaluated smoothed data for the solubility of this liquid in water in the range $273 < T/K < 343$. The data were represented by

$$w = 0.26479\text{E}{-}2 - 1.5847\text{E}{-}5(T/K) + 2.4477\text{E}{-}8(T/K)^2$$

where w is mass fraction. The mole fraction solubility is

$$x = 2.8772\text{E}{-}4 - 1.6828\text{E}{-}6(T/K) + 2.6597\text{E}{-}9(T/K)^2$$

The k_H data were calculated with this equation and solute vapour pressure data from Gmehling et al. [34]:

$t/°C$	T/K	Solubility, x	$k_H/\text{mol kg}^{-1}\,\text{Pa}^{-1}$
0	273.2	2.649E−5	2.39E−6
5	278.2	2.541E−5	1.69E−6
10	283.2	2.447E−5	1.22E−6
15	288.2	2.365E−5	8.80E−7
20	293.2	2.297E−5	6.57E−7
25	298.2	2.242E−5	4.94E−7
30	303.2	2.200E−5	3.76E−7

$t/°C$	T/K	Solubility, x	$k_H/\text{mol kg}^{-1}\,\text{Pa}^{-1}$
35	308.2	2.172E−5	2.91E−7
40	313.2	2.157E−5	2.28E−7
45	318.2	2.155E−5	1.81E−7
50	323.2	2.167E−5	1.46E−7
55	328.2	2.192E−5	1.19E−7
60	333.2	2.230E−5	9.86E−8
65	338.2	2.281E−5	8.25E−8
70	343.2	2.346E−5	6.98E−8

In these calculations, it has been assumed that the concentration of the solute was within the region where Henry's law is valid. In both tables, calculations were performed with solute vapour pressure data from Gmehling $et\ al.$ [34]. It is seen that k_H calculated from solubility agrees fairly, though not exactly, with experimental measurement data.

Solute		
Formula	Name	CAS Registry Number
$C_2HBrClF_3$	Halothane	151-67-7
Solvent: Water		

Available experimental measurements of k_H of this gas are as follows:

Original measurements					
Ref.	$t/°C$	T/K	Limiting solubility ratio	Method	$k_H/\text{mol kg}^{-1}\,\text{Pa}^{-1}$
69	4	277.2	Bunsen = 4.28	DIRECT	18.8E−7
69	10	283.2	Bunsen = 2.92	DIRECT	12.9E−7
69	20	293.2	Bunsen = 1.60	DIRECT	7.05E−7
69	25	298.2	Bunsen = 1.20	DIRECT	5.28E−7
70	25	298.2	$k_{WA} = 1.47$	DIRECT	5.83E−7
69	30	303.2	Bunsen = 0.92	DIRECT	4.05E−7
70	30	303.2	$k_{WA} = 1.18$	DIRECT	4.68E−7
70	37	310.2	$k_{WA} = 0.874$	DIRECT	3.39E−7
71	37	310.2	Ostwald = 0.80	CRIT EVAL	3.10E−7

In these calculations, it has been assumed that the concentration of the solute was within the region where Henry's law is valid. These data may be represented by the equation

$$\ln(k_H/\text{mol kg}^{-1}\,\text{Pa}^{-1}) = -30.0893 + 4679.4/(T/K)$$

The standard error of estimate of k_H by this equation is 7%. This equation generates RECOMMENDED data in the temperature range 277–310 K.

Solute		
Formula	Name	CAS Registry Number
C_2HBrF_4	Teflurane	124-72-1
Solvent: Water		

There is one estimate [71] of k_H of this gas in water. In a critical evaluation of existing data at 37 °C, a value for the Ostwald coefficient of 0.32 was found. Thus $k_H = 1.24E{-}7 \, \text{mol} \, \text{kg}^{-1} \, \text{Pa}^{-1}$; this a RECOMMENDED value.

Solute		
Formula	Name	CAS Registry Number
C_2HCl_3	Trichloroethene	79-01-6
Solvent: Water		

Available experimental measurements of k_H of this liquid in water are as follows:

			Original measurements		k_H/mol
Ref.	t/°C	T/K	Limiting solubility ratio	Method	$\text{kg}^{-1} \, \text{Pa}^{-1}$
61	10	283.2	$(1/k_H)/\text{atm} \, \text{m}^3 \, \text{mol}^{-1} = 3.78E{-}3$	EPICS	2.61E−6
61	18	291.2	$(1/k_H)/\text{atm} \, \text{m}^3 \, \text{mol}^{-1} = 6.32E{-}3$	EPICS	1.56E−6
36	20	293.2	$(1/k_H)/\text{atm} \, \text{m}^3 \, \text{mol}^{-1} = 7.0E{-}3$	GC	1.41E−6
72	20	293.2	$(1/k_H)/\text{Pa} = 4.313E7$	GC	1.29E−6
73	20	293.2	$\gamma^\infty = 5450$	DIRECT	1.42E−6
74	20	293.2	$(1/k_H)/\text{kPa} = 42.E3$	DIRECT	1.32E−6
64	20	293.2	$(1/k_H)/\text{MPa} = 42.3$	PURGE	1.31E−6
75	20	293.2	$\gamma^\infty = 5410$	DIRECT	1.43E−6
76	25	298.2	$k_{AW} = 0.473$	DIRECT	8.53E−7
66	25	298.2	$(1/k_H)/\text{atm} \, \text{m}^3 \, \text{mol}^{-1} = 1.03E{-}2$	HEADSPACE	9.58E−7
61	25	298.2	$(1/k_H)/\text{atm} \, \text{m}^3 \, \text{mol}^{-1} = 9.58E{-}3$	EPICS	1.03E−6
63	25	298.2	$(k_H)/\text{kPa} \, \text{m}^3 \, \text{mol}^{-1} = 1.144$	PURGE	8.74E−7
77	25	298.2	$\gamma^\infty = 5358$	VP/AS	1.12E−6

Ref.	$t/°C$	T/K	Original measurements Limiting solubility ratio	Method	$k_H/\text{mol kg}^{-1}\text{Pa}^{-1}$
36	30	303.2	$(1/k_H)/\text{atm m}^3\,\text{mol}^{-1} = 1.14\text{E}{-}2$	GC	8.66E−7
72	30	303.2	$(1/k_H)/\text{Pa} = 7.64\text{E}7$	GC	7.27E−7
66	30	303.2	$(1/k_H)/\text{atm m}^3\,\text{mol}^{-1} = 1.31\text{E}{-}2$	HEADSPACE	7.53E−7
67	30	303.2	$k_{AW} = 0.427$	EPICS	9.29E−7
73	30	303.2	$\gamma^\infty = 6061$	DIRECT	7.83E−7
74	30	303.2	$(1/k_H)/\text{kPa} = 64.\text{E}3$	DIRECT	8.67E−7
36	30	303.2	$\gamma^\infty = 5034$	GC	9.42E−7
61	35	308.2	$(1/k_H)/\text{atm m}^3\,\text{mol}^{-1} = 1.49\text{E}{-}2$	EPICS	6.62E−7
63	35	308.2	$1/k_H)/\text{kPa m}^3\,\text{mol}^{-1} = 1.6739$	PURGE	5.97E−7
36	40	313.2	$(1/k_H)/\text{atm m}^3\,\text{mol}^{-1} = 1.73\text{E}{-}2$	GC	5.70E−7
72	40	313.2	$(1/k_H)/\text{Pa} = 1.1532\text{E}8$	GC	4.81E−7
66	40	313.2	$(1/k_H)/\text{atm m}^3\,\text{mol}^{-1} = 1.65\text{E}{-}2$	HEADSPACE	5.98E−7
67	40	313.2	$k_{AW} = 0.693$	EPICS	5.54E−7
73	40	313.2	$\gamma^\infty = 5943$	DIRECT	5.05E−7
74	40	313.2	$(1/k_H)/\text{kPa} = 106.\text{E}3$	DIRECT	5.23E−7
36	40	313.2	$\gamma^\infty = 4973$	GC	6.03E−7
72	50	323.2	$(1/k_H)/\text{Pa} = 1.7073\text{E}8$	GC	3.25E−7
66	50	323.2	$(1/k_H)/\text{atm m}^3\,\text{mol}^{-1} = 2.63\text{E}{-}2$	HEADSPACE	3.75E−7
67	50	323.2	$k_{AW} = 0.922$	EPICS	4.04E−7
73	50	323.2	$\gamma^\infty = 5893$	DIRECT	3.36E−7
67	60	333.2	$k_{AW} = 1.273$	EPICS	2.83E−7

In these calculations, solute vapour pressure data were taken from Gmehling et al. [34]; it has been assumed that the concentration of the solute was within the region where Henry's law is valid. These data may be represented by the equation

$$\ln(k_H/\text{mol kg}^{-1}\,\text{Pa}^{-1}) = -28.051 + 4259.1/(T/K)$$

The standard error of estimate of k_H by this equation is 16%. This equation generates RECOMMENDED data in the range 283–333 K.

Horvath et al. [68] presented critically evaluated smoothed data for the solubility of this liquid in water in the range $273 < T/K < 333$. The data were represented by

$$w = 1.4049\text{E}{-}2 - 8.2223\text{E}{-}5(T/K) + 1.3218\text{E}{-}7(T/K)^2$$

where w is mass fraction. The mole fraction solubility is

$$x = 1.9262\text{E}{-}3 - 1.1273\text{E}{-}5(T/K) + 1.8123\text{E}{-}8(T/K)^2$$

The k_H data were calculated from this equation and solute vapour pressure data of Gmehling et al. [34].

$t/°C$	T/K	Solubility, x	$k_H/\text{mol kg}^{-1}\,\text{Pa}^{-1}$
0	273.2	1.991E−4	4.85E−6
5	278.2	1.927E−4	3.45E−6
10	283.2	1.872E−4	2.50E−6
15	288.2	1.826E−4	1.85E−6
20	293.2	1.789E−4	1.39E−6
25	298.2	1.762E−4	1.06E−6
30	303.2	1.743E−4	8.24E−7
35	308.2	1.733E−4	6.50E−7
40	313.2	1.733E−4	5.21E−7
45	318.2	1.741E−4	4.23E−7
50	323.2	1.759E−4	3.49E−7
55	328.2	1.785E−4	2.92E−7
60	333.2	1.821E−4	2.47E−7

In these calculations, it has been assumed that the concentration of the solute was within the region where Henry's law is valid. It is seen that k_H calculated from solubility agrees well with experimental values.

Solute		
Formula	Name	CAS Registry Number
C_2H_2	Ethyne	74-86-2
Solvent: Water		

Wilhelm *et al.* [6] presented critically evaluated smoothed data for the solubility of this gas in water at a partial pressure of gas of 101.3 kPa in the range $274 < T/K < 343$. The data were represented by

$$(R/\text{cal mol}^{-1}\,\text{K}^{-1})\ln x = -311.014 + 16215.8/(T/K) + 42.5305\ln(T/K)$$

or

$$\ln x = -156.550 + 8160.0/(T/K) + 21.4020\ln(T/K)$$

and hence

$$\ln(k_H/\text{mol kg}^{-1}\,\text{Pa}^{-1}) = -164.060 + 8160.04/(T/K) + 21.4020\ln(T/K)$$

$t/°C$	T/K	Solubility, x	$k_H/\text{mol kg}^{-1}\,\text{Pa}^{-1}$
0	273.2	13.46E−4	7.37E−7
5	278.2	11.60E−4	6.35E−7

$t/°C$	T/K	Solubility, x	$k_H/\text{mol kg}^{-1}\,\text{Pa}^{-1}$
10	283.2	10.12E−4	5.54E−7
15	288.2	8.924E−4	4.89E−7
20	293.2	7.956E−4	4.36E−7
25	298.2	7.164E−4	3.92E−7
30	303.2	6.512E−4	3.57E−7
35	308.2	5.971E−4	3.27E−7
40	313.2	5.521E−4	3.02E−7
45	318.2	5.146E−4	2.82E−7
50	323.2	4.831E−4	2.65E−7
55	328.2	4.567E−4	2.50E−7
60	333.2	4.346E−4	2.38E−7
65	338.2	4.162E−4	2.28E−7
70	343.2	4.009E−4	2.20E−7

These data are RECOMMENDED.

Solute		
Formula	Name	CAS Registry Number
$C_2H_2Cl_2$	1,1-Dichloroethene	75-35-4
Solvent: Water		

Available experimental measurements of k_H of this liquid are as follows:

			Original measurements		
Ref.	$t/°C$	T/K	Limiting solubility ratio	Method	$k_H/\text{mol kg}^{-1}\,\text{Pa}^{-1}$
78	2	272.2	$k_{AW} = 0.379$	EPICS	11.5E−7
78	4	279.2	$k_{AW} = 0.438$	EPICS	9.84E−7
78	10	283.2	$k_{AW} = 0.547$	EPICS	7.77E−7
61	10	283.2	$(1/k_H)/\text{atm m}^3\,\text{mol}^{-1} = 1.27\text{E−}2$	EPICS	7.77E−7
78	18	291.2	$k_{AW} = 0.824$	EPICS	5.01E−7
61	18	291.2	$(1/k_H)/\text{atm m}^3\,\text{mol}^{-1} = 1.91\text{E−}2$	EPICS	5.17E−7
36	20	293.2	$\gamma^\infty = 1894$	GC	4.42E−7
36	20	293.2	$(1/k_H)/\text{atm m}^3\,\text{mol}^{-1} = 2.29\text{E−}2$	GC	4.31E−7
61	25	298.2	$(1/k_H)/\text{atm m}^3\,\text{mol}^{-1} = 2.61\text{E−}2$	EPICS	3.78E−7
78	25	298.2	$k_{AW} = 1.086$	EPICS	3.71E−7
78	30	303.2	$\gamma^\infty = 1930$	GC	3.00E−7

			Original measurements		
Ref.	$t/°C$	T/K	Limiting solubility ratio	Method	$k_H/\text{mol kg}^{-1}\text{Pa}^{-1}$
36	30	303.2	$(1/k_H)/\text{atm m}^3\text{mol}^{-1} = 3.37E{-}2$	GC	2.93E$-$7
61	35	308.2	$(1/k_H)/\text{atm m}^3\text{mol}^{-1} = 3.66E{-}2$	EPICS	2.70E$-$7
36	40	313.2	$\gamma^\infty = 1936$	GC	2.12E$-$7
36	40	313.2	$(1/k_H)/\text{atm m}^3\text{mol}^{-1} = 4.75E{-}2$	GC	2.08E$-$7

In these calculations, it has been assumed that the concentration of the solute was within the region where Henry's law is valid. These data may be represented by the equation

$$\ln(k_H/\text{mol kg}^{-1}\text{Pa}^{-1}) = -27.392 + 3775.2/(T/K)$$

The standard error of estimate of k_H by this equation is 5 %. This equation generates RECOMMENDED data in the temperature range 275–313 K.

Horvath et al. [68] presented critically evaluated smoothed data for the solubility of this liquid in water in the range $278 < T/K < 363$. The data were represented by the equation

$$w = 6.27413E{-}2 - 3.8257E{-}4(T/K) + 6.04607E{-}7(T/K)^2$$

where w is mass fraction. In mole fractions,

$$x = 1.16427E{-}2 - 7.0992E{-}5(T/K) + 1.12195E{-}7(T/K)^2$$

Data for k_H were calculated from this equation and from vapour pressure data [34]:

$t/°C$	T/K	Solubility, x	$k_H/\text{mol kg}^{-1}\text{Pa}^{-1}$
10	283.2	5.38E$-$4	6.68E$-$7
20	293.2	4.75E$-$4	3.96E$-$7
30	303.2	4.35E$-$4	2.50E$-$7
40	313.2	4.15E$-$4	1.70E$-$7
50	323.2	4.19E$-$4	1.25E$-$7

In these calculations, it has been assumed that the concentration of the solute was within the region where Henry's law is valid. The k_H data calculated from solubility are consistently lower than those measured directly; this may be due to uncertainty in the published critical evaluation [68].

Solute		
Formula	Name	CAS Registry Number
$C_2H_2Cl_2$	*cis*-1,2-Dichloroethene	156-59-2
Solvent: Water		

Available experimental measurements of k_H of this liquid in water are as follows:

			Original measurements		k_H/mol
Ref.	$t/°C$	T/K	Limiting solubility ratio	Method	$\text{kg}^{-1}\,\text{Pa}^{-1}$
61	10	283.2	$(1/k_H)/\text{atm m}^3\,\text{mol}^{-1} = 1.73\text{E}{-}3$	EPICS	5.74E−6
61	17.5	290.7	$(1/k_H)/\text{atm m}^3\,\text{mol}^{-1} = 2.65\text{E}{-}3$	EPICS	3.72E−6
64	20	293.2	$(1/k_H)/\text{MPa} = 17.5$	PURGE	3.17E−6
75	20	293.2	$\gamma^\infty = 856$	DIRECT	2.99E−6
74	20	293.2	$(1/k_H)/\text{kPa} = 18.4\text{E}3$	DIRECT	3.01E−6
36	20	293.2	$(1/k_H)/\text{atm m}^3\,\text{mol}^{-1} = 3.2\text{E}{-}3$	GC	3.08E−6
36	20	293.2	$\gamma^\infty = 819$	GC	3.12E−6
61	25	298.2	$(1/k_H)/\text{atm m}^3\,\text{mol}^{-1} = 4.08\text{E}{-}3$	EPICS	2.42E−6
74	30	303.2	$(1/k_H)/\text{kPa} = 2.88\text{E}4$	DIRECT	1.93E−6
36	30	303.2	$(1/k_H)/\text{atm m}^3\,\text{mol}^{-1} = 4.9\text{E}{-}3$	GC	2.01E−6
36	30	303.2	$\gamma^\infty = 803$	GC	2.08E−6
61	35	308.2	$(1/k_H)/\text{atm m}^3\,\text{mol}^{-1} = 5.4\text{E}{-}3$	EPICS	1.81E−6
74	40	313.2	$(1/k_H)/\text{kPa} = 4.17\text{E}4$	DIRECT	1.33E−6
36	40	313.2	$(1/k_H)/\text{atm m}^3\,\text{mol}^{-1} = 7.3\text{E}{-}3$	GC	1.35E−6
36	40	313.2	$\gamma^\infty = 807$	GC	1.40E−6

In these calculations, solute vapour pressure data from Gmehling *et al.* [34] were used; also, it has been assumed that the concentration of the solute was within the region where Henry's law is valid. These data may be represented by the equation

$$\ln(k_H/\text{mol kg}^{-1}\,\text{Pa}^{-1}) = -26.171 + 3961.6/(T/K)$$

The standard error of estimate of k_H by this equation is 4 %. This equation generates RECOMMENDED data in the range 283–313 K.

Horvath *et al.* [68] presented critically evaluated smoothed data for the solubility of this liquid in water in the range $283 < T/K < 313$. The data were represented by

$$w = 27.7353\text{E}{-}2 - 0.178316\text{E}{-}2(T/K) + 2.93282\text{E}{-}6(T/K)^2$$

where w is mass fraction. The mole fraction solubility is

$$x = 5.1567E-2 - 3.31532E-4(T/K) + 5.45282E-7(T/K)^2$$

The k_H data were calculated from this equation and solute vapour pressure from Gmehling $et\ al.$ [34]. In these calculations, it has been assumed that the concentration of the solute was within the region where Henry's law is valid.

$t/°C$	T/K	Solubility, x	$k_H/\text{mol kg}^{-1}\,\text{Pa}^{-1}$
10	283.2	1.410E−3	5.70E−6
15	288.2	1.310E−3	4.19E−6
20	293.2	1.238E−3	3.16E−6
25	298.2	1.192E−3	2.45E−6
30	303.2	1.174E−3	1.96E−6
35	308.2	1.184E−3	1.62E−6
40	313.2	1.220E−3	1.37E−6

It is seen that the k_H data calculated from solubility are in good agreement with those measured experimentally.

Solute		
Formula	Name	CAS Registry Number
$C_2H_2Cl_2$	$trans$-1,2-Dichloroethene	156-60-5
Solvent: Water		

Available experimental measurements of k_H of this liquid in water are as follows:

			Original measurements		k_H/mol
Ref.	$t/°C$	T/K	Limiting solubility ratio	Method	$\text{kg}^{-1}\,\text{Pa}^{-1}$
61	10	283.2	$(1/k_H)/\text{atm m}^3\,\text{mol}^{-1} = 4.2E-3$	EPICS	23.5E−7
61	17.5	290.65	$(1/k_H)/\text{atm m}^3\,\text{mol}^{-1} = 6.6E-3$	EPICS	15.0E−7
64	20	293.2	$(1/k_H)/\text{MPa} = 42.3$	PURGE	13.1E−7
75	20	293.2	$\gamma^\infty = 1200$	DIRECT	12.8E−7
36	20	293.2	$(1/k_H)/\text{atm m}^3\,\text{mol}^{-1} = 7.9E-3$	GC	12.5E−7
72	20	293.2	$(1/k_H)/\text{Pa} = 4.42E6$	GC	12.6E−7
36	20	293.2	$\gamma^\infty = 1202$	GC	12.8E−7
73	20	293.2	$\gamma^\infty = 1216$	DIRECT	12.6E−7
61	25	298.2	$(1/k_H)/\text{atm m}^3\,\text{mol}^{-1} = 9.8E-3$	EPICS	10.0E−7

			Original measurements		k_H/mol
Ref.	$t/°C$	T/K	Limiting solubility ratio	Method	$\text{kg}^{-1}\,\text{Pa}^{-1}$
79	26	299.2	$(1/k_H)/\text{kPa}\,\text{m}^3\,\text{mol}^{-1} = 1.02$	EPICS	9.80E−7
74	30	303.2	$(1/k_H)/\text{kPa} = 68.8\text{E}3$	DIRECT	8.07E−7
36	30	303.2	$(1/k_H)/\text{atm}\,\text{m}^3\,\text{mol}^{-1} = 1.18\text{E}{-2}$	GC	8.36E−7
36	30	303.2	$\gamma^\infty = 1202$	GC	12.8E−7
73	30	303.2	$\gamma^\infty = 1435$	DIRECT	7.20E−7
79	35	308.2	$(1/k_H)/\text{kPa}\,\text{m}^3\,\text{mol}^{-1} = 1.59$	EPICS	6.29E−7
61	35	308.2	$(1/k_H)/\text{atm}\,\text{m}^3\,\text{mol}^{-1} = 1.38\text{E}{-2}$	EPICS	7.60E−7
74	40	313.2	$(1/k_H)/\text{kPa} = 10.3\text{E}4$	DIRECT	5.39E−7
36	40	313.2	$(1/k_H)/\text{atm}\,\text{m}^3\,\text{mol}^{-1} = 1.77\text{E}{-2}$	GC	5.58E−7
72	40	313.2	$(1/k_H)/\text{Pa} = 11.6\text{E}7$	GC	4.80E−7
36	40	313.2	$\gamma^\infty = 1243$	GC	5.75E−7
73	40	313.2	$\gamma^\infty = 1477$	DIRECT	4.84E−7
79	46	319.2	$(1/k_H)/\text{kPa}\,\text{m}^3\,\text{mol}^{-1} = 2.09$	EPICS	4.78E−7
72	50	323.2	$(1/k_H)/\text{Pa} = 16.5\text{E}7$	GC	3.36E−7
73	50	323.2	$\gamma^\infty = 1509$	DIRECT	3.37E−7

In these calculations, solute vapour pressure data from Gmehling *et al.* [34] were used; in addition, it has been assumed that the concentration of the solute was within the region where Henry's law is valid. These data may be represented by the equation

$$\ln(k_H/\text{mol}\,\text{kg}^{-1}\,\text{Pa}^{-1}) = -27.814 + 4191.9/(T/K)$$

The standard error of estimate of k_H by this equation is 8 %. This equation generates RECOMMENDED data in the range 283–323 K.

Horvath *et al.* [68] presented critically evaluated smoothed data for the solubility of this liquid in water in the temperature range $283 < T/K < 313$. The data were represented by the equation

$$w = 7.803906\text{E}{-2} - 4.5457\text{E}{-4}(T/K) + 6.96755\text{E}{-7}(T/K)^2$$

where w is mass fraction. The mole fraction solubility is

$$x = 1.45093\text{E}{-2} - 8.4516\text{E}{-5}(T/K) + 1.29543\text{E}{-7}(T/K)^2$$

The k_H data were calculated from this equation with solute vapour pressure data from Gmehling *et al.* [34]:

$t/°C$	T/K	Solubility, x	$k_H/\text{mol}\,\text{kg}^{-1}\,\text{Pa}^{-1}$
10	283.2	9.640E−4	2.28E−6
15	288.2	9.115E−4	1.73E−6
20	293.2	8.655E−4	1.33E−6

$t/°C$	T/K	Solubility, x	$k_H/\text{mol kg}^{-1}\,\text{Pa}^{-1}$
25	298.2	8.260E−4	1.04E−6
30	303.2	7.930E−4	8.19E−7
35	308.2	7.664E−4	6.56E−7
40	313.2	7.463E−4	5.33E−7

In these calculations, it has been assumed that the concentration of the solute was within the region where Henry's law is valid. It is seen that k_H data calculated from solubility agree with experimental measurement within experimental uncertainty.

Solute		
Formula	Name	CAS Registry Number
$C_2H_2Cl_4$	1,1,1,2-Tetrachloroethane	630-20-6
Solvent: Water		

Horvath et al. [68] presented critically evaluated smoothed data for the solubility of this liquid in water in the temperature range $273 < T/K < 323$. The data were represented by the equation

$$w = 2.17896E{-2} - 1.3966E{-4}(T/K) + 2.35307E{-7}(T/K)^2$$

where w is mass fraction. The mole fraction solubility is given by

$$x = 2.33667E{-3} - 1.4977E{-5}(T/K) + 2.52339E{-8}(T/K)^2$$

The k_H data were calculated from this equation with vapour pressure data from Boublik et al. [80]:

$t/°C$	T/K	Solubility, x	$k_H/\text{mol kg}^{-1}\,\text{Pa}^{-1}$
0	273.2	1.284E−4	2.19E−5
5	278.2	1.231E−4	1.48E−5
10	283.2	1.190E−4	1.03E−5
15	288.2	1.162E−4	0.730E−5
20	293.2	1.147E−4	0.531E−5
25	298.2	1.144E−4	0.396E−5
30	303.2	1.154E−4	0.302E−5
35	308.2	1.177E−4	0.236E−5
40	313.2	1.212E−4	0.188E−5

$t/°C$	T/K	Solubility, x	$k_H/\text{mol kg}^{-1}\,\text{Pa}^{-1}$
45	318.2	1.260E−4	0.152E−5
50	323.2	1.320E−4	0.126E−5

In these calculations, it has been assumed that the concentration of the solute was within the region where Henry's law is valid. These data may be represented by the equation

$$\ln(k_H/\text{mol kg}^{-1}\,\text{Pa}^{-1}) = -29.198 + 5017.3/(T/K)$$

The standard error of estimate of k_H by this equation is 9 %. This equation generates RECOMMENDED data in the range 273–323 K.

Solute		
Formula	Name	CAS Registry Number
$C_2H_3Cl_3$	1,1,1-Trichloroethane	71-55-6
Solvent: Water		

Available experimental measurements of k_H of this liquid in water are as follows:

			Original measurements		k_H/mol
Ref.	$t/°C$	T/K	Limiting solubility ratio	Method	$\text{kg}^{-1}\,\text{Pa}^{-1}$
61	17.5	290.7	$(1/k_H)/\text{atm m}^3\,\text{mol}^{-1} = 8.22\text{E}{-7}$	EPICS	8.22E−7
64	20	293.2	$(1/k_H)/\text{MPa} = 70.0$	PURGE	7.83E−7
75	20	293.2	$\gamma^\infty = 5880$	DIRECT	2.07E−7
74	20	293.2	$(1/k_H)/\text{kPa} = 76.6\text{E3}$	DIRECT	7.25E−7
36	20	293.2	$(1/k_H)/\text{atm m}^3\,\text{mol}^{-1} = 1.26\text{E}{-2}$	GC	7.83E−7
36	20	293.2	$\gamma^\infty = 5245$	GC	7.96E−7
81	20	293.2	$\gamma^\infty = 5660$	DIRECT	7.37E−7
28	20	293.2	$k_{AW} = 0.71$	VP/AS	5.78E−7
76	25	298.2	$k_{AW} = 0.836$	DIRECT	4.83E−7
66	25	298.2	$(1/k_H)/\text{atm m}^3\,\text{mol}^{-1} = 1.76\text{E}{-2}$	HEADSPACE	5.61E−7
79	26	299.2	$(1/k_H)/\text{kPa m}^3\,\text{mol}^{-1} = 1.76$	EPICS	5.61E−7
74	30	303.2	$(1/k_H)/\text{kPa} = 11.1\text{E4}$	DIRECT	5.00E−7
36	30	303.2	$(1/k_H)/\text{atm m}^3\,\text{mol}^{-1} = 2.0\text{E}{-2}$	GC	4.93E−7
66	30	303.2	$(1/k_H)/\text{atm m}^3\,\text{mol}^{-1} = 2.18\text{E}{-2}$	HEADSPACE	4.53E−7
67	30	303.2	$k_{AW} = 0.750$	EPICS	5.29E−7
36	30	303.2	$\gamma^\infty = 5324$	GC	5.06E−7

Ref.	$t/°C$	T/K	Original measurements Limiting solubility ratio	Method	k_H/mol $\text{kg}^{-1}\,\text{Pa}^{-1}$
79	35	308.2	$(1/k_H)/\text{kPa}\,\text{m}^3\,\text{mol}^{-1} = 2.41$	EPICS	4.15E−7
36	35	308.2	$(1/k_H)/\text{atm}\,\text{m}^3\,\text{mol}^{-1} = 2.35\text{E}{-}2$	GC	4.20E−7
61	35	308.2	$(1/k_H)/\text{atm}\,\text{m}^3\,\text{mol}^{-1} = 2.44\text{E}{-}2$	EPICS	3.96E−7
36	35	308.2	$\gamma^\infty = 5097$	GC	4.29E−7
81	35	308.2	$\gamma^\infty = 6210$	DIRECT	3.52E−7
74	40	313.2	$(1/k_H)/\text{kPa} = 16.4\text{E}4$	DIRECT	3.38E−7
36	40	313.2	$(1/k_H)/\text{atm}\,\text{m}^3\,\text{mol}^{-1} = 2.81\text{E}{-}2$	GC	3.51E−7
66	40	313.2	$(1/k_H)/\text{atm}\,\text{m}^3\,\text{mol}^{-1} = 2.64\text{E}{-}2$	HEADSPACE	3.74E−7
67	40	313.2	$k_{AW} = 1.147$	EPICS	3.35E−7
36	40	313.2	$\gamma^\infty = 4986$	GC	3.59E−7
66	45	318.2	$(1/k_H)/\text{atm}\,\text{m}^3\,\text{mol}^{-1} = 3.55\text{E}{-}2$	HEADSPACE	2.78E−7
66	50	323.2	$(1/k_H)/\text{atm}\,\text{m}^3\,\text{mol}^{-1} = 4.11\text{E}{-}2$	HEADSPACE	2.40E−7
67	50	323.2	$k_{AW} = 1.467$	EPICS	2.54E−7
81	50	323.2	$\gamma^\infty = 5850$	DIRECT	2.10E−7
67	60	333.2	$k_{AW} = 1.948$	EPICS	1.85E−7

In these calculations, solute vapour pressure data from Gmehling *et al.* [34] were used; also, it has been assumed that the concentration of the solute was within the region where Henry's law is valid. These data may be represented by the equation

$$\ln(k_H/\text{mol}\,\text{kg}^{-1}\,\text{Pa}^{-1}) = -27.223 + 3856.9/(T/K)$$

The standard error of estimate of k_H by this equation is 6%. This equation generates RECOMMENDED data in the range 293–323 K.

Horvath *et al.* [68] presented critically evaluated smoothed data for the solubility of this liquid in water in the range $273 < T/K < 323$. The data were represented by the equation

$$w = 1.09092\text{E}{-}2 - 6.52776\text{E}{-}5(T/K) + 1.10747\text{E}{-}7(T/K)^2$$

where w is mass fraction. The mole fraction solubility is

$$x = 1.47331\text{E}{-}3 - 8.81587\text{E}{-}6(T/K) + 1.49566\text{E}{-}8(T/K)^2$$

The k_H data were calculated from this equation and solute vapour pressure data of Gmehling *et al.* [34]:

$t/°C$	T/K	Solubility, x	$k_H/\text{mol}\,\text{kg}^{-1}\,\text{Pa}^{-1}$
0	273.2	1.812E−4	2.05E−6
5	278.2	1.783E−4	1.55E−6
10	283.2	1.762E−4	1.19E−6

$t/°C$	T/K	Solubility, x	$k_H/\text{mol kg}^{-1}\,\text{Pa}^{-1}$
15	288.2	1.749E−4	9.23E−7
20	293.2	1.743E−4	7.28E−7
25	298.2	1.744E−4	5.82E−7
30	303.2	1.753E−4	4.72E−7
35	308.2	1.770E−4	3.87E−7
40	313.2	1.793E−4	3.21E−7
45	318.2	1.825E−4	2.70E−7
50	323.2	1.864E−4	2.29E−7

In these calculations, it has been assumed that the concentration of the solute was within the region where Henry's law is valid. The k_H calculated from solubility are in good agreement with those measured experimentally.

Solute		
Formula	Name	CAS Registry Number
$C_2H_3Cl_3$	1,1,2-Trichloroethane	79-00-5
Solvent: Water		

Available experimental measurements of k_H of this liquid in water are as follows:

	Original measurements				$k_H/\text{mol kg}^{-1}\,\text{Pa}^{-1}$
Ref.	$t/°C$	T/K	Limiting solubility ratio	Method	
74	20	293.2	$(1/k_H)/\text{kPa} = 3.73\text{E}3$	DIRECT	14.9E−6
64	20	293.2	$(1/k_H)/\text{MPa} = 3.67$	PURGE	15.1E−6
79	26	299.2	$(1/k_H)/\text{kPa m}^3\,\text{mol}^{-1} = 0.082$	EPICS	12.2E−6
36	30	303.2	$(1/k_H)/\text{atm m}^3\,\text{mol}^{-1} = 1.1\text{E}{-3}$	GC	8.97E−6
74	35	308.2	$(1/k_H)/\text{kPa} = 7.55\text{E}3$	DIRECT	7.35E−6
81	35	308.2	$\gamma^\infty = 1520$	DIRECT	7.39E−6
36	40	313.2	$(1/k_H)/\text{atm m}^3\,\text{mol}^{-1} = 1.7\text{E}{-3}$	GC	5.81E−6
74	50	323.2	$(1/k_H)/\text{kPa} = 1.3\text{E}4$	DIRECT	4.27E−6
81	50	323.2	$\gamma^\infty = 1430$	DIRECT	3.83E−6

In these calculations, the solute vapour pressure data were taken from Gmehling *et al.* [34]; also, it has been assumed that the concentration of the solute was within the region where Henry's law is valid. These data may be represented by the equation

$$\ln(k_H/\text{mol kg}^{-1}\,\text{Pa}^{-1}) = -25.472 + 4212.1/(T/\text{K})$$

The standard error of estimate of k_H by this equation is 3%. This equation generates RECOMMENDED values in the range 293–323 K.

Horvath et al. [68] presented critically evaluated smoothed data for the solubility of this liquid in water in the range $273 < T/\text{K} < 328$. The data were represented by

$$w = 2.89796\text{E}{-2} - 1.8585\text{E}{-4}(T/\text{K}) + 3.48961\text{E}{-7}(T/\text{K})^2$$

where w is mass fraction. The mole fraction solubility is

$$x = 3.91375\text{E}{-3} - 2.5099\text{E}{-5}(T/\text{K}) + 4.71279\text{E}{-8}(T/\text{K})^2$$

The k_H data were calculated from this equation with solute vapour pressure data from Gmehling et al. [34]:

$t/°C$	T/K	Solubility, x	$k_H/\text{mol kg}^{-1}\,\text{Pa}^{-1}$
0	273.2	5.743E−4	5.13E−5
5	278.2	5.787E−4	3.69E−5
10	283.2	5.855E−4	2.70E−5
15	288.2	5.946E−4	2.02E−5
20	293.2	6.061E−4	1.53E−5
25	298.2	6.200E−4	1.18E−5
30	303.2	6.362E−4	0.925E−6
35	308.2	6.548E−4	0.735E−6
40	313.2	6.757E−4	0.592E−6
45	318.2	6.990E−4	0.482E−6
50	323.2	7.247E−4	0.397E−6
55	328.2	7.527E−4	0.331E−6

In these calculations, it has been assumed that the concentration of the solute was within the region where Henry's law is valid. The k_H data calculated from solubility are seen to be close to those measured experimentally.

Solute		
Formula	Name	CAS Registry Number
C_2H_3N	Acetonitrile	75-05-8
Solvent: Water		

Available experimental measurements of k_H of this liquid in water are as follows:

Original measurements					
Ref.	$t/°C$	T/K	Limiting solubility ratio	Method	$k_H/\text{mol kg}^{-1}\text{Pa}^{-1}$
43	0	273.2	$k_{WA} = 3.7\text{E}3$	DIRECT	16.3E−4
82	6	279.2	$(1/k_H)/\text{atm} = 0.395$	DIRECT	13.9E−4
82	12	285.2	$(1/k_H)/\text{atm} = 0.587$	DIRECT	9.34E−4
82	20	293.2	$(1/k_H)/\text{atm} = 0.808$	DIRECT	6.78E−4
43	25	298.2	$k_{WA} = 1.2\text{E}3$	DIRECT	4.84E−4
48	25	298.2	$\gamma^\infty = 10.1$	INDIRECT	4.52E−4
82	30	303.2	$(1/k_H)/\text{atm} = 1.292$	DIRECT	4.24E−4

It was assumed that the concentration of the solute was within the region where Henry's law is valid. These data may be represented by the equation

$$\ln(k_H/\text{mol kg}^{-1}\text{Pa}^{-1}) = -21.218 + 4061.2/(T/K)$$

The standard error of estimate of k_H by this equation is 6 %. This equation generates RECOMMENDED data.

Solute		
Formula	Name	CAS Registry Number
$C_2H_3NO_5$	Peroxyacetyl nitrate	2278-22-0
Solvent: Water		

Selected experimental measurements of k_H for this liquid are given in the following table. The data of Kames *et al.* [83] were generated from their equation

$$k_H/\text{M atm}^{-1} = 10^{-9.04}\exp[6513/(T/K)]$$

for the temperature range $274 < T/K < 297$.

Original measurements				
$t/°C$	T/K	Limiting solubility ratio	Method	$k_H/\text{mol kg}^{-1}\text{Pa}^{-1}$
0	273.2	$k_H/\text{M atm}^{-1} = 20.7$	DIRECT	20.3E−5
5	278.2	$k_H/\text{M atm}^{-1} = 13.5$	DIRECT	13.3E−5
10	283.2	$k_H/\text{M atm}^{-1} = 8.90$	DIRECT	8.78E−5
15	288.2	$k_H/\text{M atm}^{-1} = 5.97$	DIRECT	5.89E−5

	Original measurements			
$t/°C$	T/K	Limiting solubility ratio	Method	$k_H/\text{mol kg}^{-1}\,\text{Pa}^{-1}$
20	293.2	$k_H/\text{M atm}^{-1} = 4.06$	DIRECT	4.01E−5
25	298.2	$k_H/\text{M atm}^{-1} = 2.80$	DIRECT	2.76E−5

In these calculations, it has been assumed that the concentration of the solute was within the region where Henry's law is valid. In the preferred units, the equation is

$$\ln(k_H/\text{mol kg}^{-1}\,\text{Pa}^{-1}) = -32.351 + 6515.8/(T/K)$$

Solute		
Formula	Name	CAS Registry Number
C_2H_4	Ethene	74-85-1
Solvent: Water		

Hayduck [84a] presented critically evaluated smoothed data for the solubility of this gas in water at a partial pressure of gas of 101.3 kPa in the range $278 < T/K < 323$. The data were represented by

$$\ln x = -66.9156 + 92.2101/(T/100\ \text{K}) + 24.3792 \ln(T/100\ \text{K})$$

and the expression for the corresponding k_H is

$$\ln(k_H/\text{mol kg}^{-1}\text{Pa}^{-1}) = -186.696 + 9.22101\text{E}3/(T/K) + 24.3792 \ln(T/K)$$

$t/°C$	T/K	Solubility, x	$k_H/\text{mol kg}^{-1}\,\text{Pa}^{-1}$
5	278.2	14.71E−5	8.06E−8
10	283.2	12.64E−5	6.93E−8
15	288.2	11.01E−5	6.03E−8
20	293.2	9.703E−5	5.32E−8
25	298.2	8.647E−5	4.74E−8
30	303.2	7.788E−5	4.27E−8
35	308.2	7.084E−5	3.88E−8
40	313.2	6.504E−5	3.56E−8
45	318.2	6.025E−5	3.30E−8
50	323.2	5.627E−5	3.08E−8

These data are RECOMMENDED.

Solute		
Formula	Name	CAS Registry Number
$C_2H_4Cl_2$	1,1-Dichloroethane	75-34-3
Solvent: Water		

Available experimental measurements of k_H of this liquid in water are as follows:

	Original measurements				k_H/mol
Ref.	$t/^\circ C$	T/K	Limiting solubility ratio	Method	$\text{kg}^{-1}\,\text{Pa}^{-1}$
78	2	275.2	$k_{AW} = 6.65\text{E}{-}2$	EPICS	6.57E−6
61	8.5	281.8	$(1/k_H)/\text{atm m}^3\,\text{mol}^{-1} = 2.48\text{E}{-}3$	EPICS	3.98E−6
61	10	283.2	$k_{AW} = 8.27\text{E}{-}2$	EPICS	5.14E−6
64	10	283.2	$(1/k_H)/\text{MPa} = 15.0$	PURGE	3.70E−6
61	17.5	290.7	$(1/k_H)/\text{atm m}^3\,\text{mol}^{-1} = 3.89\text{E}{-}3$	EPICS	2.54E−6
78	18	291.2	$k_{AW} = 0.153$	EPICS	2.70E−6
64	20	293.2	$(1/k_H)/\text{MPa} = 25.6$	PURGE	2.17E−6
75	20	293.2	$\gamma^\infty = 1100$	DIRECT	2.07E−6
74	20	293.2	$(1/k_H)/\text{kPa} = 26.5\text{E}3$	DIRECT	2.09E−6
36	20	293.2	$(1/k_H)/\text{atm m}^3\,\text{mol}^{-1} = 4.6\text{E}{-}3$	GC	2.15E−6
36	20	293.2	$\gamma^\infty = 1046$	GC	2.17E−6
81	20	293.2	$\gamma^\infty = 1080$	DIRECT	2.11E−6
78	25	298.2	$k_{AW} = 0.206$	EPICS	1.96E−6
61	25	298.2	$(1/k_H)/\text{atm m}^3\,\text{mol}^{-1} = 5.62\text{E}{-}3$	EPICS	1.76E−6
64	30	303.2	$(1/k_H)/\text{MPa} = 38.5$	PURGE	1.44E−6
36	30	303.2	$(1/k_H)/\text{atm m}^3\,\text{mol}^{-1} = 7.0\text{E}{-}3$	GC	1.41E−6
36	30	303.2	$\gamma^\infty = 1034$	GC	1.44E−6
64	35	308.2	$(1/k_H)/\text{MPa} = 46.5$	PURGE	1.19E−6
74	35	308.2	$(1/k_H)/\text{kPa} = 55.3\text{E}3$	DIRECT	1.00E−6
61	35	308.2	$(1/k_H)/\text{atm m}^3\,\text{mol}^{-1} = 8.1\text{E}{-}3$	EPICS	1.22E−6
81	35	308.2	$\gamma^\infty = 1120$	DIRECT	1.09E−6
64	40	313.2	$(1/k_H)/\text{MPa} = 54.8$	PURGE	1.01E−6
36	40	313.2	$\gamma^\infty = 1017$	GC	0.996E−6
64	45	318.2	$(1/k_H)/\text{MPa} = 63.5$	PURGE	0.874E−6
74	45	318.2	$(1/k_H)/\text{kPa} = 68.1\text{E}3$	DIRECT	0.815E−6
64	50	323.2	$(1/k_H)/\text{MPa} = 72.3$	PURGE	0.768E−6
81	50	323.2	$\gamma^\infty = 1080$	DIRECT	0.655E−6

In these calculations, solute vapour pressure data were taken from Gmehling *et al.* [34]; also, it has been assumed that the concentration of the solute was within the region where Henry's law is valid. These data may be represented by the equation

$$\ln(k_H/\text{mol kg}^{-1}\,\text{Pa}^{-1}) = -26.791 + 4050.1(T/\text{K})$$

The standard error of estimate of k_H by this equation is 8%. This equation generates RECOMMENDED data in the range 273–323 K.

Solute		
Formula	Name	CAS Registry Number
$C_2H_4Cl_2$	1,2-Dichloroethane	107-06-2
Solvent: Water		

Available experimental k_H data for this liquid in water are as follows:

			Original measurements		k_H/mol
Ref.	$t/°C$	T/K	Limiting solubility ratio	Method	$\text{kg}^{-1}\,\text{Pa}^{-1}$
78	2	275.2	$k_{AW} = 1.4\text{E}{-}2$	EPICS	3.12E−5
78	6	279.2	$k_{AW} = 1.83\text{E}{-}2$	EPICS	2.35E−5
78	10	283.2	$k_{AW} = 1.68\text{E}{-}2$	EPICS	2.53E−5
78	18	291.2	$k_{AW} = 3.05\text{E}{-}2$	EPICS	1.35E−5
75	20	293.2	$\gamma^\infty = 585$	DIRECT	1.15E−5
74	20	293.2	$(1/k_H)/\text{kPa} = 5.09\text{E3}$	DIRECT	1.09E−5
36	20	293.2	$(1/k_H)/\text{atm m}^3\,\text{mol}^{-1} = 1.0\text{E}{-}3$	GC	9.87E−6
81	20	293.2	$\gamma^\infty = 626$	DIRECT	1.07E−5
28	20	293.2	$k_{WA} = 26.4$	VP/AS	1.08E−5
78	25	298.2	$k_{AW} = 4.12\text{E}{-}2$	EPICS	9.79E−6
87	25	298.2	$\gamma^\infty = 511$	HEADSPACE	1.03E−5
36	30	303.2	$(1/k_H)/\text{atm m}^3\,\text{mol}^{-1} = 1.5\text{E}{-}3$	GC	6.58E−6
36	30	303.2	$\gamma^\infty = 610$	GC	6.84E−6
74	35	308.2	$(1/k_H)/\text{kPa} = 10.3\text{E3}$	DIRECT	5.39E−6
36	35	308.2	$(1/k_H)/\text{atm m}^3\,\text{mol}^{-1} = 1.8\text{E}{-}3$	GC	5.48E−6
36	35	308.2	$\gamma^\infty = 600$	GC	5.57E−6
81	35	308.2	$\gamma^\infty = 604$	DIRECT	5.54E−6
36	40	313.2	$(1/k_H)/\text{atm m}^3\,\text{mol}^{-1} = 2.2\text{E}{-}3$	GC	4.49E−6
36	40	313.2	$\gamma^\infty = 587$	GC	4.57E−6
74	50	323.2	$(1/k_H)/\text{kPa} = 17.7\text{E3}$	DIRECT	3.14E−6
81	50	323.2	$\gamma^\infty = 552$	DIRECT	3.23E−6

In these calculations, vapour pressure data were taken from Gmehling *et al.* [34]; also, it has been assumed that the concentration of the solute was within the region where Henry's law is valid. These data may be represented by the equation

$$\ln(k_H/\text{mol kg}^{-1}\,\text{Pa}^{-1}) = -26.0965 + 4319.8/(T/\text{K})$$

The standard error of estimate of k_H by this equation is 7%. This equation generates RECOMMENDED data in the range 273–323 K.

Horvath *et al.* [68] presented critically evaluated smoothed data for the solubility of this liquid in water in the range $273 < T/\text{K} < 373$. The data were represented by the equation

$$w = 17.9147\text{E}{-}2 - 0.11684\text{E}{-}2(T/\text{K}) + 2.0003\text{E}{-}6(T/\text{K})^2$$

where w is mass fraction. The equation for mole fraction is

$$x = 3.26012\text{E}{-}2 - 2.1263\text{E}{-}4(T/\text{K}) + 3.6402\text{E}{-}7(T/\text{K})^2$$

The k_H data were calculated with this equation and vapour pressure as previously:

$t/^\circ\text{C}$	T/K	Solubility, x	$k_H/\text{mol kg}^{-1}\,\text{Pa}^{-1}$
0	273.2	1.681E−3	3.33E−5
10	283.2	1.580E−3	1.78E−5
20	293.2	1.552E−3	1.04E−5
30	303.2	1.596E−3	6.66E−6
40	313.2	1.714E−3	4.60E−6
50	323.2	1.904E−3	3.40E−6
60	333.2	2.167E−3	2.65E−6
70	343.2	2.503E−3	2.15E−6
80	353.2	2.912E−3	1.80E−6
90	363.2	3.393E−3	1.54E−6
100	373.2	3.948E−3	1.34E−6

In these calculations, it has been assumed that the concentration of the solute was within the region where Henry's law is valid. The k_H values from solubility and from direct measurement are in good agreement.

Solute		
Formula	Name	CAS Registry Number
C_2H_5I	Iodoethane	75-03-6
Solvent: Water		

Horvath *et al.* [68] presented critically evaluated smoothed data for the solubility of this liquid in water in the temperature range $273 < T/K < 303$. The data were represented by the equation

$$w = 8.5757E{-}2 - 5.5568E{-}4(T/K) + 9.43918E{-}7(T/K)^2$$

where w is mass fraction. The mole fraction solubility is

$$x = 9.9060E{-}3 - 6.4188E{-}5(T/K) + 1.09034E{-}7(T/K)^2$$

The k_H data were calculated from this equation with vapour pressure data from Gmehling *et al.* [85]:

$t/°C$	T/K	Solubility, x	$k_H/\mathrm{mol\,kg^{-1}\,Pa^{-1}}$
0	273.2	5.079E−4	5.43E−6
5	278.2	4.877E−4	4.00E−6
10	283.2	4.727E−4	3.02E−6
15	288.2	4.633E−4	2.32E−6
20	293.2	4.593E−4	1.82E−6
25	298.2	4.608E−4	1.46E−6
30	303.2	4.677E−4	1.20E−6

In these calculations, it has been assumed that the concentration of the solute was within the region where Henry's law is valid. These data may be represented by the equation

$$\ln(k_H/\mathrm{mol\,kg^{-1}\,Pa^{-1}}) = -27.963 + 4182.7/T/K$$

The standard error of estimate of k_H by this equation is 2 %. This equation generates RECOMMENDED data in the temperature interval 273–303 K.

Solute		
Formula	Name	CAS Registry Number
$C_2H_5NO_2$	Ethyl nitrate	625-58-1
Solvent: Water		

The sole report of k_H of this liquid is by Kames and Schurath [41], who used headspace chromatography. The reported datum is $k_H/\mathrm{M\,atm^{-1}} = 2.18$, or $k_H/\mathrm{mol\,kg^{-1}\,Pa^{-1}} = 2.15E{-}5$ at 20 °C.

Solute		
Formula	Name	CAS Registry Number
$C_2H_5NO_4$	2-Nitrooxyethanol	16051-48-2
Solvent: Water		

Shepson *et al.* [86] measured k_H for this liquid in water in the temperature range $279 < T/K < 304$ by the direct method. The data were expressed by the equation

$$\ln(k_H/\text{M atm}^{-1}) = -18.438 + 8647.5/(T/K)$$

The standard error of estimate of k_H by this equation is approximately 10%.

Original measurements			
$t/°C$	T/K	Limiting solubility ratio	$k_H/\text{mol kg}^{-1}\,\text{Pa}^{-1}$
5	278.2	$k_H/\text{M atm}^{-1} = 3.11\text{E}5$	3.07
10	283.2	$k_H/\text{M atm}^{-1} = 1.79\text{E}5$	1.77
15	288.2	$k_H/\text{M atm}^{-1} = 1.06\text{E}5$	1.04
20	293.2	$k_H/\text{M atm}^{-1} = 6.33\text{E}4$	0.625
25	298.2	$k_H/\text{M atm}^{-1} = 3.86\text{E}4$	0.381
30	303.2	$k_H/\text{M atm}^{-1} = 2.39\text{E}4$	0.236

In these calculations, it has been assumed that the concentration of the solute was within the region where Henry's law is valid. There are no other independent measurements of k_H for comparison.

Solute		
Formula	Name	CAS Registry Number
C_2H_6	Ethane	74-84-0
Solvent: Water		

Hayduk [84b] presented critically evaluated smoothed data for the solubility of this gas in water at a partial pressure of 101.3 kPa in the temperature range $273 < T/K < 323$. The data were represented by

$$\ln x = -90.82250 + 126.9559/(T/100\,K) + 34.74128\ln(T/100\,K)$$

The corresponding equation for k_H is

$$\ln(k_H/\text{mol kg}^{-1}\,\text{Pa}^{-1}) = -258.322 + 1.269559\text{E}4/(T/\text{K}) + 34.74128\ln(T/\text{K})$$

$t/°C$	T/K	Solubility, x	$k_H/\text{mol kg}^{-1}\,\text{Pa}^{-1}$
0	273.2	7.994E−5	4.38E−8
5	278.2	6.510E−5	3.57E−8
10	283.2	5.400E−5	2.96E−8
15	288.2	4.556E−5	2.50E−8
20	293.2	3.907E−5	2.14E−8
25	298.2	3.401E−5	1.86E−8
30	303.2	3.002E−5	1.64E−8
35	308.2	2.687E−5	1.47E−8
40	313.2	2.434E−5	1.33E−8
45	318.2	2.232E−5	1.22E−8
50	323.2	2.069E−5	1.13E−8

These are RECOMMENDED data.

Solute		
Formula	Name	CAS Registry Number
C_2H_6O	Ethanol	64-17-5
Solvent: Water		

Available experimental measurements of k_H of this liquid are given in the table.

			Original measurements		
Ref.	$t/°C$	T/K	Limiting solubility ratio	Method	$k_H/\text{mol kg}^{-1}\,\text{Pa}^{-1}$
43	0	273.2	$k_{AW} = 3.03\text{E}-5$	DIRECT	14.5E−3
45	10	283.2	$\gamma^\infty = 4.38$	DIRECT	4.08E−3
45	20	293.2	$\gamma^\infty = 4.81$	DIRECT	1.97E−3
44	20	293.2	$\gamma^\infty = 6.52$	GC	1.43E−3
46	24	297.2	$\gamma^\infty = 474$	HEADSPACE	1.58E−3
43	25	298.2	$k_{AW} = 2.13\text{E}-4$	DIRECT	1.90E−3
59	25	298.2	$\gamma^\infty = 3.73$	HEADSPACE	1.88E−3
87	25	298.2	$\gamma^\infty = 3.74$	HEADSPACE	1.87E−3
47	25	298.2	$\gamma^\infty = 3.55$	DIRECT	1.98E−3
88	25	298.2	$k_{AW} = 1.77\text{E}-4$	HEADSPACE	2.28E−3

			Original measurements		
Ref.	$t/°C$	T/K	Limiting solubility ratio	Method	$k_H/\text{mol kg}^{-1}\,\text{Pa}^{-1}$
89	25	298.2	$\gamma^\infty = 3.76$	DIRECT	1.86E−3
48	25	298.2	$\gamma^\infty = 3.80$	INDIRECT	1.85E−3
49	25	298.2	$k_{AW} = 3.33E-4$	PURGE	1.20E−3
51	25	298.2	$\gamma^\infty = 3.55$	DIRECT	1.98E−3
52	25	298.2	$\gamma^\infty = 4.03$	GC	1.74E−3
53	25	298.2	$\gamma^\infty = 3.27$	GC	2.14E−3
44	30	303.2	$\gamma^\infty = 6.15$	GC	0.864E−3
45	40	313.2	$\gamma^\infty = 5.17$	DIRECT	0.600E−3
44	40	313.2	$\gamma^\infty = 5.50$	GC	0.564E−3
55	60	333.2	$k_{AW} = 1.88E-3$	HEADSPACE	0.192E−3
57	63	336.2	$\gamma^\infty = 4.436$	GC	0.234E−3
58	100	373.2	$\gamma^\infty = 5.81$	EBULLIOMETRY	0.423E−3

These data may be represented by the equation

$$\ln(k_H/\text{mol kg}^{-1}\text{Pa}^{-1}) = -25.659 + 5749.9/(T/K)$$

The standard error of estimate of k_H by this equation is 18%. This equation generates RECOMMENDED data in the range 273–373 K.

	Solute	
Formula	Name	CAS Registry Number
C_2H_6OS	Dimethyl sulfoxide	67-68-5
Solvent: Water		

There is a single source of measurements for k_H of this liquid in water [90].

		Original measurements		
$t/°C$	T/K	Limiting solubility ratio	Method	$k_H/\text{mol kg}^{-1}\,\text{Pa}^{-1}$
26	299.2	$\gamma^\infty = 0.37$	EBULLIOMETRY	1.77
35	308.2	$\gamma^\infty = 0.60$	EBULLIOMETRY	0.573
45	318.2	$\gamma^\infty = 0.02$	EBULLIOMETRY	0.887
55	328.2	$\gamma^\infty = 0.135$	EBULLIOMETRY	0.716
65	338.2	$\gamma^\infty = 0.17$	EBULLIOMETRY	0.325

In these calculations, it has been assumed that the concentration of the solute was within the region where Henry's law is valid. These data may be represented by the equation

$$\ln(k_H/\text{mol kg}^{-1}\text{Pa}^{-1}) = -10.5141 - 3241.5/(T/K)$$

The standard error of estimate of k_H by this equation is 30 %.

Solute		
Formula	Name	CAS Registry Number
C_2H_6S	Dimethyl sulfide	75-18-3
Solvent: Water		

Available experimental measurements for k_H of this liquid in water are as follows:

			Original measurements		k_H/mol
Ref.	$t/°C$	T/K	Limiting solubility ratio	Method	$\text{kg}^{-1}\,\text{Pa}^{-1}$
91	−1	272.2	$(1/k_H)/\text{atm M}^{-1} = 0.6$	DIRECT	16.4E−6
91	4	277.2	$(1/k_H)/\text{atm M}^{-1} = 0.74$	DIRECT	13.3E−6
91	5	278.2	$(1/k_H)/\text{atm M}^{-1} = 0.74$	DIRECT	13.3E−6
92	5	278.2	$k_{AW} = 0.039$	−	11.1E−6
91	10	283.2	$(1/k_H)/\text{atm M}^{-1} = 0.95$	DIRECT	10.4E−6
91	14	287.2	$(1/k_H)/atm\,\text{M}^{-1} = 1.17$	DIRECT	8.44E−6
92	15	288.2	$k_{AW} = 0.059$	−	7.08E−6
91	18	291.2	$(1/k_H)/\text{atm M}^{-1} = 1.41$	HEADSPACE	7.00E−6
93	18	291.2	$k_{AW} = 0.058$	−	7.13E−6
92	25	298.2	$k_{AW} = 0.086$	−	4.69E−6
91	25	298.2	$(1/k_H)/\text{atm M}^{-1} = 1.79$	DIRECT	5.51E−6
60	25	298.2	$k_{WA} = 13.42$	HEADSPACE	5.42E−6
59	25	298.2	$k_H/\text{atm}^{-1} = 8.61\text{E}-3$	DIRECT	4.72E−6
91	31	304.2	$(1/k_H)/\text{atm M}^{-1} = 2.35$	DIRECT	4.20E−6

In these calculations, it has been assumed that the concentration of the solute was within the region where Henry's law is valid. These data may be represented by the equation

$$\ln(k_H/\text{mol kg}^{-1}\text{Pa}^{-1}) = -24.207 + 3588.7/(T/K)$$

The standard error of estimate of k_H by this equation is 6 %. This equation generates RECOMMENDED values in the temperature range 272–304 K.

Solute		
Formula	Name	CAS Registry Number
$C_3H_2ClF_5O$	Isoflurane	26675-46-7
Solvent: Water		

Available data for k_H of this gas in water are as follows:

Original measurements					
Ref.	$t/°C$	T/K	Limiting solubility ratio	Method	$k_H/\text{mol kg}^{-1}\,\text{Pa}^{-1}$
69	25	298.2	Bunsen = 1.08	DIRECT	4.76E−7
69	37	310.2	Bunsen = 0.54	DIRECT	2.38E−7
71	37	310.2	Ostwald = 0.62	CRIT EVAL	2.40E−7
94	37	310.2	k_{WA} = 0.626	DIRECT	2.42E−7

A RECOMMENDED value of $2.40\text{E}{-}7\,\text{mol kg}^{-1}\,\text{Pa}^{-1}$ may be concluded at 37 °C.

Solute		
Formula	Name	CAS Registry Number
$C_3H_2ClF_5O$	Enflurane	13838-16-9
Solvent: Water		

There are two reports of k_H for this gas:

Original measurements					
Ref.	$t/°C$	T/K	Limiting solubility ratio	Method	$k_H/\text{mol kg}^{-1}\,\text{Pa}^{-1}$
71	37	310.2	Ostwald = 0.78	CRIT EVAL	3.02E−7
94	37	310.2	k_{WA} ≐ 0.754	DIRECT	2.97E−7

A value of $3.0\text{E}{-}7\,\text{mol kg}^{-1}\,\text{Pa}^{-1}$ may be taken as a RECOMMENDED value at 37 °C.

Solute		
Formula	Name	CAS Registry Number
C_3H_3Br	Propargyl bromide	106-96-7
Solvent: Water		

The k_H of this liquid was measured by Yates and Gan [95] by headspace chromatography in the range $275 < T/K < 313$. The authors' equation was

$$k_{AW} = -0.0154 + 636\exp[-2756/(T/K)]$$

and the data are as follows:

	Original measurements		
$t/°C$	T/K	Limiting solubility ratio	$k_H/\text{mol kg}^{-1}\text{Pa}^{-1}$
5	278.2	$k_{AW} = 1.62E-2$	26.6E−6
10	283.2	$k_{AW} = 2.23E-2$	19.0E−6
15	288.2	$k_{AW} = 2.92E-2$	14.2E−6
20	293.2	$k_{AW} = 3.71E-2$	11.0E−6
25	298.2	$k_{AW} = 4.61E-2$	8.77E−6
30	303.2	$k_{AW} = 5.62E-2$	7.05E−6
35	308.2	$k_{AW} = 6.76E-2$	5.77E−6
40	313.2	$k_{AW} = 8.04E-2$	4.78E−2

In these calculations, it has been assumed that the concentration of the solute was within the region where Henry's law is valid. These data may be represented by the equation

$$\ln(k_H/\text{mol kg}^{-1}\text{Pa}^{-1}) = -25.807 + 4231.1/(T/K)$$

The standard error of estimate of k_H by this equation is 4 %.

Solute		
Formula	Name	CAS Registry Number
C_3H_3N	Acrylonitrile	107-13-1
Solvent: Water		

Available experimental measurements of the limiting solubility constant of this liquid in water are given in the table. The k_H data were calculated with vapour pressure from Gmehling *et al.* [34].

Ref.	T/K	Quantity	Value	$10^5 k_H/\text{mol kg}^{-1}\,\text{Pa}^{-1}$	Method
96	298.15	γ^∞	48.4	7.83	GC
97	303.15	γ^∞	41.8	7.30	PURGE
96	308.15	γ^∞	44.4	5.59	GC
97	308.15	γ^∞	41.6	5.96	PURGE
96	313.15	γ^∞	45.2	4.50	GC
97	313.15	γ^∞	41.5	4.90	PURGE
97	318.15	γ^∞	41.4	4.04	PURGE
96	323.15	γ^∞	42.0	3.31	GC
96	333.15	γ^∞	38.4	2.54	GC

These data may be represented by the equation

$$\ln(k_H/\text{mol kg}^{-1}\text{Pa}^{-1}) = -20.771 + 3388.7/(T/K); \quad r = 0.993$$

This equation generates RECOMMENDED values.

Solute		
Formula	Name	CAS Registry Number
$C_3H_4Cl_2F_2O$	Methoxyflurane	76-38-0
Solvent: Water		

Available experimental measurements of k_H for this gas in water are as follows:

		Original measurements			
Ref.	$t/°C$	T/K	Limiting solubility ratio	Method	$k_H/\text{mol kg}^{-1}\,\text{Pa}^{-1}$
69	20	293.2	Bunsen = 8.08	DIRECT	3.56E−6
70	23	296.2	k_{WA} = 8.02	DIRECT	3.26E−6
69	30	303.2	Bunsen = 5.48	DIRECT	2.41E−6
70	30	303.2	k_{WA} = 6.08	DIRECT	2.41E−6
69	37	310.2	Bunsen = 3.80	DIRECT	1.67E−6
70	37	310.2	k_{WA} = 4.32	DIRECT	1.68E−6
71	37	310.2	Ostwald = 4.5	CRIT EVAL	1.75E−6
94	37	310.2	k_{WA} = 4.33	DIRECT	1.68E−6

These data may be represented by the equation

$$\ln(k_H/\text{mol kg}^{-1}\,\text{Pa}^{-1}) = -26.567 + 4122.0/(T/\text{K})$$

The standard error of estimate of k_H by this equation is 3 %. This equation generates RECOMMENDED values.

Solute		
Formula	Name	CAS Registry Number
C_3H_6	Propene	115-07-1
Solvent: Water		

Wilhelm *et al.* [6] presented critically evaluated smoothed data for the solubility of this gas in water at 101.3 kPa partial pressure in the range $293 < T/\text{K} < 363$. The data were represented by the equation

$$(R/\text{cal mol}^{-1}\text{K}^{-1})\ln x = 199.656 - 3940.90/(T/\text{K}) - 35.8336\ln(T/\text{K})$$

or

$$\ln x = 1.00470\text{E}2 - 1.983\text{E}3/(T/\text{K}) - 18.0320\ln(T/\text{K})$$

The equation for k_H is therefore

$$\ln(k_H/\text{mol kg}^{-1}\text{Pa}^{-1}) = 92.9604 - 1.983\text{E}3/(T/\text{K}) - 18.0320\ln(T/\text{K})$$

$t/°C$	T/K	Solubility, x	$k_H/\text{mol kg}^{-1}\,\text{Pa}^{-1}$
20	293.2	1.616E−4	8.85E−8
25	298.2	1.334E−4	7.31E−8
30	303.2	1.103E−4	6.04E−8
35	308.2	9.133E−5	5.00E−8
40	313.2	7.572E−5	4.15E−8
45	318.2	6.286E−5	3.44E−8
50	323.2	5.226E−5	2.86E−8
55	328.2	4.350E−5	2.38E−8
60	333.2	3.627E−5	1.99E−8
65	338.3	3.027E−5	1.66E−8
70	343.2	2.531E−5	1.39E−8
75	348.2	2.118E−5	1.16E−8
80	353.2	1.776E−5	9.73E−9
85	358.2	1.490E−5	8.16E−9
90	363.2	1.253E−5	6.86E−9

In these calculations, it has been assumed that the concentration of the solute was within the region where Henry's law is valid. These data are RECOM-MENDED values.

Solute		
Formula	Name	CAS Registry Number
C_3H_6O	Acetone	67-64-1
Solvent: Water		

About 0.2 % of acetone exists as a *gem*-diol in aqueous solution. For most practical purposes acetone can be assumed to be unchanged when it is dissolved in Water (see Chapter 9).

Available experimental measurements of k_H of this liquid in water are given in the table. Solute vapour pressure data were taken from Gmehling *et al.* [98].

			Original measurements		
Ref.	$t/°C$	T/K	Limiting solubility ratio	Method	$k_H/\text{mol kg}^{-1}\,\text{Pa}^{-1}$
43	0	273.2	$k_{AW} = 4.0E{-}4$	DIRECT	16.7E$-$4
82	9	282.2	$(1/k_H)/\text{atm} = 0.691$	DIRECT	7.93E$-$4
82	16	289.2	$(1/k_H)/\text{atm} = 1.121$	DIRECT	4.89E$-$4
46	24	297.2	$\gamma^\infty = 7.96$	GC	2.37E$-$4
99	25	298.2	$k_{AW} = 1.64E{-}3$	HEADSPACE	2.47E$-$4
100	25	298.2	$k_{AW} = 1.6E{-}3$	DIRECT	2.52E$-$4
52	25	298.2	$\gamma^\infty = 7.31$	GC	2.47E$-$4
101	25	298.2	$k_{AW} = 1.61E{-}3$	HEADSPACE	2.50E$-$4
50	25	298.2	$\gamma^\infty = 6.90$	DIRECT	2.62E$-$4
87	25	298.2	$\gamma^\infty = 7.01$	HEADSPACE	2.58E$-$4
47	30	303.2	$\gamma^\infty = 7.70$	DIRECT	1.90E$-$4
102	45	318.2	$\gamma^\infty = 8.3$	DIRECT	0.980E$-$4

These data may be represented by the equation

$$\ln(k_H/\text{mol kg}^{-1}\,\text{Pa}^{-1}) = -27.221 + 5657.5/(T/K)$$

The standard error of estimate of k_H by this equation is 7 %. This equation generates RECOMMENDED values.

Solute		
Formula	Name	CAS Registry Number
C_3H_8	Propane	74-98-6
Solvent: Water		

Hayduk [103] presented critically evaluated smoothed data for the solubility of this gas in water at a gas partial pressure of 101.3 kPa in the range $273 < T/K < 308$. The data were represented by

$$\ln x = -102.044 + 144.345/(T/100\,\mathrm{K}) + 39.4740\ln(T/100\,\mathrm{K})$$

The corresponding equation for k_H is

$$\ln(k_H\,\mathrm{mol\,kg^{-1}\,Pa^{-1}}) = -291.338 + 1.44345\mathrm{E}4(T/\mathrm{K}) + 39.4740\ln(T/\mathrm{K})$$

$t/°C$	T/K	Solubility, x	$k_H/\mathrm{mol\,kg^{-1}\,Pa^{-1}}$
0	273.2	7.236E−5	3.96E−8
5	278.2	5.727E−5	3.14E−8
10	283.2	4.627E−5	2.53E−8
15	288.2	3.813E−5	2.09E−8
20	293.2	3.200E−5	1.75E−8
25	298.2	2.732E−5	1.50E−8
30	303.2	2.370E−5	1.30E−8
35	308.2	2.088E−5	1.14E−8
40	313.2	1.865E−5	1.02E−8
45	318.2	1.689E−5	0.925E−8
50	323.2	1.550E−5	0.849E−8
55	328.2	1.438E−5	0.788E−8
60	333.2	1.350E−5	0.740E−8
65	338.2	1.281E−5	0.702E−8
70	343.2	1.228E−5	0.672E−8
75	348.2	1.188E−5	0.651E−8

These are RECOMMENDED k_H data.

Solute		
Formula	Name	CAS Registry Number
C_3H_8O	1-Propanol	71-23-8
Solvent: Water		

Available experimental measurements of k_H of this liquid in water are given in the table.

Original measurements					$k_H/\text{mol kg}^{-1}\text{Pa}^{-1}$
Ref.	$t/°C$	T/K	Limiting solubility ratio	Method	
43	0	273.2	$k_{AW} = 3.33\text{E}{-}5$	DIRECT	13.2E$-$3
44	20	293.2	$\gamma^\infty = 24$	GC	1.26E$-$3
43	25	298.2	$k_{AW} = 3.03\text{E}{-}4$	DIRECT	1.33E$-$3
46	25	298.2	$\gamma^\infty = 17.2$	GC	1.35E$-$3
47	25	298.2	$\gamma^\infty = 11.2$	DIRECT	1.80E$-$3
48	25	298.2	$\gamma^\infty = 14.17$	INDIRECT	1.42E$-$3
51	25	298.2	$\gamma^\infty = 11.2$	DIRECT	1.80E$-$3
52	25	298.2	$\gamma^\infty = 15.0$	GC	1.34E$-$3
53	25	298.2	$\gamma^\infty = 10.9$	GC	1.85E$-$3
104a	25	298.2	$(1/k_H)/\text{mmHg} = 291.$	DIRECT	1.43E$-$3
54	25	298.2	$(1/k_H)/\text{mmHg} = 283.$	HEADSPACE	1.47E$-$3
87	25	298.2	$\gamma^\infty = 13.36$	HEADSPACE	1.51E$-$3
105	25	298.2	$\gamma^\infty = 13.35$	INDIRECT	1.51E$-$3
44	30	303.2	$\gamma^\infty = 26.$	GC	0.607E$-$3
44	40	313.2	$\gamma^\infty = 22.$	GC	0.393E$-$3
57	69	342.2	$\gamma^\infty = 14.412$	GC	0.133E$-$3

In these calculations, it has been assumed that the concentration of the solute was within the region where Henry's law is valid. These data may be represented by a $\ln(k_H)$ vs inverse temperature plot; such a plot shows a distinct curvature A linear regression of all the data gives the equation

$$\ln(k_H/\text{mol kg}^{-1}\,\text{Pa}^{-1}) = -27.507 + 6246.8/(T/K)$$

The standard error of estimate of k_H by this equation is 18%. This equation generates RECOMMENDED data in the range 273–343 K.

Solute		
Formula	Name	CAS Registry Number
C_3H_8O	2-Propanol	67-63-0
Solvent: Water		

Available experimental measurements of the HLC of this liquid are given in the table.

Ref.	T/K	Quantity	Unit	Value	$10^3 \, \text{mol} \, \text{kg}^{-1}$ Pa^{-1}	Method
43	273.15	k_{AW}	–	3.57E−5	12.3	DIRECT
45	288.15	γ^∞	–	12.9	1.39	DIRECT
43	298.15	k_{AW}	–	3.23E−4	1.25	DIRECT
104a	298.15	$1/k_H$	mmHg	339	1.23	DIRECT
54	298.15	$1/k_H$	mmHg	335	1.24	HEADSPACE
52	298.15	γ^∞	–	7.75	1.18	GC
106	313.15	k_{AW}	–	1.21E−3	0.317	HEADSPACE
56	318.15	γ^∞	–	8.8	0.356	EBULLIOMETRY
106	333.15	k_{AW}	–	3.50E−3	0.103	HEADSPACE
56	337.15	γ^∞	–	9.5	0.130	EBULLIOMETRY
106	343.15	k_{AW}	–	5.59E−3	0.0627	HEADSPACE
56	349.15	γ^∞	–	11.0	0.0657	EBULLIOMETRY
106	353.15	k_{AW}	–	8.55E−3	0.0398	HEADSPACE
107	353.15	γ^∞	–	13.62	0.0447	EBULLIOMETRY
56	358.15	γ^∞	–	11.6	0.0428	EBULLIOMETRY
107	363.15	γ^∞	–	13.68	0.0294	EBULLIOMETRY
107	373.15	γ^∞	–	14.00	0.0198	EBULLIOMETRY

These data may be represented by the equation

$$\ln(k_H/\text{mol} \, \text{kg}^{-1} \, \text{Pa}^{-1}) = -27.480 + 6179.2/(T/K); r = 0.991$$

This equation generates RECOMMENDED values.

Solute		
Formula	Name	CAS Registry Number
C_4Cl_6	Hexachloro-1,3-butadiene	87-68-3
Solvent: Water		

Available experimental measurements of k_H are as follows:

	Original measurements			
Ref.	$t/°C$	T/K	Limiting solubility ratio	Method	$k_H/\text{mol} \, \text{kg}^{-1} \, \text{Pa}^{-1}$
78	2	275.2	$k_{AW} = 0.157$	GC	2.78E−6
78	6	279.2	$k_{AW} = 0.256$	GC	1.68E−6
78	10	283.2	$k_{AW} = 0.297$	GC	1.43E−6
78	18	291.2	$k_{AW} = 0.440$	GC	0.939E−6

			Original measurements		
78	25	298.2	$k_{AW} = 0.624$	GC	0.646E−6
37	25	298.2	$k_{AW} = 0.603$	EPICS	0.669E−6
37	35	308.2	$k_{AW} = 0.681$	EPICS	0.573E−6
37	45	318.2	$k_{AW} = 0.818$	EPICS	0.462E−6
37	55	328.2	$k_{AW} = 1.032$	EPICS	0.355E−6
37	60	333.2	$k_{AW} = 1.146$	EPICS	0.315E−6

In these calculations, it has been assumed that the concentration of the solute was within the region where Henry's law is valid. Solute vapour pressure data were taken from Gmehling *et al.* [34]. The k_H data may be represented by the equation

$$\ln(k_H/\text{mol kg}^{-1}\text{Pa}^{-1}) = -24.313 + 3075.2/(T/\text{K})$$

The standard error of estimate of k_H by this equation is 13 %. This equation generates RECOMMENDED data.

Solute		
Formula	Name	CAS Registry Number
$C_4H_5F_3O$	Fluoxene	406-90-6
Solvent: Water		

Available experimental measurements of k_H for this gas are as follows:

			Original measurements		
Ref.	$t/^{\circ}\text{C}$	T/K	Limiting solubility ratio	Method	$k_H/\text{mol kg}^{-1}\text{Pa}^{-1}$
69	25	298.2	Bunsen = 1.24	DIRECT	5.46E−7
70	25	298.2	$k_{WA} = 1.35$	DIRECT	5.45E−7
70	31	304.2	$k_{WA} = 1.01$	DIRECT	3.99E−7
69	37	310.2	Bunsen = 0.71	DIRECT	3.13E−7
71	37	310.2	Ostwald = 0.85	CRIT EVAL	3.30E−7
70	37	310.2	$k_{WA} = 0.84$	DIRECT	3.26E−7
108	37	310.2	$k_{WA} = 0.84$	DIRECT	3.26E−7

These data may be represented by the equation

$$\ln(k_H/\text{mol kg}^{-1}\text{Pa}^{-1}) = -27.909 + 4018.5/(T/\text{K}).$$

The standard error of estimate of k_H by this equation is 3 %. This equation generates RECOMMENDED values in the range 298–310 K.

Solute		
Formula	Name	CAS Registry Number
C_4H_6O	Methyl vinyl ketone	78-94-4
Solvent: Water		

There is one report [109] of k_H of this liquid in water:

		Original measurements		
$t/°C$	T/K	Limiting solubility ratio	Method	$k_H/\text{mol kg}^{-1}\,\text{Pa}^{-1}$
5	278.2	$k_{AW} = 2.9E{-}4$	DIRECT	14.9E−4
10	283.2	$k_{AW} = 5.6E{-}4$	DIRECT	7.59E−4
15	288.2	$k_{AW} = 8.0E{-}4$	DIRECT	5.22E−4
20	293.2	$k_{AW} = 1.2E{-}3$	DIRECT	3.42E−4
25	298.2	$k_{AW} = 1.9E{-}3$	DIRECT	2.12E−4

(A recent measurement at 25 °C [110] was equivalent to $k_H/\text{mol kg}^{-1}\,\text{Pa}^{-1} = $ 4.05E−4 and is not consistent with the tabulated results.) The tabulated data may be represented by the equation

$$\ln(k_H/\text{mol kg}^{-1}\,\text{Pa}^{-1}) = -34.620 + 7798.9/(T/K)$$

The standard error of estimate of k_H by this equation is 5 %. This equation generates RECOMMENDED values in the range 278–298 K.

Solute		
Formula	Name	CAS Registry Number
C_4H_8	1-Butene	106-98-9
Solvent: Water		

There are three reports of k_H of this gas in water:

			Original measurements		
Ref.	$t/°C$	T/K	Limiting solubility ratio	Method	$k_H/mol\,kg^{-1}\,Pa^{-1}$
40	25	298.2	$\gamma^\infty = 1.40E4$	VP/AS	1.34E$-$8
38	25	298.2	$(1/k_H)/atm = 1.36E4$	–	4.03E$-$8
111	25	298.2	$k_{WA} = 9.77E{-}2$	–	3.94E$-$8

The method of calculation in Refs 38 and 111 was not stated; the data were probably derived from aqueous solubility and vapour pressure.

Solute		
Formula	Name	CAS Registry Number
C_4H_8O	Methyl ethyl ketone	78-93-3
Solvent: Water		

Available experimental measurements of k_H for this liquid in water are given in the table. Solute vapour pressures were taken from Gmehling et al. [98].

			Original measurements		k_H/mol
Ref.	$t/°C$	T/K	Limiting solubility ratio	Method	$kg^{-1}\,Pa^{-1}$
43	0	273.2	$k_{AW} = 4.35E{-}4$	DIRECT	10.1E$-$4
112	10	283.2	$k_H/M\,atm^{-1} = 49.0$	DIRECT	4.84E$-$4
112	25	298.2	$k_H/M\,atm^{-1} = 19.8$	DIRECT	1.95E$-$4
100	25	298.2	$k_{AW} = 1.9E{-}3$	DIRECT	2.12E$-$4
113	30	303.2	$k_{AW} = 3.9E{-}3$	HEADSPACE	1.02E$-$4
112	35	308.2	$k_H/M\,atm^{-1} = 10.9$	DIRECT	1.08E$-$4
112	45	318.2	$k_H/M\,atm^{-1} = 7.1$	DIRECT	0.701E$-$4
114	70	343.2	$\gamma^\infty = 26.7$	EBULLIOMETRY	0.282E$-$4
115	70	343.2	$\gamma^\infty = 29.5$	DIRECT	0.225E$-$4
114	80	353.2	$\gamma^\infty = 28.5$	EBULLIOMETRY	0.190E$-$4
114	90	363.2	$\gamma^\infty = 30.2$	EBULLIOMETRY	0.132E$-$4
114	100	373.2	$\gamma^\infty = 31.8$	EBULLIOMETRY	0.0936E$-$4

In these calculations, it has been assumed that the concentration of the solute was within the region where Henry's law is valid. These data may be represented by the equation

$$\ln(k_H/mol\,kg^{-1}\,Pa^{-1}) = -23.889 + 4571.7/(T/K)$$

The standard error of estimate of k_H by this equation is 17%. This equation generates RECOMMENDED data.

Solute		
Formula	Name	CAS Registry Number
$C_4H_8O_2$	Ethyl acetate	141-78-6
Solvent: Water		

Selected experimental measurements of k_H for this liquid at 25 °C are as follows:

	Original measurements		
Ref.	Limiting solubility ratio	Method	$k_H/\text{mol kg}^{-1}\,\text{Pa}^{-1}$
104b	$(1/k_H)/\text{mmHg} = 5574$	DIRECT	7.47E−5
116	$\gamma^\infty = 68.6$	DIRECT	6.41E−4
116	$\gamma^\infty = 56.8$	VP/AS	7.75E−5
117	$k_{AW} = 6.94\text{E}{-}3$	HEADSPACE	5.81E−5

Getzen *et al.* [118] presented critically evaluated smoothed data for the solubility of this liquid in water in the range $273 < T/K < 303$. The k_H data calculated from these solubilities are as follows:

$t/°C$	T/K	Solubility, x	$k_H/\text{mol kg}^{-1}\,\text{Pa}^{-1}$
0	273.2	2.17E−2	3.70E−4
10	283.2	1.91E−2	1.83E−4
20	293.2	1.74E−2	9.81E−5
25	298.2	1.61E−2	7.08E−5
30	303.2	1.54E−2	5.34E−5

In both tables, it has been assumed that the concentration of the solute was within the region where Henry's law is valid; in the latter table, that the activity of the equilibrium ethyl acetate was unity. Solute vapour pressure data were taken from Gmehling *et al.* [119].

Solute		
Formula	Name	CAS Registry Number
C_4H_{10}	Isobutane	75-28-5
Solvent: Water		

Hayduk [103] presented critically evaluated smoothed data for the solubility of this gas in water at a partial pressure of gas of 101.3 kPa in the range $278 < T/K < 318$. The data were represented by the equation

$$\ln x = -129.714 + 183.044/(T/100\,\text{K}) + 53.4651\ln(T/100\,\text{K})$$

The corresponding equation for k_H is

$$\ln(k_H/\text{mol kg}^{-1}\,\text{Pa}^{-1}) = -383.440 + 1.83044\text{E}4/(T/\text{K}) + 53.4651\ln(T/\text{K})$$

$t/^\circ\text{C}$	T/K	Solubility, x	$k_H/\text{mol kg}^{-1}\,\text{Pa}^{-1}$
5	278.2	9.983E−5	5.47E−8
10	283.2	8.096E−5	4.43E−8
15	288.2	6.723E−5	3.68E−8
20	293.2	5.708E−5	3.13E−8
25	298.2	4.948E−5	2.71E−8
30	303.2	4.373E−5	2.40E−8
35	308.2	3.937E−5	2.16E−8
40	313.2	3.606E−5	1.98E−8
45	318.2	3.356E−5	1.84E−8

These are RECOMMENDED data.

Solute		
Formula	Name	CAS Registry Number
C_4H_{10}	Butane	106-97-8
Solvent: Water		

Hayduk [103] presented critically evaluated smoothed data for the solubility of this gas in water at a gas partial pressure of 101.3 kPa in the range $273 < T/K < 353$. The data were represented by the equation

$$\ln x = -102.029 + 146.040/(T/100\,\text{K}) + 38.7599\ln(T/100\,\text{K})$$

and the corresponding expression for k_H is

$$\ln(k_H/\text{mol kg}^{-1}\,\text{Pa}^{-1}) = -288.034 + 1.4604\text{E}4/(T/\text{K}) + 38.7599\ln(T/\text{K})$$

$t/°C$	T/K	Solubility, x	$k_H/\text{mol kg}^{-1}\,\text{Pa}^{-1}$
0	273.2	6.665E−5	3.65E−8
5	278.2	5.150E−5	2.82E−8
10	283.2	4.065E−5	2.23E−8
15	288.2	3.273E−5	1.79E−8
20	293.2	2.686E−5	1.47E−8
25	298.2	2.244E−5	1.23E−8
30	303.2	1.906E−5	1.04E−8
35	308.2	1.644E−5	0.901E−8
40	313.2	1.440E−5	0.789E−8
45	318.2	1.278E−5	0.700E−8
50	323.2	1.150E−5	0.630E−8
55	328.2	1.047E−5	0.574E−8
60	333.2	0.9651E−5	0.529E−8
65	338.2	0.8992E−5	0.493E−8
70	343.2	0.8465E−5	0.464E−8
75	348.2	0.8048E−5	0.441E−8
80	353.2	0.7723E−5	0.423E−8

These are RECOMMENDED data.

Solute		
Formula	Name	CAS Registry Number
$C_4H_{10}O$	1-Butanol	71-36-3
Solvent: Water		

Available experimental measurements of k_H of this liquid are given in the following table.

			Original measurements		k_H/mol
Ref.	$t/°C$	T/K	Limiting solubility ratio	Method	$\text{kg}^{-1}\,\text{Pa}^{-1}$
43	0	273.2	$k_{AW} = 3.85\text{E}-5$	DIRECT	114.E−4
43	25	298.2	$k_{AW} = 3.23\text{E}-5$	DIRECT	12.5E−4
54	25	298.2	$(1/k_H)/\text{mmHg} = 358.$	HEADSPACE	11.6E−4
87	25	298.2	$\gamma^\infty = 50.2$	HEADSPACE	11.9E−4

Original measurements					k_H/mol kg^{-1} Pa^{-1}
Ref.	t/°C	T/K	Limiting solubility ratio	Method	
48	25	298.2	$\gamma^\infty = 53.33$	INDIRECT	11.2E−4
52	25	298.2	$\gamma^\infty = 53.7$	GC	11.1E−4
53	25	298.2	$\gamma^\infty = 45.1$	GC	13.2E−4
100	25	298.2	$k_{AW} = 3.6E-4$	DIRECT	11.2E−4
120	25	298.2	$(1/k_H)/\text{kPa} = 31.3$	DIRECT	17.7E−4
104a	25	298.2	$(1/k_H)/\text{mmHg} = 359.$	DIRECT	11.67E−4
113	30	303.2	$k_{AW} = 7.46E-4$	HEADSPACE	5.32E−4
106	40	313.2	$k_{AW} = 1.55E-3$	HEADSPACE	2.48E−4
121	50	323.2	$\gamma^\infty = 78.7$	DIRECT	1.58E−4
106	60	333.2	$k_{AW} = 4.2E-3$	HEADSPACE	0.860E−4
106	70	343.2	$k_{AW} = 6.49E-3$	HEADSPACE	0.505E−4
122	70	343.2	$\gamma^\infty = 67.8$	EBULLIOMETRY	0.615E−4
114	70	343.2	$\gamma^\infty = 59.3$	EBULLIOMETRY	0.704E−4
106	80	353.2	$k_{AW} = 1.01E-2$	HEADSPACE	0.337E−4
122	80	353.2	$\gamma^\infty = 46.5$	EBULLIOMETRY	0.552E−4
114	80	353.2	$\gamma^\infty = 57.2$	EBULLIOMETRY	0.449E−4
114	90	363.2	$\gamma^\infty = 55.5$	EBULLIOMETRY	0.294E−4
122	99	372.2	$\gamma^\infty = 27.1$	EBULLIOMETRY	0.412E−4
114	100	373.2	$\gamma^\infty = 54.0$	EBULLIOMETRY	0.199E−4

In these calculations, it has been assumed that the concentration of the solute was within the region where Henry's law is valid. These data may be represented by the equation

$$\ln(k_H/\text{mol kg}^{-1}\,\text{Pa}^{-1}) = -27.756 + 6255.0/(T/\text{K})$$

The standard error of estimate of k_H by this equation is 24 %. This equation generates RECOMMENDED values.

Solute		
Formula	Name	CAS Registry Number
$C_4H_{10}O$	2-Butanol	78-92-2
Solvent: Water		

Available experimental measurements of k_H of this liquid in water are as follows:

Original measurements				k_H/mol
Ref.	$t/°C$	T/K	Limiting solubility ratio	Method	$\text{kg}^{-1}\,\text{Pa}^{-1}$
43	0	273.2	$k_{AW} = 4.35\text{E}{-}5$	DIRECT	10.1E−3
120	20	293.2	$(1/k_H)/\text{kPa} = 36.0$	DIRECT	1.54E−3
54	25	298.2	$(1/k_H)/\text{mmHg} = 424.$	HEADSPACE	0.983E−3
43	25	298.2	$k_{AW} = 3.7\text{E}{-}4$	DIRECT	1.09E−3
104a	25	298.2	$(1/k_H)/\text{mmHg} = 431.$	DIRECT	0.967E−3
52	25	298.2	$\gamma^\infty = 22.4$	GC	1.01E−3
121	50	323.2	$\gamma^\infty = 33.5$	DIRECT	0.15E−3

In these calculations, solute vapour pressure data were taken from Gmehling *et al.* [98]; also, it has been assumed that the concentration of the solute was within the region where Henry's law is valid. These data may be represented by the equation

$$\ln(k_H/\text{mol}\,\text{kg}^{-1}\,\text{Pa}^{-1}) = -31.820 + 7432.5/(T/K)$$

The standard error of estimate of k_H by this equation is 3 %. This equation generates RECOMMENDED data.

Solute		
Formula	Name	CAS Registry Number
C_5H_8	Isoprene	78-79-5
Solvent: Water		

There are three reports of k_H of this gas in water at 25 °C:

	Original measurements		
Ref.	Limiting solubility ratio	Method	$k_H/\text{mol}\,\text{kg}^{-1}\,\text{Pa}^{-1}$
40	$\gamma^\infty = 5.89\text{E}3$	VP/AS	1.29E−7
111	$k_{WA} = 0.316$	–	1.28E−7
123	$(1/k_H)/\text{kPa}\,\text{mol}^3\,mol^{-1} = 7.78$	CRIT EVAL	1.29E−7

In these calculations, it has been assumed that the concentration of the solute was within the region where Henry's law is valid. The method of calculation in Ref. 111 was not stated; probably the datum was derived from aqueous solubility and vapour pressure.

Solute		
Formula	Name	CAS Registry Number
C_6Cl_6	Hexachlorobenzene	118-74-1
Solvent: Water		

There are two measurements for the solubility of this solid in water at 25 °C [125,126]. The k_H were calculated from the solubility data and the vapour pressure of solid hexachlorobenzene at 25 °C: 1.82E−3 Pa [124]:

Original data			
Ref.	Solubility, S	Solubility, x	$k_H/\text{mol kg}^{-1}\,\text{Pa}^{-1}$
1	$S/M = 1.8\text{E}-8$	3.2E−10	9.89E−6
2	$S/M = 1.65\text{E}-8$	2.97E−10	9.07E−6

These results are in fair agreement with the following available experimental measurements of k_H:

Original measurements					
Ref.	$t/°C$	T/K	Limiting solubility ratio	Method	$k_H/\text{mol kg}^{-1}\,\text{Pa}^{-1}$
127	20	293.2	$(1/k_H)/\text{atm m}^3\,\text{mol}^{-1} = 4.8\text{E}-4$	PURGE	2.06E−5
128	22	295.2	$(1/k_H)/\text{Pa m}^3\,\text{mol}^{-1} = 46.6$	PURGE	2.15E−5
128	24	295.2	$(1/k_H)/\text{Pa m}^3\,\text{mol}^{-1} = 52.5$	PURGE	1.90E−5
128	35	308.2	$(1/k_H)/\text{Pa m}^3\,\text{mol}^{-1} = 88.3$	PURGE	1.13E−5
128	50.5	323.7	$(1/k_H)/\text{Pa m}^3\,\text{mol}^{-1} = 217.2$	PURGE	0.460E−5

In these calculations, it has been assumed that the concentration of the solute was within the region where Henry's law is valid.

Solute		
Formula	Name	CAS Registry Number
C_6HCl_5	Pentachlorobenzene	608-93-5
Solvent: Water		

There are two measurements for the solubility of this solid in water at 25 °C. The k_H data were calculated by the simplified VP/AS method, using the vapour pressure of the supercooled liquid [129] and $\Delta_{fus}S° = 56\,\text{J mol}^{-1}\,\text{K}^{-1}$:

Ref.	Solubility, S	Solubility, x	$k_H/\mathrm{mol\,kg^{-1}\,Pa^{-1}}$
130	$S/M = 2.2E{-}6$	$3.97E{-}8$	$3.56E{-}5$
126	$S/M = 3.32E{-}3$	$5.98E{-}8$	$2.40E{-}5$

In these calculations, it has been assumed that the concentration of the solute was within the region where Henry's law is valid. These data are not in good agreement with experimental measurements of k_H:

			Original measurements		
Ref.	$t/°C$	T/K	Limiting solubility ratio	Method	$k_H/\mathrm{mol\,kg^{-1}\,Pa^{-1}}$
128	15	288.2	$(1/k_H)/\mathrm{Pa\,m^3\,mol^{-1}} = 37.4$	PURGE	$26.7E{-}6$
127	20	293.2	$(1/k_H)/\mathrm{atm\,m^3\,mol^{-1}} = 7.1E{-}4$	PURGE	$13.9E{-}6$
128	20	293.2	$(1/k_H)/\mathrm{Pa\,m^3\,mol^{-1}} = 49.4$	PURGE	$20.2E{-}6$
128	22	295.2	$(1/k_H)/\mathrm{Pa\,m^3\,mol^{-1}} = 68.1$	PURGE	$14.7E{-}6$
128	24	297.2	$(1/k_H)/\mathrm{Pa\,m^3\,mol^{-1}} = 66.7$	PURGE	$15.0E{-}6$
128	35	308.2	$(1/k_H)/\mathrm{Pa\,m^3\,mol^{-1}} = 124.1$	PURGE	$8.05E{-}6$
128	50.5	323.7	$(1/k_H)/\mathrm{Pa\,m^3\,mol^{-1}} = 276.2$	PURGE	$3.62E{-}6$

In these calculations, it has been assumed that the concentration of the solute was within the region where Henry's law is valid. These data from the PURGE method may be represented by the equation

$$\ln(k_H/\mathrm{mol\,kg^{-1}\,Pa^{-1}}) = -27.873 + 4965.9/(T/K)$$

The standard error of estimate of k_H by this equation is 9%. This equation generates RECOMMENDED data in the range 288–323 K.

	Solute	
Formula	Name	CAS Registry Number
C_6HCl_5O	Pentachlorophenol	87-86-5
Solvent: Water		

Horvath and Getzen [130a] presented critically evaluated data for the solubility of this solid in water at 25 °C. A tentative value of $x = 6.8E{-}7$ was given. This corresponds to $\gamma^\infty = 1.47E6$. If the vapour pressure of the supercooled liquid be taken as 3.55 Pa [40], the calculated k_H for this compound at 25 °C is $1.06E{-}5\ \mathrm{mol\,kg^{-1}\,Pa^{-1}}$.

There are two other estimates of k_H for this compound:

Ref.	T/K	Quantity	Unit	Value	$10^3\,\text{mol}\,\text{kg}^{-1}\,\text{Pa}^{-1}$	Method
33	293.15	$1/k_H$	$\text{Pa}\,\text{m}^3\,\text{mol}^{-1}$	0.044	23	VP/AS
40	298.15	γ^∞	–	6010.	2.6	VP/AS

The data are not entirely consistent, but an order-of-magnitude estimate for k_H can be made as $1.8\text{E}-3\,\text{mol}\,\text{kg}^{-1}\,\text{Pa}^{-1}$.

Solute		
Formula	Name	CAS Registry Number
$C_6H_2Cl_4$	1,2,3,4-Trichlorobenzene	634-66-2
Solvent: Water		

There are two measurements for the solubility of this solid in water [126,130]. Values of k_H were calculated by the simplified VP/AS method from these data, the vapour pressure of the supercooled liquid [131], viz. 8.00 Pa, and $\Delta_{\text{fus}}S^\circ = 56\,\text{J}\,\text{mol}^{-1}\text{K}^{-1}$:

	Original measurements			
Ref.	$t/^\circ\text{C}$	T/K	Solubility, S	$k_H/\text{mol}\,\text{kg}^{-1}\,\text{Pa}^{-1}$
130	25	298.2	$S/M = 2.0\text{E}-5$	$4.10\text{E}-6$
126	25	298.2	$S/M = 5.65\text{E}-5$	$1.15\text{E}-5$

These data are in fair agreement with other measurements of k_H:

		Original measurements			
Ref.	$t/^\circ\text{C}$	T/K	Limiting solubility ratio	Method	$k_H/\text{mol}\,\text{kg}^{-1}\,\text{Pa}^{-1}$
128	20	293.2	$(1/k_H)/\text{Pa}\,\text{m}^3\,\text{mol}^{-1} = 52$	PURGE	$1.92\text{E}-5$
128	22	295.2	$(1/k_H)/\text{Pa}\,\text{m}^3\,\text{mol}^{-1} = 68.1$	PURGE	$1.47\text{E}-5$
128	24	297.2	$(1/k_H)/\text{Pa}\,\text{m}^3\,\text{mol}^{-1} = 70.9$	PURGE	$1.41\text{E}-5$
128	35	308.2	$(1/k_H)/\text{Pa}\,\text{m}^3\,\text{mol}^{-1} = 127.9$	PURGE	$7.82\text{E}-6$
128	50.5	324.7	$(1/k_H)/\text{Pa}\,\text{m}^3\,\text{mol}^{-1} = 276.2$	PURGE	$3.62\text{E}-6$

In these calculations, it has been assumed that the concentration of the solute was within the region where Henry's law is valid.

The data in this table may be represented by the equation

$$\ln(k_H/\text{mol kg}^{-1}\,\text{Pa}^{-1}) = -27.898 + 4972.5/(T/\text{K})$$

The standard error of estimate of k_H by this equation is 4%. This equation generates RECOMMENDED k_H data in the range 293–325 K.

Solute		
Formula	Name	CAS Registry Number
$C_6H_2Cl_4$	1,2,3,5-Tetrachlorobenzene	634-90-2
Solvent: Water		

There are two measurements for the solubility of this solid in water at 25 °C [126,130a]. The k_H values were calculated by the simplified VP/AS method from these data, the vapour pressure of the supercooled liquid [129] and $\Delta_{fus}S° = 56\,\text{J mol}^{-1}\,\text{K}^{-1}$:

Ref.	Solubility, S	Solubility, x	$k_H/\text{mol kg}^{-1}\,\text{Pa}^{-1}$
1	$S/M = 1.6E{-}5$	2.89E−7	2.11E−6
2	$S/M = 1.34E{-}5$	2.41E−7	1.76E−6

In these calculations, it has been assumed that the concentration of the solute was within the region where Henry's law is valid. These data are not in good agreement with experimental measured data:

Original measurements					
Ref.	$t/°C$	T/K	Limiting solubility ratio	Method	$k_H/\text{mol kg}^{-1}\,\text{Pa}^{-1}$
128	20	293.2	$(1/k_H)/\text{Pa m}^3\,\text{mol}^{-1} = 99$	PURGE	1.01E−5
132	25	298.2	$(1/k_H)/\text{Pa m}^3\,\text{mol}^{-1} = 160$	PURGE	0.625E−5

Solute		
Formula	Name	CAS Registry Number
$C_6H_2Cl_4$	1,2,4,5-Tetrachlorobenzene	95-94-3
Solvent: Water		

There are two reported measurements of the solubility of this solid in water at 25 °C [126,130]. The k_H values were calculated by means of the simplified VP/AS method, using a vapour pressure of the supercooled liquid of 28.0 Pa [129] and $\Delta_{fus}S^\circ = 56\,J\,mol^{-1}K^{-1}$:

Ref.	Solubility, S	Solubility, x	$k_H/mol\,kg^{-1}\,Pa^{-1}$
130	$S/M = 2.8E{-}6$	5.06E−8	2.78E−7
126	$S/M = 10.9E{-}6$	19.6E−8	1.08E−6

In these calculations, it has been assumed that the concentration of the solute was within the region where Henry's law is valid. There is one measured k_H datum [127]: $(1/k_H)/atm\,m^3\,mol^{-1} = 1.0E{-}3$, or $k_H/mol\,kg^{-1}\,Pa^{-1} = 9.87E{-}6$ at 20 °C.

Solute		
Formula	Name	CAS Registry Number
$C_6H_3Cl_3$	1,2,3-Trichlorobenzene	87-61-6
Solvent: Water		

There are two reports of the solubility of this solid in water at 25 °C [126,130]. The k_H values were calculated by the simplified VP/AS method with these data, a vapour pressure of the supercooled liquid of 60.2 Pa [129] and $\Delta_{fus}S^\circ = 56\,J\,mol^{-1}\,K^{-1}$:

Ref.	Solubility, S	Solubility, x	$k_H/mol\,kg^{-1}\,Pa^{-1}$
130	$S/M = 1.7E{-}4$	3.07E−6	5.86E−6
126	$S/M = 0.676E{-}4$	2.14E−6	2.14E−6

These data are not in good agreement with experimental measurements of k_H:

	Original measurements			
Ref.	$t/°C$	T/K	Limiting solubility ratio	Method	$k_H/mol\,kg^{-1}\,Pa^{-1}$
127	20	293.2	$(1/k_H)/atm\,m^3\,mol^{-1} = 8.9E{-}4$	PURGE	1.11E−5
128	20	293.2	$(1/k_H)/Pa\,m^3\,mol^{-1} = 72$	PURGE	1.39E−5
132	25	298.2	$(1/k_H)/Pa\,m^3\,mol^{-1} = 127$	PURGE	0.787E−5

In these calculations, it has been assumed that the concentration of the solute was within the region where Henry's law is valid.

Solute		
Formula	Name	CAS Registry Number
$C_6H_3Cl_3$	1,2,4-Trichlorobenzene	120-82-1
Solvent: Water		

There are two reported measurements for the solubility of this liquid in water at 25 °C [126,130a]. The k_H values were calculated from these data with the vapour pressure at this temperature, 29.4 Pa [133].

Ref.	Solubility, S	Solubility, x	$k_H/\text{mol kg}^{-1}\,\text{Pa}^{-1}$
130a	$S/M = 2.09E{-}4$	3.76E$-$6	7.10E$-$6
126	$S/M = 2.54E{-}4$	4.58E$-$6	8.65E$-$6

These are in fair agreement with measured experimental data at 20 °C:

	Original measurements		
Ref.	Limiting solubility ratio	Method	$k_H/\text{mol kg}^{-1}\,\text{Pa}^{-1}$
127	$(1/k_H)/\text{atm m}^3\,\text{mol}^{-1} = 1.2E{-}3$	PURGE	8.22E$-$6
128	$(1/k_H)/\text{Pa m}^3\,\text{mol}^{-1} = 101$	PURGE	9.90E$-$6

In these calculations, it has been assumed that the concentration of the solute was within the region where Henry's law is valid.

Solute		
Formula	Name	CAS Registry Number
$C_6H_3Cl_3$	1,3,5-Trichlorobenzene	108-70-3
Solvent: Water		

There are two reported measurements of the solubility of this solid in water at 25 °C [126,130]. The k_H values were calculated by the simplified VP/AS method with these data, the vapour pressure of the supercooled liquid 64.0 Pa [129] and $\Delta_{\text{fus}}S° = 56\,\text{J mol}^{-1}\,\text{K}^{-1}$:

Ref.	Solubility, S	Solubility, x	$k_H/\text{mol kg}^{-1}\,\text{Pa}^{-1}$
130	$S/M = 3.6E{-}5$	6.51E$-$7	1.36E$-$6
126	$S/M = 2.27E{-}5$	4.09E$-$7	8.48E$-$7

There are two measurements of k_H of this solid in water [65,128]:

			Original measurements		
Ref.	$t/°C$	T/K	Limiting solubility ratio	Method	$k_H/\text{mol kg}^{-1}\,\text{Pa}^{-1}$
128	20	293.2	$(1/k_H)/\text{Pa m}^3\,\text{mol}^{-1} = 192$	PURGE	0.521E−5
65	22	295.2	$k_{WA} = 87$	DIRECT	3.55E−5

In these calculations, it has been assumed that the concentration of the solute was within the region where Henry's law is valid. The great disparity in these results is unexplained.

Solute		
Formula	Name	CAS Registry Number
$C_6H_3Cl_3$	2,4,5-Trichlorophenol	95-95-4
Solvent: Water		

Horvath and Getzen [130a] presented critically evaluated data for the solubility of this solid in water at 25 °C. A tentative value was given as $x = 9.0E - 5$. A value for k_H was calculated by the simplified VP/AS method, using a vapour pressure of the supercooled liquid of 28.6 Pa [40] and $\Delta_{fus}S° = 56\,\text{J mol}^{-1}\,\text{K}^{-1}$. The calculated datum is 4.66E−4 mol kg^{-1} Pa^{-1}.

Hwang *et al.* [40] made the same kind of estimate, stating that $\gamma^\infty = 3.93E3$, or $k_H = 4.94E - 4\,\text{mol kg}^{-1}\,\text{Pa}^{-1}$ at the same temperature. An order-of-magnitude estimate for k_H of this compound is therefore 4.8E−4 mol kg^{-1} Pa^{-1}.

Solute		
Formula	Name	CAS Registry Number
$C_6H_3Cl_3O$	2,4,6-Trichlorophenol	88-06-2
Solvent: Water		

There are two estimates for k_H of this compound:

			Original measurements		
Ref.	$t/°C$	T/K	Limiting solubility ratio	Method	$k_H/\text{mol kg}^{-1}\,\text{Pa}^{-1}$
134	18	291.2	$(1/k_H)/\text{atm m}^3\,\text{mol}^{-1} = 1.3E{-}6$	VP/AS	7.59E−3
135	20	293.2	$(1/k_H)/\text{mmHg M}^{-1} = 3.92$	VP/AS	1.91E−3

This provides only an order-of-magnitude estimate: $k_H/\text{mol}\,\text{kg}^{-1}\,\text{Pa}^{-1} = 7.5\text{E}{-}3$.

Solute		
Formula	Name	CAS Registry Number
$C_6H_4Cl_2$	1,2-Dichlorobenzene	95-50-1
Solvent: Water		

Horvath and Getzen [130a] presented critically evaluated smoothed data for the solubility of this liquid in water in the range $273 < T/\text{K} < 333$. The data were represented by the equation

$$x = 2.35679\text{E}{-}3 - 2.21985\text{E}{-}5(T/\text{K}) + 6.92512\text{E}{-}8(T/\text{K})^2$$
$$- 7.07945\text{E}{-}11(T/\text{K})^3$$

where x is mole fraction. The k_H data in the following table were calculated from this equation and vapour pressure from Gmehling $et\ al.$ [34]:

$t/^\circ$C	T/K	Solubility, x	$k_H/\text{mol}\,\text{kg}^{-1}\,\text{Pa}^{-1}$
0	273.2	1.738E−5	42.4E−6
5	278.2	1.659E−05	26.3E−6
10	283.2	1.630E−5	17.1E−6
15	288.2	1.647E−5	11.7E−6
20	293.2	1.705E−5	8.31E−6
25	298.2	1.798E−5	6.13E−6
30	303.2	1.920E−5	4.65E−6
35	308.2	2.066E−5	3.60E−6
40	313.2	2.232E−5	2.84E−6
45	318.2	2.411E−5	2.27E−6
50	323.2	2.600E−5	1.83E−6
55	328.2	2.790E−5	1.49E−6
60	333.2	2.978E−5	1.22E−6

In these calculations, it has been assumed that the concentration of the solute was within the region where Henry's law is valid. Experimental measurements of k_H are given in the following table; solute vapour pressure data were taken from Gmehling $et\ al.$ [34]:

			Original measurements		
Ref.	$t/°C$	T/K	Limiting solubility ratio	Method	$k_H/\text{mol kg}^{-1}\text{Pa}^{-1}$
89	15	288.2	$(1/k_H)/\text{kPa m}^3\text{mol}^{-1} = 0.1201$	EPICS	8.33E$-$6
64	20	293.2	$(1/k_H)/\text{atm m}^3\text{mol}^{-1} = 12.\text{E}-4$	PURGE	8.22E$-$6
65	22	295.2	$k_{WA} = 23$	DIRECT	9.38E$-$6
48	25	298.2	$\gamma^\infty = 6.82\text{E}4$	INDIRECT	4.52E$-$6
132	25	298.2	$(1/k_H)/\text{Pa m}^3\text{mol}^{-1} = 195$	PURGE	5.13E$-$6
37	35	308.2	$k_{AW} = 0.121$	EPICS	3.23E$-$6
89	35	308.2	$(1/k_H)/\text{kPa m}^3\text{mol}^{-1} = 0.3199$	EPICS	3.13E$-$6
37	45	318.2	$k_{AW} = 0.179$	EPICS	2.11E$-$6
37	55	328.2	$k_{AW} = 0.242$	EPICS	1.52E$-$6
37	60	333.2	$k_{AW} = 0.280$	EPICS	1.29E$-$6

In these calculations, it has been assumed that the concentration of the solute was within the region where Henry's law is valid. The combined data of both tables may be represented by the equation

$$\ln(k_H/\text{mol kg}^{-1}\,\text{Pa}^{-1}) = -29.527 + 5258.8/(T/K)$$

The standard error of estimate of k_H by this equation is 12%. This equation generates RECOMMENDED data in the temperature range 273–333 K.

	Solute	
Formula	Name	CAS Registry Number
$C_6H_4Cl_2$	1,3-Dichlorobenzene	541-73-1
Solvent: Water		

Horvath and Getzen [130a] presented critically evaluated smoothed data for the solubility of this liquid in the temperature range $283 < T/K < 333$. The data were represented by the equation

$$x = 3.39249\text{E}-3 - 3.20585\text{E}-5(T/K) + 1.00454\text{E}-7(T/K)^2$$
$$- 1.03797\text{E} - 10(T/K)^3$$

where x is mole fraction. The k_H values were calculated from this equation with vapour pressure from Rohac et al. [133].

$t/°C$	T/K	Solubility, x	$k_H/\text{mol kg}^{-1}\,\text{Pa}^{-1}$
10	283.2	1.260E−5	8.11E−6
15	288.2	1.221E−5	5.44E−6
20	293.2	1.235E−5	3.87E−6
25	298.2	1.296E−5	2.90E−6
30	303.2	1.394E−5	2.25E−6
35	308.2	1.523E−5	1.81E−6
40	313.2	1.674E−5	1.47E−6
45	318.2	1.840E−5	1.22E−6
50	323.2	2.013E−5	1.01E−6
55	328.2	2.185E−5	0.841E−6
60	333.2	2.349E−5	0.700E−6

In these calculations, it has been assumed that the concentration of the solute was within the region where Henry's law is valid. Experimental measurements of k_H are as follows:

			Original measurements		
Ref.	$t/°C$	T/K	Limiting solubility ratio	Method	$k_H/\text{mol kg}^{-1}\,\text{Pa}^{-1}$
64	20	293.2	$(1/k_H)/\text{MPa} = 11.7$	PURGE	4.75E−6
127	20	293.2	$(1/k_H)/\text{atm m}^3\,\text{mol}^{-1} = 18.E-4$	PURGE	5.48E−6
37	25	298.2	$k_{AW} = 0.100$	EPICS	4.04E−6
37	35	308.2	$k_{AW} = 0.176$	EPICS	2.22E−6
37	45	318.2	$k_{AW} = 0.252$	EPICS	1.50E−6
37	55	328.2	$k_{AW} = 0.337$	EPICS	1.09E−6
37	60	333.2	$k_{AW} = 0.376$	EPICS	0.961E−6

In these calculations, it has been assumed that the concentration of the solute was within the region where Henry's law is valid. The combined data of both tables may be represented by the equation

$$\ln(k_H/\text{mol kg}^{-1}\,\text{Pa}^{-1}) = -27.076 + 4319.9/(T/K)$$

The standard error of estimate of k_H by this equation is 13 %. This equation generates RECOMMENDED data in the temperature range 283–333 K.

Solute		
Formula	Name	CAS Registry Number
$C_6H_4Cl_2$	1,4-Dichlorobenzene	106-46-7
Solvent: Water		

Horvath and Getzen [130a] presented critically evaluated smoothed data for the solubility of this liquid in water in the range $328 < T/K < 348$. The data were represented by the equation

$$x = 1.7125\text{E}{-}3 - 1.05183\text{E}{-}5(T/\text{K}) + 1.63787\text{E}{-}8(T/\text{K})^2$$

where x is mole fraction. The k_H values were calculated from this equation with vapour pressure from Gmehling *et al.* [34].

$t/{\circ}C$	T/K	Solubility, x	$k_H/\text{mol kg}^{-1}\,\text{Pa}^{-1}$
55	328.2	2.462E$-$5	10.4E$-$7
60	333.2	2.618E$-$5	8.60E$-$7
65	338.2	2.857E$-$5	7.35E$-$7
70	343.2	3.177E$-$5	6.45E$-$7
75	348.2	3.579E$-$5	5.78E$-$7

In these calculations, it has been assumed that the concentration of the solute was within the region where Henry's law is valid. Experimentally measured k_H data are as follows:

			Original measurements		
Ref.	$t/{\circ}C$	T/K	Limiting solubility ratio	Method	$k_H/\text{mol kg}^{-1}\,\text{Pa}^{-1}$
64	20	293.2	$(1/k_H)/\text{MPa} = 10.3$	PURGE	5.39E$-$6
127	20	293.2	$(1/k_H)/\text{atm m}^3\,\text{mol}^{-1} = 15\text{E}{-}4$	PURGE	6.58E$-$6
132	25	298.2	$(1/k_H)/\text{Pa m}^3\,\text{mol}^{-1} = 244$	PURGE	4.10E$-$6
37	25	298.2	$k_{AW} = 0.079$	EPICS	5.11E$-$6
37	35	308.2	$k_{AW} = 0.149$	EPICS	2.62E$-$6
37	45	318.2	$k_{AW} = 0.232$	EPICS	1.63E$-$6
37	55	328.2	$k_{AW} = 0.304$	EPICS	1.21E$-$6
37	60	333.2	$k_{AW} = 0.343$	EPICS	1.05E$-$6

In these calculations, it has been assumed that the concentration of the solute was within the region where Henry's law is valid. The combined data of both tables may be represented by

$$\ln(k_H/\text{mol kg}^{-1}\,\text{Pa}^{-1}) = -27.230 + 4447.9/(T/\text{K})$$

The standard error of estimate of k_H by this equation is 9%. This equation generates RECOMMENDED data in the range 293–348 K.

Solute		
Formula	Name	CAS Registry Number
$C_6H_4Cl_2O$	2,4-Dichlorophenol	120-83-2
Solvent: Water		

There is one report of the experimental measurement of k_H of this solid in water [136]:

		Original measurements		
$t/°C$	T/K	Limiting solubility ratio	Method	$k_H/mol\,kg^{-1}\,Pa^{-1}$
50	323.2	$(1/k_H)/kPa = 48.2$	EBULLIOMETRY	11.5E−4
60	333.2	$(1/k_H)/kPa = 94.0$	EBULLIOMETRY	5.90E−4
70	343.2	$(1/k_H)/kPa = 170.3$	EBULLIOMETRY	3.26E−4
80	353.2	$(1/k_H)/kPa = 296.6$	EBULLIOMETRY	1.87E−4
90	363.2	$(1/k_H)/kPa = 486.9$	EBULLIOMETRY	1.14E−4

In these calculations, it has been assumed that the concentration of the solute was within the region where Henry's law is valid. These data may be represented by the equation

$$\ln(k_H/mol\,kg^{-1}\,Pa^{-1}) = -27.776 + 6781.6/(T/K)$$

The standard error of estimate of k_H by this equation is 2 %. The extrapolated value of k_H at 25 °C is 6.54E−3 mol kg^{-1} Pa^{-1}. This datum and those in the table are not recommended, however, because other k_H data in Tabai *et al.* [136] are inconsistent with other independently measured k_H data for this substance.

Solute		
Formula	Name	CAS Registry Number
C_6H_5Br	Bromobenzene	108-86-1
Solvent: Water		

Horvath and Getzen [130a] presented critically evaluated smoothed data for the solubility of this liquid in water in the range $283 < T/K < 313$. The data were represented by the equation

$$x = 2.09901E-4 - 1.55679E-6(T/K) + 3.43454E-9(T/K)^2$$

where x is mole fraction. The following k_H data were calculated from this equation and vapour pressure data from Gmehling *et al.* [34]:

$t/^{\circ}C$	T/K	Solubility, x	$k_H/mol\,kg^{-1}\,Pa^{-1}$
10	283.2	4.445E−5	12.2E−6
15	288.2	4.648E−5	8.98E−6
20	293.2	4.868E−5	6.74E−6
25	298.2	5.105E−5	5.14E−6
30	303.2	5.359E−5	3.97E−6
35	308.2	5.631E−5	3.12E−6
40	313.2	5.919E−5	2.47E−6

In these calculations, it has been assumed that the concentration of the solute was within the region where Henry's law is valid. Experimental measurements of k_H are as follows:

			Original measurements		k_H/mol
Ref.	$t/^{\circ}C$	T/K	Limiting solubility ratio	Method	$kg^{-1}\,Pa^{-1}$
64	10	283.2	$(1/k_H)/MPa = 5.01$	PURGE	11.1E−6
64	20	293.2	$(1/k_H)/MPa = 6.08$	PURGE	6.08E−6
48	25	298.2	$\gamma^{\infty} = 2.24E−4$	INDIRECT	4.49E−6
38	25	298.2	$(1/k_H)/atm = 115.9$	VP/AS	4.72E−6
37	25	298.2	$k_{AW} = 0.075$	EPICS	5.38E−6
111	25	298.2	$k_{AW} = 0.085$	VP/AS	4.75E−6
123	25	298.2	$(1/k_H)/kPa\,m^3\,mol^{-1} = 0.21$	CRIT EVAL	4.76E−6
132	25	298.2	$(1/k_H)/Pa\,m^3\,mol^{-1} = 250$	PURGE	4.00E−6
79	30	303.2	$(1/k_H)/kPa\,m^3\,mol^{-1} = 0.256$	EPICS	3.91E−6
64	30	303.2	$(1/k_H)/MPa = 15.3$	PURGE	3.63E−6
79	35	308.2	$(1/k_H)/kPa\,m^3\,mol^{-1} = 0.332$	EPICS	3.01E−6
37	35	308.2	$k_{AW} = 0.141$	EPICS	2.77E−6
37	45	318.2	$k_{AW} = 0.192$	EPICS	1.97E−6
64	50	323.2	$(1/k_H)/MPa = 32.8$	PURGE	1.69E−6
37	55	328.2	$k_{AW} = 0.230$	EPICS	1.62E−6
37	60	333.2	$k_{AW} = 0.292$	EPICS	1.24E−6

In these calculations, it has been assumed that the concentration of the solute was within the region where Henry's law is valid. The combined data of both tables may be represented by the equation

$$\ln(k_H/mol\,kg^{-1}\,Pa^{-1}) = -26.283 + 4195.0/(T/K)$$

The standard error of estimate of k_H by this equation is 7 %. This equation generates RECOMMENDED data in the temperature range 283–333 K.

Solute		
Formula	Name	CAS Registry Number
C_6H_5Cl	Chlorobenzene	108-90-7
Solvent: Water		

Horvath and Getzen [130a] presented critically evaluated smoothed data for the solubility of this liquid in water in the range $283 < T/K < 363$. The data were represented by the equation

$$x = 1.81349E{-}3 - 4.84607E{-}6(T/K) - 2.99435E{-}8(T/K)^2$$
$$+ 8.95052E{-}11(T/K)^3$$

The k_H values were calculated from these data with vapour pressure data from Gmehling *et al.* [34].

$t/°C$	T/K	Solubility, x	$k_H/\text{mol kg}^{-1}\,\text{Pa}^{-1}$
10	283.2	7.252E−5	6.05E−6
15	288.2	7.230E−5	4.45E−6
20	293.2	7.446E−5	3.42E−6
25	298.2	7.906E−5	2.74E−6
30	303.2	8.617E−5	2.28E−6
35	308.2	9.584E−5	1.95E−6
40	313.2	10.82E−5	1.71E−6
45	318.2	12.32E−5	1.53E−6
50	323.2	14.10E−5	1.39E−6
55	328.2	16.20E−5	1.27E−6
60	333.2	18.52E−5	1.17E−6
65	338.2	21.17E−5	1.09E−6
70	343.2	24.13E−5	1.01E−6

In these calculations, it has been assumed that the concentration of the solute was within the region where Henry's law is valid. These data may be represented by the equation

$$\ln(k_H/\text{mol kg}^{-1}\,\text{Pa}^{-1}) = -20.930 + 2440.7/(T/K)$$

The standard error of estimate of k_H by this equation is 10%. This equation generates RECOMMENDED data in the range 283–343 K.

Solute		
Formula	Name	CAS Registry Number
C_6H_5ClO	2-Chlorophenol	95-57-8
Solvent: Water		

Horvath and Getzen [130a] presented critically evaluated data for the solubility of this liquid in water at 25 °C. A tentative value was given as $x = 2.9E{-}3$. On the assumption that the vapour pressure of the liquid at this temperature is 130.5 Pa (extrapolated [136], the calculated k_H is $1.23E{-}3\,mol\,kg^{-1}\,Pa^{-1}$.

Other reported estimates and experimental measurements are as follows:

			Original measurements		k_H/mol
Ref.	$t/°C$	T/K	Limiting solubility ratio	Method	$kg^{-1}\,Pa^{-1}$
137	25	298.2	$(1/k_H)/Pa\,m^3\,mol^{-1} = 0.688$	VP/AS	1.45E$-$3
136	50	323.2	$(1/k_H)/kPa = 66.3$	EBULLIOMETRY	0.837E$-$3
136	60	333.2	$(1/k_H)/kPa = 114.8$	EBULLIOMETRY	0.484E$-$3
136	70	343.2	$(1/k_H)/kPa = 191.1$	EBULLIOMETRY	0.290E$-$3
136	80	353.2	$(1/k_H)/kPa = 300.3$	EBULLIOMETRY	0.185E$-$3
136	90	363.2	$(1/k_H)/kPa = 453.9$	EBULLIOMETRY	0.122E$-$3

In these calculations, it has been assumed that the concentration of the solute was within the region where Henry's law is valid. These data may be represented by the equation

$$\ln(k_H/mol\,kg^{-1}\,Pa^{-1}) = -6.7952 + 327.96/(T/K).$$

The standard error of estimate of k_H by this equation is 2 %. This equation generates RECOMMENDED data in the range 298–363 K.

Solute		
Formula	Name	CAS Registry Number
C_6H_5ClO	3-Chlorophenol	108-43-0
Solvent: Water		

Horvath and Getzen [130a] presented critically evaluated data for the solubility of this solid in water at 25 °C. A tentative value was given as $x = 3.2E{-}3$. On the assumption that the vapour pressure of the supercooled liquid at this temperature is 41.5 Pa (extrapolated [136] and $\Delta_{fus}S° = 56\,J\,mol^{-1}\,K^{-1}$, k_H was calculated by the simplified VP/AS method as $7.36E{-}3\,mol\,kg^{-1}\,Pa^{-1}$.

Other reported estimates and experimental measurements are as follows:

			Original measurements		k_H/mol
Ref.	$t/°C$	T/K	Limiting solubility ratio	Method	$kg^{-1}\,Pa^{-1}$
137	25	298.2	$(1/k_H)/Pa\,m^3\,mol^{-1} = 0.205$	VP/AS	4.88E−3
136	50	323.2	$(1/k_H)/kPa = 8.5$	EBULLIOMETRY	6.53E−3
136	60	333.2	$(1/k_H)/kPa = 16.0$	EBULLIOMETRY	3.47E−3
136	70	343.2	$(1/k_H)/kPa = 28.1$	EBULLIOMETRY	1.98E−3
136	80	353.2	$(1/k_H)/kPa = 46.5$	EBULLIOMETRY	1.19E−3
136	90	363.2	$(1/k_H)/kPa = 76.9$	EBULLIOMETRY	0.722E−3

In these calculations, it has been assumed that the concentration of the solute was within the region where Henry's law is valid. The data from Ref. 136 may be represented by

$$\ln(k_H/mol\,kg^{-1}\,Pa^{-1}) = -24.953 + 6432.8/(T/K)$$

The standard error of estimate of k_H by this equation is 1%. This equation generates RECOMMENDED data in the range 323–363 K.

Solute		
Formula	Name	CAS Registry Number
C_6H_5ClO	4-Chlorophenol	106-48-9
Solvent: Water		

Horvath and Getzen [130a] presented critically evaluated data for the solubility of this solid in water at 25 °C. A tentative value was given as $x = 3.9E{-}3$. On the assumption that the vapour pressure of the supercooled liquid at this temperature is 26.6 Pa (extrapolated [136] and $\Delta_{fus}S° = 56\,J\,mol^{-1}\,K^{-1}$, k_H was calculated by the simplified VP/AS method as $1.22E{-}2\,mol\,kg^{-1}\,Pa^{-1}$.

Other reported estimates and experimental measurements are as follows:

			Original measurements		k_H/mol
Ref.	$t/°C$	T/K	Limiting solubility ratio	Method	$kg^{-1}\,Pa^{-1}$
137	25	298.2	$(1/k_H)/Pa\,m^3\,mol^{-1} = 9.52E{-}2$	VP/AS	1.05E−2
136	50	323.2	$(1/k_H)/kPa = 0.5$	EBULLIOMETRY	11.0E−2
136	60	333.2	$(1/k_H)/kPa = 2.7$	EBULLIOMETRY	2.06E−2

Ref. $t/°C$ T/K	Original measurements		k_H/mol $kg^{-1} Pa^{-1}$
	Limiting solubility ratio	Method	
136 70 343.2	$(1/k_H)/kPa = 6.1$	EBULLIOMETRY	9.10E−3
136 80 353.2	$(1/k_H)/kPa = 12.9$	EBULLIOMETRY	4.30E−3
136 90 363.2	$(1/k_H)/kPa = 24.7$	EBULLIONMETRY	2.25E−3

In these calculations, it has been assumed that the concentration of the solute was within the region where Henry's law is valid. The data from Ref. 136 may be represented by

$$\ln(k_H/mol\,kg^{-1}\,Pa^{-1}) = -36.742 + 11059/(T/K)$$

The standard error of estimate of k_H by this equation is 21 %. There is a large discrepancy between the data at high and low temperatures, and recommended values cannot be given.

Solute		
Formula	Name	CAS Registry Number
$C_7H_7Cl_3NO_3PS$	Chlorpyrifos methyl	5598-13-0
Solvent: Water		

There are two estimates of k_H for this solid:

Ref. $t/°C$ T/K	Original measurements		k_H/mol $kg^{-1} Pa^{-1}$
	Limiting solubility ratio	Method	
33 20 293.2	$(1/k_H)/Pa\,m^3\,mol^{-1} = 0.34$	VP/AS	2.94E−3
123 25 298.2	$(1/k_H)/kPa\,m^3\,mol^{-1} = 3.4E{-}4$	CRIT EVAL	3.29E−3

In these calculations, it has been assumed that the concentration of the solute was within the region where Henry's law is valid. These data permit only an order-of-magnitude estimate of k_H at ambient temperature: $3.1E{-}3\,mol\,kg^{-1}\,Pa^{-1}$.

Solute		
Formula	Name	CAS Registry Number
C_7H_8	Toluene	108-88-3
Solvent: Water		

Shaw [138] presented critically evaluated smoothed data for the solubility of this liquid in water in the range $273 < T/K < 328$. The data may be represented by

$$x = 3.16638 - 1.58395E{-}7(T/K) + 2.36013E{-}10(T/K)^2$$

$t/°C$	T/K	Solubility, x	$k_H/\text{mol kg}^{-1}\,\text{Pa}^{-1}$
0	273.2	1.382E−4	8.57E−6
5	278.2	1.289E−4	5.84E−6
10	283.2	1.216E−4	4.07E−6
15	288.2	1.163E−4	2.92E−6
20	293.2	1.133E−4	2.16E−6
25	298.2	1.128E−4	1.65E−6
30	303.2	1.149E−4	1.31E−6
35	308.2	1.198E−4	1.07E−6
40	313.2	1.278E−4	0.900E−6
45	318.2	1.388E−4	0.781E−6
50	323.2	1.533E−4	0.694E−6
55	328.2	1.712E−4	0.628E−6

In these calculations, it has been assumed that the concentration of the solute was within the region where Henry's law is valid. When these data are plotted as $\ln k_H$ vs inverse temperature, there is a definite curvature. However, the data may be represented by the equation

$$\ln(k_H/\text{mol kg}^{-1}\,\text{Pa}^{-1}) = -27.551 + 4286.9/(T/K)$$

The standard error of estimate of k_H by this equation is 10%. This equation generates RECOMMENDED data in the range 273–328 K.

Solute		
Formula	Name	CAS Registry Number
$C_7H_{12}ClN_5$	Simazine	122-34-9
Solvent: Water		

There are two estimates of k_H for this solid. Although no temperature was stated in Ref. 139, it was assumed to be 23 °C.

			Original measurements		
Ref.	$t/°C$	T/K	Limiting solubility ratio	Method	$k_H/\text{mol}\,\text{kg}^{-1}\,\text{Pa}^{-1}$
33	20	293.2	$(1/k_H)/\text{Pa}\,\text{m}^3\,\text{mol}^{-1} = 3.4\text{E}{-4}$	VP/AS	2.94
139	23	296.2	$k_{AW} = 2.5\text{E}{-8}$	VP/AS	1.62

In these calculations, it has been assumed that the concentration of the solute was within the region where Henry's law is valid. These data allow only an order-of-magnitude estimate of $2.3\,\text{mol}\,\text{kg}^{-1}\,\text{Pa}^{-1}$ for k_H of this compound at ambient temperature.

	Solute	
Formula	Name	CAS Registry Number
C_8Cl_8	Octachlorostyrene	29082-74-4
Solvent: Water		

There is one reported measurement of k_H of this liquid [127] by the PURGE method. The datum, at 20 °C, is $(1/k_H)/\text{atm}\,\text{m}^3\,\text{mol}^{-1} = 1.3\text{E}{-4}$ or, in preferred units, $(1/k_H)/\text{mol}\,\text{kg}^{-1}\,\text{Pa}^{-1} = 7.6\text{E}{-5}$.

	Solute	
Formula	Name	CAS Registry Number
C_8H_8	Styrene	100-42-5
Solvent: Water		

Available experimental data for k_H of this liquid in water are as follows [37]:

		Original measurements		
$t/°C$	T/K	Limiting solubility ratio	Method	$k_H/\text{mol}\,\text{kg}^{-1}\,\text{Pa}^{-1}$
25	298.2	$k_{AW} = 0.100$	EPICS	4.04E−6
35	308.2	$k_{AW} = 0.168$	EPICS	2.32E−6
45	318.2	$k_{AW} = 0.230$	EPICS	1.64E−6
55	328.2	$k_{AW} = 0.308$	EPICS	1.20E−6
60	333.2	$k_{AW} = 0.351$	EPICS	1.03E−6

In these calculations, it has been assumed that the concentration of the solute was within the region where Henry's law is valid. These data may be represented by the equation

$$\ln(k_H/\mathrm{mol\,kg^{-1}\,Pa^{-1}}) = -25.190 + 3787.9/(T/\mathrm{K})$$

The standard error of estimate of k_H by this equation is 4%. This equation generates RECOMMENDED data in the range 298–333 K.

Solute		
Formula	Name	CAS Registry Number
$C_8H_9Cl_3O_3\,PS$	Dicapthon	2463-84-5
Solvent: Water		

There are two estimates of k_H for this solid:

			Original measurements		k_H/mol
Ref.	$t/^\circ C$	T/K	Limiting solubility ratio	Method	$\mathrm{kg^{-1}\,Pa^{-1}}$
33	20	293.2	$(1/k_H)/\mathrm{Pa\,m^3\,mol^{-1}} = 2.4E{-}2$	VP/AS	4.16E−2
123	25	298.2	$(1/k_H)/k\mathrm{Pa\,m^3\,mol^{-1}} = 2.28E{-}5$	CRIT EVAL	4.39E−2

In these calculations, it has been assumed that the concentration of the solute was within the region where Henry's law is valid. These data permit only an order-of-magnitude estimate of k_H for this compound at ambient temperature: $4.3E{-}2\,\mathrm{mol\,kg^{-1}\,Pa^{-1}}$.

Solute		
Formula	Name	CAS Registry Number
C_8H_{10}	o-Xylene	95-47-6
Solvent: Water		

Available experimental measurements of k_H of this liquid in water are as follows:

Ref.	$t/°C$	T/K	Original measurements		k_H/mol $kg^{-1}\,Pa^{-1}$
			Limiting solubility ratio	Method	
140	0	273.2	$k_{AW} = 2.82E{-}2$	DIRECT	15.6E$-$6
78	2	275.2	$k_{AW} = 5.82E{-}2$	EPICS	7.51E$-$6
78	6	279.2	$k_{AW} = 5.07E{-}2$	EPICS	8.50E$-$6
78	10	283.2	$k_{AW} = 6.6E{-}2$	EPICS	6.44E$-$6
141	15	288.2	$(1/k_H)/Pa = 1.66E7$	VP/AS	3.35E$-$6
78	18	291.2	$k_{AW} = 0.134$	EPICS	3.08E$-$6
140	23	296.2	$k_{AW} = 0.122$	DIRECT	3.33E$-$6
37	25	298.2	$k_{AW} = 0.172$	EPICS	2.35E$-$6
141	25	298.2	$(1/k_H)/Pa = 2.92E7$	VP/AS	1.90E$-$6
78	25	298.2	$k_{AW} = 0.173$	EPICS	2.33E$-$6
48	25	298.2	$\gamma^\infty = 3.25E4$	INDIRECT	1.92E$-$6
66	25	298.2	$(1/k_H)/atm\,m^3\,mol^{-1} = 4.99E{-}3$	HEADSPACE	1.98E$-$6
66	30	303.2	$(1/k_H)/atm\,m^3\,mol^{-1} = 6.29E{-}3$	HEADSPACE	1.57E$-$6
37	35	308.2	$k_{AW} = 0.242$	EPICS	1.61E$-$6
106	40	313.2	$k_{AW} = 0.410$	HEADSPACE	0.937E$-$6
37	45	318.2	$k_{AW} = 0.309$	EPICS	1.22E$-$6
66	45	318.2	$(1/k_H)/atm\,m^3\,mol^{-1} = 1.06E{-}2$	HEADSPACE	0.931E$-$6
66	50	323.2	$(1/k_H)/atm\,m^3\,mol^{-1} = 1.16E{-}2$	HEADSPACE	0.851E$-$6
37	55	328.2	$k_{AW} = 0.415$	EPICS	0.884E$-$6
37	60	333.2	$k_{AW} = 0.459$	EPICS	0.787E$-$6
106	60	333.2	$k_{AW} = 0.763$	HEADSPACE	0.473E$-$6
106	70	343.2	$k_{AW} = 0.990$	HEADSPACE	0.354E$-$6
106	80	353.2	$k_{AW} = 1.01$	HEADSPACE	0.337E$-$6

In these calculations, it has been assumed that the concentration of the solute was within the region where Henry's law is valid. These data may be represented by the equation

$$\ln(k_H/mol\,kg^{-1}\,Pa^{-1}) = -27.007 + 4196.7/(T/K)$$

The standard error of estimate of k_H by this equation is 20%. This equation generates RECOMMENDED data in the range 273–353 K.

Solute		
Formula	Name	CAS Registry Number
C_8H_{10}	*m*-Xylene	108-38-3
Solvent: Water		

Shaw [138] presented critically evaluated smoothed data for the solubility of this liquid in water in the range $273 < T/K < 343$. The mole fraction data may be represented by the equation

$$x = 1.06889E{-}3 - 7.12319E{-}6(T/K) + 1.22026E{-}8(T/K)^2; r = 0.980$$

The k_H values were calculated from this equation and vapour pressure from Gmehling *et al.* [85]:

T/K	$10^5 x$	$10^6 k_H$/mol kg^{-1} Pa^{-1}
273.15	3.364	8.61
283.15	3.029	3.87
293.15	2.938	1.97
298.15	2.984	1.50
303.15	3.091	1.17
313.15	3.489	0.767
323.15	4.130	0.552
333.15	5.015	0.423
343.15	6.145	0.337

A plot of $\ln k_H$ vs $1/T$ is slightly curved; a linear plot represents the data by

$$\ln(k_H/\text{mol kg}^{-1}\,\text{Pa}^{-1}) = -27.569 + 4267.5/(T/K); \ r = 0.987$$

This equation generates RECOMMENDED values.

Solute		
Formula	Name	CAS Registry Number
C_8H_{10}	*p*-Xylene	106-42-3
Solvent: Water		

Shaw [138] presented critically evaluated smoothed data for the solubility of this liquid in water in the range $273 < T/K < 363$. The data may be represented by the equation

$$x = 3.31258E{-}3 - 2.96082E{-}5(T/K) + 8.70521E{-}8(T/K)^2$$
$$- 8.26822E{-}11(T/K)^3$$

$t/°C$	T/K	Solubility, x	$k_H/\text{mol kg}^{-1}\text{Pa}^{-1}$
0	273.2	3.505E−5	8.40E−6
10	283.2	3.133E−5	3.77E−6
20	293.2	3.097E−5	1.98E−6
30	303.2	3.347E−5	1.20E−6
40	313.2	3.835E−5	0.804E−6
50	323.2	4.509E−5	0.577E−6
60	333.2	5.322E−5	0.431E−6
70	343.2	6.222E−5	0.329E−6
80	353.2	7.161E−5	0.254E−6
90	363.2	8.089E−5	0.198E−6

Other experimental measurements of k_H are as follows:

			Original measurements		$k_H/\text{mol kg}^{-1}\text{Pa}^{-1}$
Ref.	$t/°C$	T/K	Limiting solubility ratio	Method	
140	0	273.2	$k_{AW} = 3.70\text{E}{-2}$	DIRECT	11.9E−6
78	2	275.2	$k_{AW} = 7.71\text{E}{-2}$	EPICS	5.67E−6
78	6	279.2	$k_{AW} = 6.81\text{E}{-2}$	EPICS	6.63E−6
78	10	283.2	$k_{AW} = 0.107$	EPICS	3.97E−6
140	13	286.2	$k_{AW} = 8.77\text{E}{-2}$	DIRECT	4.80E−6
78	18	291.2	$k_{AW} = 0.193$	EPICS	2.14E−6
78	23	296.2	$k_{AW} = 0.156$	DIRECT	2.60E−6
63	25	298.2	$(1/k_H)/\text{kPa m}^3\text{mol}^{-1} = 0.9516$	EPICS	1.05E−6
48	25	298.2	$\gamma^{\infty} = 3.79E4$	INDIRECT	1.25E−6
37	25	298.2	$k_{AW} = 0.271$	EPICS	1.48E−6
66	30	303.2	$(1/k_H)/\text{atm m}^3\text{mol}^{-1} = 7.29\text{E}{-3}$	HEADSPACE	1.35E−6

In these calculations, it has been assumed that the concentration of the solute was within the region where Henry's law is valid. The combined data of both tables may be represented by the equation

$$\ln(k_H/\text{mol kg}^{-1}\text{Pa}^{-1}) = -27.407 + 4240.0/(T/K)$$

The standard error of estimate of k_H by this equation is 22 %. This equation generates RECOMMENDED data in the range 273–363 K.

Solute		
Formula	Name	CAS Registry Number
C_8H_{10}	Ethylbenzene	100-41-4
Solvent: Water		

Shaw [138] presented critically evaluated smoothed data for the solubility of this liquid in water in the range $273 < T/\text{K} < 343$. The data are given in the following table.

$t/°\text{C}$	T/K	Solubility, x^a	$k_H/\text{mol kg}^{-1}\,\text{Pa}^{-1}$
0	273.2	3.4E−5	7.46E−6
10	283.2	3.1E−5	3.42E−6
20	293.2	3.07E−5	1.80E−6
25	298.2	2.87E−5	1.26E−6
30	303.2	3.2E−5	1.06E−6
40	313.2	3.39E−5	0.656E−6
50	323.2	3.7E−5	0.438E−6
60	333.2	4.2E−5	0.315E−6
70	343.2	4.8E−5	0.236E−6

^{a}Data are quoted precisely from Shaw [138]; no fitting equation was given.

Available experimental measurements of k_H of this liquid are as follows:

			Original measurements		k_H/mol
Ref.	$t/°\text{C}$	T/K	Limiting solubility ratio	Method	$\text{kg}^{-1}\,\text{Pa}^{-1}$
78	2	275.2	$k_{AW} = 7.86\text{E}{-2}$	EPICS	5.56E−6
142	4.5	277.7	$k_{WA} = 12.3$	HEADSPACE	5.33E−6
78	6	279.2	$k_{AW} = 8.37\text{E}{-2}$	EPICS	5.15E−6
142	6	279.2	$k_{WA} = 10.7$	HEADSPACE	4.61E−6
142	7	280.2	$k_{WA} = 9.91$	HEADSPACE	4.25E−6
142	9	282.2	$k_{WA} = 8.82$	HEADSPACE	3.76E−6
143	10	283.2	$(1/k_H)/\text{atm m}^3\,\text{mol}^{-1} = 3.02\text{E}{-3}$	HEADSPACE	3.27E−6
142	12	285.2	$k_{WA} = 7.15$	HEADSPACE	3.02E−6
144	13	286.2	$\gamma^\infty = 2.4E4$	GC	3.78E−6
141	15	288.2	$(1/k_H)/\text{Pa} = 2.55E7$	VP/AS	2.18E−6
143	15	288.2	$(1/k_H)/\text{atm m}^3\,\text{mol}^{-1} = 4.22\text{E}{-3}$	HEADSPACE	2.34E−6
142	15	288.2	$k_{WA} = 5.85$	HEADSPACE	2.44E−6
142	18	291.2	$k_{WA} = 5.04$	HEADSPACE	2.08E−6
143	20	293.2	$(1/k_H)/\text{atm m}^3\,\text{mol}^{-1} = 5.75\text{E}{-3}$	HEADSPACE	1.72E−6
142	20	293.2	$k_{WA} = 4.36$	HEADSPACE	1.79E−6
76	25	298.2	$k_{AW} = 0.361$	DIRECT	1.12E−6
78	25	298.2	$k_{AW} = 0.270$	EPICS	1.49E−6
66	25	298.2	$(1/k_H)/\text{atm m}^3\,\text{mol}^{-1} = 7.78\text{E}{-3}$	HEADSPACE	1.27E−6
48	25	298.2	$\gamma^\infty = 3.34E4$	INDIRECT	1.31E−6
141	25	298.2	$(1/k_H)/\text{Pa} = 4.43E7$	VP/AS	1.25E−6
143	25	298.2	$(1/k_H)/\text{atm m}^3\,\text{mol}^{-1} = 7.84\text{E}{-3}$	HEADSPACE	1.26E−6

			Original measurements		k_H/mol
Ref.	$t/°C$	T/K	Limiting solubility ratio	Method	$kg^{-1} Pa^{-1}$
66	25	298.2	$(1/k_H)/atm\,m^3\,mol^{-1} = 7.78E{-}3$	HEADSPACE	1.27E$-$6
123	25	298.2	$(1/k_H)/atm\,m^3\,mol^{-1} = 8.43E{-}3$	CRIT EVAL	1.17E$-$6
37	25	298.2	$k_{AW} = 0.281$	EPICS	1.43E$-$6
77	25	298.2	$\gamma^\infty = 3.12E4$	VP/AS	1.40E$-$6
66	30	303.2	$(1/k_H)/atm\,m^3\,mol^{-1} = 1.02E{-}2$	HEADSPACE	0.968E$-$6
143	30	303.2	$(1/k_H)/atm\,m^3\,mol^{-1} = 1.03E{-}2$	HEADSPACE	0.958E$-$6
141	35	308.2	$(1/k_H)/Pa = 7.44E7$	VP/AS	0.746E$-$6
66	40	313.2	$(1/k_H)/atm\,m^3\,mol^{-1} = 1.64E{-}2$	HEADSPACE	0.602E$-$6
141	45	318.2	$(1/k_H)/Pa = 1.12E8$	VP/AS	0.496E$-$6

The combined data of both tables may be represented by the equation

$$\ln(k_H/mol\,kg^{-1}\,Pa^{-1}) = -29.658 + 4828.3/(T/K)$$

The standard error of estimate of k_H by this equation is 10 %. This equation generates RECOMMENDED data in the range 273–343 K.

Solute		
Formula	Name	CAS Registry Number
$C_8H_{10}NO_5PS$	Parathion methyl	298-00-0
Solvent: Water		

There are six reports for k_H of this solid in water. Although no temperature is stated in Refs 139 and 147, the results were calculated on the assumption that the temperature is 23 °C.

			Original measurements		k_H/mol
Ref.	$t/°C$	T/K	Limiting solubility ratio	Method	$kg^{-1} Pa^{-1}$
145	20	293.2	$(1/k_H)/atm\,m^3\,mol^{-1} = 1.0E{-}7$	PURGE	9.87E$-$2
146	20	293.2	$k_{AW} = 1.57E{-}6$	WET WALL	2.61E$-$2
33	20	293.2	$(1/k_H)/Pa\,m^3\,mol^{-1} = 1.0E{-}7$	VP/AS	4.76E$-$2
147	23	296.2	$(1/k_H)/mmHg\,ppm^{-1} = 1.96E{-}7$	VP/AS	0.145
139	23	296.2	$k_{AW} = 4.4E{-}6$	VP/AS	9.23E$-$2
148	25	298.2	$(1/k_H)/kPa\,m^3\,mol^{-1} = 2.0E{-}5$	CRIT EVAL	5.00E$-$2

In these calculations, it has been assumed that the concentration of the solute was within the region where Henry's law is valid. These data allow only an order-of-magnitude estimate of k_H of this compound at ambient temperature: $6.3E-2 \, mol \, kg^{-1} \, Pa^{-1}$.

Solute		
Formula	Name	CAS Registry Number
$C_8H_{14}ClN_5$	Atrazine	1912-24-9
Solvent: Water		

Available estimates of k_H for this solid are as follows:

			Original measurements			
Ref.	$t/°C$	T/K	Limiting solubility ratio		Method	$k_H/mol \, kg^{-1} \, Pa^{-1}$
139	23	296.2	$k_{AW} = 2.0E-7$		VP/AS	2.03
33	20	293.2	$(1/k_H)/Pa \, m^3 \, mol^{-1} = 2.9E-4$		VP/AS	3.45

In these calculations, it has been assumed that the concentration of the solute was within the region where Henry's law is valid. Although no temperature was stated in Ref. 139, the result was calculated with an assumed temperature of $23 \, °C$. An order-of-magnitude estimate for k_H is $2.8 \, mol \, kg^{-1} \, Pa^{-1}$.

Solute		
Formula	Name	CAS Registry Number
$C_8H_{24}O_4Si_4$	Octamethylcyclotetrasiloxane	556-67-2
Solvent: Water		

There is only one report of k_H for this liquid [149], in which the direct method was used. The reported datum is $k_{AW} = 3.4$ at $20 \, °C$, or $k_H/mol \, kg^{-1} \, Pa^{-1} = 1.2E-7$.

Solute		
Formula	Name	CAS Registry Number
$C_9H_{11}Cl_3NO_3$ PS	Chlorpyrifos	2921-88-2
Solvent: Water		

There are three reports of k_H of this solid:

	Original measurements				k_H/mol
Ref.	$t/^{\circ}C$	T/K	Limiting solubility ratio	Method	$kg^{-1} Pa^{-1}$
146	20	293.2	$k_{AW} = 1.30E{-}4$	WET WALL	31.6E−4
33	20	293.2	$(1/k_H)/Pa\,m^3\,mol^{-1} = 1.75$	VP/AS	5.71E−4
139	23	296.2	$k_{AW} = 5.0E{-}4$	VP/AS	8.12E−4

In these calculations, it has been assumed that the concentration of the solute was within the region where Henry's law is valid. Although no temperature was stated in Ref. 139, the result was calculated for a temperature of 23 °C. An order-of-magnitude estimate of k_H of this compound is $6.9\,mol\,kg^{-1}\,Pa^{-1}$.

Solute		
Formula	Name	CAS Registry Number
C_9H_{12}	1,2,3-Trimethylbenzene	526-73-8
Solvent: Water		

Shaw [138] presented critically evaluated data for the solubility of this liquid in water in the temperature range $299 < T/K < 318$. The k_H values were calculated from these data and vapour pressure from Gmehling et al. [85].

$t/^{\circ}C$	T/K	Solubility, x	$k_H/mol\,kg^{-1}\,Pa^{-1}$
15	288.2	9.0E−6	5.13E−6
25	298.2	10.E−6	2.80E−6
35	308.2	11.E−6	1.60E−6
45	318.2	13.E−6	1.04E−6

Experimental measurements are as follows:

			Original measurements		k_H/mol
Ref.	$t/^\circ C$	T/K	Limiting solubility ratio	Method	$\text{kg}^{-1}\,\text{Pa}^{-1}$
48	15	288.2	$(1/k_H)/\text{Pa} = 1.34\text{E}{-}7$	VP/AS	4.14E−6
48	25	298.2	$(1/k_H)/\text{Pa} = 2.45\text{E}{-}7$	VP/AS	2.27E−6
77	25	298.2	$\gamma^\infty = 1.0\text{E}5$	VP/AS	2.80E−6
123	25	298.2	$(1/k_H)/k\text{Pa}\,\text{m}^3\,\text{mol}^{-1} = 0.323$	CRIT EVAL	3.10E−6
48	45	318.2	$(1/k_H)/\text{Pa} = 3.92\text{E}7$	VP/AS	1.42E−6
48	55	328.2	$(1/k_H)/\text{Pa} = 5.88\text{E}7$	VP/AS	0.944E−6

In these calculations, it has been assumed that the concentration of the solute was within the region where Henry's law is valid. The combined data of both tables may be represented by the equation

$$\ln(k_H/\text{mol}\,\text{kg}^{-1}\,\text{Pa}^{-1}) = -28.980 + 4816.6/(T/\text{K})$$

The standard error of estimate of k_H by this equation is 11 %. This equation generates RECOMMENDED values for the range $288 < T/\text{K} < 328$.

Solute		
Formula	Name	CAS Registry Number
C_9H_{12}	1,2,4-Trimethylbenzene	95-63-6
Solvent: Water		

Shaw [138] presented critically evaluated smoothed data for the solubility of this liquid in water in the range $288 < T/\text{K} < 318$.

$t/^\circ C$	T/K	Solubility, x	$k_H/\text{mol}\,\text{kg}^{-1}\,\text{Pa}^{-1}$
15	288.2	7.8E−6	3.20E−6
25	298.2	8.4E−6	1.72E−6
35	308.2	9.3E−6	1.01E−6
45	318.2	10.E−6	0.603E−6

Available experimental measurements of k_H of this liquid in water are as follows:

			Original measurements		k_H/mol
Ref.	$t/°C$	T/K	Limiting solubility ratio	Method	$\text{kg}^{-1}\,\text{Pa}^{-1}$
141	25	298.2	$(1/k_H)/\text{Pa} = 3.44\text{E}7$	VP/AS	1.61E$-$6
111	25	298.2	$k_{WA} = 0.63$	VP/AS	1.72E$-$6
123	25	298.2	$(1/k_H)/\text{kPa}\,\text{m}^3\,\text{mol}^{-1} = 0.590$	CRIT EVAL	1.69E$-$6
79	25	298.2	$(1/k_H)/\text{kPa}\,\text{m}^3\,\text{mol}^{-1} = 0.704$	EPICS	1.42E$-$6
141	35	308.2	$(1/k_H)/\text{Pa} = 5.79\text{E}7$	VP/AS	0.930E$-$6
37	35	308.2	$k_{AW} = 0.251$	EPICS	1.55E$-$6
141	45	318.2	$(1/k_H)/\text{Pa} = 9.24\text{E}7$	VP/AS	0.601E$-$6
37	45	318.2	$k_{AW} = 0.329$	EPICS	1.15E$-$6
37	55	328.2	$k_{AW} = 0.438$	EPICS	0.837E$-$6
37	60	333.2	$k_{AW} = 0.516$	EPICS	0.700E$-$6

In these calculations, it has been assumed that the concentration of the solute was within the region where Henry's law is valid. The combined data of both tables may be represented by the equation

$$\ln(k_H/\text{mol}\,\text{kg}^{-1}\,\text{Pa}^{-1}) = -23.644 + 3084.8/(T/K)$$

The standard error of estimate of k_H by this equation is 20%. This equation generates RECOMMENDED values in the range 288–333 K.

Solute		
Formula	Name	CAS Registry Number
C_9H_{12}	1,3,5-Trimethylbenzene	108-67-8
Solvent: Water		

Shaw [138] presented critically evaluated smoothed data for the solubility of this liquid in water in the range $288 < T/K < 318$. The k_H data were calculated from these data and solute vapour pressure from Gmehling et al. [85].

$t/°C$	T/K	Solubility, x	$k_H/\text{mol}\,\text{kg}^{-1}\,\text{Pa}^{-1}$
15	288.2	6.9E$-$6	2.36E$-$6
25	298.2	7.4E$-$6	1.27E$-$6
35	308.2	8.2E$-$6	0.752E$-$6
45	318.2	8.7E$-$6	0.447E$-$6

Available experimental measurements of k_H are as follows:

			Original measurements		k_H/mol
Ref.	$t/°C$	T/K	Limiting solubility ratio	Method	$kg^{-1}\,Pa^{-1}$
141	15	288.2	$(1/k_H)/Pa = 2.84E7$	VP/AS	1.95E−6
141	25	298.2	$(1/k_H)/Pa = 4.93E7$	VP/AS	1.13E−6
48	25	298.2	$\gamma^\infty = 1.28E5$	INDIRECT	1.34E−6
123	25	298.2	$(1/k_H)/kPa\,m^3\,mol^{-1} = 0.60$	CRIT EVAL	1.67E−6
37	25	298.2	$k_{AW} = 0.272$	EPICS	1.48E−6
141	35	308.2	$(1/k_H)/Pa = 8.14E7$	VP/AS	0.682E−6
37	35	308.2	$k_{AW} = 0.346$	EPICS	1.13E−6
141	45	318.2	$(1/k_H)/Pa = 1.33E8$	VP/AS	0.417E−6
37	45	318.2	$k_{AW} = 0.419$	EPICS	0.902E−6
37	55	328.2	$k_{AW} = 0.549$	EPICS	0.668E−6
37	60	333.2	$k_{AW} = 0.642$	EPICS	0.562E−6

In these calculations, it has been assumed that the concentration of the solute was within the region where Henry's law is valid. Data of the two tables are not consistent; an average value at 25 °C may be taken to be $k_H/\text{mol}\,kg^{-1}\,Pa^{-1} = 1.4E{-}6$.

	Solute	
Formula	Name	CAS Registry Number
C_9H_{12}	Isopropylbenzene	98-82-8
Solvent: Water		

Available experimental measurements of k_H of this liquid are as follows:

			Original measurements		
Ref.	$t/°C$	T/K	Limiting solubility ratio	Method	$k_H/\text{mol}\,kg^{-1}\,Pa^{-1}$
141	15	288.2	$(1/k_H)/Pa = 3.50E7$	VP/AS	15.9E−7
141	25	298.2	$(1/k_H)/Pa = 6.47E7$	VP/AS	8.58E−7
48	25	298.2	$\gamma^\infty = 9.91E4$	INDIRECT	9.20E−7
37	25	298.2	$k_{AW} = 0.407$	EPICS	9.91E−7
141	35	308.2	$(1/k_H)/Pa = 10.6E7$	VP/AS	5.24E−7
37	35	308.2	$k_{AW} = 0.467$	EPICS	6.85E−7
141	45	318.2	$(1/k_H)/Pa = 16.6E{-}7$	VP/AS	3.34E−7
37	45	318.2	$k_{AW} = 0.552$	EPICS	6.85E−7

			Original measurements		
Ref.	$t/°C$	T/K	Limiting solubility ratio	Method	$k_H/\text{mol kg}^{-1}\,\text{Pa}^{-1}$
37	55	328.2	$k_{AW} = 0.698$	EPICS	5.25E−7
37	60	333.2	$k_{AW} = 0.781$	EPICS	4.62E−7

In these calculations, it has been assumed that the concentration of the solute was within the region where Henry's law is valid. The data are not consistent enough for plotting or fitting.

Solute		
Formula	Name	CAS Registry Number
$C_9H_{12}NO_5\,PS$	Fenitrothion	122-14-5
Solvent: Water		

There are three reports of k_H of this solid:

			Original measurements		k_H/mol
Ref.	$t/°C$	T/K	Limiting solubility ratio	Method	$\text{kg}^{-1}\,\text{Pa}^{-1}$
145	20	293.2	$(1/k_H)/\text{atm m}^3\,\text{mol}^{-1} = 9.3E−7$	PURGE	1.06E−2
145	20	293.2	$(1/k_H)/\text{atm m}^3\,\text{mol}^{-1} = 6.6E−7$	VP/AS	1.50E−2
123	25	298.2	$(1/k_H)/\text{kPa m}^3\,\text{mol}^{-1} = 6.65E−5$	CRIT EVAL	1.50E−2

In these calculations, it has been assumed that the concentration of the solute was within the region where Henry's law is valid. These data permit only an order-of-magnitude estimate of k_H for this compound at ambient temperature: $1.4E−2\,\text{mol kg}^{-1}\,\text{Pa}^{-1}$.

13 HENRY'S LAW CONSTANTS FOR ORGANIC COMPOUNDS ($C_{10}-C_{20}$)

Solute		
Formula	Name	CAS Registry Number
$C_{10}Cl_{12}$	Mirex	2385-85-5
Solvent: Water		

There are three reports of k_H for this solid in water:

Original measurements				
Ref.	$t/°C$	T/K	Limiting solubility ratio	Method	$k_H/mol\,kg^{-1}\,Pa^{-1}$
69	20	293.2	$(1/k_H)/Pa\,m^3\,mol^{-1} = 1013$	VP/AS	0.987E−6
33	20	293.2	$(1/k_H)/Pa\,m^3\,mol^{-1} = 839.37$	VP/AS	1.19E−6
150	22	295.2	$(1/k_H)/atm\,m^3\,mol^{-1} = 5.16E−4$	PURGE	19.1E−6

In these calculations, it has been assumed that the concentration of the solute was within the region where Henry's law is valid. The value from Ref. 150 is not consistent with the other two. An order-of-magnitude estimate of $1.1E−6\,mol\,kg^{-1}\,Pa^{-1}$ may be given for k_H of this compound.

Solute		
Formula	Name	CAS Registry Number
$C_{10}H_8$	Naphthalene	91-20-3
Solvent: Water		

Available experimental measurements of k_H of this solid in water are as follows:

Original measurements					$k_H/mol\,kg^{-1}\,Pa^{-1}$
Ref.	$t/°C$	T/K	Limiting solubility ratio	Method	
151	8	281.2	$k_{AW} = 6.85E−3$	DIRECT	6.25E−5
152	9	282.2	$(1/k_H)/Pa\,m^3\,mol^{-1} = 15.3$	PURGE	6.49E−5
152	15	288.2	$(1/k_H)/Pa\,m^3\,mol^{-1} = 21.4$	PURGE	4.67E−5
151	17	290.2	$k_{AW} = 1.1E−2$	DIRECT	3.77E−5
152	20	293.2	$(1/k_H)/Pa\,m^3\,mol^{-1} = 33.2$	PURGE	3.01E−5
153	20	293.2	$(1/k_H)/Pa\,m^3\,mol^{-1} = 45.$	PURGE	2.22E−5
151	22	295.2	$k_{AW} = 1.49E−2$	DIRECT	2.73E−5
151	25	298.2	$k_{AW} = 1.89E−2$	DIRECT	2.14E−5
132	25	298.2	$(1/k_H)/Pa\,m^3\,mol^{-1} = 44.6$	PURGE	2.24E−5
154	25	298.2	$k_{AW} = 2.26E−2$	PURGE	1.79E−5
152	30	303.2	$(1/k_H)/Pa\,m^3\,mol^{-1} = 58.6$	PURGE	1.71E−5
37	35	308.2	$k_{AW} = 3.4E−2$	EPICS	1.19E−5
152	36	309.2	$(1/k_H)/Pa\,m^3\,mol^{-1} = 79.1$	PURGE	1.26E−5
37	45	318.2	$k_{AW} = 7.4E−2$	EPICS	0.545E−5

			Original measurements		k_H/mol
Ref.	$t/°C$	T/K	Limiting solubility ratio	Method	$\text{kg}^{-1}\,\text{Pa}^{-1}$
37	55	328.2	$k_{AW} = 9.0\text{E}{-}2$	EPICS	0.448E−5
37	60	333.2	$k_{AW} = 0.107$	EPICS	0.377E−5

In these calculations, it has been assumed that the concentration of the solute was within the region where Henry's law is valid. These data may be represented by the equation

$$\ln(k_H/\text{mol kg}^{-1}\,\text{Pa}^{-1}) = -28.605 + 5329.3(T/K)$$

The standard error of estimate of k_H by this equation is 15%. This equation generates RECOMMENDED values in the range 281–333 K.

Solute		
Formula	Name	CAS Registry Number
$C_{10}H_{14}NO_5PS$	Parathion	56-38-2
Solvent: Water		

There are four reports of k_H of this solid in water. No temperature was stated in Refs 139 and 147, and a temperature of 23 °C was assumed.

			Original measurements		k_H/mol
Ref.	$t/°C$	T/K	Limiting solubility ratio	Method	$\text{kg}^{-1}\,\text{Pa}^{-1}$
33	20	293.2	$(1/k_H)/\text{Pa}\,\text{m}^3\,\text{mol}^{-1} = 1.2\text{E}{-}2$	VP/AS	8.33E−2
147	23	296.2	$(1/k_H)/\text{mmHg ppm}^{-1} = 2.52\text{E}{-}7$	VP/AS	10.0E−2
139	23	296.2	$k_{AW} = 9.5\text{E}{-}6$	VP/AS	4.27E−2
123	25	298.2	$(1/k_H)/\text{kPa}\,\text{m}^3\,\text{mol}^{-1} = 1.23\text{E}{-}4$	CRIT EVAL	0.813E−2(?)

In these calculations, it has been assumed that the concentration of the solute was within the region where Henry's law is valid. These data allow only an order-of-magnitude estimate for k_H. The mean of the three more recent data (Refs 33, 139 and 147) is $7.5\text{E}{-}2\,\text{mol kg}^{-1}\,\text{Pa}^{-1}$.

Solute		
Formula	Name	CAS Registry Number
$C_{10}H_{16}$	α-Pinene	80-56-8
Solvent: Water		

Available experimental measurements of k_H of this liquid in water are as follows:

			Original measurements		
Ref.	$t/°C$	T/K	Limiting solubility ratio	Method	$k_H/\mathrm{mol\,kg^{-1}\,Pa^{-1}}$
155	6	279.2	$\gamma^\infty = 3.32E6$	VP/AS	28.0E−8
155	24	297.2	$\gamma^\infty = 3.03E6$	VP/AS	3.37E−8
156	25	298.2	$\gamma^\infty = 1.5E6$	VP/AS	6.29E−8
157	37	310.2	$k_{AW} = 8.33$	–	4.66E−8

In these calculations, it has been assumed that the concentration of the solute was within the region where Henry's law is valid. The quality of the data is mediocre.

Solute		
Formula	Name	CAS Registry Number
$C_{10}H_{16}$	β-Pinene	127-91-3
Solvent: Water		

Available experimental measurements of k_H for this liquid in water are as follows:

			Original measurements		
Ref.	$t/°C$	T/K	Limiting solubility ratio	Method	$k_H/\mathrm{mol\,kg^{-1}\,Pa^{-1}}$
156	25	298.2	$\gamma^\infty = 6.86E5$	VP/AS	20.5E−8
157	37	310.2	$k_{AW} = 8.33$	VP/AS	4.66E−8

In these calculations, it has been assumed that the concentration of the solute was within the region where Henry's law is valid.

Solute		
Formula	Name	CAS Registry Number
$C_{10}H_{16}$	Δ^3-Carene	13466-78-9
Solvent: Water		

There is one report of the limiting solubility constant of this liquid at 37 °C (from VP/AS estimation [157]). The original datum is $k_{AW} = 2.44$, or k_H/mol $kg^{-1} Pa^{-1} = 1.59E-7$.

Solute		
Formula	Name	CAS Registry Number
$C_{10}H_{16}$	Limonene	138-86-3
Solvent: Water		

Available experimental measurements of k_H of this liquid in water are given in the table. Solute vapour pressure data were taken from Gmehling *et al.* [85]

	Original measurements				k_H/mol
Ref.	$t/°C$	T/K	Limiting solubility ratio	Method	$kg^{-1} Pa^{-1}$
158	0	273.2	$(1/k_H)$/mmHg M^{-1} = 5.61E3	VP/AS	74.6E−6
158	5	278.2	$(1/k_H)$/mmHg M^{-1} = 7.39E3	VP/AS	1.01E−6
155	6	279.2	γ^∞ = 1.75E6	VP/AS	1.72E−6
155	24	297.2	γ^∞ = 1.35E6	VP/AS	0.204E−6
158	25	298.2	$(1/k_H)$/mmHg M^{-1} = 2.05E4	VP/AS	0.365E−6
156	25	298.2	γ^∞ = 3.7E5	VP/AS	0.542E−6
157	37	310.2	k_{AW} = 0.556	VP/AS	0.698E−6

The data are inconsistent.

Solute		
Formula	Name	CAS Registry Number
$C_{10}H_{18}O$	Linalool	78-70-6
Solvent: Water		

There are three estimates of k_H of this liquid in water. The result for Ref. 156 was calculated on the assumption that the vapour pressure of the pure liquid is 21.2 Pa (24 °C [155]).

			Original measurements		
Ref.	$t/°C$	T/K	Limiting solubility ratio	Method	$k_H/\text{mol kg}^{-1}\text{Pa}^{-1}$
155	6	279.2	$\gamma^\infty = 1.56E4$	VP/AS	0.474E-4
155	24	297.2	$\gamma^\infty = 1.00E4$	VP/AS	2.62E-4
156	25	298.2	$\gamma^\infty = 5.49E3$	VP/AS	4.77E-4

In these calculations, it has been assumed that the concentration of the solute was within the region where Henry's law is valid. There is not good agreement among the data, but an order-of-magnitude estimate of $3.7E-4 \text{ mol kg}^{-1}\text{Pa}^{-1}$ can be given for k_H of this compound at ambient temperature.

	Solute	
Formula	Name	CAS Registry Number
$C_{10}H_{18}O$	α-Terpineol	98-55-5
Solvent: Water		

There are three estimates of k_H for this liquid. The result for Ref. 156 was calculated assuming the vapour pressure of the pure liquid is 5.69 Pa (24 °C [155]).

			Original measurements		
Ref.	$t/°C$	T/K	Limiting solubility ratio	Method	$k_H/\text{mol kg}^{-1}\text{Pa}^{-1}$
155	6	279.2	$\gamma^\infty = 2.52E4$	VP/AS	2.56E-3
155	24	297.2	$\gamma^\infty = 1.21E4$	VP/AS	0.806E-3
156	25	298.2	$\gamma^\infty = 4.53E3$	VP/AS	2.15E-3

In these calculations, it has been assumed that the concentration of the solute was within the region where Henry's law is valid. The data are not in good agreement, and so only an order-of-magnitude estimate of k_H can be given: $1.5E-3 \text{ mol kg}^{-1}\text{Pa}^{-1}$.

Solute		
Formula	Name	CAS Registry Number
$C_{10}H_{19}O_6PS_2$	Malathion	121-75-5
Solvent: Water		

There are five reports of k_H of this solid in water. Although there is no statement of temperature in Refs 139, 147 and 159, the results were calculated on the assumption that the temperature was 23 °C.

			Original measurements		k_H/mol kg^{-1} Pa^{-1}
Ref.	$t/°C$	T/K	Limiting solubility ratio	Method	
33	20	293.2	$(1/k_H)/Pa\,m^3\,mol^{-1} = 2.3E{-}3$	VP/AS	0.435
147	23	296.2	$(1/k_H)/mmHg\,ppm^{-1} = 5.68E{-}8$	VP/AS	0.400
139	23	296.2	$k_{AW} = 2.4E{-}6$	VP/AS	0.169
159	23	296.2	$(1/k_H)/atm\,m^3\,mol^{-1} = 3.9E{-}8$	VP/AS	0.253
123	25	298.2	$(1/k_H)/kPa\,m^3\,mol^{-1} = 3.8E{-}5$	CRIT EVAL	0.0263(?)

In these calculations, it has been assumed that the concentration of the solute was within the region where Henry's law is valid. An order-of-magnitude estimate of k_H for this compound at ambient temperature is 0.2 mol kg^{-1} Pa^{-1}.

Solute		
Formula	Name	CAS Registry Number
$C_{11}H_{10}$	1-Methylnaphthalene	90-12-0
Solvent: Water		

Available experimental measurements of k_H of this liquid in water are as follows:

			Original measurements		k_H/mol kg^{-1} Pa^{-1}
Ref.	$t/°C$	T/K	Limiting solubility ratio	Method	
160	4	277.2	$(1/k_H)/Pa\,m^3\,mol^{-1} = 10.1$	PURGE	9.90E$-$5
157	8	281.2	$k_{WA} = 166$	DIRECT	7.10E$-$5
160	11	284.2	$(1/k_H)/Pa\,m^3\,mol^{-1} = 17.4$	PURGE	5.75E$-$5

			Original measurements		k_H/mol
Ref.	$t/°C$	T/K	Limiting solubility ratio	Method	$kg^{-1} Pa^{-1}$
151	17	290.2	$k_{WA} = 110$	DIRECT	4.56E−5
160	18	291.2	$(1/k_H)/Pa\, m^3\, mol^{-1} = 29.2$	PURGE	3.42E−5
151	23	296.2	$k_{WA} = 73$	DIRECT	2.96E−5
160	25	298.2	$(1/k_H)/Pa\, m^3\, mol^{-1} = 47.8$	PURGE	2.09E−5
161	25	298.2	$k_{AW} = 2.1E{-}2$	PURGE	1.92E−5
151	30	303.2	$k_{WA} = 51$	DIRECT	2.02E−5
160	31	304.2	$(1/k_H)/Pa\, m^3\, mol^{-1} = 71.7$	PURGE	1.39E−5

In these calculations, it has been assumed that the concentration of the solute was within the region where Henry's law is valid. These data may be represented by the equation

$$\ln(k_H/mol\, kg^{-1}\, Pa^{-1}) = -31.061 + 6060.0/(T/K)$$

The standard error of estimate of k_H by this equation is 10%. This equation generates RECOMMENDED values in the range 277–304 K.

Solute		
Formula	Name	CAS Registry Number
$C_{11}H_{10}$	2-Methylnaphthalene	91-57-6
Solvent: Water		

Available experimental data of k_H for this liquid in water are as follows:

			Original measurements		k_H/mol
Ref.	$t/°C$	T/K	Limiting solubility ratio	Method	$kg^{-1} Pa^{-1}$
160	4	277.2	$(1/k_H)/Pa\, m^3\, mol^{-1} = 13.0$	PURGE	7.69E−5
160	11	284.2	$(1/k_H)/Pa\, m^3\, mol^{-1} = 21.0$	PURGE	4.76E−5
160	18	291.2	$(1/k_H)/Pa\, m^3\, mol^{-1} = 33.2$	PURGE	3.01E−5
153	20	293.2	$(1/k_H)/Pa\, m^3\, mol^{-1} = 46.0$	PURGE	2.17E−5
160	25	298.2	$(1/k_H)/Pa\, m^3\, mol^{-1} = 51.2$	PURGE	1.95E−5
161	25	298.2	$k_{AW} = 2.5E{-}2$	WET WALL	1.61E−5
160	31	304.2	$(1/k_H)/Pa\, m^3\, mol^{-1} = 73.3$	PURGE	1.36E−5

In these calculations, it has been assumed that the concentration of the solute was within the region where Henry's law is valid. These data may be represented by the equation

$$\ln(k_H/\text{mol kg}^{-1}\,\text{Pa}^{-1}) = -29.799 + 5629.3/(T/\text{K})$$

The standard error of estimate of k_H by this equation is 8 %. This equation generates RECOMMENDED data in the range 277–304 K.

Solute		
Formula	Name	CAS Registry Number
$C_{12}Cl_8O_2$	Octachlorodibenzodioxin	3268-87-9
Solvent: Water		

An estimate of k_H of this solid at 25 °C from vapour pressure and solubility was reported by Shiu et al. [162]. The datum is $(1/k_H)/\text{Pa m}^3\,\text{mol}^{-1} = 0.683$, or $k_H/\text{mol kg}^{-1}\,\text{Pa}^{-1} = 1.46\text{E}{-3}$.

Solute		
Formula	Name	CAS Registry Number
$C_{12}Cl_{10}$	Decachloro-PCB	2051-24-3
Solvent: Water		

An estimate for k_H of this solid was reported by Shiu and Mackay [163] from vapour pressure and aqueous solubility. The datum is $20.84\,\text{Pa m}^3\,\text{mol}^{-1}$ at 25 °C, or $k_H/\text{mol kg}^{-1}\,\text{Pa}^{-1} = 4.80\text{E}{-5}$.

Solute		
Formula	Name	CAS Registry Number
$C_{12}H_3Cl_5O_2$	1,2,3,4,7-Pentachlorodi-benzodioxin	39227-61-7
Solvent: Water		

An estimate of k_H was made by Shiu *et al.* [162] from vapour pressure and aqueous solubility. The reported datum is $(1/k_H)/\text{Pa m}^3 \text{ mol}^{-1} = 0.264$ at 25 °C, or $k_H/\text{mol kg}^{-1} \text{Pa}^{-1} = 3.79\text{E}{-}3$.

Solute		
Formula	Name	CAS Registry Number
$C_{12}H_4Cl_4O_2$	1,2,3,4-Tetrachlorodi-benzodioxin	30746-58-8
Solvent: Water		

There are two reports of k_H for this solid in water at 25 °C:

	Original measurements		
Ref.	Limiting solubility ratio	Method	$k_H/\text{mol kg}^{-1} \text{Pa}^{-1}$
162	$(1/k_H)/\text{Pa m}^3 \text{ mol}^{-1} = 3.77$	VP/AS	2.65E−4
164	$(1/k_H)/\text{Pa m}^3 \text{ mol}^{-1} = 2.02$	DIRECT	4.95E−4

In these calculations, it has been assumed that the concentration of the solute was within the region where Henry's law is valid. The mean value, $k_H/\text{mol kg}^{-1} \text{Pa}^{-1} = 3.8\text{E}{-}4$, is a RECOMMENDED value.

Solute		
Formula	Name	CAS Registry Number
$C_{12}H_4Cl_6$	2,2′,4,4′,5,5′-PCB	35065-27-1
Solvent: Water		

Available experimental data for k_H of this solid in water are as follows:

			Original measurements		k_H/mol
Ref.	$t/°C$	T/K	Limiting solubility ratio	Method	$\text{kg}^{-1} \text{Pa}^{-1}$
165	20	293.2	$(1/k_H)/\text{atm m}^3 \text{ mol}^{-1} = 0.99\text{E}{-}4$	VP/AS	9.95E−5
166	25	298.2	$(1/k_H)/\text{atm m}^3 \text{ mol}^{-1} = 1.31\text{E}{-}4$	PURGE	7.48E−5
167	25	298.2	$(1/k_H)/\text{atm m}^3 \text{ mol}^{-1} = 0.23\text{E}{-}4$	WET WALL	42.9E−5
163	25	298.2	$(1/k_H)/\text{Pa m}^3 \text{ mol}^{-1} = 42.9$	VP/AS	2.33E−5

In these calculations, it has been assumed that the concentration of the solute was within the region where Henry's law is valid. The data of Refs 163, 165 and 166 permit an order-of-magnitude estimate for k_H of 6.6E−5 mol kg^{-1} Pa^{-1}.

Solute		
Formula	Name	CAS Registry Number
$C_{12}H_5Cl_3O_2$	1,2,4-Trichlorodibenzo-dioxin	39227-58-2
Solvent: Water		

There are two reports of k_H for this solid in water at 25 °C:

	Original measurements		
Ref.	Limiting solubility ratio	Method	k_H/mol kg^{-1} Pa^{-1}
162	$(1/k_H)$/Pa m^3 mol^{-1} = 3.84	VP/AS	2.60E−4
164	$(1/k_H)$/Pa m^3 mol^{-1} = 3.84	DIRECT	2.60E−4

In these calculations, it has been assumed that the concentration of the solute was within the region where Henry's law is valid. The value k_H/mol kg^{-1} Pa^{-1} = 2.60E−4 is a RECOMMENDED value.

Solute		
Formula	Name	CAS Registry Number
$C_{12}H_5Cl_5$	2,2′,4,5,5′-PCB	37680-73-2
Solvent: Water		

Available experimental data for k_H of this solid in water are as follows:

			Original measurements		k_H/mol kg^{-1} Pa^{-1}
Ref.	t/°C	T/K	Limiting solubility ratio	Method	
165	20	293.2	$(1/k_H)$/atm m^3 mol^{-1} = 1.79E−4	VP/AS	5.51E−5
166	25	298.2	$(1/k_H)$/atm m^3 mol^{-1} = 2.51E−4	PURGE	3.93E−5
168	25	298.2	$(1/k_H)$/atm m^3 mol^{-1} = 2.3E−4	VP/AS	4.29E−5
167	25	298.2	$(1/k_H)$/atm m^3 mol^{-1} = 0.90E−4	WET WALL	11.0E−5
163	25	298.2	$(1/k_H)$/Pa m^3 mol^{-1} = 35.48	VP/AS	2.82E−5

In these calculations, it has been assumed that the concentration of the solute was within the region where Henry's law is valid. The data of Refs 163, 165, 166 and 168 permit an order-of-magnitude estimate for k_H of 4.E−5 mol kg^{-1} Pa^{-1}.

Solute		
Formula	Name	CAS Registry Number
$C_{12}H_5Cl_5$	2,2′,4,5,5′-PCB	37680-73-2
Solvent: Water		

Available experimental data for k_H of this solid in water are as follows:

			Original measurements		k_H/mol kg^{-1} Pa^{-1}
Ref.	$t/°C$	T/K	Limiting solubility ratio	Method	
165	20	293.2	$(1/k_H)$/atm m^3 mol^{-1} = 1.79E−4	VP/AS	5.51E−5
166	25	298.2	$(1/k_H)$/atm m^3 mol^{-1} = 2.51E−4	PURGE	3.93E−5
168	25	298.2	$(1/k_H)$/atm m^3 mol^{-1} = 2.3E−4	VP/AS	4.29E−5
167	25	298.2	$(1/k_H)$/atm m^3 mol^{-1} = 0.90E−4	WET WALL	11.0E−5
163	25	298.2	$(1/k_H)$/Pa m^3 mol^{-1} = 35.48	VP/AS	2.82E−5

In these calculations, it has been assumed that the concentration of the solute was within the region where Henry's law is valid. The data of Refs 163, 165, 166 and 168 permit an order-of-magnitude estimate for k_H of 4.E−5 mol kg^{-1} Pa^{-1}.

Solute		
Formula	Name	CAS Registry Number
$C_{12}H_6Cl_2O_2$	2,7-Dichlorodibenzo-dioxin	33857-26-0
Solvent: Water		

There are two reports of k_H for this solid in water at 25 °C:

	Original measurements		
Ref.	Limiting solubility ratio	Method	k_H/mol kg^{-1} Pa^{-1}
162	$(1/k_H)$/Pa m^3 mol^{-1} = 8.11	VP/AS	1.23E−4
164	$(1/k_H)$/Pa m^3 mol^{-1} = 5.96	DIRECT	1.68E−4

In these calculations, it has been assumed that the concentration of the solute was within the region where Henry's law is valid. The mean value, 1.42E−4 mol $kg^{-1} Pa^{-1}$, can be RECOMMENDED.

Solute		
Formula	Name	CAS Registry Number
$C_{12}H_6Cl_4$	2,2′,3,3′-PCB	38444-93-8
Solvent: Water		

Available experimental measurements of k_H of this solid in water are as follows:

Original measurements					k_H/mol $kg^{-1} Pa^{-1}$
Ref.	$t/°C$	T/K	Limiting solubility ratio	Method	
165	20	293.2	$(1/k_H)$/atm m^3 mol^{-1} = 1.61E−4	VP/AS	6.13E−5
127	20	293.2	$(1/k_H)$/atm m^3 mol^{-1} = 1.2E−4	PURGE	8.22E−5
167	25	298.2	$(1/k_H)$/atm m^3 mol^{-1} = 1.0E−4	WET WALL	9.87E−5
166	25	298.2	$(1/k_H)$/atm m^3 mol^{-1} = 2.02E−4	PURGE	4.89E−5
163	25	298.2	$(1/k_H)$/Pa m^3 mol^{-1} = 21.94	VP/AS	4.56E−5

In these calculations, it has been assumed that the concentration of the solute was within the region where Henry's law is valid. An order-of-magnitude estimate of k_H of this compound is k_H/mol $kg^{-1} Pa^{-1}$ = 6.7E−5.

Solute		
Formula	Name	CAS Registry Number
$C_{12}H_6Cl_4$	2,2′,5,5′-PCB	35693-99-3
Solvent: Water		

Available experimental data for k_H of this solid in water are as follows:

Original measurements					k_H/mol $kg^{-1} Pa^{-1}$
Ref.	$t/°C$	T/K	Limiting solubility ratio	Method	
165	20	293.2	$(1/k_H)$/atm m^3 mol^{-1} = 2.38E−4	VP/AS	4.15E−5
168	25	298.2	$(1/k_H)$/atm m^3 mol^{-1} = 4.20E−4	VP/AS	2.325E−5

Original measurements					k_H/mol
Ref.	$t/°C$ T/K	Limiting solubility ratio		Method	$\mathrm{kg^{-1}\,Pa^{-1}}$
167	25 298.2	$(1/k_H)/\mathrm{atm\,m^3\,mol^{-1}} = 2.0E{-}4$		WET WALL	4.93E−5
166	25 298.2	$(1/k_H)/\mathrm{atm\,m^3\,mol^{-1}} = 3.432E{-}4$		PURGE	2.89E−5
163	25 298.2	$(1/k_H)/\mathrm{Pa\,m^3\,mol^{-1}} = 48$		VP/AS	2.08E−5

In these calculations, it has been assumed that the concentration of the solute was within the region where Henry's law is valid. These data permit an order-of-magnitude estimate of $k_H/\mathrm{mol\,kg^{-1}\,Pa^{-1}} = 3.3E{-}5$.

Solute		
Formula	Name	CAS Registry Number
$C_{12}H_7Cl_3$	2,2′,5-PCB	37680-65-2
Solvent: Water		

There are six reports for k_H of this solid in water:

Original measurements					k_H/mol
Ref.	$t/°C$ T/K	Limiting solubility ratio		Method	$\mathrm{kg^{-1}\,Pa^{-1}}$
127	20 293.2	$(1/k_H)/\mathrm{atm\,m^3\,mol^{-1}} = 2.0E{-}4$		PURGE	4.93E−5
169	23 296.2	$k_{AW} = 4.1E{-}2$		PURGE	0.991E−5
163	25 298.2	$(1/k_H)/\mathrm{Pa\,m^3\,mol^{-1}} = 92.21$		VP/AS	1.08E−5
167	25 298.2	$(1/k_H)/\mathrm{atm\,m^3\,mol^{-1}} = 2.5E{-}4$		WET WALL	3.94E−5
165	25 298.2	$(1/k_H)/\mathrm{atm\,m^3\,mol^{-1}} = 2.99E{-}4$		VP/AS	3.30E−5

In these calculations, it has been assumed that the concentration of the solute was within the region where Henry's law is valid. An order-of-magnitude estimate for this compound is $k_H/\mathrm{mol\,kg^{-1}\,Pa^{-1}} = 2.8$ at ambient temperature.

Solute		
Formula	Name	CAS Registry Number
$C_{12}H_7Cl_3$	3,4,4′-PCB	38444-90-5
Solvent: Water		

Available experimental data for k_H of this solid in water are as follows:

Original measurements					k_H/mol
Ref.	t/°C	T/K	Limiting solubility ratio	Method	kg^{-1} Pa^{-1}
165	20	293.2	$(1/k_H)$/atm m^3 mol$^{-1} = 1.52E-4$	VP/AS	6.49E−5
167	25	298.2	$(1/k_H)$/atm m^3 mol$^{-1} = 1.0E-4$	WET WALL	9.87E−5
163	25	298.2	$(1/k_H)$/Pa m^3 mol$^{-1} = 84$	VP/AS	1.19E−5

In these calculations, it has been assumed that the concentration of the solute was within the region where Henry's law is valid. These data permit an order-of-magnitude estimate for k_H of 6.E−5 mol kg^{-1} Pa^{-1}.

Solute		
Formula	Name	CAS Registry Number
$C_{12}H_8$	Acenaphthylene	208-96-8
Solvent: Water		

Available experimental measurements of k_H of this solid in water are as follows:

Original measurements					k_H/mol
Ref.	t/°C	T/K	Limiting solubility ratio	Method	kg^{-1} Pa^{-1}
160	4	277.2	$(1/k_H)$/Pa m^3 mol$^{-1} = 2.38$	PURGE	42.0E−5
160	11	294.2	$(1/k_H)$/Pa m^3 mol$^{-1} = 4.27$	PURGE	23.4E−5
160	18	291.2	$(1/k_H)$/Pa m^3 mol$^{-1} = 7.46$	PURGE	13.4E−5
160	25	298.2	$(1/k_H)$/Pa m^3 mol$^{-1} = 12.7$	PURGE	7.87E−5
170	25	298.2	$k_{AW} = 4.6E-3$	WET WALL	8.77E−5
160	31	304.2	$(1/k_H)$/Pa m^3 mol$^{-1} = 19.6$	PURGE	5.10E−5

In these calculations, it has been assumed that the concentration of the solute was within the region where Henry's law is valid. These data may be represented by the equation

$$\ln(k_H/\text{mol kg}^{-1}\,\text{Pa}^{-1}) = -31.867 + 6727.4/(T/\text{K})$$

The standard error of estimate of k_H by this equation is 17 %. This equation generates RECOMMENDED data in the range 277–304 K.

Solute		
Formula	Name	CAS Registry Number
$C_{12}H_8Cl_2$	2,2'-PCB	13029-08-8
Solvent: Water		

Available experimental data for k_H of this solid in water are as follows:

		Original measurements			
Ref.	$t/°C$	T/K	Limiting solubility ratio	Method	$k_H/\text{mol kg}^{-1}\,\text{Pa}^{-1}$
165	20	293.2	$(1/k_H)/\text{atm m}^3\,\text{mol}^{-1} = 2.98\text{E}{-}4$	VP/AS	3.31E−5
166	25	298.2	$(1/k_H)/\text{atm m}^3\,\text{mol}^{-1} = 3.37\text{E}{-}4$	PURGE	2.93E−5
163	25	298.2	$(1/k_H)/\text{Pa m}^3\,\text{mol}^{-1} = 59.17$	VP/AS	1.69E−5

In these calculations, it has been assumed that the concentration of the solute was within the region where Henry's law is valid. These data permit an order-of-magnitude estimate for k_H of 2.6E−5 mol kg^{-1} Pa^{-1}.

Solute		
Formula	Name	CAS Registry Number
$C_{12}H_8Cl_2$	4,4'-PCB	2050-68-2
Solvent: Water		

Available experimental measurements of k_H of this solid in water at 25 °C are as follows:

	Original measurements		
Ref.	Limiting solubility ratio	Method	$k_H/\text{mol kg}^{-1}\,\text{Pa}^{-1}$
163	$(1/k_H)/\text{Pa m}^3\,\text{mol}^{-1} = 17.0$	VP/AS	5.88E−5
166	$(1/k_H)/\text{atm m}^3\,\text{mol}^{-1} = 1.99\text{E}{-}4$	PURGE	4.95E−5
170	$(1/k_H)/\text{atm m}^3\,\text{mol}^{-1} = 3.9\text{E}{-}3$	WET WALL	0.253E−5
40	$\gamma^\infty = 1.21\text{E}7$	VP/AS	0.167E−5

In these calculations, it has been assumed that the concentration of the solute was within the region where Henry's law is valid. The data are not consistent.

Solute		
Formula	Name	CAS Registry Number
$C_{12}H_8Cl_6O$	Dieldrin	60-57-1
Solvent: Water		

There are three reports of k_H for this solid in water:

			Original measurements		k_H/mol kg^{-1} Pa^{-1}
Ref.	$t/°C$	T/K	Limiting solubility ratio	Method	
171	20	293.2	$(1/k_H)$/atm m^3 mol^{-1} = 2.9E−5	PURGE	3.40E−4
33	20	293.2	$(1/k_H)$/Pa m^3 mol^{-1} = 1.12	VP/AS	8.93E−4
123	25	298.2	$(1/k_H)$/kPa m^3 mol^{-1} = 1.1E−3	CRIT EVAL	9.09E−4

In these calculations, it has been assumed that the concentration of the solute was within the region where Henry's law is valid. An order-of-magnitude estimate of k_H of this compound is 7.1E−4 mol kg^{-1} Pa^{-1}.

Solute		
Formula	Name	CAS Registry Number
$C_{12}H_{10}$	Acenaphthene	83-32-9
Solvent: Water		

Available experimental measurements of k_H of this solid in water are as follows:

			Original measurements		
Ref.	$t/°C$	T/K	Limiting solubility ratio	Method	k_H/mol kg^{-1} Pa^{-1}
160	4	277.2	$(1/k_H)$/Pa m^3 mol^{-1} = 3.52	PURGE	28.4E−5
160	11	284.2	$(1/k_H)$/Pa m^3 mol^{-1} = 6.29	PURGE	15.9E−5
160	18	291.2	$(1/k_H)$/Pa m^3 mol^{-1} = 10.9	PURGE	9.17E−5
160	25	298.2	$(1/k_H)$/Pa m^3 mol^{-1} = 18.5	PURGE	5.41E−5
148	25	298.2	$(1/k_H)$/atm m^3 mol^{-1} = 1.4E−4	PURGE	6.76E−5
132	25	298.2	$(1/k_H)$/Pa m^3 mol^{-1} = 16.2	PURGE	6.17E−5
123	25	298.2	$(1/k_H)$/kPa m^3 mol^{-1} = 2.4E−2	PURGE	4.17E−5
160	31	304.2	$(1/k_H)$/Pa m^3 mol^{-1} = 28.6	PURGE	3.50E−5

In these calculations, it has been assumed that the concentration of the solute was within the region where Henry's law is valid. These data may be represented by the equation

$$\ln(k_H/\mathrm{mol\,kg^{-1}\,Pa^{-1}}) = -31.499 + 6466.4/(T/\mathrm{K})$$

The standard error of estimate of k_H by this equation is 8 %. This equation generates RECOMMENDED data in the range 277–304 K.

Solute		
Formula	Name	CAS Registry Number
$C_{12}H_{21}N_2O_3$ PS	Diazinon	333-41-5
Solvent: Water		

There are five reports of k_H for this solid in water:

			Original measurements		k_H/mol
Ref.	$t/^{\circ}\mathrm{C}$	T/K	Limiting solubility ratio	Method	$\mathrm{kg^{-1}\,Pa^{-1}}$
33	20	293.2	$(1/k_H)/\mathrm{Pa\,m^3\,mol^{-1}} = 6.7\mathrm{E}{-}2$	VP/AS	1.49E−2
147	23	296.2	$(1/k_H)/\mathrm{mmHg\,ppm^{-1}} = 4.05\mathrm{E}{-}6$	VP/AS	0.609E−2
139	23	296.2	$k_{AW} = 6.0\mathrm{E}{-}5$	VP/AS	0.677E−2
170	23	296.2	$k_{AW} = 4.6\mathrm{E}{-}6$	WET WALL	8.82E−2
172	23	296.2	$k_{AW} = 4.9\mathrm{E}{-}6$	FOG	8.28E−2

In these calculations, it has been assumed that the concentration of the solute was within the region where Henry's law is valid. Although no temperature was stated in Refs. 139 and 147, the results were calculated with an assumed temperature of 23 °C. An order-of-magnitude estimate for k_H of this compound is 4.E−2 mol kg^{-1} Pa^{-1}.

Solute		
Formula	Name	CAS Registry Number
$C_{13}H_{10}$	Fluorene	86-73-7
Solvent: Water		

Available experimental measurements of k_H of this solid in water are as follows:

Ref.	$t/^\circ C$	T/K	Original measurements — Limiting solubility ratio	Method	$k_H/mol\ kg^{-1}\ Pa^{-1}$
160	4	277.2	$(1/k_H)/Pa\,m^3\,mol^{-1} = 2.05$	PURGE	4.88E−4
160	11	284.2	$(1/k_H)/Pa\,m^3\,mol^{-1} = 3.54$	PURGE	2.82E−4
160	18	291.2	$(1/k_H)/Pa\,m^3\,mol^{-1} = 5.96$	PURGE	1.68E−4
153	20	293.2	$(1/k_H)/Pa\,m^3\,mol^{-1} = 6.5$	PURGE	1.54E−4
132	25	298.2	$(1/k_H)/Pa\,m^3\,mol^{-1} = 9.75$	PURGE	1.03E−4
160	25	298.2	$(1/k_H)/Pa\,m^3\,mol^{-1} = 9.81$	PURGE	1.02E−4
160	25	298.2	$(1/k_H)/kPa\,m^3\,mol^{-1} = 8.5E\text{-}3$	CRIT EVAL	1.18E−4
160	31	304.2	$(1/k_H)/Pa\,m^3\,mol^{-1} = 14.8$	PURGE	0.676E−4

In these calculations, it has been assumed that the concentration of the solute was within the region where Henry's law is valid. These data may be represented by the equation

$$\ln(k_H/mol\,kg^{-1}\,Pa^{-1}) = -29.402 + 6036.2/(T/K)$$

The standard error of estimate of k_H by this equation is 4 %. This equation generates RECOMMENDED values in the range 277–304 K.

Solute		
Formula	Name	CAS Registry Number
$C_{13}H_{16}F_3N_3O_4$	Trifluralin	1582-09-8
Solvent: Water		

There are five reports of k_H for this solid in water. Although no temperature was stated in Ref. 147, it was assumed to be 23 °C.

Ref.	$t/^\circ C$	T/K	Original measurements — Limiting solubility ratio	Method	$k_H/mol\ kg^{-1}\ Pa^{-1}$
146	20	293.2	$k_{AW} = 4.23E\text{-}3$	WET WALL	0.970E−4
33	20	293.2	$(1/k_H)/Pa\,m^3\,mol^{-1} = 4.02$	VP/AS	2.49E−4
147	23	296.2	$(1/k_H)/mmHg\,ppm^{-1} = 1.08E\text{-}4$	VP/AS	2.37E−4
172	23	296.2	$k_{AW} = 2.1E\text{-}3$	FOG	1.93E−4
172	23	296.2	$k_{AW} = 2.4E\text{-}3$	WET WALL	1.69E−4

In these calculations, it has been assumed that the concentration of the solute was within the region where Henry's law is valid. These results allow only an order-of-magnitude estimate for k_H of this compound at ambient temperature: $1.9E-4\,\mathrm{mol\,kg^{-1}\,Pa^{-1}}$.

Solute		
Formula	Name	CAS Registry Number
$C_{14}H_9Cl_5$	p,p'-DDT	50-29-3
Solvent: Water		

There are three reports of k_H of this solid in water:

			Original measurements		k_H/mol $\mathrm{kg^{-1}\,Pa^{-1}}$
Ref.	$t/°C$	T/K	Limiting solubility ratio	Method	
33	20	293.2	$(1/k_H)/\mathrm{Pa\,m^3\,mol^{-1}} = 2.36$	VP/AS	4.24E−4
172	23	296.2	$k_{AW} = 5.3E{-4}$	FOG	7.66E−4
172	23	296.2	$k_{AW} = 3.5E{-4}$	WET WALL	11.6E−4
123	25	298.2	$(1/k_H)/\mathrm{kPa\,m^3\,mol^{-1}} = 5.3E{-3}$	CRIT EVAL	1.87E−4

In these calculations, it has been assumed that the concentration of the solute was within the region where Henry's law is valid. Although no temperature was stated in Ref. 172, the results were calculated at an assumed temperature of 23 °C. An order-of-magnitude estimate of k_H of this compound is $6.E-4\,\mathrm{mol\,kg^{-1}\,Pa^{-1}}$.

Solute		
Formula	Name	CAS Registry Number
$C_{14}H_{10}$	Anthracene	120-12-7
Solvent: Water		

Available experimental measurements for k_H of this solid in water are as follows:

Ref.	$t/°C$	T/K	Original measurements Limiting solubility ratio	Method	k_H/mol $kg^{-1} Pa^{-1}$
160	4	277.2	$(1/k_H)/Pa\,m^3\,mol^{-1} = 1.25$	PURGE	8.00E−4
152	11	284.2	$(1/k_H)/Pa\,m^3\,mol^{-1} = 2.12$	PURGE	4.72E−4
152	15	288.2	$(1/k_H)/Pa\,m^3\,mol^{-1} = 3.44$	PURGE	2.91E−4
160	18	291.2	$(1/k_H)/Pa\,m^3\,mol^{-1} = 3.50$	PURGE	2.86E−4
154	25	298.2	$k_{AW} = 2.66E−3$	PURGE	1.52E−4
173	25	298.2	$\gamma^\infty = 2.35E7$	VP/AS	1.64E−4
123	25	298.2	$(1/k_H)/kPa\,m^3\,mol^{-1} = 6.0E−3$	CRIT EVAL	1.67E−4
160	25	298.2	$(1/k_H)/Pa\,m^3\,mol^{-1} = 5.64$	PURGE	1.77E−4
160	31	302.1	$(1/k_H)/Pa\,m^3\,mol^{-1} = 8.36$	PURGE	1.20E−4

In these calculations, it has been assumed that the concentration of the solute was within the region where Henry's law is valid. These data may be represented by the equation

$$\ln(k_H/mol\,kg^{-1}\,Pa^{-1}) = -27.620 + 5654.3/(T/K)$$

The standard error of estimate of k_H by this equation is 8 %. This equation generates RECOMMENDED values in the range 277–302 K.

	Solute	
Formula	Name	CAS Registry Number
$C_{14}H_{10}$	Phenanthrene	85-01-8
Solvent: Water		

Available experimental measurements of k_H of this solid in water are as follows:

Ref.	$t/°C$	T/K	Original measurements Limiting solubility ratio	Method	k_H/mol $kg^{-1} Pa^{-1}$
152	6	279.2	$(1/k_H)/Pa\,m^3\,mol^{-1} = 1.81$	PURGE	5.52E−4
152	15	288.2	$(1/k_H)/Pa\,m^3\,mol^{-1} = 3.06$	PURGE	3.27E−4
160	18	291.2	$(1/k_H)/Pa\,m^3\,mol^{-1} = 2.65$	PURGE	3.77E−4
152	20	293.2	$(1/k_H)/Pa\,m^3\,mol^{-1} = 3.66$	PURGE	2.73E−4
123	25	298.2	$(1/k_H)/kPa\,m^3\,mol^{-1} = 4.0E−3$	CRIT EVAL	2.50E−4
132	25	298.2	$(1/k_H)/Pa\,m^3\,mol^{-1} = 3.61$	PURGE	2.77E−4

			Original measurements		k_H/mol
Ref.	$t/°C$	T/K	Limiting solubility ratio	Method	$kg^{-1} Pa^{-1}$
148	25	298.2	$(1/k_H)/atm\, m^3\, mol^{-1} = 3.93E-5$	PURGE	2.51E-4
160	25	298.2	$(1/k_H)/Pa\, m^3\, mol^{-1} = 4.29$	PURGE	2.33E-4
152	26	299.2	$(1/k_H)/Pa\, m^3\, mol^{-1} = 4.73$	PURGE	2.11E-4
152	30	303.2	$(1/k_H)/Pa\, m^3\, mol^{-1} = 5.54$	PURGE	1.81E-4
160	31	304.2	$(1/k_H)/Pa\, m^3\, mol^{-1} = 6.38$	PURGE	1.57E-4
152	35	308.2	$(1/k_H)/Pa\, m^3\, mol^{-1} = 7.90$	PURGE	1.27E-4

In these calculations, it has been assumed that the concentration of the solute was within the region where Henry's law is valid. These data may be represented by the equation

$$\ln(k_H/mol\, kg^{-1}\, Pa^{-1}) = -22.366 + 4169.5/(T/K)$$

The standard error of estimate of k_H by this equation is 9%. This equation generates RECOMMENDED values in the range 279–308 K.

	Solute	
Formula	Name	CAS Registry Number
$C_{14}H_{20}ClNO_2$	Alachlor	15972-60-8
Solvent: Water		

There are four reports of k_H of this solid in water. No temperature was stated in Ref. 139, and its result was calculated on the assumption that the temperature was 23 °C.

			Original measurements		k_H/mol
Ref.	$t/°C$	T/K	Limiting solubility ratio	Method	$kg^{-1} Pa^{-1}$
1	20	293.2	$(1/k_H)/Pa\, m^3\, mol^{-1} = 6.2E-3$	VP/AS	0.161
2	23	296.2	$k_{AW} = 4.5E-7$	FOG	0.902
3	23	296.2	$k_{AW} = 43.4E-7$	WET WALL	1.19
4	23	296.2	$k_{AW} = 1.3E-6$	VP/AS	0.312

Ref.	T/K	Quantity	Unit	Value	$Mol\,kg^{-1}\,Pa^{-1}$	Method
33	293.15	$1/k_H$	$Pa\,m^3\,mol^{-1}$	6.2E−3	0.161	VP/AS
172	296.15	k_{AW}	–	4.5E−7	0.902	FOG
170	296.15	k_{AW}	–	3.4E−7	1.19	WET WALL
139	296.15	k_{AW}	–	1.3E−6	0.312	VP/AS

In these calculations, it has been assumed that the concentration of the solute was within the region where Henry's law is valid. The data are inconsistent.

Solute		
Formula	Name	CAS Registry Number
$C_{15}H_{22}ClNO_2$	Metolachlor	51218-45-2
Solvent: Water		

Available estimates of k_H for this solid in water are given in the table. No temperature was stated in Ref. 139, and it was assumed to be 23 °C.

	Original measurements				k_H/mol
Ref.	$t/°C$	T/K	Limiting solubility ratio	Method	$kg^{-1}\,Pa^{-1}$
174	20	293.2	$(1/k_H)/atm\,m^3\,mol^{-1} = 9.0E{-}9$	INDIRECT	1.10
139	23	296.2	$k_{AW} = 3.7E{-}7$	VP/AS	1.10
174	25	298.2	$(1/k_H)/atm\,m^3\,mol^{-1} = 1.39E{-}8$	INDIRECT	0.710
174	30	303.2	$(1/k_H)/atm\,m^3\,mol^{-1} = 3.8E{-}8$	INDIRECT	0.260
174	40	313.2	$(1/k_H)/atm\,m^3\,mol^{-1} = 1.98E{-}7$	INDIRECT	0.0500

In these calculations, it has been assumed that the concentration of the solute was within the region where Henry's law is valid. These data may be represented by the equation

$$\ln(k_H/mol\,kg^{-1}\,Pa^{-1}) = -51.5808 + 15235(T/K)$$

The standard error of estimate of k_H by this equation is 26 %. This equation generates RECOMMENDED data in the range 293–313 K.

Solute		
Formula	Name	CAS Registry Number
$C_{16}H_{10}$	Fluoranthene	206-44-0
Solvent: Water		

Available experimental measurements of k_H of this solid in water are as follows:

Original measurements					
Ref.	$t/°C$	T/K	Limiting solubility ratio	Method	$k_H/mol\,kg^{-1}\,Pa^{-1}$
160	4	277.2	$(1/k_H)/Pa\,m^3\,mol^{-1} = 0.56$	PURGE	14.7E−4
160	11	284.2	$(1/k_H)/Pa\,m^3\,mol^{-1} = 0.87$	PURGE	11.5E−4
160	18	291.2	$(1/k_H)/Pa\,m^3\,mol^{-1} = 1.32$	PURGE	7.58E−4
153	20	293.2	$(1/k_H)/Pa\,m^3\,mol^{-1} = 1.11$	PURGE	9.09E−4
160	25	298.2	$(1/k_H)/Pa\,m^3\,mol^{-1} = 1.96$	PURGE	5.10E−4
160	31	302.2	$(1/k_H)/Pa\,m^3\,mol^{-1} = 2.72$	PURGE	3.68E−4

In these calculations, it has been assumed that the concentration of the solute was within the region where Henry's law is valid. These data may be represented by the equation

$$\ln(k_H/mol\,kg^{-1}\,Pa^{-1}) = -23.705 + 4824.6/(T/K)$$

The standard error of estimate of k_H by this equation is 13%. This equation generates RECOMMENDED data in the temperature range 277–302 K.

Solute		
Formula	Name	CAS Registry Number
$C_{16}H_{10}$	Pyrene	129-00-0
Solvent: Water		

Available experimental measurements of k_H of this solid in water are as follows:

Ref.	$t/°C$	T/K	Limiting solubility ratio	Method	$k_H/\mathrm{mol}\ \mathrm{kg}^{-1}\ \mathrm{Pa}^{-1}$
			Original measurements		
160	4	277.2	$(1/k_H)/\mathrm{Pa}\,\mathrm{m}^3\,\mathrm{mol}^{-1} = 0.43$	PURGE	23.3E−4
160	11	284.2	$(1/k_H)/\mathrm{Pa}\,\mathrm{m}^3\,\mathrm{mol}^{-1} = 0.69$	PURGE	14.5E−4
160	18	291.2	$(1/k_H)/\mathrm{Pa}\,\mathrm{m}^3\,\mathrm{mol}^{-1} = 1.10$	PURGE	9.09E−4
153	20	293.2	$(1/k_H)/\mathrm{Pa}\,\mathrm{m}^3\,\mathrm{mol}^{-1} = 1.46$	PURGE	6.85E−4
132	25	298.2	$(1/k_H)/\mathrm{Pa}\,\mathrm{m}^3\,\mathrm{mol}^{-1} = 1.21$	PURGE	8.26E−4
123	25	298.2	$(1/k_H)/\mathrm{kPa}\,\mathrm{m}^3\,\mathrm{mol}^{-1} = 1.2\mathrm{E}{-3}$	CRIT EVAL	8.33E−4
160	31	304.2	$(1/k_H)/\mathrm{Pa}\,\mathrm{m}^3\,\mathrm{mol}^{-1} = 2.45$	PURGE	4.08E−4

In these calculations, it has been assumed that the concentration of the solute was within the region where Henry's law is valid. These data may be represented by the equation

$$\ln(k_H/\mathrm{mol}\ \mathrm{kg}^{-1}\ \mathrm{Pa}^{-1}) = -23.401 + 4794.8/(T/K)$$

The standard error of estimate of k_H by this equation is 13 %. This equation generates RECOMMENDED values in the range 277–304 K.

Solute		
Formula	Name	CAS Registry Number
$C_{18}H_{12}$	Benz[a]anthracene	56-55-3
Solvent: Water		

Available experimental measurements of k_H of this solid in water are as follows:

Ref.	$t/°C$	T/K	Limiting solubility ratio	Method	$k_H/\mathrm{mol}\ \mathrm{kg}^{-1}\ \mathrm{Pa}^{-1}$
			Original measurements		
160	4	277.2	$(1/k_H)/\mathrm{Pa}\,\mathrm{m}^3\,\mathrm{mol}^{-1} = 0.15$	PURGE	6.67E−3
160	11	284.2	$(1/k_H)/\mathrm{Pa}\,\mathrm{m}^3\,\mathrm{mol}^{-1} = 0.31$	PURGE	3.23E−3
160	18	291.2	$(1/k_H)/\mathrm{Pa}\,\mathrm{m}^3\,\mathrm{mol}^{-1} = 0.63$	PURGE	1.59E−3
160	25	298.2	$(1/k_H)/\mathrm{Pa}\,\mathrm{m}^3\,\mathrm{mol}^{-1} = 1.22$	PURGE	0.820E−3
154	25	298.2	$k_{AW} = 3.28\mathrm{E}{-4}$	PURGE	1.23E−3
160	31	304.2	$(1/k_H)/\mathrm{Pa}\,\mathrm{m}^3\,\mathrm{mol}^{-1} = 2.11$	PURGE	0.474E−3

In these calculations, it has been assumed that the concentration of the solute was within the region where Henry's law is valid. These data may be represented by the equation

$$\ln(k_H/\mathrm{mol\,kg^{-1}\,Pa^{-1}}) = -33.349 + 7854.5/(T/\mathrm{K})$$

The standard error of estimate of k_H by this equation is 11%. This equation generates RECOMMENDED data in the range 277–304 K.

Solute		
Formula	Name	CAS Registry Number
$C_{18}H_{12}$	Chrysene	218-01-9
Solvent: Water		

There is one available report of measurement of k_H of this solid in water [160]:

Original measurements				
$t/°C$	T/K	Limiting solubility ratio	Method	$k_H/\mathrm{mol\,kg^{-1}\,Pa^{-1}}$
4	277.2	$(1/k_H)/\mathrm{Pa\,m^3\,mol^{-1}} = 0.02$	PURGE	50.0E−3
11	284.2	$(1/k_H)/\mathrm{Pa\,m^3\,mol^{-1}} = 0.07$	PURGE	14.3E−3
18	291.2	$(1/k_H)/\mathrm{Pa\,m^3\,mol^{-1}} = 0.19$	PURGE	5.26E−3
25	298.2	$(1/k_H)/\mathrm{Pa\,m^3\,mol^{-1}} = 0.53$	PURGE	1.89E−3
31	304.2	$(1/k_H)/\mathrm{Pa\,m^3\,mol^{-1}} = 1.20$	PURGE	0.833E−3

In these calculations, it has been assumed that the concentration of the solute was within the region where Henry's law is valid. These data may be represented by the equation

$$\ln(k_H/\mathrm{mol\,kg^{-1}\,Pa^{-1}}) = -48.819 + 12\,688/(T/\mathrm{K})$$

The standard error of estimate of k_H by this equation is 3%. This equation generates RECOMMENDED data in the range 277–304 K.

Solute		
Formula	Name	CAS Registry Number
$C_{20}H_{12}$	Benzo[a]fluorene	238-84-6
Solvent: Water		

There is one report of the measurement of k_H of this solid in water [160]:

		Original measurements		
$t/°C$	T/K	Limiting solubility ratio	Method	$k_H/\text{mol kg}^{-1}\,\text{Pa}^{-1}$
4	277.2	$(1/k_H)/\text{Pa m}^3\,\text{mol}^{-1} = 0.88$	PURGE	11.4E−4
11	284.2	$(1/k_H)/\text{Pa m}^3\,\text{mol}^{-1} = 1.30$	PURGE	7.69E−4
18	291.2	$(1/k_H)/\text{Pa m}^3\,\text{mol}^{-1} = 1.89$	PURGE	5.29E−4
25	298.2	$(1/k_H)/\text{Pa m}^3\,\text{mol}^{-1} = 3.70$	PURGE	3.70E−4
31	304.2	$(1/k_H)/\text{Pa m}^3\,\text{mol}^{-1} = 2.76$	PURGE	2.76E−4

In these calculations, it has been assumed that the concentration of the solute was within the region where Henry's law is valid. These data may be represented by the equation

$$\ln(k_H/\text{mol kg}^{-1}\,\text{Pa}^{-1}) = -22.756 + 4429.5/(T/K)$$

The standard error of estimate of k_H by this equation is 0.02 %. This equation generates RECOMMENDED data in the range 277–304 K.

Solute		
Formula	Name	CAS Registry Number
$C_{20}H_{12}$	Benzo[a]pyrene	50-32-8
Solvent: Water		

There are two estimates for k_H of this solid in water at 25 °C:

	Original measurements		
Ref.	Limiting solubility ratio	Method	$k_H/\text{mol kg}^{-1}\,\text{Pa}^{-1}$
40	$\gamma^\infty = 1.17E8$	VP/AS	6.09E−5
173	$\gamma^\infty = 3.75E8$	VP/AS	1.90E−5

In these calculations, it has been assumed that the concentration of the solute was within the region where Henry's law is valid.

REFERENCES

1. Clever, H. L. (ed.), *Solubility Data Series, Vol. 4, Argon*, Pergamon Press, Oxford, 1980.

2. Young, C. L. (ed.), *Solubility Data Series, Vol. 12, Sulfur dioxide, Chlorine, Fluorine and Chlorine Oxides*, Pergamon Press, Oxford, 1983.

3. Cargill, R. W. (ed.), *Solubility Data Series, Vol. 43, Carbon Monoxide*, Pergamon Press, Oxford, 1990.

4. Scharlin, P. (ed.), *Solubility Data Series, Vol. 62, Carbon Dioxide in Water and Aqueous Solutions*, Oxford University Press, Oxford, 1996.

5. Carroll, J. J.; Slupsky, J. D.; Mather, A. E., *J. Phys. Chem. Ref. Data*, 1991, **20**, 1201–9.

6. Wilhelm, E.; Battino, R.; Wilcock, R. J., *Chem. Rev.*, 1977, **77**, 219–62.

7. Thompson, H. W.; Kearton, C. F.; Lamb, S. A., *J. Chem. Soc.*, 1934, 1033–7.

8. Ferm, R. J., *Chem. Rev.*, 1957, **57**, 621–40.

9. (a) Young, C. L. (ed.), *Solubility Data Series, Vol. 5/6, Hydrogen and Deuterium*, Pergamon Press, Oxford, 1981; (b) Young, C. L. (ed.), *Solubility Data Series, Vol. 8, Oxides of Nitrogen*, Pergamon Press, Oxford, 1981; (c) Hwang, H.; Dasgupta, P. K., *Environ. Sci. Technol.*, 1985, **19**, 255–8.

10. O'Sullivan, D. W.; Lee, M.; Noone, B. C.; Heikes, B. G., *J. Phys. Chem.*, 1996, **100**, 3241–7.

11. Martin, L. R.; Damschen, D. E., *Atmos. Environ.*, 1981, **15**, 1615–21.

12. Zhou, X.; Lee, Y.-N., *J. Phys. Chem.*, 1992, **96**, 265–72.

13. Lind, J. A.; Kok, G. L., *J. Geophys. Res.*, 1986, **91**, 7889–95., *Erratum*: 1994, **99**, 21119.

14. Staffelbach, T. A.; Kok, G. L., *J. Geophys. Res.*, 1993, **98**, 12713–7.

15. Yoshizumi, K.; Aoki, K.; Nouchi, I.; Kobayashi, T.; Kamakura, S.; Tajima, M., *Atmos. Environ.*, 1984, **18**, 395–401.

16. Clever, H. L. (ed.), *Solubility Data Series, Vol. 1, Helium and Neon*, Pergamon Press, Oxford, 1979.

17. Clever, H. L. (ed.), *Solubility Data Series, Vol. 2, "Krypton, Xenon and Radon"*, Pergamon Press, Oxford, 1979.

18. Battino, R.; Rettich, T. R.; Tominaga, T., *J. Phys. Chem. Ref. Data*, 1984, **13**, 563–600.

19. Battino, R. (ed.), *Solubility Data Series, Vol. 7, Oxygen and Ozone*, Pergamon Press, Oxford, 1981.

20. Mroczek, E. K., *J. Chem. Eng. Data*, 1997, **42**, 116–19.

21. Ashton, J. T.; Dawe, R. A.; Miller, K. W.; Smith, E. B.; Stickings, B. J., *J. Chem. Soc. A*, 1968, 1793–6.

22. Cosgrove, B. A.; Walkley, J., *J. Chromatogr.*, 1981, **216**, 161–7.

23. O'Connell, W. L., *Trans. Am. Inst. Min. Metall. Petrol.*, 1963, **226**, 126–32.

24. Weigl, Z., *Predict II, Estimation and Prediction of Properties of Pure Organic Compounds and Their Mixtures. A Computer System*, Ch and T, Warsaw, 1993.

25. Kudchadker, A. P.; Kudchadker, S. A.; Shukla, R. P.; Patnaik, P. R., *J. Phys. Chem. Ref. Data*, 1979, **8**, 499–517.

26. van Arkel, A. E.; Vies, S. E., *Recl. Trav. Chim. Pays-Bas*, 1936, **55**, 407–11.

27. Gross, P. M.; Saylor, J. H., *J. Am. Chem. Soc.*, 1931, **53**, 1744–51.

28. Pearson, C. R.; McConnell, G., *Proc. R. Soc. London, Ser. B*, 1975, **189**, 305–32.

29. Roberts, P. V.; Dändliker, P. G., *Environ. Sci. Technol.*, 1983, **17**, 484–9.

30. Horvath, A. L.; Getzen, F. W., *Solubility Data Series, Vol. 60, Halogenated Methanes with Water*, Oxford University Press, Oxford, 1995.

31. Yurttaş, L.; Holste, J. C.; Hall, K. R.; Gammon, B. E.; Marsh, K. N., *Fluid Phase Equil.*, 1990, **59**, 217–23.

32. Kawamoto, K.; Urano, K., *Chemosphere*, 1989, **18**, 1987–96.

33. Suntio, L. R.; Shiu, W.-Y.; Mackay, D.; Seiber, J. N.; Glotfelty, D., *Rev. Environ. Contam. Toxicol.*, 1988, **103**, 1–59.

34. Gmehling, J.; Onken, U.; Arlt, W., *Chemistry Data Series, Vol. I/8, Vapor–liquid Equilibrium Data Collection*, Deutsche Gesellschaft für Apparatewesen, Frankfurt/Main, 1984.
35. Nicholson, B. C.; Maguire, B. P.; Bursill, D. B., *Environ. Sci. Technol.*, 1984, **18**, 518–21.
36. Tse, G.; Orbey, H.; Sandler, S. I., *Environ. Sci. Technol.*, 1992, **26**, 2017–22.
37. Kondoh, H.; Nakajima, T., *Kankyo Kagaku*, 1997, **7**, 81–9.
38. Yaws, C.; Yang, H.-C.; Pan, X., *Chem. Eng. (N.Y.)*, 1991, **98**, 179–85.
39. Tewari, Y. B.; Miller, M. M.; Wasik, S. P.; Martirre, D. E., *J. Chem. Eng. Data*, 1982, **27**, 451–4.
40. Hwang, Y.-L.; Olson, J. D.; Keller, G. E., *Ind. Eng. Chem. Res.*, 1992, **31**, 1759–68.
41. Kames, J.; Schurath, U., *J. Atmos. Chem.*, 1992, **15**, 79–95.
42. Clever, H. L.; Young, C. L. (eds), *Solubility Data Series, Vol. 27/28, Methane*, Pergamon Press, Oxford, 1987.
43. Snider, J. R.; Dawson, G. A., *J. Geophys. Res.*, 1985, **90**, 3797–805.
44. Pecsar, R. E.; Martin, J. J., *Anal. Chem.*, 1966, **38**, 1661–9.
45. Pividal, K. A.; Birtigh, A.; Sandler, S. I., *J. Chem. Eng. Data*, 1992, **37**, 484–7.
46. Shaffer, J. R.; Daubert, G. A., *Anal. Chem.*, 1969, **41**, 1585–9.
47. Sorrentino, G.; Voilley, A.; Richon, D., *AIChE J.*, 1986, **32**, 1988–93.
48. Li, J.; Carr. P. W., *Anal. Chem.*, 1993, **65**, 1443–50.
49. Burnett, M. G., *Anal. Chem.*, 1963, **35**, 1567–70.
50. Bader, M. S. H.; Gasem, K. A. M., *Chem. Eng. Commun.*, 1996, **140**, 41–72.
51. Richon, D.; Sorrentino, F.; Voilley, A., *Ind. Eng. Chem., Process Des. Dev.*, 1985, **24**, 1160–5.
52. Landau, I.; Nelfer, A. J.; Locke, D. C., *Ind. Eng. Chem. Res.*, 1991, **30**, 1900–6.
53. Lebert, A.; Richon, D., *J. Agric. Food Chem.*, 1984, **32**, 1156–61.
54. Rytting, J. H.; Huston, L. P.; Higuchi, T., *J. Pharm. Sci.*, 1978, **67**, 615–8.
55. Chai, X. S.; Zhu, J. Y., *J. Chromatogr. A*, 1998, **799**, 207–14.
56. Bergmann, D. L.; Eckert, C. A., *Fluid Phase Equil.*, 1991, **63**, 141–50.
57. Katsanos, N. A.; Karaiskakis, G.; Agathonos, P., *J. Chromatogr.*, 1986, **349**, 369–76.
58. Ochi, K.; Kojima, K., *J. Chem. Eng. Jpn.*, 1987, **20**, 6–10.
59. DeBruyn, W. J.; Swartz, E.; Hu, J. H.; Shorter, J. A.; Davidovits, P.; Worsnop, D. R.; Zahniser, M. S.; Kolb, C. E., *J. Geophys. Res.*, 1995, **100**, 7245–51.
60. Przyjazny, A.; Janicki, W.; Chrzanowski, W.; Staszewski, R., *J. Chromatogr.*, 1983, **280**, 244–60.
61. Gossett, J. M., *Environ. Sci. Technol.*, 1987, **21**, 202–8.
62. Peng, J.; Wann, A., *Environ. Sci. Technol.*, 1997, **31**, 2998–3003.
63. Park, S.-J.; Han, S.-D.; Ryu, S.-A., *Hwahak Konghak*, 1997, **35**, 915–20.
64. Hovorka, S.; Dohnal, V., *J. Chem. Eng. Data*, 1997, **42**, 924–33.
65. Hellmann, H., *Fresenius' Z. Anal. Chem.*, 1987, **328**, 475–9.
66. Robbins, G. A.; Wang, S.; Stuart, J. D., *Anal. Chem.*, 1993, **65**, 3113–18.
67. Vane, L. M.; Giroux, E. L., *J. Chem. Eng. Data*, 2000, **45**, 38–47.
68. Horvath, A. L.; Getzen, F. W.; Maczynska, Z., *J. Phys. Chem. Ref. Data*, 1999, **28**, 395–624.
69. Smith, R. A.; Porter, E. G.; Miller, K. W., *Biochim. Biophys. Acta*, 1981, **645**, 327–38.
70. Regan, M. J.; Eger, E. I., *Anesthesiology*, 1967, **28**, 689–700.
71. Steward, A.; Allott, P. R.; Cowles, A. L.; Mapleson, W. W., *Br. J. Anesthesiol.*, 1973, **45**, 282–93.
72. Khalfaoui, B.; Newsham, D. M. T., *J. Chromatogr. A*, 1994, **673**, 85–92.
73. Cooling, M. R.; Khalfaoui, B.; Newsham, D. M. T., *Fluid Phase Equil.*, 1992, **81**, 217–29.

74. Wright, D. A.; Sandler, S. I.; DeVoll, D., *Environ. Sci. Technol.*, 1992, **26**, 1828–31.
75. Sandler, S. I.; Orbey, H., *Fluid Phase Equil.*, 1993, **82**, 63–9.
76. Turner, L. H.; Chiew, Y. C.; Ahlert, R. C.; Kosson, D. S., *AIChE J.*, 1996, **42**, 1772–88.
77. Wasik, S. P.; Tewari, Y. B.; Miller, M. M.; Martire, D. E., *Octanol–Water Partition Coefficients and Aqueous Solubilities of Organic Compounds*, Report NBSIR-81-2406, US Department of Commerce, Washington, DC, 1981.
78. DeWulf, J.; van Langenhave, H.; Everaert, P., *J. Chromatogr. A*, 1999, **830**, 353–63.
79. Hansen, K. C.; Zhou, Z.; Yaws, C. L.; Aminabhavi, T. M., *J. Chem. Educ.*, 1995, **72**, 93–6.
80. Boublik, T.; Fried, V.; Hala, E., *The Vapor Pressures of Pure Substances*, Elsevier, Amsterdam, 1973.
81. Barr, R. S.; Newsham, D. M. T., *Fluid Phase Equil.*, 1987, **35**, 189–205.
82. Benkelberg, H.-J.; Hamm, S.; Warneck, P., *J. Atmos. Chem.*, 1995, **20**, 17–34.
83. Kames, J.; Schweighoefer, S.; Schurath, U., *J. Atmos. Chem.*, 1991, **12**, 169–80.
84. (a) Hayduk, W., *Solubility Data Series*, Vol. 57, *Ethene*, Oxford University Press, Oxford, 1994; (b) Hayduk, W., *Solubility Data Series*, Vol. 9, *Ethane*, Pergamon Press, Oxford, 1982.
85. Gmehling, J.; Onken, U.; Arlt, W., *Chemistry Data Series*, Vol. I/7, *Vapor–liquid Equilibrium Data Collection*, Deutsche Gesellschaft für Apparatewesen, Frankfurt/Main, 1980.
86. Shepson, P. B.; Mackay, E.; Muthurama, K., *Environ. Sci. Technol.*, 1996, **30**, 3618–23.
87. Dallas, A. J. *PhD Thesis*, University of Minnesota, Minneapolis, MN, 1993.
88. Rohrschneider, L., *Anal. Chem.*, 1973, **45**, 1241–7.
89. Park, J. H.; Hussam, A.; Couasnon, P.; Fritz, D.; Carr, P. W., *Anal. Chem.*, 1987, **59**, 1970–6.
90. Trampe, D. B.; Eckert, C. A., *AIChE J.*, 1993, **39**, 1045–50.
91. Dacey, J. W. H.; Wakeham, S. G.; Howes, B. L., *Geophys. Res. Lett.*, 1984, **11**, 991–4.
92. Aneja, V. P.; Overton, J. H., *Chem. Eng. Commun.*, 1990, **98**, 199–209.
93. Wong, P. K.; Wang, Y. H., *Chemosphere*, 1997, **35**, 535–44.
94. Lerman, J.; Willis, M. M.; Gregory, G. A.; Eger, E. I., *Anesthesiology*, 1983, **59**, 554–8.
95. Yates, S. R.; Gan, J., *J. Agric. Food Chem.*, 1998, **46**, 755–61.
96. Belfer, A. J.; Locke, D. C.; Landau, I., *Anal. Chem.*, 1990, **62**, 347–9.
97. Bao, J.-B.; Han, S.-J., *Phase Equilib.*, 1995, **112**, 307–17.
98. Gmehling, J.; Onken, U.; Arlt, W., *Chemistry Data Series*, Vol. I/2b, *Vapor–liquid Equilibrium Data Collection*, Deutsche Gesellschaft für Apparatewesen, Frankfurt/Main, 1978.
99. Vitenberg, A. G.; Ioffe, B. V.; Dimitrova, Z. St.; Butaeva, I. L., *J. Chromatogr.*, 1975, **112**, 319–27.
100. Buttery, R. G.; Ling, L. C.; Guadagni, D. G., *J. Agric. Food Chem.*, 1969, **17**, 385–9.
101. Vitenberg, A. G.; Ioffe, B. V.; Borisov, V. N., *Chromatographia*, 1974, **7**, 610–9.
102. Hartwick, R. P.; Howat, C. S., *J. Chem. Eng. Data*, 1995, **40**, 738–45.
103. Hayduk, W., *Solubility Data Series*, Vol. 24, *Propane, Butane and 2-Methylpropane*, Pergamon Press, Oxford, 1986.
104. (a) Butler, J. A. V.; Ramachandran, C. N.; Thomson, D. W., *J. Chem. Soc.*, 1935, 280–5; (b) Butler, J. A. V.; Ramachandran, C. N., *J. Chem. Soc.*, 1935, 952–5.
105. Djerki, R. A.; Laub, R. J., *J. Liq. Chromatogr.*, 1988, **11**, 585–62.
106. Kolb, B.; Welter, C.; Bichler, C., *Chromatographia*, 1992, **34**, 235–40.

107. Slocum, E. W.; Dodge, B. F., *AIChE J.*, 1964, **10**, 364–8.

108. Munson, E. S.; Saidman, L. J.; Eger, E. I., *Anesthesiology*, 1964, **25**, 638–40.

109. Allen, J. M.; Balcavage, W. X.; Ramachandran, B. R.; Shrout, A. L., *Environ. Toxicol. Chem.*, 1998, **17**, 1216–21.

110. Iraci, L. T.; Baker, B. M.; Tyndall, G. S.; Orlando, J. J., *J. Atmos. Chem.*, 1999, **33**, 321–30.

111. Hine, J.; Mookerjee, P. K., *J. Org. Chem.*, 1975, **40**, 292–8.

112. Zhou, X.; Mopper, K., *Environ. Sci. Technol.*, 1990, **24**, 1864–9.

113. Friant, S. L.; Suffet, I. H., *Anal. Chem.*, 1979, **51**, 2167–72.

114. Tochigi, K.; Kojima, K., *J. Chem. Eng. Jpn.*, 1976, **9**, 267–73.

115. Zou, M.; Prausnitz, J. M., *J. Chem. Eng. Data*, 1987, **32**, 34–7.

116. LeThanh, M.; Lamer, T.; Voilley, A.; Jose, J., *J. Chim. Phys. Phys.-Chim. Biol.*, 1993, **90**, 545–60.

117. Kieckbusach, T. G.; King, C. J., *J. Chromatogr. Sci.*, 1979, **17**, 273–80.

118. Getzen, F.; Hefter, G.; Maczynski, A. (eds), Solubility Data Series, *Vol. 48, Esters with Water. Part I*, Pergamon Press, Oxford, 1992.

119. Gmehling, J.; Onken, U.; Grenzheuser, P., *Chemistry Data Series, Vol. I/5, Vapor–liquid Equilibrium Data Collection* , Deutsche Gesellschaft für Apparatewesen, Frankfurt/Main, 1983.

120. Sagert, N. H.; Lau, D. W., *J. Chem. Eng. Data*, 1986, **31**, 475–8.

121. Fischer, K.; Gmehling, J., *J. Chem. Eng. Data*, 1994, **39**, 309–15.

122. Loblen, G. M.; Prausnitz, J. M., *Ind. Eng. Chem. Fundam.*, 1982, **21**, 109–13.

123. Mackay, D.; Shiu, W. Y., *J. Phys. Chem. Ref. Data*, 1981, **10**, 1175–99.

124. Wania, F.; Shiu, W.-Y.; Mackay, D., *J. Chem. Eng. Data*, 1994, **39**, 572–7.

125. Weil, L.; Dure, G.; Quentin, K.-E., *Wasser Abwasser Forsch.*, 1974, **7**, 169–75.

126. Miller, M. M.; Wasik, S. P.; Huang, G.-L.; Shiu, W.-Y.; Mackay, D., *Environ. Sci. Technol.*, 1985, **19**, 522–9.

127. Oliver, B. G., *Chemosphere*, 1985, **14**, 1087–106.

128. ten Hulscher, T. E. M.; van der Velde, L. E.; Bruggeman, W. A., *Environ. Toxicol. Chem.*, 1992, **11**, 1595–603.

129. Stull, D. R., *Ind. Eng. Chem.*, 1947, **39**, 517–40.

130. Yalkowsky, S. H.; Orr, R. J.; Valvani, S. C., *Ind. Eng. Chem. Fundam.*, 1979, **18**, 351–3.

130a. Horvath, A. L.; Getzen, F. W., 1985. *Solubility Data Series, Vol. 20, Halogenated Benzenes, Toluenes and Phenols with Water*, Pergamon Press, Oxford, 1985.

131. Hinckley, D. A.; Bidleman, T. F.; Foreman, W. T., Tuschall, J. R., *J. Chem. Eng. Data*, 1990, **35**, 232–7.

132. Shiu, W.-Y.; Mackay, D., *J. Chem. Eng. Data*, 1997, **42**, 27–30.

133. Rohac, V.; Ruzicka, V.; Ruzicka, K., *J. Chem. Eng. Data*, 1998, **43**, 770–5.

134. Leuenberger, C.; Ligocki, M. P.; Pankow, J. F., *Environ. Sci. Technol.*, 1985, **19**, 1053–8.

135. Yoshida, K.; Shigeoka, T.; Yamauchi, F., *Chemosphere*, 1987, **16**, 2531–44.

136. Tabai, S.; Rogalski, M.; Solimando, R.; Malanowski, S. K., *J. Chem. Eng. Data*, 1997, **42**, 1147–50.

137. Shiu, W.-Y.; Ma, K.-C.; Varhanickova, D.; Mackay, D., *Chemosphere*, 1994, **29**, 1155–224.

138. Shaw, D. G. (ed.), *Solubility Data Series, Vol. 37/38, Hydrocarbons with Water and Seawater*, Pergamon Press, Oxford, 1989.

139. Glotfelty, D. E.; Seiber, J. N.; Liljedahl, L. A., *Nature*, 1987, **325**, 602–5,

140. Wasik, S. P.; Tsang, W., *J. Phys. Chem.*, 1970, **74**, 2970–6.

141. Sanemasa, I.; Araki, M.; Degucgi, T.; Nagai, H., *Bull. Chem. Soc. Jpn.*, 1982, **55**, 1054–62.

142. Brown, R. L.; Wasik, S. P., *J. Res. Natl. Bur. Stand. A*, 1974, **78**, 453–60.

143. Perlinger, J. A.; Eisenreich, S. J.; Capel, P. D., *Environ. Sci. Technol.*, 1993, **27**, 928–37.
144. Karger, B. L.; Sewell, P. A.; Castells, R. C. Hartkopf, A., *J. Colloid Interface Sci.*, 1971, **35**, 328–39.
145. Metcalf, C. D.; McLeese, D. W.; Zitko, V., *Chemosphere*, 1980, **9**, 151–5.
146. Rice, C. P.; Chernyak, S. M.; McConnell, L. L., *J. Agric. Food Chem.*, 1997, **45**, 2291–8.
147. Sanders, P. F.; Seiber, J. N., *Chemosphere*, 1983, **12**, 999–1012.
148. Mackay, D.; Shiu, W. Y.; Sutherland, R. P., *Environ. Sci. Technol.*, 1979, **13**, 333–7.
149. Hamelink, J. L.; Simon, P. B.; Silberhorn, E. M., *Environ. Sci. Technol.*, 1996, **30**, 1946–52.
150. Yin, C.; Hassett, J. D., *Environ. Sci. Technol.*, 1986, **20**, 1213–17.
151. Schwarz, F. P.; Wasik, S. P., *J. Chem. Eng. Data*, 1977, **22**, 270–3.
152. Alaee, M.; Whittal, R. M.; Strachan, W. M. J., *Chemosphere*, 1996, **32**, 1153–64.
153. DeMaagd, P. G.-J.; ten Hulscher, D. T. E. M.; van den Heuvel, H.; Opperhuizen, A.; Sijm, D. T. H., *Environ. Toxicol. Chem.*, 1998, **17**, 251–7.
154. Southworth, G. R., *Bull. Environ. Contam. Toxicol.*, 1979, **21**, 507–14.
155. Li, J.; Perdue, E. M.; Pavlostathis, S. G.; Araujo, R., *Environ. Int.*, 1998, **24**, 353–8.
156. Fichan, I.; Larroche, C.; Gros, J. B., *J. Chem. Eng. Data*, 1999, **44**, 56–62.
157. Falk, A.; Gullstrand, E.; Löf, A.; Wigaeus-Hjelm, E., *Br. J. Ind. Med.*, 1990, **47**, 62–4.
158. Massaldi, H. A.; King, C. J., *J. Chem. Eng. Data*, 1973, **18**, 393–7.
159. Cotham, W. E.; Bidleman, T. F., *J. Agric. Food Chem.*, 1989, **37**, 824–8.
160. Bamford, H. A.; Poster, D. L.; Baker, J. E., *Environ. Toxicol. Chem.*, 1999, **18**, 1905–12.
161. Altschuh, J.; Brüggemann, R.; Santl, H.; Eichinger, G.; Piringer, O. G., *Chemosphere*, 1999, **39**, 1871–7.
162. Shiu, W. Y.; Doucette, W.; Gobas, F. A. P. C.; Andren, A.; Mackay, D., *Environ. Sci. Technol.*, 1988, **22**, 651–8.
163. Shiu, W. Y.; Mackay, D., *J. Phys. Chem. Ref. Data*, 1986, **15**, 911–29.
164. Santl, H.; Brandsch, R.; Gruber, L., *Chemosphere*, 1994, **29**, 2209–14.
165. Murphy, T. J.; Mullin, M. D.; Meyer, J. A., *Environ. Sci. Technol.*, 1987, **21**, 155–62.
166. Dunnivant, F. M.; Elzerman, A. W., *Chemosphere*, 1988, **17**, 525–41.
167. Brunner, S.; Hornung, E.; Santl, H.; Wolff, E.; Piringer, O. G.; Altschuh, J.; Brüggemann, R., *Environ. Sci. Technol.*, 1990, **24**, 1751–4.
168. Westcott, J. W.; Simon, C. G.; Bidleman, T. F., *Environ. Sci. Technol.*, 1981, **15**, 1375–8.
169. Atlas, E.; Foster, R.; Giam, C. S., *Environ. Sci. Technol.*, 1981, **16**, 283–6.
170. Fendinger, N. J.; Glotfelty, D. E., *Environ. Sci. Technol.*, 1988, **22**, 1289–93.
171. Slater, R. M.; Spedding, D. J., *Arch. Environ. Contam. Toxicol.*, 1981, **10**, 25–33.
172. Fendinger, N. J.; Glotfelty, D. E.; Freeman, H. P., *Environ. Sci. Technol.*, 1989, **23**, 1528–31.
173. Haines, R. I. S.; Sandler, S. I., *J. Chem. Eng. Data*, 1995, **40**, 833–6.
174. Lau, Y. L.; Liu, D. L. S.; Pacepavicius, G. J.; Maguire, R. J., *J. Environ. Sci. Health B*, 1995, **30**, 605–20.

CHAPTER 12

Henry's Law Constants for Dissolution in Seawater

James Sangster

Sangster Research Laboratories, Montreal, Canada

Solubility data for many solutes in pure water are presented in Chapter 11. These data are valuable in a number of environmental and industrial contexts. On a world scale, however, fresh water is rare: about 97 % of the water of the planet is salt [1]. It is therefore important to be aware of the properties of solutes in seawater. On a world scale, the oceans represent a reservoir of vast capacity for the containment and transport of dissolved species.

1 SEAWATER

The salt content of seawater varies somewhat over the earth (the Caspian and Dead 'seas' do not figure in this account). For the present purposes, a 'standard' seawater can be chosen, whose principal solutes at $25\,°C$ are [2]

Salt	Molarity
NaCl	0.457
$MgSO_4$	0.028
$MgCl_2$	0.028
$CaCl_2$	0.010

Seawater is thus a dilute solution of common salts, most of which is sodium chloride. The solids content of standard seawater – the salinity – from the table is $\sim$3.5 mass%, or 35 parts per thousand (usually designated by ppt or ‰).

For the calculations which follow, it is necessary to be cognizant of the density of both water and seawater as a function of temperature. The density of pure water

Chemicals in the Atmosphere – Solubility, Sources and Reactivity.   Edited by P. G. T. Fogg and J. Sangster
© 2003 IUPAC   ISBN: 0-471-98651-8

between 0 and 40 °C may be represented by [2]

$$d_0/\,\text{kg dm}^{-3} = 1.00120 - 1.94790\text{E}{-}4(t/^\circ\text{C}) \tag{12.1}$$

within a standard error of estimate of less than 0.1 %. The density of standard seawater, in the same temperature range, is [3]

$$d/\text{kg dm}^{-3} = 1.02917 - 2.35214\text{E}{-}4(t/^\circ\text{C}) \tag{12.2}$$

within a standard error of estimate of less than 0.1 %.

2 PHYSICAL CHEMISTRY OF SOLUTES IN SEAWATER

The solubility of neutral solutes in aqueous salt solutions was reviewed briefly by Harned and Owen [4]. Experimentally, it was found that for most (but not all) neutral solutes in aqueous electrolytes, the solubility of the solute is decreased by the presence of the salt. Alternatively, the activity coefficient of the solute is increased by the addition of the salt; this is known as 'salting out'.

A complete theory of this effect is not available, but the order of magnitude of the experimental results can be accounted for by consideration of coulombic forces in the solution. A simplified theory was given by Debye and McAulay [5]. The theory postulates a pure solvent of dielectric constant D_0, containing ions of radius b_i and charge e_i at concentrations n_i per unit volume. The theory considers the electrical work of discharging the ions in the pure solvent and recharging them in a medium of dielectric constant D (i.e. water with dissolved neutral species). The change in Gibbs energy (or chemical potential, $\Delta\mu$) of the neutral solute due to the influence of the electrolyte is

$$\Delta\mu = kT \ln\, (f/f_0) = -(\beta/2D_0)\sum_i (n_i e_i^2/b_i) \tag{12.3}$$

where

        k = Planck's constant
        T = absolute temperature
f and f_0 = activity coefficients of the neutral solute in the electrolyte and
               pure water solutions, respectively,

and the sum is over all ions in the solution. The quantity β represents the effect of dissolved neutral solute on the dielectric constant of water:

$$D = D_0(1 + \beta n_s) \tag{12.4}$$

where n_s is the concentration of neutral solute. For solutes considered here, β is negative, $f/f_0 > 1$, and the solute is salted out.

Equation (12.3) may be converted to a more recognizable form. For individual ions, we may put $e_i = z_i e$, where e is the electronic charge and z_i is the number and sign of electronic charges carried by the ion i (e.g., for chloride ion $z_i =$

-1, for calcium ion $z_i = +2$, etc.). Furthermore, b_i may be considered to be a constant, b. Then, Equation (12.3) becomes

$$kT \ln (f/f_0) = -(\beta/2bD_0) \sum_i (n_i z_i^2 e^2) \tag{12.5}$$

$$= -(\beta e^2/bD_0) \sum_i (n_i z_i^2)/2 \tag{12.6}$$

Multiplying Equation (12.6) by Avogadro's number,

$$RT \ln (f/f_0) = -(\beta e^2/bD_0) \sum_i (c_i z_i^2)/2 \tag{12.7}$$

where c_i is the molar concentration of ions i. Now, $\sum_i (c_i z_i^2)/2$ is defined as the ionic strength I on the molarity scale, and hence

$$RT \ln (f/f_0) = -(\beta e^2/bD_0)\, I \tag{12.8}$$

or

$$\ln (f/f_0) = -(\beta e^2/RTbD_0)\, I \tag{12.9}$$

If we consider a pure solute in equilibrium with both its saturated aqueous and salt solutions, at concentrations c_0 and c, respectively, its activity is unity by definition and

$$1 = c_0 f_0 = cf \tag{12.10}$$

or

$$f/f_0 = c_0/c \tag{12.11}$$

From Equations (12.9) and (12.11), we have

$$\ln (c_0/c) = -(\beta e^2/RTbD_0)\, I \tag{12.12}$$

Equation (12.12) may be recognized as a form of the Setchenov equation, an empirical relation between the ratio of solute solubilities and concentration of electrolyte [6].

Equation (12.12) has been used with concentrations expressed as molarity, molality or mole fractions [7]. As mentioned above, the constant $-(\beta e^2/dD_0 RT)$ is a positive quantity for the solutes considered here and is called the Setchenov coefficient. For 1:1 salts such as NaCl, $I = 1\,M$; for $CaCl_2$ and Na_2SO_4, $I = 3\,M$; for $MgSO_4$, $I = 4\,M$.

The quantity $RT \ln(f/f_0)$ is the Gibbs energy of transfer of solute from pure water to salt solution. In this form, it becomes a useful thermodynamic property as input parameter for various models for the fate of organic pollutants in the environment.

3 HENRY'S LAW CONSTANT

This quantity is defined as (Chapter 3)

$$k_{HSW} = \lim_{m \to 0} (m/p) \tag{12.13}$$

where m is the molality of dissolved solute and p is the partial pressure (fugacity) of pure solute in equilibrium with dissolved solute. If we let k_{HSW} in Equation (12.13) refer to seawater as solvent and $k_{HW} = (m_0/p)$ refer to pure water as solvent, the Setchenov ratio may be written as

$$k_{HW}/k_{HSW} = m_0/m \tag{12.14}$$

As in Chapter 11, m_0 and m are interpreted as saturation molalities, with the assumption that Henry's law holds up to saturation.

In the examination of experimental data for Henry's law in seawater which follows, it will be seen that Setchenov ratios conform to Equation (12.12), i.e. $m < m_0$. The apparent Setchenov coefficients for organic compounds are generally larger than those for simple gases. Also, Equation (12.12) predicts that the Setchenov coefficient decreases with temperature; for simple gases, this is seen to be true. For organic compounds, however, the temperature dependence may be either positive or negative. The reasons for this are not accounted for in the simple theoretical treatment given here.

4 TREATMENT OF DATA

Volume-Based and Mass-Based Solubility

Data for the solubility of a number of unreactive solutes in seawater are examined in this chapter, together with data for pure water as solvent, in order to obtain data for the Setchenov ratio k_{HW}/k_{HSW}. In the examples, the solubility was reported by the original investigators as Bunsen coefficients, molarity Setchenov coefficients, air–water partition coefficients, mole fraction, molality and molarity.

The following consideration could complicate the interpretation and presentation of the final results. The Bunsen, air–water partition coefficient, molarity Setchenov coefficient and molarity solubility are volume-based, whereas mole fraction and molality solubilities are mass-based. The quantity of solvent in the mass-based quantities is independent of temperature, whereas that in the other quantities is not. The densities of water and seawater change with temperature, and hence, for example, the mass of $1\,dm^3$ of solvent changes with temperature. Since we have chosen to present final results in terms of molality k_H and ratios thereof, it should first be ascertained to what extent changes in density due to temperature affect the final result.

It can be assumed that the neutral solute is at a low enough concentration that the density and mass fraction is unaffected by the solute. The conversion between molality and molarity is straightforward. From the above, the mass of $1\,dm^3$ of water is d_0 kg. The ratio $(m/M)_W$ is therefore $1/d_0$. For seawater,

$(m/M)_{SW}$ is $1/(0.965d)$, since 3.5 mass% of seawater is dissolved solids. Since, according to Equation (12.14) we are always concerned with the ratio m_0/m, it is the behaviour of the ratio $R = (m/M)_W/(m/M)_{SW}$ with temperature which is pertinent here:

$t/°C$	0	10	20	30	40
R	0.99196	0.99162	0.99128	0.99094	0.99060

Hence, although the quantities $(m/M)_W$ and $(m/M)_{SW}$ individually may vary significantly from unity with temperature, the ratio R is practically invariant, owing to cancellation of effects. Hence, in what follows, molality and molarity will be considered as identical at all temperatures.

Definitions

Although these were given in Chapter 3, they are stated here again for convenience. The Bunsen coefficient is the volume of gas (reduced to STP) dissolved in $1 \, dm^3$ of solvent at the temperature and pressure of measurement. The quantity of gas represented by $1 \, dm^3$, reduced to STP, is $4.4614E-2 \, mol$. The air–(sea)water partition coefficient, $k_{A(S)W}$, is the ratio $[solute]_{air}/[solute]_{(sea)water}$, at the temperature and pressure of measurement.

5 EVALUATIONS

In the course of this work, it was observed that, for a given solute, the measurements of its solubility in water outnumbered the measurements of its solubility in seawater. Inasmuch as the quantity sought, k_{HW}/k_{HSW}, requires two data, the choice of which datum to use for the numerator may become crucial. This is because the Setchenov ratio is of the order of 1.2, whereas the presence of evident systematic error in seawater solubility measurements sometimes entirely obscures this water–seawater difference.

Hence, in the following evaluations, where a single investigator measured solubilities in both water and seawater by the same method, that investigator's data for *both* solvents were used to calculate the Setchenov ratio. In this way, the sometimes large systematic error was accounted for, and the final results were consistent with general trends for the solutes examined.

Since the solubilities of many simple gases were reported as Bunsen coefficients, a conversion example is given here. As explained above, molality and molarity are treated as equivalent. Representing the Bunsen coefficient as α, the molal solubility in water is

$$m_0/ \, mol \, kg^{-1} = 4.4614E-2\alpha_0 \tag{12.15}$$

where $\alpha_0 [=] \, dm^3 \, (STP) \, dm^{-3}$. Hence

$$k_{HW}/mol \, kg^{-1} \, Pa^{-1} = 4.4614E-2\alpha_0/p \tag{12.16}$$

For seawater, the corresponding equations are

$$m/\text{mol kg}^{-1} = 4.4614\text{E}-2\alpha \tag{12.17}$$

and

$$k_{\text{HSW}}/\text{mol kg}^{-1}\,\text{Pa}^{-1} = 4.4614\text{E}-2\alpha/p \tag{12.18}$$

The air–water partition coefficient for pure water as solvent is designated as k_{AW} and for seawater k_{ASW}. Using c for concentrations, we have $k_{\text{AW}} = c_A/m_W$ and $k_{\text{ASW}} = c_A/m_{\text{SW}}$, since for aqueous solutions we may use molarity and molality indiscriminately. The Setchenov ratio is therefore

$$k_{\text{HW}}/k_{\text{HSW}} = k_{\text{ASW}}/k_{\text{AW}} \tag{12.19}$$

In the calculations, as many significant figures are retained from basic quantities to allow for three significant figures for the Setchenov ratio, irrespective of the number of significant figures in the original report of experimental results.

6 SOLUBILITY DATA

Solute		
Formula	Name	CAS Registry Number
Ar	Argon	7440-37-1
Solvent: Water and seawater		

Clever [8] presented critically evaluated smoothed data for the solubility of this gas in water and seawater at a partial pressure of 101.3 kPa in the temperature range $273 < T/K < 308$. The data were represented by the Bunsen coefficient:

$$\ln[\alpha/\text{dm}^3\,(\text{STP})\,\text{dm}^{-3}] = -686.1811 + 21\,561.99/(T/K)$$
$$+ 115.5947\ln(T/K) - 0.161717/(T/K)$$
$$+ S[5.887\text{E}-3 - 3.4803/(T/K)]$$

or

$$\ln(k_H/\text{mol kg}^{-1}\,\text{Pa}^{-1}) = -700.817 + 21561.99/(T/K) + 115.5947\ln(T/K)$$
$$- 0.161717/(T/K) + S[5.887\text{E}-3 - 3.4803/(T/K)]$$

for salinity $0 < S/\text{‰} < 39$. The standard error of estimate for α by this equation is $\sim$0.25 % [8].

		Original data				
		α/dm^3 (STP) dm^{-3}		k_H/mol kg^{-1} Pa^{-1}		
t/°C	T/K	Water ($S = 0$)	Seawater ($S = 35$)	Water	Seawater	k_{HW}/k_{HSW}
0	273.2	5.348E−2	4.208E−2	2.355E−8	1.853E−8	1.27
5	278.2	4.693E−2	3.722E−2	2.068E−8	1.640E−8	1.26
10	283.2	4.171E−2	3.334E−2	1.656E−8	1.471E−8	1.25
15	288.2	3.750E−2	3.020E−2	1.505E−8	1.333E−8	1.24
20	293.2	3.406E−2	2.763E−2	1.381E−8	1.221E−8	1.23
25	298.2	3.122E−2	2.550E−2	1.277E−8	1.128E−8	1.22
30	303.2	2.886E−2	2.373E−2	1.190E−8	0.985E−8	1.22
35	308.2	2.686E−2	2.223E−2	1.116E−8	0.930E−8	1.21

These are RECOMMENDED data for the temperature range $273 < T/\mathrm{K} < 308$.

Solute		
Formula	Name	CAS Registry Number
CO	Carbon monoxide	630-08-0
Solvent: Water and seawater		

Cargill [9] presented critically evaluated smoothed data for the solubility of this gas in water and seawater at a partial pressure of 101.3 kPa in the temperature range $283 < T/\mathrm{K} < 303$. The data were represented by the Bunsen coefficient:

$$\begin{aligned}\ln[\alpha/\mathrm{dm}^3\,(\mathrm{STP})\,\mathrm{dm}^{-3}] = {} & -47.6148 + 69.5068[100/(T/\mathrm{K})] \\ & + 18.7397\ln[(T/\mathrm{K})/100] + S\{4.5657\mathrm{E}{-2} \\ & - 4.0721\mathrm{E}{-2}[(T/\mathrm{K})/100] + 7.97\mathrm{E}{-3}[(T/\mathrm{K})100]^2\}\end{aligned}$$

or

$$\begin{aligned}\ln(k_H/\mathrm{mol\,kg^{-1}\,Pa^{-1}}) = {} & -62.2406 + 69.5068[100/(T/\mathrm{K})] \\ & + 18.7397\ln[(T/\mathrm{K})/100] + S\{4.5657\mathrm{E}{-2} \\ & - 4.0721\mathrm{E}{-2}[(T/\mathrm{K})/100] + 7.97\mathrm{E}{-3}[(T/\mathrm{K})100]^2\}\end{aligned}$$

for salinity $0 < S/‰ < 40$. The standard error of estimate for α by this equation is ~0.4 % [9].

		Original data				
		α/dm^3 (STP) dm^{-3}		k_H/mol kg^{-1} Pa^{-1}		
t/°C	T/K	Water $(S = 0)$	Seawater $(S = 35)$	Water	Seawater	k_{HW}/k_{HSW}
0	273.2	3.552E−2	2.868E−2	1.562E−8	1.272E−8	1.23
5	278.2	3.158E−2	2.564E−2	1.390E−8	1.138E−8	1.22
10	283.2	2.836E−2	2.320E−2	1.250E−8	1.031E−8	1.21
15	288.2	2.572E−2	2.121E−2	1.134E−8	9.435E−9	1.20
20	293.2	2.353E−2	1.959E−2	1.039E−8	8.727E−9	1.19
25	298.2	2.170E−2	1.828E−2	9.591E−9	8.150E−9	1.18
30	303.2	2.018E−2	1.721E−2	8.926E−9	7.683E−9	1.16

These are RECOMMENDED data for the temperature range $273 < T/K < 303$.

Solute		
Formula	Name	CAS Registry Number
CO_2	Carbon dioxide	124-38-9
Solvent: Water and seawater		

Scharlin [10] presented critically evaluated smoothed data for the solubility of this gas in water and seawater at a partial pressure of 101.3 kPa in the temperature range $273 < T/K < 313$. The data were represented by the Henry's law constant:

$$\ln(k_H/\text{mol dm}^{-3} \text{atm}^{-1}) = -58.0931 + 90.5069[100/(T/\text{K})] + 22.294$$
$$\times \ln[(T/\text{K})/100] + S\{2.7766\text{E}-2 - 2.5888\text{E}-2$$
$$\times [(T/\text{K})/100] + 5.0578\text{E}-3[(T/\text{K})/100]^2\}$$

or

$$\ln(k_H/\text{mol kg}^{-1} \text{Pa}^{-1}) = -69.6192 + 90.5069[100/(T/\text{K})] + 22.294$$
$$\times \ln[(T/\text{K})/100] + S\{2.7766\text{E}-2 - 2.5888\text{E}-2$$
$$\times [(T/\text{K})/100] + 5.0578\text{E}-3[(T/\text{K})/100]^2\}$$

for salinity $0 < S/‰ < 40$. The standard error of estimate for α by this equation is ~0.3 % [10].

		Original data				
		k_H/mol/dm^{-3} atm^{-1}		k_H/mol kg^{-1} Pa^{-1}		
t/°C	T/K	Water ($S = 0$)	Seawater ($S = 35$)	Water	Seawater	k_{HW}/k_{HSW}
0	273.2	7.743E−2	6.457E−2	7.647E−7	6.373E−7	1.20
5	278.2	6.396E−2	5.353E−2	6.323E−7	5.287E−7	1.20
10	283.2	5.357E−2	4.503E−2	5.300E−7	4.451E−7	1.19
15	288.2	4.545E−2	3.841E−2	4.501E−7	3.800E−7	1,18
20	293.2	3.904E−2	3.320E−2	3.869E−7	3.287E−7	1.18
25	298.2	3.392E−2	2.905E−2	3.365E−7	2.879E−7	1.17
30	303.2	2.979E−2	2.571E−2	2.958E−7	2.550E−7	1.16
35	308.2	2.643E−2	2.302E−2	2.626E−7	2.285E−7	1.15
40	313.2	2.368E−2	2.082E−2	2.355E−7	2.069E−7	1.14

These are RECOMMENDED data for the temperature range $273 < T/\mathrm{K} < 313$.

Solute		
Formula	Name	CAS Registry Number
H_2	Hydrogen	1333-74-0
Solvent: Water and seawater		

Young [11] presented critically evaluated smoothed data for the solubility of this gas in water and seawater at a partial pressure of 101.3 kPa in the temperature range $273 < T/\mathrm{K} < 303$. The data were represented by the Bunsen coefficient:

$$\begin{aligned}\ln[\alpha/\mathrm{dm}^3\,(\mathrm{STP})\,\mathrm{dm}^{-3}] = {} & -47.9848 + 65.0369[100/(T/\mathrm{K})] + 20.1709 \\ & \times \ln[(T/\mathrm{K})/100] + S\{-8.2225\mathrm{E}{-2} + 4.9564 \\ & \times [(T/\mathrm{K})/100] - 7.8689\mathrm{E}{-3}[(T/\mathrm{K})/100]^2\}\end{aligned}$$

or

$$\begin{aligned}\ln(k_H/\mathrm{mol\,kg^{-1}\,Pa^{-1}}) = {} & -62.6206 + 65.0369[100/(T/\mathrm{K})] + 20.1709 \\ & \times \ln[(T/\mathrm{K})/100] + S\{-8.2225\mathrm{E}{-2} + 4.9564 \\ & \times [(T/\mathrm{K})/100] - 7.8689\mathrm{E}{-3}[(T/\mathrm{K})/100]^2\}\end{aligned}$$

for salinity $0 < S/‰ < 40$. The standard error of estimate for α by this equation is ~0.5 % [11].

		Original data				
		α/dm^3 (STP) dm^{-3}		$k_H/mol\,kg^{-1}\,Pa^{-1}$		
$t/°C$	T/K	Water $(S = 0)$	Seawater $(S = 35)$	Water	Seawater	k_{HW}/k_{HSW}
0	273.2	2.012E−2	1.657E−2	8.848E−9	7.393E−9	1.20
5	278.2	1.891E−2	1.574E−2	8.323E−9	7.032E−9	1.18
10	283.2	1.792E−2	1.506E−2	7.897E−9	6.737E−9	1.17
15	288.2	1.712E−2	1.451E−2	7.553E−9	6.497E−9	1.16
20	293.2	1.649E−2	1.406E−2	7.279E−9	6.304E−9	1.15
25	298.2	1.599E−2	1.370E−2	7.064E−9	6.152E−9	1.15
30	303.2	1.513E−2	1.343E−2	6.901E−9	6.034E−9	1.14

These are RECOMMENDED data for the temperature range $273 < T/K < 303$.

Solute		
Formula	Name	CAS Registry Number
Kr	Krypton	7439-90-9
Solvent: Water and seawater		

Clever [12] presented critically evaluated smoothed data for the solubility of this gas in water and seawater at a partial pressure of 101.3 kPa in the temperature range $273 < T/K < 313$. The data [13] were represented by the Bunsen coefficient:

$$\ln[\alpha/dm^3\,(STP)\,dm^{-3}] = -57.2596 + 87.4242[100/(T/K)] + 22.9332$$
$$\times \ln(T/K)/100] + S\{-8.723E{-3} - 2.793E{-3}$$
$$\times [(T/K)/100] + 1.2398E{-3}[(T/K)/100]^2\}$$

or

$$\ln(k_H/mol\,kg^{-1}\,Pa^{-1}) = -71.8954 + 87.4242[100/(T/K)] + 22.9332$$
$$\times \ln[(T/K)/100] + S\{-8.723E{-3} - 2.793E{-3}$$
$$\times [(T/K)/100] + 1.2398E{-3}[(T/K)/100]^2\}$$

for salinity $0 < S/\text{‰} < 40$.

		Original data				
		α/dm^3 (STP) dm^{-3}		$k_H/mol\,kg^{-1}\,Pa^{-1}$		
$t/°C$	T/K	Water $(S=0)$	Seawater $(S=35)$	Water	Seawater	k_{HW}/k_{HSW}
0	273.2	0.1098	8.562E−2	4.828E−8	3.663E−8	1.32
5	278.2	9.363E−2	7.353E−2	4.121E−8	3.149E−8	1.31
10	283.2	8.086E−2	6.398E−2	3.563E−8	2.743E−8	1.30
15	288.2	7.070E−2	5.636E−2	3.118E−8	2.420E−8	1.29
20	293.2	6.252E−2	5.023E−2	2.760E−8	2.159E−8	1.28
25	298.2	5.588E−2	4.525E−2	2.470E−8	1.947E−8	1.27
30	303.2	5.045E−2	4.119E−2	2.232E−8	1.774E−8	1.26
35	308.2	4.598E−2	3.785E−2	2.036E−8	1.633E−8	1.25
40	313.2	4.228E−2	3.511E−2	1.874E−8	1.516E−8	1.24

These are RECOMMENDED data for the temperature range $273 < T/K < 313$.

Solute		
Formula	Name	CAS Registry Number
N_2	Nitrogen	7727-37-9
Solvent: Water and seawater		

Battino [14] presented critically evaluated smoothed data for the solubility of this gas in water and seawater at a partial pressure of 101.3 kPa in the temperature range $273 < T/K < 313$. The data were represented by the Bunsen coefficient:

$$\ln[\alpha/dm^3\,(STP)\,dm^{-3}] = 59.7745 - 76.7685[100/(T/K)]$$
$$- 88.327\ln[(T/K)/100] + 19.5287[(T/K)/100]$$
$$+ S\{7.1485E-3 - 3.9793E-2[100/(T/K)]\}$$

or

$$\ln(k_H/mol\,kg^{-1}\,Pa^{-1}) \doteq 45.1387 - 76.7685[100/(T/K)]$$
$$- 88.327\ln[(T/K)/100] + 19.5287[(T/K)/100]$$
$$+ S\{7.1485E-3 - 3.9793E-2[100/(T/K)]\}$$

for salinity $0 < S/\%_o < 40$. The standard error of estimate for α is $\sim0.4\%$ [14].

$t/°C$	T/K	α/dm^3 (STP) dm^{-3}		$k_H/mol\,kg^{-1}\,Pa^{-1}$		k_{HW}/k_{HSW}
		Water $(S=0)$	Seawater $(S=35)$	Water	Seawater	
0	273.2	2.368E−2	1.826E−2	1.041E−8	8.097E−9	1.29
5	278.2	2.099E−2	1.634E−2	9.239E−9	7.251E−9	1.27
10	283.2	1.881E−2	1.477E−2	8.262E−9	6.562E−9	1.26
15	288.2	1.703E−2	1.349E−2	7.511E−9	6.000E−9	1.25
20	293.2	1.558E−2	1.245E−2	6.880E−9	5.543E−9	1.24
25	298.2	1.441E−2	1.160E−2	6.369E−9	5.173E−9	1.23
30	303.2	1.347E−2	1.092E−2	5.957E−9	4.876E−9	1.22
35	308.2	1.271E−2	1.039E−2	5.629E−9	4.643E−9	1.21
40	313.2	1.212E−2	9.980E−3	5.373E−9	4.465E−2	1.20

(Original data)

These are RECOMMENDED data for the temperature range $273 < T/K < 313$.

Solute		
Formula	Name	CAS Registry Number
N_2O	Nitrous oxide	10024-97-2
Solvent: Water and seawater		

Young [15] presented critically evaluated smoothed data for the solubility of this gas in water and seawater at a partial pressure of 101.3 kPa in the temperature range $273 < T/K < 313$. The data were represented by Henry's law constant:

$$\ln(k_H/mol\,dm^{-3}\,atm^{-1}) = -62.7062 + 97.3006[100/(T/K)] + 24.1406$$
$$\times \ln[(T/K)/100)] + S\{-5.842E-2 + 3.3193E-2$$
$$\times [(T/K)100] - 5.1313E-3[(T/K)/100]^2\}$$

or

$$\ln(k_H/mol\,kg^{-1}\,Pa^{-1}) = -74.2323 + 97.3006[100/(T/K)] + 24.1406$$
$$\times \ln[(T/K)/100)] + S\{-5.842E-2 + 3.3193E-2$$
$$\times [(T/K)100] - 5.1313E-3[(T/K)/100]^2\}$$

where S is the salinity in ‰.

		Original data				
		k_H/mol/dm^{-3} atm^{-1}		k_H/mol kg^{-1} Pa^{-1}		
$t/°C$	T/K	Water ($S = 0$)	Seawater ($S = 35$)	Water	Seawater	k_{HW}/k_{HSW}
0	273.2	5.908E−2	4.783E−2	5.830E−7	4.720E−7	1.24
5	278.2	4.826E−2	3.940E−2	4.763E−7	3.889E−7	1.22
10	283.2	4.001E−2	3.292E−2	3.948E−7	3.249E−7	1.22
15	288.2	3.363E−2	2.786E−2	3.319E−7	2.750E−7	1.21
20	293.2	2.865E−2	2.387E−2	2.827E−7	2.356E−7	1.20
25	298.2	2.470E−2	2.068E−2	2.438E−7	2.041E−7	1.20
30	303.2	2.154E−2	1.182E−2	2.126E−2	1.788E−7	1.19
35	308.2	1.900E−2	1.603E−2	1.875E−7	1.582E−7	1.19
40	313.2	1.693E−2	1.431E−2	1.670E−7	1.412E−7	1.18

These are RECOMMENDED data for the temperature range $273 < T/K < 313$.

Solute		
Formula	Name	CAS Registry Number
O_2	Oxygen	7782-44-7
Solvent: Water and seawater		

Battino [16] presented critically evaluated smoothed data for the solubility of this gas in water and seawater at a partial pressure of 101.3 kPa in the temperature range $283 < T/K < 308$. The data were represented by the Bunsen coefficient:

$$\ln[\alpha/\text{cm}^3\,(\text{STP})\,\text{dm}^{-3}] = -1268.9782 + 36\,063.19/(T/K) + 220.1832\ln(T/K)$$
$$- 0.351299(T/K) + S[6.229\text{E}{-}3 - 3.5912/(T/K)]$$
$$+ 3.44\text{E}{-}6S^2$$

or

$$\ln(k_H/\text{mol kg}^{-1}\,\text{Pa}^{-1}) = -1290.4900 + 36\,063.19/(T/K) + 220.1832\ln(T/K)$$
$$- 0.351299(T/K) + S[6.229\text{E}{-}3 - 3.5912/(T/K)]$$
$$+ 3.44\text{E}{-}6S^2$$

for salinity $0 < S/‰ < 40$. The standard error of estimate for α by this equation is $\sim 0.1\%$ [16].

$t/°C$	T/K	α/dm^3 (STP) dm^{-3}		k_H/mol kg^{-1} Pa^{-1}		k_{HW}/k_{HSW}
		Water $(S=0)$	Seawater $(S=35)$	Water	Seawater	
10	283.2	7.876	6.311	3.580E−9	2.868E−9	1.25
15	288.2	7.040	5.684	3.200E−9	2.583E−9	1.24
20	293.2	6.349	5.165	2.886E−9	2.348E−9	1.23
25	298.2	6.770	4.728	2.623E−9	2.149E−9	1.22
30	303.2	5.276	4.353	2.398E−9	1.978E−9	1.21
35	308.2	4.847	4.026	2.203E−9	1.830E−9	1.20

These are RECOMMENDED data for the temperature range $283 < T/K < 308$.

Solute		
Formula	Name	CAS Registry Number
CCl_4	Tetrachloromethane	56-23-5
Solvent: Water and seawater		

Hunter-Smith et al. [17] measured the air–solvent partition coefficient of this gas in the temperature range $278 < T/K < 306$ by the headspace chromatographic method [18]. The solvents were water and seawater ($S = 35‰$) and the data were presented only as parameters of fitting equations. For water as solvent

$$\ln k_{AW} = -2918/(T/K) + 9.77 \qquad (12.20)$$

and for seawater

$$\ln k_{ASW} = -3230/(T/K) + 11.27 \qquad (12.21)$$

From Equations (12.20) and (12.21)

$$k_{ASW}/k_{AW} = \exp[1.50 - 312/(T/K)] \qquad (12.22)$$

The result for k_{HW}/k_{HSW} follows from Equation (12.19):

	Original data			
$t/°C$	T/K	k_{AW}	k_{ASW}	k_{HW}/k_{HSW}
5	278.2	0.4873	0.7115	1.46
10	283.2	0.5865	0.8734	1.49

Original data			
15	288.2	0.7012	1.065	1.52
20	293.2	0.8334	1.289	1.55
25	298.2	0.9847	1.550	1.57
30	303.2	1.157	1.853	1.60

Solute		
Formula	Name	CAS Registry Number
CCl$_3$F	Fluorotrichloro- methane	75-69-4
Solvent: Water and seawater		

Hunter-Smith *et al.* [17] measured the air–solvent partition coefficient of this gas in the temperature range $278 < T/\mathrm{K} < 306$ by the headspace chromatographic method [18]. The solvents were water and seawater ($S = 35\%o$) and the data were presented only as parameters of fitting equations. For water as solvent

$$\ln k_{AW} = -2372/(T/\mathrm{K}) + 9.25 \tag{12.23}$$

and for seawater

$$\ln k_{ASW} = -2652/(T/\mathrm{K}) + 10.50 \tag{12.24}$$

From Equations (12.23) and (12.24)

$$k_{ASW}/k_{AW} = \exp[1.25 - 280/(T/\mathrm{K})] \tag{12.25}$$

Original data				
$t/°\mathrm{C}$	T/K	k_{AW}	k_{ASW}	k_{HW}/k_{HSW}
5	278.2	2.062	2.631	1.28
10	283.2	2.397	3.113	1.30
15	288.2	2.772	3.662	1.32
20	293.2	3.190	4.284	1.34
25	298.2	3.653	4.986	1.39
30	303.2	4.165	5.774	1.41

Solute		
Formula	Name	CAS Registry Number
CCl_2F_2	Dichlorodifluoro-methane	75-71-8
Solvent: Water and seawater		

The solubility of this gas was measured [19,20] by the direct method in the temperature range $288 < T/K < 303, 318$ at a partial pressure of the gas of 101.3 kPa. For water as solvent [19] the data are

$$\ln x_0 = -68.904 + 98.951[(T/K)/100] + 23.7031 \ln[(T/K)/100]$$

Original data			
$t/°C$	T/K	Solubility, x_0	Solubility, m_0
15	288.2	7.667E−5	4.148E−3
20	293.2	6.389E−5	3.478E−3
25	298.2	5.358E−5	2.952E−3
30	303.2	4.648E−5	2.580E−3
35	308.2	4.097E−5	2.274E−3
40	313.2	3.538E−5	1.964E−3
45	318.2	3.117E−5	1.730E−3

and for seawater ($S/‰ = 35$ [20])

Original data		
$t/°C$	T/K	Solubility, m
15	288.2	3.232E−3
20	293.2	2.783E−3
25	298.2	2.475E−3
30	303.2	2.271E−3

Thus, from Equation (12.14)

$t/°C$	T/K	k_{HW}/k_{HSW}
15	288.2	1.28
20	293.2	1.25
25	298.2	1.20
30	303.2	1.14

Solute		
Formula	Name	CAS Registry Number
$CClF_3$	Chlorotrifluoro- methane	75-72-9
Solvent: Water and seawater		

The solubility of this gas was measured [19,20] by the direct method in the temperature range $288 < T/K < 303$, 318 at a partial pressure of the gas of 101.3 kPa. For water as solvent [19] the data are

$$\ln x_0 = -112.372 + 155.260/[(T/K)/100] + 45.1236\ln[(T/K)/100]$$

Original data			
$t/°C$	T/K	Solubility, x_0	Solubility, m_0
15	288.2	2.166E−5	1.202E−3
20	293.2	1.884E−5	1.046E−3
25	298.2	1.680E−5	9.326E−4
30	303.2	1.492E−5	8.282E−4
35	308.2	1.354E−5	7.516E−4
40	313.2	1.258E−5	6.983E−4
45	318.2	1.185E−5	6.578E−4

and for seawater ($S/‰ = 35$ [20])

Original data		
$t/°C$	T/K	Solubility, m
15	288.2	9.581E−4
20	293.2	8.554E−4
25	298.2	7.858E−4
30	303.2	7.414E−4

Thus, from Equation (12.14)

$t/°C$	T/K	k_{HW}/k_{HSW}
15	288.2	1.25
20	293.2	1.22
25	298.2	1.19
30	303.2	1.12

Solute		
Formula	Name	CAS Registry Number
CF_4	Tetrafluoromethane	75-73-0
Solvent: Water and seawater		

The solubility of this gas was measured [20,21] by the direct method in the temperature range $288 < T/K < 303$, 318 at a partial pressure of the gas of 101.3 kPa. For water as solvent [21] the data are

$$\ln x_0 = -178.693 + 245.958/[(T/K)/100] + 76.6351 \ln[(T/K)/100]$$

	Original data		
$t/^{\circ}C$	T/K	Solubility, x_0	Solubility, m_0
15	288.2	4.880E−6	2.709E−4
20	293.2	4.337E−6	2.407E−4
25	298.2	3.769E−6	2.092E−4
30	303.2	3.485E−6	1.935E−4
35	308.2	3.303E−6	1.835E−4
40	313.2	3.150E−6	1.749E−4
45	318.2	3.071E−6	1.705E−4

and for seawater ($S/\%o = 35$ [20])

	Original data	
$t/^{\circ}C$	T/K	Solubility, m
15	288.2	2.130E−4
20	293.2	1.836E−4
25	298.2	1.608E−4
30	303.2	1.430E−4

Thus, from Equation (12.14)

$t/^{\circ}C$	T/K	k_{HW}/k_{HSW}
15	288.2	1.27
20	293.2	1.31
25	298.2	1.30
30	303.2	1.35

Solute		
Formula	Name	CAS Registry Number
CH$_4$	Methane	74-82-8
Solvent: Water and seawater		

Clever and Young [22] presented critically evaluated smoothed data for the solubility of this gas in water and seawater at a partial pressure of 101.3 kPa in the temperature range $273 < T/K < 303$. The data were represented by the Bunsen coefficient:

$$\ln[\alpha/dm^3\,(STP)\,dm^{-3}] = -68.8862 + 101.4956[100/(T/K)] + 28.7314$$
$$\times \ln[(T/K)/100] + S\{-7.6146E{-}2 + 4.397E{-}2$$
$$\times [(T/K)/100] - 6.8672E{-}3[(T/K)/100]^2\}$$

or

$$\ln(k_H/mol\,kg^{-1}\,Pa^{-1}) = -83.5220 + 101.4956[100/(T/K)] + 28.7314$$
$$\times \ln[(T/K)/100] + S\{-7.6146E{-}2 + 4.397E{-}2$$
$$\times [(T/K)/100] - 6.8672E{-}3[(T/K)/100]^2\}$$

for salinity $0 < S/\text{‰} < 40$. The standard error of estimate for α by this equation is $\sim0.4\%$ [22].

		Original data				
		α/dm^3 (STP) dm^{-3}		k_H/mol kg^{-1} Pa^{-1}		
$t/°$C	T/K	Water $(S=0)$	Seawater $(S=35)$	Water	Seawater	k_{HW}/k_{HSW}
0	273.2	5.739E−2	4.450E−2	2.524E−8	1.986E−8	1.27
5	278.2	4.955E−2	3.883E−2	2.181E−8	1.735E−8	1.26
10	283.2	4.340E−2	3.434E−2	1.913E−8	1.536E−8	1.25
15	288.2	3.853E−2	3.074E−2	1.700E−8	1.377E−8	1.23
20	293.2	3.464E−2	2.783E−2	1.529E−8	1.248E−8	1.23
25	298.2	3.151E−2	2.547E−2	1.393E−8	1.143E−8	1.22
30	303.2	2.898E−2	2.353E−2	1.282E−8	1.058E−8	1.21

These are RECOMMENDED data for the temperature range $273 < T/K < 303$.

Solute		
Formula	Name	CAS Registry Number
C_2Cl_4	Tetrachloroethene	127-18-4
Solvent: Water and seawater		

Peng and Wan [23,24] measured Henry's law constants for this liquid in the temperature range $288 < T/K < 318$ by the headspace chromatographic method [17,18]. The solvents were water and aqueous NaCl (salinity 36‰), which was assumed here to be equivalent to seawater. The data for water as solvent were

$$\log k_{HW}/\text{atm}^{-1} = 1822/(T/K) - 9.06 \tag{12.26}$$

and for seawater

$$\log k_{HSW}/\text{atm}^{-1} = 1795/(T/K) - 9.09 \tag{12.27}$$

Hence

$$\log(k_{HW}/k_{HSW}) = 27/(T/K) + 0.03 \tag{12.28}$$

	Original data			
$t/°C$	T/K	k_{HW}	k_{HSW}	k_{HW}/k_{HSW}
15	288.2	1.828E−3	1.375E−3	1.33
20	293.2	1.426E−3	1.077E−3	1.32
25	298.2	1.122E−3	8.501E−4	1.32
30	303.2	8.897E−4	6.764E−4	1.32
35	308.2	7.108E−4	5.422E−4	1.31
40	313.2	5.720E−4	4.377E−4	1.31
45	318.2	4.634E−4	3.557E−4	1.30

Solute		
Formula	Name	CAS Registry Number
C_2HCl_3	Trichloroethene	79-01-6
Solvent: Water and seawater		

Peng and Wan [23,24] measured air-solvent partition coefficients for this liquid in the temperature range $288 < T/K < 318$ by the headspace chromatographic method [17,18]. The solvents were water and aqueous NaCl (salinity 36‰), which was assumed here to be equivalent to seawater. For water as solvent, the data were

$$\log k_{HW}/\text{atm}^{-1} = 1642/(T/K) - 8.19 \qquad (12.29)$$

and for seawater

$$\log k_{HSW}/\text{atm}^{-1} = 1736/(T/K) - 8.62 \qquad (12.30)$$

Hence

$$\log(k_H/k_{HSW}) = 0.43 - 94/(T/K) \qquad (12.31)$$

		Original data		
$t/°C$	T/K	k_{HW}	k_{HSW}	k_{HW}/k_{HSW}
15	288.2	3.217E−3	2.533E−3	1.27
20	293.2	2.572E−3	1.999E−3	1.33
25	298.2	2.072E−3	1.591E−3	1.35
30	303.2	1.681E−3	1.275E−3	1.27
35	308.2	1.373E−3	1.030E−3	1.31
40	313.2	1.129E−3	8.371E−4	1.33
45	318.2	9.339E−4	6.850E−4	1.35

Solute		
Formula	Name	CAS Registry Number
$C_2H_3Cl_3$	1,1,1-Trichloroethane	71-55-6
Solvent: Water and seawater		

Hunter-Smith et al. [17] measured the air–solvent partition coefficient of this gas in the temperature range $278 < T/K < 306$ by the headspace chromatographic method [18]. The solvents were water and seawater ($S/‰ = 35$) and the data were presented only as parameters of fitting equations. For water as solvent

$$\ln k_{AW} = -2915/(T/K) + 9.15 \qquad (12.32)$$

and for seawater

$$\ln k_{ASW} = -3905/(T/K) + 13.04 \qquad (12.33)$$

From Equations (12.32) and (12.33)

$$k_{ASW}/k_{AW} = \exp[3.89 - 990/(T/K)] \qquad (12.34)$$

The result for k_{HW}/k_{HSW} follows from Equation (12.19):

Original data				
$t/°C$	T/K	k_{AW}	k_{ASW}	k_{HW}/k_{HSW}
5	278.2	0.2650	0.3691	1.39
10	283.2	0.3188	0.4729	1.48
15	288.2	0.3812	0.6007	1.58
20	293.2	0.4529	0.7569	1.67
25	298.2	0.5351	0.9463	1.77
30	303.2	0.6287	1.174	1.87

Solute		
Formula	Name	CAS Registry Number
C_2H_4	Ethene	74-85-1
Solvent: Water and seawater		

Hayduk [25] presented critically evaluated smoothed data for the solubility of this gas in aqueous electrolyte solutions. The author reported solubilities in the form of Setchenov coefficients for ethene in NaCl solutions in the form

$$\log(m_0/m) = k_{smm}m_{NaCl} \qquad (12.35)$$

for the temperature range $288 < T/K < 313$ and NaCl molality range $0 < m_{NaCl} < 4$. The temperature dependence of k_{smm} could be represented by

$$k_{smm}/\text{kg mol}^{-1} = 0.534629 - 1.97858\text{E}{-}3(T/K) + 2.09116\text{E}{-}6(T/K)^2 \qquad (12.36)$$

For the present purposes, seawater is represented by 0.7 m NaCl. From Equations (12.35), (12.36) and (12.14):

Original data			
$t/°C$	T/K	$k_{smm}/\text{kg mol}^{-1}$	k_{HW}/k_{HSW}
15	288.2	0.13813	1.25
20	293.2	0.13432	1.24
25	298.2	0.13060	1.23
30	303.2	0.12700	1.22
35	308.2	0.12350	1.22
40	313.2	0.12010	1.21

These are RECOMMENDED data in the temperature range $288 < T/K < 313$.

Solute		
Formula	Name	CAS Registry Number
C_2H_6	Ethane	74-84-0
Solvent: Water and seawater		

Hayduk [26] presented critically evaluated smoothed data for the solubility of this gas in aqueous electrolyte solutions. The author reported solubilities in the form of Setchenov coefficients for ethane in NaCl solutions in the form (Bunsen coefficients):

$$\log(\alpha_0/\alpha) = k_{sm\alpha} m_{NaCl} \tag{12.37}$$

for the temperature range $273 < T/K < 348$ and NaCl molality range $0 < m_{NaCl} < 4$. The temperature dependence of k_{smm} could be represented by

$$\log(k_{sm\alpha}/\text{kg mol}^{-1}) = -1.50078 + 220.679/(T/K) \tag{12.38}$$

For the present purposes, seawater is represented by 0.7 m NaCl. From Equations (12.37), (12.38), (12.16) and (12.18):

$$\log(k_{HW}/k_{HSW}) = 0.7 \times 10^{[-1.50078+220.679/(T/K)]} \tag{12.39}$$

$t/°C$	T/K	k_{HW}/k_{HSW}
0	273.2	1.38
5	278.2	1.36
10	283.2	1.35
15	288.2	1.33
20	293.2	1.32
25	298.2	1.31
30	303.2	1.30
35	308.2	1.29
40	313.2	1.28
45	318.2	1.27
50	323.2	1.27
55	328.2	1.26
60	333.2	1.25
65	338.2	1.24
70	343.2	1.24
75	348.3	1.23

Solute		
Formula	Name	CAS Registry Number
C_2H_6S	Dimethyl sulfide	75-18-3
Solvent: Water and seawater		

Wong and Wang [27] measured the air–solvent partition coefficients of this liquid by the modified EPICS procedure of [28]. The result for water as solvent were obtained at 291.2 K only; for seawater as solvent, the temperature range was $291 < T/K < 317$. The salinity of the seawater was not specified [27], but it is assumed here to be 35‰. Using the equivalence $k_H = 1/k_{AW}RT$ (Chapter 3), we have

$t/°C$	T/K	k_{AW}	k_{ASW}	$k_{HW}/\text{mol kg}^{-1}\text{Pa}^{-1}$	$k_{HSW}/\text{mol kg}^{-1}\text{Pa}^{-1}$
	Original data				
18	291.2	0.058	0.069	7.12E−6	5.986E−6
25	298.2	–	0.094	–	4.291E−6
35	308.2	–	0.149	–	2.619E−6
44	317.2	–	0.211	–	1.797E−6

The temperature dependence of k_H for dimethyl sulfide in water from Chapter 11 is

$$\ln(k_H/\text{mol kg}^{-1}\text{Pa}^{-1}) = -24.207 + 3588.7/(T/K) \qquad (12.40)$$

Equation (12.40) can be normalized to Wong and Wang's datum at 291 K:

$$\ln(k_H/\text{mol kg}^{-1}\text{Pa}^{-1}) = -24.176 + 3588.7/(T/K) \qquad (12.41)$$

$t/°C$	T/K	$k_{HW}/\text{mol kg}^{-1}\text{Pa}^{-1}$	$k_{HSW}/\text{mol kg}^{-1}\text{Pa}^{-1}$	k_{HW}/k_{HSW}
18	291.2	7.121E−6	5.986E−6	1.19
25	298.2	5.334E−6	4.291E−6	1.24
35	308.2	3.610E−6	2.619E−6	1.37
44	317.2	2.594E−6	1.797E−6	1.44

Solute		
Formula	Name	CAS Registry Number
C_6H_6	Benzene	71-43-2
Solvent: Water and seawater		

Peng and Wan [23,24] measured Henry's law constants for this liquid in the temperature range $288 < T/K < 318$ by the headspace chromatographic method [17,18]. The solvents were water and aqueous NaCl (salinity 36‰), which was assumed here to be equivalent to seawater. For water as solvent, the data were given by

$$\log k_{HW}/\text{atm}^{-1} = 1397/(T/K) - 7.15 \tag{12.42}$$

and for seawater

$$\log k_{HSW}/\text{atm}^{-1} = 1448/(T/K) - 7.44 \tag{12.43}$$

Hence

$$\log(k_{HW}/k_{HSW}) = 0.29 - 51/(T/K) \tag{12.44}$$

		Original data		
$t/°C$	T/K	k_{HW}/atm^{-1}	k_{HSW}/atm^{-1}	k_{HW}/k_{HSW}
15	288.2	4.981E−3	3.840E−3	1.30
20	293.2	4.118E−3	3.152E−3	1.31
25	298.2	3.426E−3	2.605E−3	1.32
30	303.2	2.868E−3	2.166E−3	1.32
35	308.2	2.414E−3	1.812E−3	1.33
40	313.2	2.044E−3	1.525E−3	1.34
45	318.2	1.739E−3	1.290E−3	1.35

Keeley et al. [29] measured the solubility of benzene in water and in aqueous NaCl solutions (0–1 M) at 25 °C. The solubilities in water and 0.7 M NaCl were 2.17E−3 and 1.55E−3 M, respectively; the Setchenov ratio was thus $2.17/1.55 = 1.40$.

Solute		
Formula	Name	CAS Registry Number
$C_6H_6Cl_6$	α-Hexachlorocyclo-hexane	319-84-6
Solvent: Water and seawater		

Kucklick et al. [30] measured Henry's law constants of this pesticide by the gas-purge method [31] in the temperature range $273 < T/K < 318$. The solvents were water and seawater (salinity 30‰). The data for water as solvent could be represented by

$$\log k_{HW}/\text{mol m}^{-3} \text{Pa}^{-1} = 2810/(T/K) - 9.31 \tag{12.45}$$

and for seawater

$$\log k_{\text{HSW}}/\text{mol m}^{-3}\,\text{Pa}^{-1} = 2969/(T/\text{K}) - 9.88 \qquad (12.46)$$

Hence

$$\log k_{\text{HW}}/k_{\text{HSW}} = 0.57 - 159/(T/\text{K}) \qquad (12.47)$$

$t/^\circ\text{C}$	T/K	$k_{\text{HW}}/\text{mol m}^{-3}\,\text{Pa}^{-1}$	$k_{\text{HSW}}/\text{mol m}^{-3}\,\text{Pa}^{-1}$	$k_{\text{HW}}/k_{\text{HSW}}$
0	273.2	9.452	9.716	0.97
5	278.2	6.175	6.197	1.00
10	283.2	4.096	4.016	1.02
15	288.2	2.755	2.642	1.04
20	293.2	1.879	1.763	1.07
25	298.2	1.298	1.192	1.09
30	303.2	0.9074	0.8170	1.11
35	308.2	0.6419	0.5667	1.13
40	313.2	0.4591	0.3977	1.16
45	318.2	0.3318	0.2822	1.18

Solute		
Formula	Name	CAS Registry Number
$C_6H_6Cl_6$	Lindane	58-89-9
Solvent: Water and seawater		

Kucklick *et al.* [30] measured Henry's law constants of this pesticide by the gas-purge method [31] in the temperature range $273 < T/\text{K} < 318$. The solvents were water and seawater (salinity 30‰). The data for water as solvent could be represented by

$$\log k_{\text{HW}}/\text{mol m}^{-3}\,\text{Pa}^{-1} = 2382/(T/\text{K}) - 7.54 \qquad (12.48)$$

and for seawater

$$\log k_{\text{HSW}}/\text{mol m}^{-3}\,\text{Pa}^{-1} = 2703/(T/\text{K}) - 8.68 \qquad (12.49)$$

Hence

$$\log k_{\text{HW}}/k_{\text{HSW}} = 1.14 - 321/(T/\text{K}) \qquad (12.50)$$

$t/^\circ\text{C}$	T/K	$k_{\text{HW}}/\text{mol m}^{-3}\,\text{Pa}^{-1}$	$k_{\text{HSW}}/\text{mol m}^{-3}\,\text{Pa}^{-1}$	$k_{\text{HW}}/k_{\text{HSW}}$
0	273.2	15.10	16.36	0.92
5	278.2	10.52	10.87	0.97

$t/°C$	T/K	$k_{HW}/mol\,m^{-3}\,Pa^{-1}$	$k_{HSW}/mol\,m^{-3}\,Pa^{-1}$	k_{HW}/k_{HSW}
10	283.2	7.431	7.320	1.02
15	288.2	5.310	4.999	1.06
20	293.2	3.838	3.459	1.11
25	298.2	2.805	2.423	1.16
30	303.2	2.071	1.718	1.21
35	308.2	1.544	1.231	1.25
40	313.2	1.162	0.8918	1.30
45	318.2	0.8828	0.6526	1.35

Solute		
Formula	Name	CAS Registry Number
C_7H_8	Toluene	108-88-3
Solvent: Water and seawater		

Peng and Wan [23,24] measured Henry's law constants for this liquid in the temperature range $288 < T/K < 318$ by the headspace chromatographic method [17,18]. The solvents were water and aqueous NaCl (salinity 36‰), which was assumed here to be equivalent to seawater. For water as solvent, the data were given by

$$\log k_{HW}/atm^{-1} = 1621/(T/K) - 7.94 \qquad (12.51)$$

and for seawater

$$\log k_{HSW}/atm^{-1} = 1565/(T/K) - 7.89 \qquad (12.52)$$

Hence

$$\log(k_{HW}/k_{HSW}) = 56/(T/K) - 0.05 \qquad (12.53)$$

Original data				
$t/°C$	T/K	k_{HW}/atm^{-1}	k_{HSW}/atm^{-1}	k_{HW}/k_{HSW}
15	288.2	4.837E−3	3.469E−3	1.39
20	293.2	3.878E−3	2.803E−3	1.38
25	298.2	3.133E−3	2.281E−3	1.37
30	303.2	2.549E−3	1.869E−3	1.36
35	308.2	2.087E−3	1.541E−3	1.35
40	313.2	1.720E−3	1.279E−3	1.34
45	318.2	1.427E−3	1.067E−3	1.34

Keeley *et al.* [29] measured the solubility of toluene in water and in aqueous NaCl solutions (0–1 M) at 25 °C. The solubilities in water and 0.7 M NaCl were 6.29E−3 and 4.59E−3 M, respectively; the Setchenov ratio was thus $6.29/4.59 = 1.37$.

REFERENCES

1. Maidment, D. R. (ed.), *Handbook of Hydrology*, McGraw-Hill, New York, 1993.
2. Lide, D. R. (ed.), *Handbook of Chemistry and Physics*, 73rd edn, CRC Press, Boca Raton, FL, 1993.
3. Washburn, E. W. (ed.), *International Critical Tables*, Vol. 3, McGraw-Hill, New York, 1928.
4. Harned, H. S.; Owen, B. B., *The Physical Chemistry of Electrolyte Solutions*, 3rd edn, Reinhold, New York, 1958.
5. Debye, P.; McAulay, J., *Phys. Z.*, 1925, **26**, 22–9.
6. Scatchard, G., *Trans. Faraday Soc.*, 1927, **23**, 454–62.
7. Clever, H. L., The Sechenov salt effect parameter, in *Solubility Data Series, Vol. 10, Nitrogen and Air*, Battino, R. (ed.), Pergamon Press, Oxford, 1982, pp. xxix–xxxvii.
8. Clever, H. L. (ed.), *Solubility Data Series, Vol. 4, Argon*, Pergamon Press, Oxford, 1980.
9. Cargill, R. W. (ed.), *Solubility Data Series, Vol. 43, Carbon Monoxide*, Pergamon Press, Oxford, 1990.
10. Scharlin, P. (ed.), *Solubility Data Series, Vol. 62, Carbon Dioxide in Water and Aqueous Electrolyte Solutions*, Oxford University Press, Oxford, 1996.
11. Young, C. L. (ed.), *Solubility Data Series, Vol. 5/6, Hydrogen and Deuterium*, Pergamon Press, Oxford, 1981.
12. Clever, H. L. (ed.), *Solubility Data Series, Vol. 2, Krypton, Xenon and Radon*, Pergamon Press, Oxford, 1979.
13. Weiss, R. F.; Keper, T. K., *J. Chem. Eng. Data*, 1978, **23**, 69–72.
14. Battino, R. (ed.), *Solubility Data Series, Vol. 10, Nitrogen and Air*, Pergamon Press, Oxford, 1982.
15. Young, C. L. (ed.), *Solubility Data Series, Vol. 8, Oxides of Nitrogen*, Pergamon Press, Oxford, 1981.
16. Battino, R. (ed.), *Solubility Data Series, Vol. 7, Oxygen and Ozone*, Pergamon Press, Oxford, 1981.
17. Hunter-Smith, R. J.; Balls, P. W.; Liss, P. S., *Tellus B*, 1983, **35**, 170–6.
18. McAuliffe, C. D., *Chem. Tech.*, 1971, **1**(1), 46–51.
19. Scharlin, P.; Battino, R., *Fluid Phase Equilib.*, 1994, **95**, 137–47.
20. Scharlin, P.; Battino, R., *J. Chem. Eng. Data*, 1995, **40**, 167–9.
21. Scharlin, P.; Battino, R., *J. Solution Chem.*, 1992, **21**, 67–91.
22. Clever, H. L.; Young, C. L. (eds), *Solubility Data Series, Vol. 27/28, Methane*, Pergamon Press, Oxford, 1987.
23. Peng, J.; Wan, A., *Environ. Sci. Technol.*, 1997, **31**, 2998–3003.
24. Peng, J.; Wan, A., *Chemosphere*, 1998, **36**, 2731–40.
25. Hayduk, W. (ed.), *Solubility Data Series, Vol. 57, Ethene*, Oxford University Press, Oxford, 1994.
26. Hayduk, W. (ed.), *Solubility Data Series, Vol. 9, Ethane*, Pergamon Press, Oxford, 1982.
27. Wong, P. K.; Wang, Y. H., *Chemosphere*, 1997, **35**, 535–44.
28. Gossett, J. M., *Environ. Sci. Technol.*, 1987, **21**, 202–8.
29. Keeley, D. F.; Hoffpauir, M. A.; Meriwether, J. R., *J. Chem. Eng. Data*, 1988, **33**, 87–9.
30. Kucklick, J. R.; Hinckley, D. A.; Bidleman, T. F., *Mar. Chem.*, 1991, **34**, 197–209.
31. Mackay, D.; Shiu, W. Y.; Sutherland, R. P., *Environ. Sci. Technol.*, 1979, **13**, 333–7.

APPENDIX I

A Selection of Sources of Additional Data Currently Available from the Internet

Peter Fogg

University of North London

NASA JPL Reports

http://jpldataeval.jpl.nasa.gov/

The two reports below contain important information about processes in the atmosphere. Reports are freely available from the Internet in Word, PDF, Post-Script and RTF formats.

The PDF files include bookmarks and hypertext links

JPL Publication No. 97-4

DeMore, W. B.; Howard, C. J.; Sander, S. P.; Ravishankara, A. R.; Golden, D. M. Kolb, C. E.; Hampson, R. F.; Molina, M. J.; Kurylo, M. J., *Chemical Kinetics and Photochemical Data for Use in Stratospheric Modeling, Evaluation Number 12*, NASA Panel for Data Evaluation Jet Propulsion Laboratory, California Institute of Technology, Pasadena, CA, 1997 (266 pages).

Gas-phase enthalpy and entropy data for species of importance in atmospheric chemistry are also included.

JPL Publication No. 00-3

Sander, S. P.; Friedl, R. R.; DeMore, W. B.; Golden, D. M.; Kurylo, M. J.; Hampson, R. F.; Huie, R. E.; Moortgat, G. K.; Ravishankara, A. R.; Kolb, C. E.; Molina, M. J., *Chemical Kinetics and Photochemical Data for Use in Stratospheric Modeling, Evaluation Number 13. Supplement to Evaluation 12: Update of Key Reactions*. NASA Panel for Data Evaluation, Jet Propulsion Laboratory, California Institute of Technology, Pasadena, CA, 2000.

Chemicals in the Atmosphere – Solubility, Sources and Reactivity.   Edited by P. G. T. Fogg and J. Sangster
© 2003 IUPAC   ISBN: 0-471-98651-8

These two files are excellent sources of kinetic and equilibrium data for important homogeneous and heterogeneous reactions in the atmosphere. Cross-section data for photochemical reactions, accommodation coefficients and uptake coefficients are included.

Compilation of Henry's Law Constants for Inorganic and Organic Species of Potential Importance in Environmental Chemistry (Version 3)

http://www.mpch-mainz.mpg.de/~sander/res/henry.html

This site contains a very extensive compilation of Henry's law constants prepared by R. Sander. It is periodically updated and extended and is currently available as either a PDF or PostScript file.

NIST Chemistry WebBook

http://webbook.nist.gov/chemistry/

This database contains physical properties of a very wide range of chemicals. Data are obtained by inputting a name, CA registry number or empirical formula of a chemical substance.

The Pesticide Properties Database

http://wizard.arsusda.gov/acsl/ppdb.html

Henry's law constants for pesticides are included in this database.

CS ChemFinder

http://chemfinder.cambridgesoft.com/

This database of properties of chemical compounds is maintained by CambridgeSoft.

Publications of IUPAC Subcommittee on Gas Kinetic Data Evaluation

The IUPAC Subcommittee on Gas Kinetic Data Evaluation for Atmospheric Chemistry has published a series of nine evaluations of kinetic data in the *Journal of Physical and Chemical Reference Data* in the period 1980–2000. Full details, recent evaluations and summary data are available on the website maintained by the Subcommittee. Recent data on the website include information about heterogeneous reactions in the atmosphere in addition to that for homogeneous gas-phase reactions. The enthalpy of formation of many species participating in reactions in the atmosphere are also available from this website (http://www.iupac-kinetic.ch.cam.ac.uk).

Mass Accommodation Coefficients

http://www.mi.uni-hamburg.de/technische_meteorologie/Meso/homepages/fmueller/mass_acc.html

This is a compilation of mass accommodation coefficients prepared by F. Müller.

EUROTRAC Reports

http://www.gsf.de/eurotrac

EUROTRAC (the European Experiment on Transport and Transformation of Environmentally Relevant Trace Constituents in the Troposphere over Europe) was involved in a cooperative venture involving scientists from 24 European countries. It ran for 8 years, finishing at the end of 1995, but has been succeeded by a follow-on project entitled EUROTRAC-2, which is ongoing. Reports on this second project are available from the Internet.

The follow-on project, EUROTRAC-2, was started on the July 1, 1996, and involves scientists from about 30 countries. It is due for completion at the end of 2002.

Preliminary reports on this project are available on the Internet.

CMD-Scenarios for Modeling of Multi-Phase Tropospheric Chemistry
http://www.chem.leeds.ac.uk/Atmospheric/MCM/mcmproj.html

The Multi-Phase Modeling Groups within the EUROTRAC-2 subproject on Chemical Mechanism Development (CMD) has developed models for the chemical changes in the atmosphere under different conditions. These have been termed Master Chemical Mechanisms

CMD-Data Review Panel
http://www.fz-juelich.de/icg/icg3/CMD_DATA_PANEL/cmddata.html

Kinetic and other data for various gas-phase, aqueous-phase and heterogenous reactions which take place in the atmosphere have been compiled and are available on the Internet.

Estimation Software from the US Environmental Protection Agency
http://www.epa.gov/oppt/exposure/docs/episuitedl.htm

The EPI Suite™ may be freely downloaded from the above site but is subject to the following statement displayed on the site:

The EPI Suite™ includes 11 programs for estimating parameters associated with the behaviour of pollutants from the structure of the compounds. Of special

relevance to the subject of this volume are the following which are included in
the suite:

- AOPWIN for estimating rate constants for oxidation of compounds by photo-
  chemically produced OH radicals and by ozone;
- HENRYWIN for estimating Henry's law constants on the basis of the method
  described by Hine and Mookerjee [1] and later developed by Meylan and
  Howard at the Syracuse Research Corporation [2] (see Chapter 6);
- HYDRO for estimating hydrolysis rate constants;
- MPBPVP for estimating melting points, boiling points and vapour pressures;
- WSKOWWIN for estimating water solubility

REFERENCES

1. Hine, J.; Mookerjee, P. K., *J. Org. Chem.*, 1975, **40**, 292–8.
2. Meylan, W.; Howard, P. H., *Environ. Toxicol. Chem.*, 1991, **10**, 1283–93.

APPENDIX II

Dissolution of Carbon Dioxide and of Oxygen and Fractionation of Oxygen Isotopes in the Environment

Peter Fogg

University of North London, UK (retired)

Molecules of the same chemical species but containing different isotopes of an element have different chemical potentials and hence a difference in equilibrium constants. Molecules containing a heavy isotope undergo mass transport less readily than a molecule containing a lighter isotope. Lighter molecules tend to react more quickly than heavier molecules. The effects on the properties of water and of carbon dioxide of different isotopes of oxygen leads to fractionation of isotopes. This fractionation is temperature dependent and is giving much information about past climatic changes.

Oxygen has three stable isotopes. The proportions in oxygen gas in the air are

^{16}O 99.759 %
^{17}O 0.037 %
^{18}O 0.204 %

Of special interest is the slight variation of the O^{18}/O^{16} ratio in compounds containing oxygen.

In an equilibrium such as

$$H_2{}^{18}O + \tfrac{1}{3}C^{16}O_3{}^{2-} \rightleftharpoons H_2{}^{16}O + \tfrac{1}{3}C^{18}O_3{}^{2-}$$

Chemicals in the Atmosphere – Solubility, Sources and Reactivity. Edited by P. G. T. Fogg and J. Sangster
© 2003 IUPAC ISBN: 0-471-98651-8

an equilibrium constant may be written as

$$K = \frac{[H_2{}^{16}O][C^{18}O_3{}^{2-}]^{\frac{1}{3}}}{[H_2{}^{18}O][C^{16}O_3{}^{2-}]^{\frac{1}{3}}}$$

Such an equilibrium constant is temperature dependent.

This equilibrium constant can be equated with the fractionation factor α:

$$\alpha = [{}^{18}O/{}^{16}O]_{CO_2}/[{}^{18}O/{}^{16}O]_{H_2O}$$

Equilibrium values of fractionation are expressed as *del values*, i.e.

$$\Delta = (\alpha - 1)$$

These are usually quoted in parts per mil (‰)

Isotopic concentrations are difficult to measure directly. However, isotopic ratios can be measured very accurately by mass spectrometry. These ratios are usually compared with the ratio in a standard. The comparison is expressed using a del notation using the lower case symbol, δ, i.e.

$$\delta = [{}^{18}O/{}^{16}O]_{sample}/[{}^{18}O/{}^{16}O]_{standard}$$

This is also expressed as parts per mil (‰).

The standard for the measurement of the isotope ratio in water is the standard mean ocean water.

Water molecules containing ^{18}O have a lower vapour pressure than molecules containing ^{16}O. This difference leads to a fractionation of -8.5 to -10‰ depending on the temperature. There is also a mass transport effect due to differential transport across the air–water interface. This contributes as additional -2 to -4‰ depending on the temperature. The overall fractionation is therefore -10.5 to -14‰. There is also a fractionation effect when water vapour condenses to liquid. In this case the heavier water tends to precipitate more readily, further reducing the residual ^{18}O water vapour in the atmosphere. Water vapour which reaches cold regions is therefore depleted in ^{18}O relative to the oceans. Water vapour condensing in polar regions show the effect the most markedly.

The oxygen isotopes in carbon dioxide cannot equilibrate with water in the gas phase. However, they very readily equilibrate when dissolved in water by the reaction

$$H_2O + CO_2 \rightleftharpoons H^+ + CO_3{}^{2-}$$

The half-life for the equilibration is about 30 s. At equilibrium the $^{18}O/^{16}O$ ratio in the CO_2 is 1.041 times that of the water at 25 °C and varies slightly with temperature.

Exchange occurs with seawater and lake water. It also takes place with water in chloroplast leaf cells of plants. The exchange with leaf water is catalysed by the enzyme carbonic anhydrase. Isotope exchange also occurs in the soil between carbon dioxide produced by biological respiratory processes and soil water.

The oxygen isotopic composition of both leaf and soil water depends on several factors. Soil water is close to the composition of the local rainfall which is depleted in ^{18}O relative to the sea. Chloroplast water tends to have a higher proportion of ^{18}O than the local soil water because ^{16}O water preferentially evaporates from the surfaces of the leaves.

Oxygen gas in the atmosphere has a greater proportion of $^{18}O/^{16}O$ than seawater. This was first observed by Dole [1] and independently by Morita [2] and is called the Dole effect. It was partially explained by Lane and Dole [3]. They postulated that during photosynthesis the oxygen produced had the same isotopic composition as water in the plant cells. However, during respiration processes ^{16}O is preferentially converted to carbon dioxide relative to ^{18}O. As a result, oxygen with an augmented proportion of ^{18}O is produced by overall biological processes.

Another factor which contributes to the Dole effect is the fractionation of oxygen gas which dissolves in cell water prior to respiration. Photochemical reactions in the stratosphere can lead to isotope exchange between O_2 and CO_2. This causes a slight reduction in the Dole effect.

The possibility of investigating paleotemperatures from partitioning of stable isotopes was first suggested by Urey [4]. The greater the quantity of water locked up in glaciers, the higher is the ratio of ^{18}O to ^{16}O in the oceans.

Stratified cores of sediment can be removed from the sea bed. These cores contain remains of ancient living matter. It is possible to date roughly the material from different layers. The $^{18}O/^{16}O$ ratio in the calcium carbonate of the shells of *Foraminifera* and other microfossils depends upon the ratio in the seawater at the time that they were formed. The ratio is also determined by the extent of fractionation during the biological processes leading to the shell formation. This depends on the temperature of the seawater in which the *Foraminifera* have grown. The water at the bottom of deeper oceans does not change temperature very much even during glaciations. The water temperature nearer the surface reflects the local air temperature.

It follows that the $^{18}O/^{16}O$ ratio in species of *Foraminifera* which live in very deep water reflect the contemporary ratio in the seawater. This depends on the mass of water locked in glaciers and gives information about major glaciations. The ratio in species which live close to the surface also depends on the contemporary ratio in the sea but also shows changes due to shorter duration changes in the local air temperature. A comparison of data from the two sources gives information about local temperature changes.

Ice cores have been extracted from polar ice caps. The appearance of the ice at different levels shows slight seasonal variation. Ice formed in different years can be distinguished and counted. Counting from layers near to the surface of the ice sheet gives an absolute chronology of layers. Layers formed during interglacial periods have relatively low $^{18}O/^{16}O$ ratios and layers formed during glacial periods have higher $^{18}O/^{16}O$ ratios. Fractionation still takes place as damp air moves towards the poles but the sea is enriched in ^{18}O during these periods. Analysis by mass spectrometry can give the oxygen isotopic ratios in water from the ice and also that in carbon dioxide trapped in the ice. Oxygen isotopic ratios in water and carbon dioxide are likely to be identical because of ready exchange of oxygen between the two. A similar fractionation of deuterium and ^1H occurs.

Data from ice cores and deep sea cores have allowed the estimation of the fluctuation in climate changes during the Pliocene and Pleistocene, a time span of about 600 000 years. These changes can be partially interpreted by the Milankovich theory [5]. This theory assumes that climatic changes are caused by changes in insolation due to the changes in earth's eccentricity, axial tilt and precession. These are associated with cycles of about 100 000, 41 000 and 23 000 years, respectively. The total effect of these cycles with different periodicities causes a complex periodicity in the amount of solar radiation received by the upper atmosphere. The effect varies with latitude. There is reasonably good correlation with terrestrial evidence of palaeo-climates [6–8].

REFERENCES

1. Dole, M., *J. Amer. Chem. Soc.*, 1935, **57**, 2731.
2. Morita, N., *J. Chem. Soc. Jpn.*, 1935, **56**, 1291.
3. Lane, G. A.; Dole, M., *Science*, 1956, **123**, 574–6.
4. Urey, H. C., *J. Chem. Soc.*, 1947, 562–581.
5. Milankovich, M. K., *Serb. Akad. Beogr. Spec. Publ.*, 1941, **132**.
6. Hays, J. D.; Imbrie, J.; Shackleton, N. J., *Science*, 1976, **194**, 1121–32.
7. Lorius, C.; Jouzel, J.; Ritz, C.; Merlivat, L.; Barkov, N. I.; Korotkevich, Y. S.; Kotlyakov, V. M., *Nature*, 1985, **316**, 591–6.
8. Winograd, I. J.; Szabo, B. J.; Coplen, T. B.; Riggs, A. C., *Science*, 1988, **242**, 1275–80.

CAS Registry Index

Italic page numbers refer to illustrations, diagrams and tables.

Chemicals in the Atmosphere – Solubility, Sources and Reactivity. Edited by P. G. T. Fogg and J. Sangster
© 2003 IUPAC ISBN: 0-471-98651-8

Subject Index

Italic page numbers refer to illustrations, diagrams and tables.

Chemicals in the Atmosphere – Solubility, Sources and Reactivity.   Edited by P. G. T. Fogg and J. Sangster
© 2003 IUPAC   ISBN: 0-471-98651-8